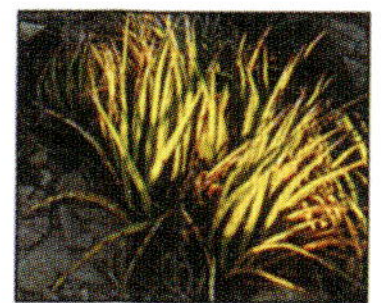

GROVER E. MURRAY
STUDIES IN THE
AMERICAN SOUTHWEST

Also in the Grover E. Murray Studies in the American Southwest

Brujerías: Stories of Witchcraft and the Supernatural in the American Southwest and Beyond
Nasario García

Cacti of Texas: A Field Guide
A. Michael Powell, James F. Weedin, and Shirley A. Powell

Cacti of the Trans-Pecos and Adjacent Areas
A. Michael Powell and James F. Weedin

Deep Time and the Texas High Plains: History and Geology
Paul H. Carlson

From Texas to San Diego in 1851: The Overland Journal of Dr. S. W. Woodhouse, Surgeon-Naturalist of the Sitgreaves Expedition
Edited by Andrew Wallace and Richard H. Hevly

Javelinas
Jane Manaster

Picturing a Different West: Vision, Illustration, and the Tradition of Austin and Cather
Janis P. Stout

Texas Quilts and Quilters: A Lone Star Legacy
Marcia Kaylakie with Janice Whittington

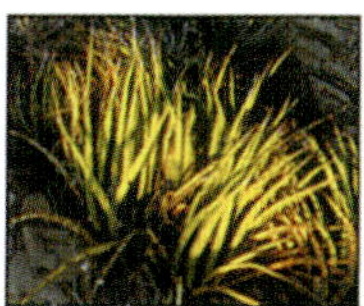

Little Big Bend

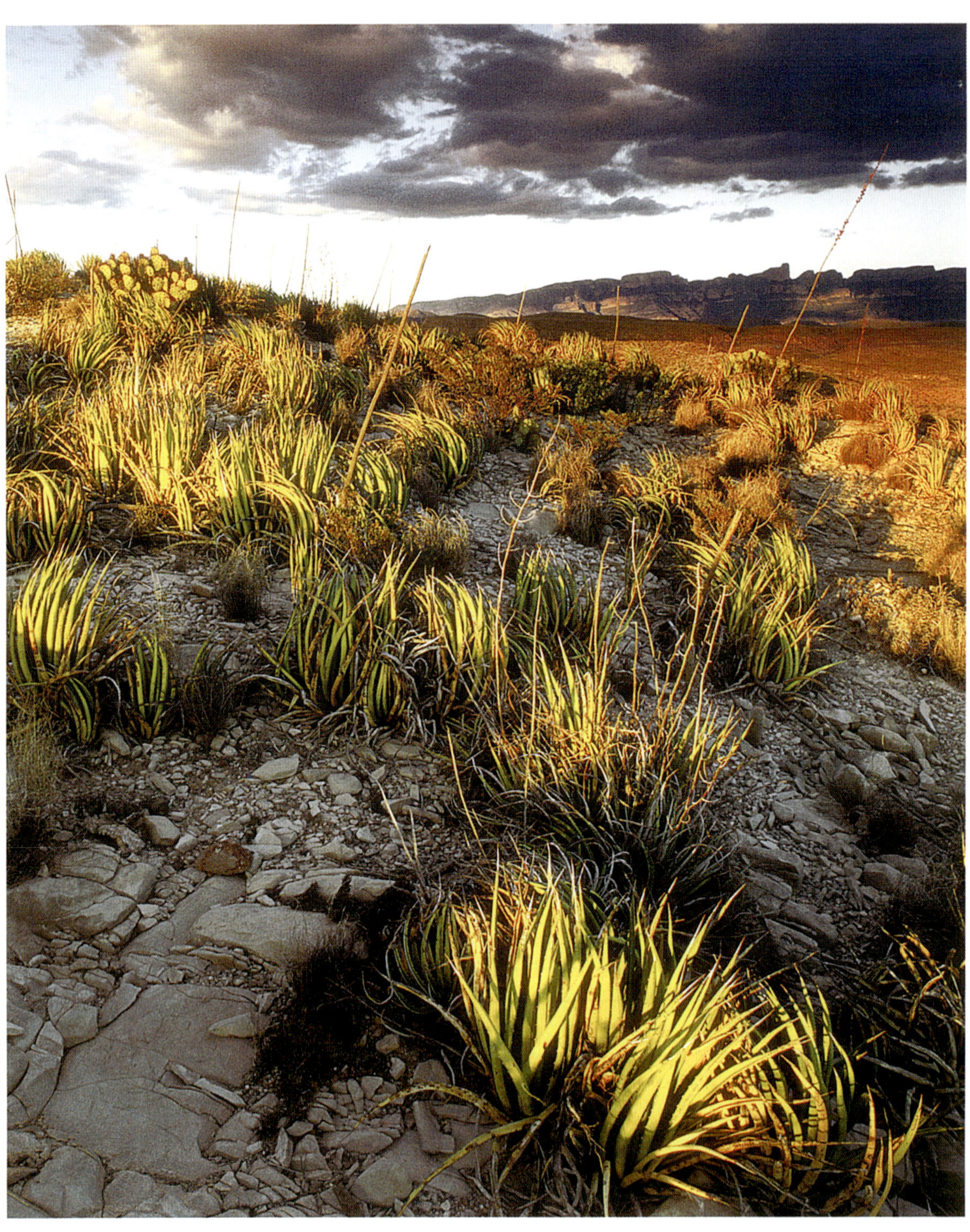

Little Big Bend

Common, Uncommon, and Rare Plants of Big Bend National Park

Roy Morey

Texas Tech University Press

This book is typeset in Adobe Minion. The paper used in this book meets the minimum requirements of ANSI/NISO Z39.48-1992 (R1997). ♾

Frontispiece: photograph of lechuguilla-covered slope with the Sierra del Carmen in the distance.

Library of Congress Cataloging-in-Publication Data

Morey, Roy.

Little Big Bend : common, uncommon, and rare plants of Big Bend National Park / by Roy Morey.

p. cm. — (Grover E. Murray studies in the American Southwest)

Includes bibliographical references and index.

ISBN-13: 978-0-89672-613-0 (pbk. : alk. paper)

ISBN-10: 0-89672-613-4 (pbk. : alk. paper)

1. Plants—Texas—Big Bend National Park—Identification. 2. Plants—Texas—Big Bend National Park—Pictorial works. 3. Big Bend National Park (Tex.) I. Title.

QK188.M67 2008

581.9764'932—dc22

2007002522

16 17 18 19 20 21 22 23 / 9 8 7 6 5 4 3 2

Texas Tech University Press

Box 41037

Lubbock, Texas 79409-1037 USA

800.832.4042

ttup@ttu.edu

www.ttupress.org

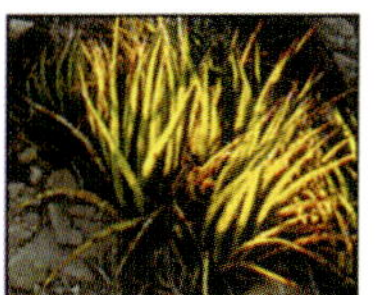

CONTENTS

Preface viii

Acknowledgments ix

About This Book xi

Big Bend, the Land of Extremes 3

How Plants Are Named 22

Checklist of Plants 24

The Plants 35

Appendix A Critically Imperiled, Imperiled, and Vulnerable Plants 291

Appendix B Selected Plants by Park Location 294

Appendix C Photo Equipment and Techniques for Plant Photography 303

Appendix D A Note on Botanical Nomenclature 304

Glossary 307

Sources 313

References 315

Index 319

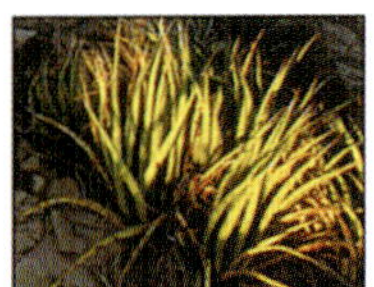

PREFACE

I FIRST BEGAN hiking in Big Bend National Park in 1986, shortly after taking up photography as a hobby. The grand views—the Chisos mountain islands towering above the desert floor like a stone fortress, the steep, vaultlike canyons encasing the Rio Grande—astounded me as they do everyone visiting Big Bend for the first time.

In those days many national parks attracted my growing photographic interest. But somehow Big Bend kept pulling me back, without my fully understanding why. The focus of my photography gradually evolved from the traditional landscape view to closer attractions nearby—wildflowers, trees and shrubs, mushrooms, lichens, rock formations, fossils. Although I often hiked long distances and explored many remote corners of the park, it was the little things right under my nose, at my feet, that provided startling evidence of the desert's bounty and intrigue. Gradually, these little discoveries became more fascinating than the broader views.

I came to realize that it was this near landscape of Big Bend, the amazing wealth of relatively small things—for example, thirteen hundred different species of plants—that kept drawing me back. Few if any other national parks offer this kind of diversity. So I became a photographic "explorer" of the close at hand, using my camera as an educational tool for capturing images of plants and then subsequently identifying and learning more about them—their botany, history, and natural beauty.

This self-teaching process became a central part of my modus operandi, and my photography evolved into a process of discovery, with on-the-job training. Nothing was known in advance, so everything surprised. My eye was never jaundiced. My goal was to discover firsthand a plant new to me—not, initially, to learn about it from a book.

The learning proved slow and personal, but gradually the process changed me. Very small plants or flowers became a special source of fascination because they often have more intricate and elaborate flower structures than their larger counterparts. I took pleasure, literally, in small things.

Photographing *Little Big Bend* was not something that could be accomplished in weeks or months; it has taken many years. During these years I have come to know something of this close landscape and its secrets. This book is the product of my continuing fascination with Big Bend biodiversity and the "little" in the Big Bend.

ACKNOWLEDGMENTS

WYNN ANDERSON, the designer, developer, and curator of the Chihuahuan Desert Gardens at the University of Texas at El Paso Centennial Museum, provided invaluable assistance and many hours of his time over a number of years, first in helping me with plant identification and then in reviewing, editing, and encouraging this work.

Mike Boren, executive director of the Big Bend Natural History Association, first suggested this book many years ago. He was undoubtedly surprised when I showed up on his doorstep years later with the first draft. Mike steered me to Texas Tech University Press.

The staff of the Science and Resource Management Division at Big Bend National Park assisted me in ways too numerous to recount. Joe Sirotnak, park botanist/ecologist, provided a thorough and constructive review of the book with many helpful suggestions and also assisted with plant identification. Betty Alex, GIS specialist and director of the park's Sensitive Plant Project, has been a guide to me for many years, offering timely information on rare plants blooming in the park and kindly examining many of my photographs. Allison Leavitt, the chief plant hunter for the Sensitive Plant Project, contributed helpful comments on an early draft and regularly provided me with tips on plant locations. Allison, who has rediscovered many rare Big Bend plants that have gone undocumented for years, has been a good friend and sometimes hiking companion.

Patty Manning, a botanist and environmental science technician who maintains the cactus garden and greenhouses at Sul Ross State University, graciously reviewed the first early draft, offering numerous helpful suggestions and her enthusiastic support for the project.

A. Michael Powell, professor emeritus of biology and director and curator of the A. Michael Powell Herbarium at Sul Ross State University, assisted in the identification of all the cacti and some other species.

Bonnie Amos, head of the biology department at Angelo State University, inspected an early draft and assisted with some difficult plant identifications.

Jackie Poole, rare plant botanist with the Texas Parks and Wildlife Department, took hours from her busy schedule to evaluate an early draft. She also helped to verify plant identifications and suggested additional species to include.

Emily Lott, a botanical consultant and research affiliate with the Plant Resources Center of the University of Texas at Austin, offered what was needed most at one time: a thorough, exhaustive review and edit to enliven my tired prose. Her spirited support for the project and my photography is much appreciated.

Tom Wendt, curator of the Plant Resources Center at the University of Texas at Austin, was a cordial host and on more than one occasion kept me from becoming hopelessly lost in the PRC's labyrinthine herbaria facilities. He also helped in the identification of some *Ephedra* and other species.

B. L. Turner, former director of the Plant Resources Center and professor emeritus at the University of Texas at Austin, enthusiastically examined an early draft and assisted in the identification of some Asteraceae and Fabaceae photographs.

James Henrickson, visiting scholar at the Plant Resources Center and an expert on Chihuahuan Desert flora, helped me with some of the most difficult plant identifications, such as smooth bur cucumber (*Sicyos glaber*), and graciously permitted me the use of an unpublished copy of *The Flora of the Chihuahuan Desert Region,* edited by himself and Marshall Johnston. As noted elsewhere, this monumental, three-volume work has proved invaluable to my efforts. It was the principal source of information on the distribution of plants in Mexico.

William Carr, botanist with the Texas Conservation Data Center of the Nature Conservancy of Texas, provided me with a January 2005 working draft of the *Annotated List of the G3/T3 and Rarer Plant Taxa of Texas.* I made frequent use of this draft in discussions of the rarity and distribution of plants, and unless otherwise indicated, it was the source of information on a plant's conservation status.

Richard Worthington, associate professor of biology and curator of the Herbarium of the University of Texas at El Paso, helped me locate and sort through many plant specimens in the herbarium and also identify some unusual species such as southwestern ringstem (*Anulocaulis leiosolenus* var. *lasianthus*).

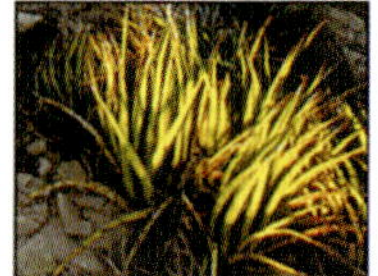

Although I received valuable assistance from these numerous individuals in the difficult identification of plants from photographs, the final responsibility for all these identifications is of course my own.

Two old friends, Tom Walker and Lonn Taylor, reviewed an early draft of the book and offered insightful suggestions on ways to make it more entertaining and useful to the reader.

My wife, Mary, spent many hours in the field with me as a hiking companion and unpaid photographer's assistant. And she allowed me to be away from home for many months, doing what I love.

ABOUT THIS BOOK

THIS BOOK is intended to give the reader a sense of the incredible plant diversity in Big Bend National Park. The choice of plants reflects the wide range of habitats within the park—desert, foothills, mountains, and moist woodlands, river canyons, and floodplain. Big Bend's three major blooming seasons—spring, summer, and fall—are all well represented.

Yucca and agave species are emphasized because in many ways they define the Big Bend landscape. Cacti receive special attention because the Chihuahuan Desert is home to more cactus species than any other North American desert, and Big Bend National Park has more cactus species than any other national park. Cactus enthusiasts from all over the world travel to Big Bend to experience this extraordinary diversity firsthand.

Little Big Bend is not a traditional botanical guide to the area's common plants. Although this guide features many species because they are characteristic of the Chihuahuan Desert environment or other unique habitats in the park, other species, such as orchids, are included precisely because they are uncommon or rare and therefore a special thrill to find. All other things equal, plants not found in other wildflower guides, or those with a limited geographic range that the reader is less likely to have encountered elsewhere, are pictured here.

This guide describes 109 species found in the United States only in Trans-Pecos Texas; 62 of these occur only in the Big Bend portion of the Trans-Pecos, and 24 of them only within Big Bend National Park. Of 252 featured species, 72 are considered "sensitive plants"; in Texas, 28 are classified as critically imperiled, 18 as imperiled, and 25 as vulnerable.

In spite of the emphasis on sensitive plants, and those with a more restricted distribution, a good cross-section of plant communities is represented. The *Inventory of Vascular Flora, Big Bend National Park* (Clelland 2001) documents 114 plant families, 531 genera, and 1,313 species in the park. This book features only 19 percent of the species, but it covers 71 plant families (more than 62 percent) and 180 genera (almost 34 percent).

Ultimately, the final choice of plants to include became a personal one. The focus is on the beauty as well as the botany of plants—on plants both interesting and artful. The emphasis is on the "little" in the Big Bend—the overlooked small plants, or inconspicuous tiny flowers of larger plants that so often go unnoticed. In a landscape so immense, these plants may be right before our eyes but seldom really seen, or they may be tucked away and quite difficult to find.

All photographs were taken by me in the park where the plants were found. My goal has been to impart to others the pleasure I have experienced in witnessing these desert survivors for the first time.

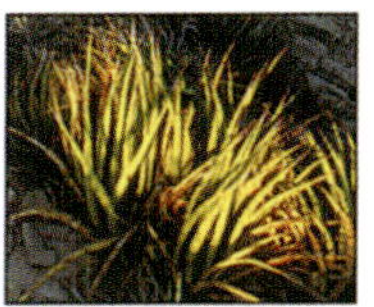

Little Big Bend

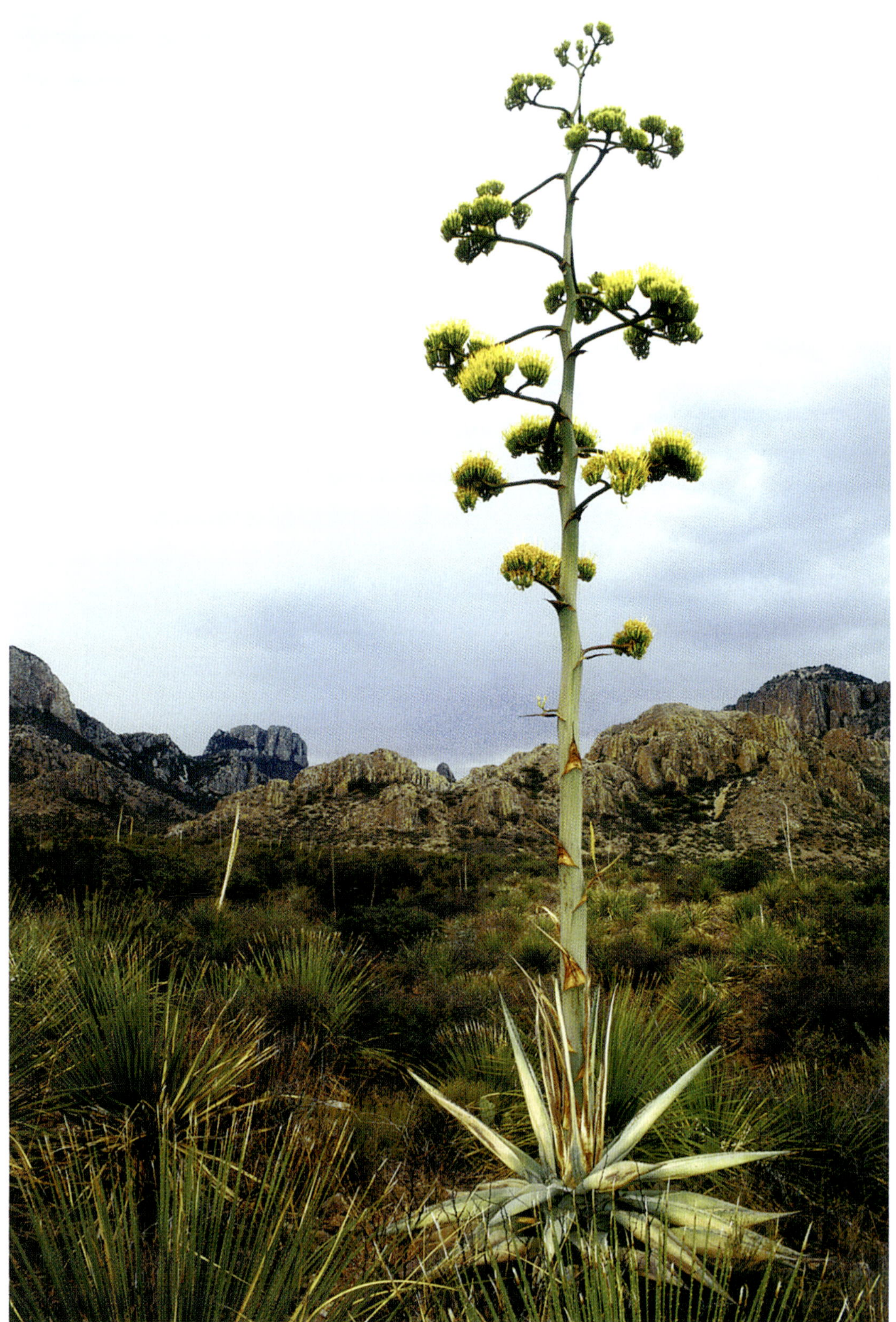

Havard agave in Green Gulch

BIG BEND, THE LAND OF EXTREMES

The Geography

Big Bend National Park is located at the southeastern edge of the Basin and Range physiographic province in the United States, a region characterized by elongated, uplifted mountain ranges running north-south alternating with sunken, arid valleys, or basins. This province extends across much of the southwestern United States, from the Sierra Nevada in eastern California to the Colorado Plateau and south through the border states to far west Texas. The Basin and Range province also reaches into northern Mexico from Baja California to Chihuahua.

The park is also situated at the northernmost limit of the Mexican Plateau, which stretches between the two great mountain ranges north of Mexico City, the Sierra Madre Occidental on the west, with elevations exceeding ten thousand feet, and the Sierra Madre Oriental on the east, with elevations surpassing thirteen thousand feet. The Sierra Madre ranges are considered southern extensions of the Rocky Mountains. The high Mexican Plateau lying between them covers 40 percent of the land area of Mexico, more than 165,000 square miles, from Chihuahua and Coahuila on the Rio Grande south to near Mexico City.

Elevation

Park elevation ranges from about 1,800 feet in the Rio Grande floodplain to 3,750 feet at Panther Junction, 5,400 feet in the Chisos Basin, and 7,825 feet at the top of Emory Peak in the Chisos Mountains. Most of the Dead Horse Mountains range reaches above 4,000 feet, and the highest point, Sue Peaks, is 5,854 feet above sea level. Rosillos Peak, at 5,445 feet, is the highest point in the Rosillos Mountains in the park's northwestern corner. Far to the southwest, Mesa de Anguila stretches from Lajitas to the mouth of Santa Elena Canyon and reaches a maximum elevation of 3,883 feet. These elevation changes contribute to a wide variation in moisture and temperature.

Rainfall

The shrub desert receives less than 10 inches of rainfall annually, while the Chisos Mountains average 18 inches. Arid regions near Mariscal Mountain and west to Castolon often get no more than 5 inches annually, while some moist Chisos Mountains canyons may receive well over 20 inches. Rainfall is highly concentrated in the summer and early fall months, which is quite appropriately considered the region's "monsoon" season. From a low of only .33 inch in March, average rainfall peaks in August at 2.4 inches (National Park Service 2006).

When the rain does come, it may

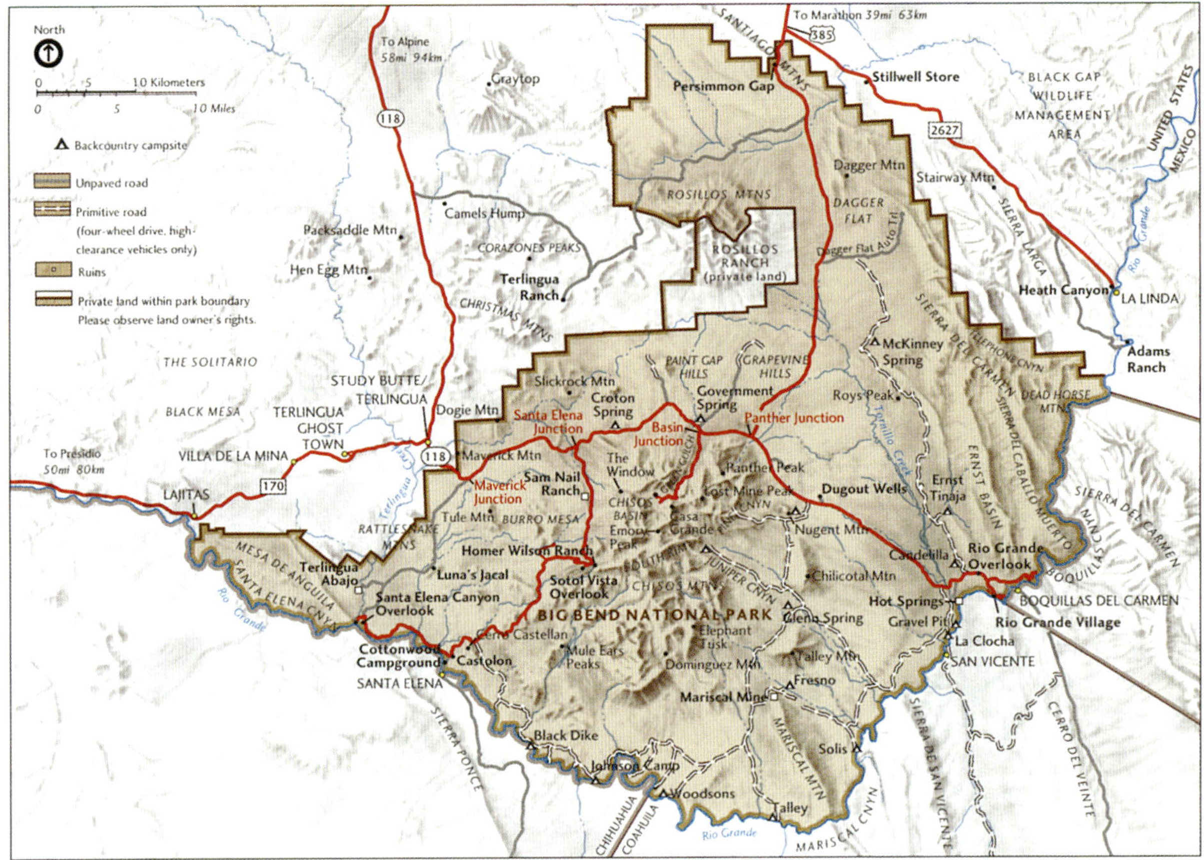

Source maps: National Park Service

come in sudden, torrential downpours, and these torrents occasionally produce life-threatening flash floods. Short and unpredictable yet intense thunderstorms originating mostly in the Gulf of Mexico bring the monsoon rains. The storms may bring with them fantastic, and also dangerous, electrical displays as well as high winds.

Temperature

Winter temperatures can be quite cool, often dropping below freezing at night. The summer is hot, characterized by daytime temperatures frequently above 100 degrees at lower elevations. Early summer, before the monsoon rains cool the desert, is the hottest part of the year, with average temperatures peaking at 94 degrees in June (National Park Service 2006).

Temperatures along the river and in the desert can be twenty-five degrees warmer than in the Chisos Mountains. In the scorching desert sun, temperatures may jump as much as fifty degrees or more from early morning to early afternoon. These are just air temperatures though. Also to be considered are the desert ground temperatures, which on a summer day can hit 180 degrees (National Park Service 2006). Animals and insects survive such life-threatening heat by taking shelter in burrows.

Natural Features

Big Bend's physical features are extreme as well, ranging from deep river canyons to mountain islands surrounded by a sea of desert.

Big Bend sits at a crossroads of geologic history and structure—a juncture of old and relatively new mountains. The new mountains, formed only 40 million to 60 million years ago, are the Santiago Mountains at the park's northern entrance and Mariscal Mountain on the Rio Grande, the park's southern boundary. They constitute a southern extension of the great Rocky Mountain range, much reduced in height over the eons, which continues into Mexico as the Sierra del Carmen and Sierra Madre. The old mountains, uplifted 275 million to 290 million years ago, are the remnants of the Ozark-Ouachita-Marathon mountain range, probably an ancient, largely eroded extension of the Appalachian Mountains of the eastern United States through Arkansas and southeastern Oklahoma into Texas. Unlike the trend of western mountains, this range runs northeast to southwest and crosses the Rocky Mountains chain in the Marathon Basin and near Persimmon Gap, the park's northern entrance.

The Rio Grande, the youngest major river in the United States, was formed in the last two million years. Known south of the United States as the Río Bravo del Norte, it delineates a 118-mile border between Big Bend National Park and the Mexican states of Chihuahua and Coahuila. The river's southeasterly flow from El Paso gradually shifts to the northeast, creating the fabled "Big Bend." In that process the river slices through three deep canyons—Santa Elena, Mariscal, and Boquillas—with steep, vaultlike limestone walls. The limestone layers were deposited during the Cretaceous period, between 145 million and 65 million years ago, when the area was covered by a shallow sea. Over millions of years, skeletons of marine animals accumulated on the ocean floor and can now be seen in the canyon walls.

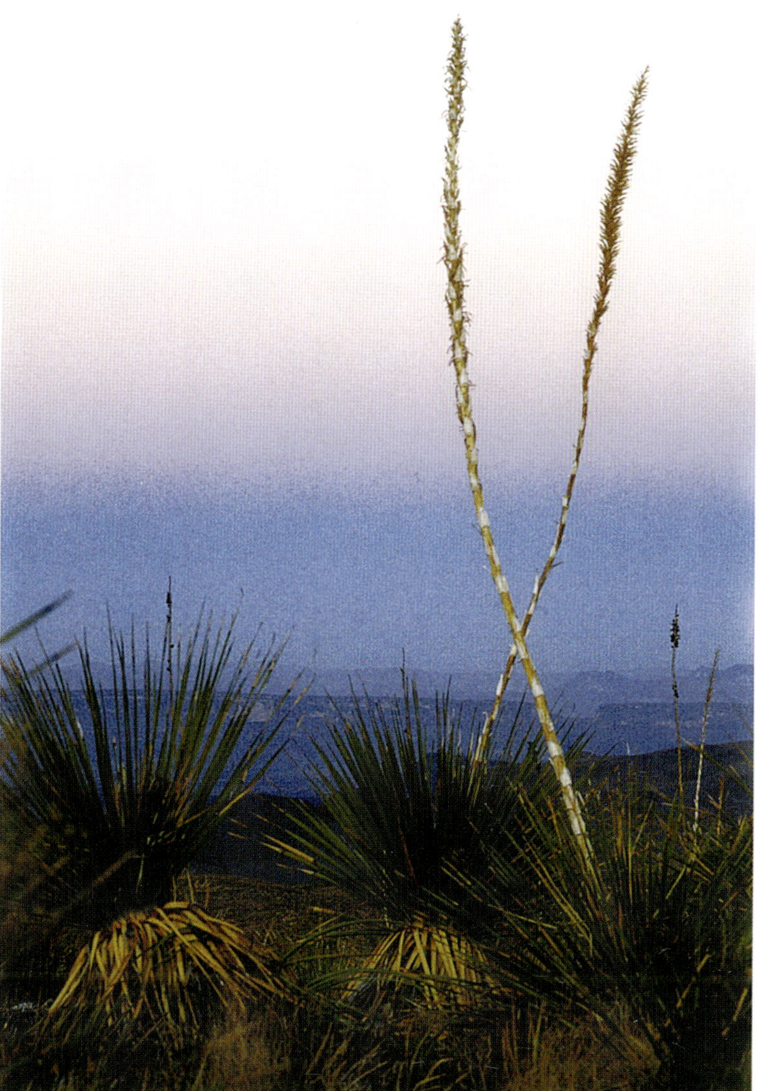

Smooth sotol at Sotol Vista

The Río Conchos, which flows northeast out of the Sierra Tarahumara of the Sierra Madre Occidental in western Mexico to join the Rio Grande at Presidio, is believed to be the prehistoric river that originally carved through Mesa de Anguila to form Santa Elena Canyon, through Mariscal Mountain to form Mariscal Canyon, and through the Sierra del Carmen to form Boquillas Canyon. Santa Elena Canyon, beginning at Lajitas in the park's southwest corner, extends east for eighteen miles to the junction with Terlingua Creek. This canyon is only thirty feet wide in places, with vertical walls fifteen hundred feet high. The most isolated and perhaps most spectacular of the three great canyons, sixteen-hundred-foot-deep Mariscal Canyon, cuts through the Lower Cretaceous limestone of Mariscal Mountain. Boquillas Canyon, also fifteen hundred feet deep, is the longest: beginning about three miles east of Rio Grande Village, it contin-

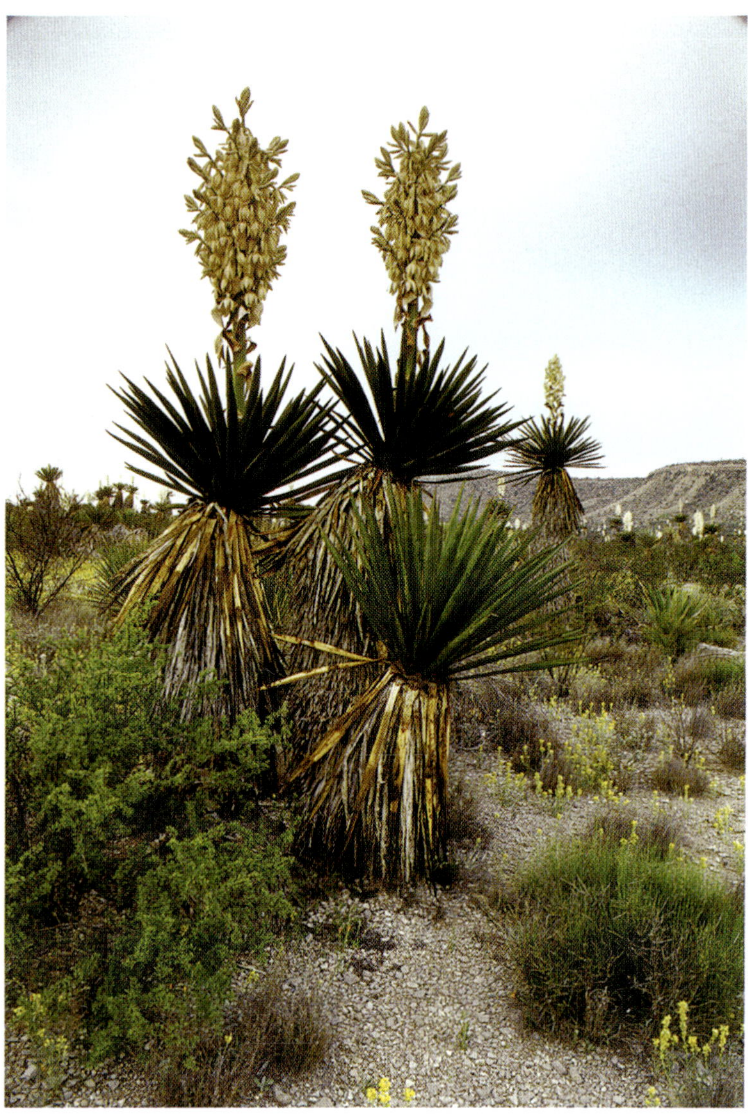

Giant dagger at Dagger Flat

ues through the Sierra del Carmen for twenty-five miles.

The massive Sierra del Carmen, looming above the Rio Grande, Rio Grande Village, and the small town of Boquillas, Mexico, on the park's eastern border, is the northernmost extension of the Sierra Madre Oriental out of Mexico and a defining feature of the Big Bend landscape. El Pico, also known in the United States as Schott Tower, is the prominent, notch-shaped peak atop the Sierra del Carmen at 7,022 feet. The Dead Horse Mountains, a name given to the Texas portion of the Sierra del Carmen, extend thirty miles northwest from Boquillas Canyon to the park's northern boundary and consist mostly of low hills alternating with dry arroyos.

Lying wholly within Big Bend National Park and surrounded by desert, the Chisos Mountains were created later, in the Tertiary period, when Big Bend was mostly floodplain and swamp. A series of volcanic eruptions 38 million to 32 million years ago, and subsequent erosion, formed this mountain island. Volcanic activity occurred in the Sierra Quemada below the South Rim, at Pine Canyon, and elsewhere. The Chisos Mountains lie at the heart of a forty-mile-wide sunken trough, dropped downward over time by block faulting that began 26 million years ago, as Mesa de Anguila on the west and the Sierra del Carmen on the east broke away and tilted upward.

The Chisos Mountains are built mostly of extrusive igneous rock that breaks through the earth's surface and lava and ash flows that can be seen in the multicolored white to maroon hills at lower elevations. Some of the Chisos Mountains, such as Nugent Mountain and Pulliam Ridge, were formed by intrusive processes in which the molten rock cooled underground, forming hard, erosion-resistant igneous chambers now exposed by the erosion of softer overlying materials. The knifelike ridges of dark rock, called "dikes," that cut across the foothills, especially on the west side of the Chisos, were also formed by now exposed magma intrusions as the covering layers of softer volcanic ash were worn away.

The Big Bend desert is part of the Chihuahuan Desert, the largest of four North American deserts, covering nearly two hundred thousand square miles. About 800 miles long and 250 miles wide, the Chihuahuan Desert reaches from San Luis Potosí in north-central Mexico across the Mexican Plateau into western Texas, southern New Mexico, and southeastern Arizona. In Mexico it is bordered by the Sierra Madre Occidental on the west, which blocks rainstorms from the Pacific Ocean, and the Sierra Madre Oriental on the east, which blocks most storms from the Gulf of Mexico. It is known as a "rain shadow" desert because it lies in the shadow of these massive mountain ranges that seldom allow moisture to penetrate onto the high desert plateau.

This primarily shrub desert is relatively green, however, because much of it lies between thirty-five hundred and five thousand feet and because rains that do penetrate the enclosing mountains come primarily in the hottest part of the year when they are most critical. Formed about eight thousand years ago, the young desert comprises alluvial flats, bajadas (low, dry, gravelly slopes formed by alluvial fans at the base of a mountain range), and small mountain ranges. Almost 80 percent of the Chihuahuan Desert consists of calcareous soils formed from beds of limestone when the area was beneath the sea. Sandy and gravelly soils cover the desert flats, but debris from mountain erosion covers the bajadas.

In the driest part of the desert in Big Bend National Park, just above the Rio Grande floodplain, a peculiar "desert pavement" composed of shallow limestone soil covered by an accumulation of interlocking rock fragments coats the desert surface. Over decades this dry pavement, nearly devoid of vegetation, has been varnished by dark clay particles and bacteria living on them. In some places small chimneylike pinnacles (for example, on the Chimneys Trail) and sandstone pillars (such as those off the Persimmon Gap Road) protrude from the barren flats. In other places colorful rain-eroded badlands (for example, the Maverick Badlands at the park's west entrance) and ever-changing sand dunes (as at Boquillas Canyon) typify the landscape.

Ecological Diversity

The unusual confluence of extremes in elevation, precipitation, temperature, and natural habitats has produced an enormous biodiversity that is a principal attraction of the park. Differences in elevation contribute to the dramatic variation in temperature and rainfall, and all three factors help create a striking diversity of habitats. The number of plant habitats is also affected by a wide variation in soil types, from gypseous, sandy, and saline in the flatter desert areas to the gravelly soils of bajadas, and the rocky soils, both limestone and volcanic, of the foothills and mountains.

The Chihuahuan Desert is characterized by small mountain ranges alternating with valleys, higher mountains of biotic islands surrounded by desert, and river valleys with their own unique floodplain habitats. These features have made the Chihuahuan Desert richer than other North American deserts. Big Bend is a perfect example, with its small mountain ranges, such as the Rosillos and Santiago, in the north part of the park; the Sierra Quemada to the south; the Rattlesnake Mountains to the southwest; the Chisos complex, forming a cooler island in the center of it all, high above the desert floor; and the valley of the Rio Grande, which delineates the southern boundary.

In Big Bend all these factors have conspired to produce a seemingly endless array of mini-habitats. The summer monsoon rains are an additional wild card, creating even more, albeit ephemeral, plant communities, such as those of intermittent streams and desert arroyos, alluvial flats, and regenerated desert springs, all with their own special mix of species. Locals in this vast country term the monsoon season the "fifth season" because a whole new group of desert annuals suddenly appear, spring-flowering plants bloom again, and the desert is more lush and green than at any other time of year.

As much as it is a crossroads of ancestral mountain ranges, Big Bend is a crossroads of biotic communities. Many plants found in the northern United States in cooler climates reach the southern limit of their range in the park, and many South or Central American and Mexican species, some tropical, reach their northern limit here. Likewise, species more commonly found in deserts farther west and in the Rocky Mountains may extend into the Big Bend, and species associated with wetter climates to the east also find their limits here.

The Chihuahuan Desert region is home to thirty-five hundred plant species, a thousand of which are found nowhere else. Although much smaller in size, Big Bend National Park houses as many as thirteen hundred species, with more kinds of cacti than in any other U.S. park. In Big Bend you will find widely spaced plants, such as creosote, adapted to an arid and sparse desert environment, and you will find delicate orchids in hot but moist mountain habitats. You will find plants armed to the teeth with thorns, prickles, and spines growing only a few feet from ferns, longspur columbine, or cardinal flower at desert springs or near willow and giant cane lining the river corridor.

The first park naturalist, Ro Wauer, divided the park into five broad vegetation zones, each supporting a different mix of plants: (1) the river floodplain–arroyo, from 1,800 to 4,000 feet; (2) the shrub desert, from 1,800 to 3,500 feet; (3) the sotol-grassland, from 3,200 to 5,500 feet; (4) the woodland, from 3,700 to 7,800 feet; and (5) the moist Chisos woodland, from 5,000 to 7,200 feet (Wauer 1980).

Plants marked with asterisks (*) in the following text are included in the detailed plant descriptions in this book. The other plants cited are not, but most of them are relatively common and can be found in many wildflower and flora guides, especially those listed in the references section of this book.

River Floodplain–Arroyo

More closely resembling dense jungle than the sparse desert nearby, the river floodplain is a long, narrow oasis for plants, many unsuited to the desert environment. A lush green belt of rapidly growing trees and shrubs lines the Rio Grande and extends its arms up countless desert arroyos.

Reaching as much as fifteen feet high, giant cane (*Arundo donax*) and common reed (*Phragmites australis*) clearly demarcate the river corridor. They are often accompanied by cattail (*Typha domingensis*) and a number of willow and baccharis shrubs, mostly Goodding willow (*Salix gooddingii*), coyote willow (*Salix exigua*), and seepwillow (*Baccharis salicifolia*), which is not a willow at all but an aster.

Invasive tamarisks also line and sometimes choke the river, pushing out the native species. The largest is athel (*Tamarix aphylla*), found primarily near Boquillas Canyon; it is easily distinguished by its scalelike leaves sheathing the stems. Guayacán* (*Guaiacum angustifolium*); tree tobacco (*Nicotiana glauca*), a slender shrub with yellow tubular flowers particularly attractive to hummingbirds; and the colony-forming aster Mexican devilweed (*Chloracantha spinosa*), with leafless green stems and dainty white and yellow flower heads, may compete with tamarisk along the river bank.

Far more useful than the tamarisks, honey mesquite (*Prosopis glandulosa*) and screwbean mesquite (*Prosopis pubescens*), with its signature corkscrew-shaped pods, provide food and protective cover for a variety of birds and animals. Huisache acacia (*Acacia farnesiana*), with round, yellow flower clusters, is prominent near Rio Grande Village. Most of the stately native cottonwoods that once lined the river were harvested long ago by farmers and miners, but Arizona cottonwood (*Populus fremontii* ssp. *mesetae*) occurs near Santa Elena and Boquillas canyons, and narrowleaf cottonwood (*Populus angustifolia*) is rare.

Desert willow* (*Chilopsis linearis*); the brushlike burrobrush (*Hymenoclea monogyra*), with needlelike leaves and abundant, papery-white flower heads; blackbrush acacia* (*Acacia rigidula*); and other species extend from the river up creek beds and dry arroyos. Scented lippia (*Lippia graveolens*), a highly aromatic shrub with creamy yellow-white flower clusters, occurs frequently along the Hot Springs Trail and also along desert washes winding into dense chaparral and rocky desert canyons miles to the north and west.

Of special interest is the rare littleleaf brongniart* (*Brongniartia minutifolia*), a feathery-foliaged legume with butterfly-shaped yellow flowers, which occupies a small strip from San Vicente to Mariscal Mountain and sandy washes leading toward Glenn Spring. The purplish-pink, globelike blooms of another uncommon shrub, Turner mimosa* (*Mimosa turneri*), may be seen at fishing camps on the Rio Grande or along the River Road. A far more plentiful but bizarre mint, bladder sage* (*Salazaria mexicana*), with characteristic paperbag-like fruits, clogs sandy arroyos from Castolon to Smoky Creek.

Of course, many of the plants prominent in floodplain-arroyo environments also reach into the shrub desert or even higher elevations. Honey mesquite is nearly ubiquitous. Catclaw acacia (*Acacia greggii*), with recurved spines and cylindrical yellow flower clusters, forms thickets along the river but also stretches up Tornillo and other creeks and skirts springs in mountain foothills. Viscid acacia (*Acacia neovernicosa*), with short, straight thorns; round, yellow flower heads; and shiny-sticky leaflets, is present near Boquillas Canyon but also at Terlingua Abaja, along Fresno Creek, and at numerous sites in the Chisos foothills. Narrowleaf moonpod* (*Acleisanthes angustifolia*) and littleleaf moonpod* (*Acleisanthes parvifolia*), with their odd membranous, winged fruits, are conspicuous in the limestone hills above Hot Springs but are widely scattered elsewhere, especially on gypseous clay soils in the Dead Horse Mountains.

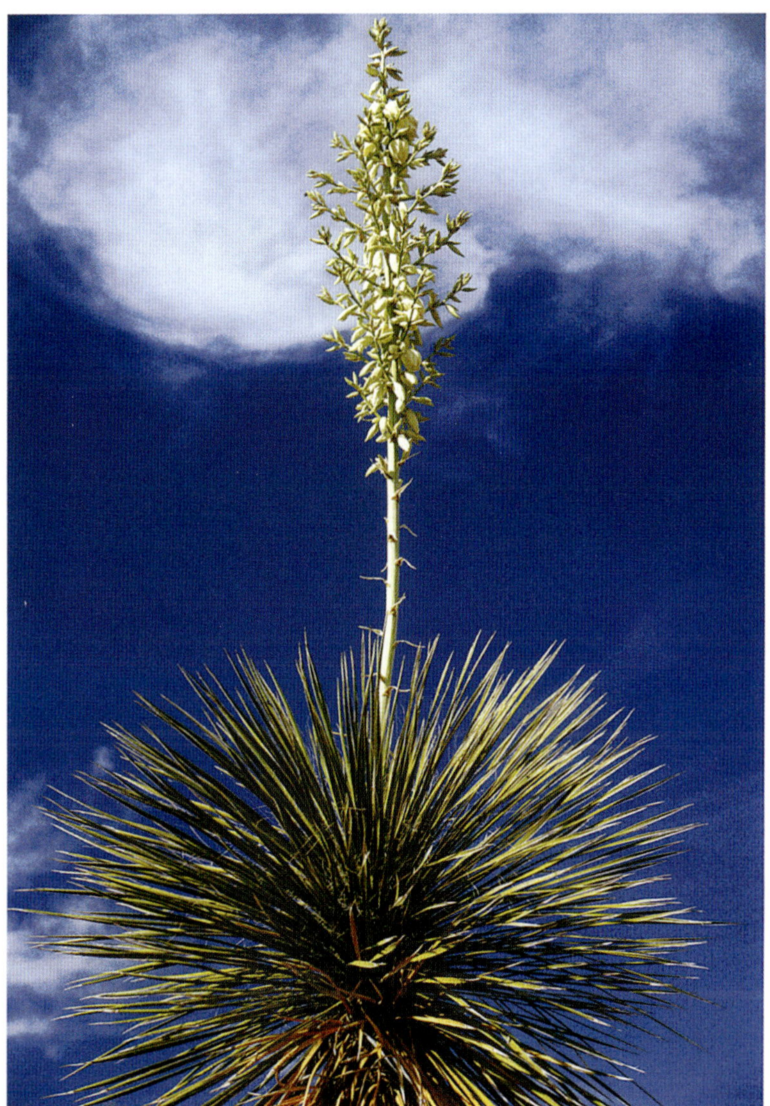

Thompson yucca

Prickly pear cacti, including the rare golden-spined prickly pear* (*Opuntia azurea* var. *aureispina*) and spiny-fruited prickly pear* (*Opuntia spinosibacca*), occupy ledges above the Rio Grande. Cylindrical-jointed cholla cacti, such as the common tasajillo or desert Christmas cholla (*Opuntia leptocaulis*), with slender joints, greenish-yellow flowers, and bright red fruits resembling Christmas decorations, are residents of the river corridor, arroyos leading from it, and the hills above. The less common candle cholla* (*Opuntia kleiniae*), with somewhat thicker joints, is most prevalent from Santa Elena Canyon to Black Dike, and the rare Big Bend cholla* (*Opuntia imbricata* var. *argentea*), with stout joints and lustrous silvery spines, is restricted to Mariscal Mountain and the surrounding creosote flats.

A number of unusual smaller cacti are found exclusively in this environment. The whiskerbrush pincushion cactus* (*Coryphantha ramillosa*), classified as Threatened under the Endangered Species Act, and the low, mounding club cholla, Big Bend devil cholla* (*Opuntia densispina*), have very circumscribed distributions along or near the Rio Grande. Two other noteworthy species reach their northern limits here. Glory of Texas cactus* (*Thelocactus bicolor* var. *bicolor*), remarkable for its luscious magenta flowers and strange twisting and curling spines, inhabits silty washes and low hills just north of the river. Populations of Potts' mammillaria cactus* (*Mammillaria pottsii*), notable for its tiny maroon flowers, are largely confined to Mesa de Anguila and Mariscal Mountain.

Shrub Desert

A predominantly shrub desert covers almost half of the park. With slightly more rainfall and cooler temperatures during parts of the year, this desert is unlike the relatively drier Sonoran and Mojave deserts to the west and more closely resembles the Great Basin Desert farther north. Even in the shrub desert portions of the park, a variety of habitats coexist. Low desert flats with mostly sandy and gravelly soils support sparse, well-spaced vegetation. This plant community is replaced by yuccas and other less drought-tolerant plants on the bajadas or low desert hills with slightly deeper, rockier soils that hold more moisture.

The most common shrub is the ubiquitous creosote bush* (*Larrea tridentata*), easily recognized by its fuzzy, round fruits and signature creosote odor. Highly adapted to the desert environment, it is a world unto itself: one grasshopper and one walking stick feed on nothing but creosote, and twenty-two species of bees rely exclusively on creosote pollen (Bowers 1993). Numerous species dwell in burrows beneath these shrubs and emerge with the spring bloom. Creosote bush is often accompanied by tarbush* (*Flourensia cernua*), a sometimes codominant shrub with leaves that produce a tarlike odor.

Stem succulents are key components of this desert landscape. Lechuguilla* (*Agave lechuguilla*) is considered an indicator species of the Chihuahuan Desert. Its presence helps to define the desert boundaries. Texas false agave,* or hechtia (*Hechtia texensis*), a member of the pineapple family, often grows alongside lechuguilla on dry limestone hills.

Ocotillo, found in other southwestern deserts; candellilla; and leatherstem are other typical plants of the desert lowlands. The thorny, whiplike branches of ocotillo* (*Fouquieria splendens*) form small "forests" near the west entrance to the park and along the River Road. The waxy stems of candelilla,* or wax plant (*Euphorbia antisyphilitica*), fill desert lowlands to the east of the park. They are still used in Mexico to manufacture wax candles. Leatherstem* (*Jatropha dioica* var. *graminea*), sometimes known as dragon's blood, is also widespread at lower elevations. Its rubbery, reddish stems are used to make a red dye.

The Chihuahuan Desert supports about four hundred species of cacti, many endangered, and about sixty-five species can be found in Big Bend National Park, including prickly pear, cholla, hedgehog, pincushion, fishhook, and barrel cactus. Many grow on low limestone ridges in the desert, but some occur along the river and others on the highest mountain slopes. Dry limestone hills near the river and throughout the Dead Horse Mountains are the preferred habitat for a number of unusual small cacti, such as the white, mushroomlike Boke's button cactus* (*Epithelantha bokei*); the white, fuzzy golf ball cactus* (*Mammillaria lasiacantha*); and the spineless, warty living rock cactus* (*Ariocarpus fissuratus* var. *fissuratus*).

The giant, waxy yellow flowers of Texas rainbow cactus* (*Echinocereus dasyacanthus*) and the spectacular tricolored flowers of the Chisos hedgehog cactus* (*Echinocereus chisoensis* var. *chisoensis*), also classified as Threatened, color the desert flats, growing in small gullies near desert pavement or in shallow limestone soils not far above the floodplain. Far less conspicuous are horse-crippler cactus* (*Echinocactus texensis*), with punishing spines only slightly above the ground, and Heyder's pincushion cactus* (*Mammillaria heyderi* var. *heyderi*), which forms spiny mats flush with the desert floor. They inhabit silty soils and sandy flats. The purple pads of Big Bend purplish prickly pear* (*Opuntia azurea* var. *parva*) and the atypical, spineless pads of blind prickly pear* (*Opuntia rufida*) are a common sight from the desert lowlands well into the grasslands and mountain foothills.

A myriad of other thorny, prickly, and spiny plants abound, including a variety of acacias and mimosas, allthorn, buckthorn, and mesquite. The thicket-forming allthorn* (*Koeberlinia spinosa*), with its maze of green, leafless thorns, is characteristic of this bristling community.

Warnock condalia (*Condalia warnockii*), a spine-tipped, dense shrub with tiny, petalless, greenish-yellow flowers and reddish-black fruit, grows on low, dry bajadas or winds along rocky desert arroyos.

Several species of Mormon tea, or ephedra, a mostly leafless shrub with greenish-yellow stems, can be quite common locally, especially at Dagger Flat, Dog Flats, and Tornillo Flat. By far the most common are boundary ephedra (*Ephedra aspera*) and longleaf ephedra (*Ephedra trifurca*). The stems of both species were boiled by Native Americans and Anglo settlers to make a mild tea sometimes used for medicinal purposes.

Salt-tolerant plants, such as the widespread fourwing saltbush (*Atriplex canescens*) with its conspicuous winged fruits, thrive on saline and gypseous desert flats. The less common obovateleaf saltbush (*Atriplex obovata*), a rounded, silvery shrub, and tubercled saltbush (*Atriplex acanthocarpa*), with prominent tubercled fruits, are mostly limited to clay soils near the park's west entrance. The low, rounded, nearly white domes of shrubby tidestromia* (*Tidestromia suffruticosa*) are largely restricted to gypseous clay and sandy soils in the Dead Horse Mountains, while the prostrate mats of fleshy tidestromia* (*Tidestromia carnosa*), an imperiled annual, cover clay dunes and saline flats far to the west in the Maverick Badlands and near Terlingua Abaja.

Threeflower goldenweed, or Terlingua stickbush* (*Xylothamia triantha*), a notoriously sticky shrub with rayless yellow flower heads, occurs in similar habitats, often on gyp-clay outcrops near Terlingua Abaja and Santa Elena Canyon. It is a Mexican species that reaches its northern limit in the park. Two larger, slightly spinescent shrubs of the nightshade (Solanaceae) family, silvery wolfberry* (*Lycium puberulum* var. *berberioides*) and the more open, less dense Berlandier wolfberry (*Lycium berlandieri*), known for their creamy white, tubular flowers and small red or orange berries, also occupy clay and alkaline flats, especially at Dog Flats and along upper Tornillo Creek.

Many shrubs span the desert floor and reach into the mountain foothills. Propellerbush* (*Janusia gracilis*), a strange half-vine, half-shrub pollinated by oil-gathering bees and producing propeller-shaped fruits, is present almost everywhere except in the higher Chisos Mountains. Skeleton goldeneye (*Viguiera stenoloba*), a brightly blooming sunflower with yellow flower heads and thin, skeletonlike leaves, and mariola (*Parthenium incanum*), a nondescript shrub with conspicuous white, woolly foliage and yellowish-white disk flowers, are widespread and characteristic Chihuahuan Desert plants found in both desert and grasslands. Much less common is mariola's close relative, guayule,* or rubber plant (*Parthenium argentatum*), which was overharvested for rubber production.

The regal, early blooming annual Big Bend bluebonnet* (*Lupinus havardii*) is the signature wildflower of this rich desert.

Sotol-Grassland

At about thirty-two hundred feet, shrub desert is gradually replaced by the sotol-grassland community, which covers another 49 percent of the park. A superb example of this community is Sotol Vista on the Ross Maxwell Scenic Drive, which provides wonderful sunset views above towering sotol stalks. These grasslands occur in upland areas of increased rainfall and deeper soils, especially on mesas and rolling hills, and extend in narrow corridors into more forested mountains to fifty-five hundred feet.

Grama and other grasses are predominant at lower elevations, while smooth sotol and beargrass extend up the hills. Curly clumps of chino grama* (*Bouteloua ramosa*) cover large expanses surrounding the Chisos. As much as fifteen feet high, the flowering stalks of smooth sotol* (*Dasylirion leiophyllum*) dominate the foothills. Beargrass* (*Nolina erumpens*), a mounding, grasslike shrub, often grows alongside smooth sotol.

Yucca, agave, acacia, and mimosa are other typical shrubby plants of the foothills. Present throughout the park at low to middle elevations, the markedly shaggy Spanish dagger* (*Yucca torreyi*) is the most common grassland yucca. The largest yucca is the massive, treelike giant dagger* (*Yucca faxoniana*), which dominates bajadas and desert plateaus at Dagger Flat and elsewhere in the northern Dead Horse Mountains. Smaller, slimmer, with much narrower leaves, Thompson yucca* (*Yucca thompsoniana*) may grow alongside giant dagger in the Dead Horse Mountains in desert scrub and foothills. The tall, slender soaptree yucca, notable for the curling, threadlike fibers of its leaf margins, is encountered on desert grasslands but is also conspicuous at lower elevations near the Rio Grande.

Easily recognized by its towering, candelabra-shaped flowering stalk, the majestic Havard agave* (*Agave havardiana*) is another key component of this community. The rare Chisos agave (*Agave glomerulifolia*), with much narrower leaves and a more slender flowering stalk, is an unusual hybrid between Havard agave and lechuguilla. Both also occur at much higher elevations in the mountains.

Roemer acacia* (*Acacia roemeriana*), with see-through oblong pods, is one of the most common acacias of the foothills. The atypical fern acacia (*Acacia angustissima*), lacking spines and prickles, spreads through underground rhizomes to form colonies in both grassland and mountain habitats. Catclaw mimosa (*Mimosa aculeaticarpa* var. *biuncifera*), with paired recurved prickles, white or pink flower spheres, and pods with recurved spines along the margins, is a prominent, thicket-forming mimosa of the grasslands. The smaller, often scrawny Emory mimosa* (*Mimosa emoryana*), with unusual yellow-bristled pods, is most prevalent on low slopes in the Grapevine and Paint Gap hills.

Two other prominent grassland legumes are feather dalea and black dalea. Feather dalea* (*Dalea formosa*), with striking plumelike, purplish flowers, inhabits gravelly flats and open slopes in desert and grassland areas. Populations of black dalea* (*Dalea frutescens*), a rounded, low shrub with purplish but not plumose flowers, stretch from the foothills into higher mountains.

Two other common grassland shrubs with showy plumose flowers or fruits are plume tiquilia and Apache plume. A rounded shrub less than one foot high, plume tiquilia* (*Tiquilia greggii*) produces small magenta flowers in woolly balls. Apache plume (*Fallugia paradoxa*), a larger shrub to six feet high, bears showy white flowers and purple-tinged, plumelike fruits said to resemble the headdress of Apache warriors.

Three species of cenizo, or purple sage, members of the figwort (Scrophulariaceae) family, serve as the desert's ornamentals, creating a gardenlike atmosphere when they spring into bloom after soaking summer rains. They are most common in the Chisos foothills and across the shrub desert to Boquillas Canyon. The largest is purple sage* (*Leucophyllum frutescens*), with silvery gray leaves and rose-colored flowers. Boquillas silverleaf* (*Leucophyllum candidum*), a smaller, more compact shrub, has nearly white leaves and purple or dark violet flowers. Big Bend silverleaf (*Leucophyllum minus*), a short, rounded shrub up to three feet high, is distinguished by its lavender to purple flowers, which have an arresting white blotch with orange dots in the throat.

Three other small shrubs—mejorana, shorthorn jefea, and desert yaupon—are plentiful on gravelly hills surrounding the Chisos. The aromatic mejorana* (*Lantana macropoda*), with pinkish flower clusters on elongated stalks, is especially abundant from Sam Nail Ranch to the Chisos. A low shrub with yellow flower heads, shorthorn jefea (*Jefea brevifolia*) is a characteristic Chihuahuan Desert plant of shrub desert and grasslands, quite prevalent in the Grapevine Hills. Desert yaupon* (*Schaefferia cuneifolia*), with short, stiff, whitish branchlets, is common along the road from Nugent Mountain to Glenn Spring.

Four small trees of this community—desert olive* (*Forestiera angustifolia*), desert sumac* (*Rhus microphylla*), desert hackberry, and Texas persimmon—can be found seeking moisture along desert arroyos and at desert springs, but they are more common in the grasslands. Desert hackberry (*Celtis pallida*) is a dense evergreen with spiny gray branches, minuscule flowers, and round orange berries. An evergreen with smooth, gray bark, Texas persimmon (*Diospyros texana*) has bell-shaped white flower clusters and black fruit.

A slender shrub with stiff, spine-tipped branches, beebrush (*Aloysia gratissima*) may form extensive thickets in the grasslands and at higher elevations. A significant bee attraction, it produces spikes of tiny, sweet-smelling white flowers. Agarito (*Mahonia trifoliolata*), identified by its three spiny, waxy, gray-green leaflets and small, cup-shaped yellow flowers, is yet another prickly shrub of this regime.

A variety of cacti share grassland habitats with larger shrubs. Tree cholla* (*Opuntia imbricata* var. *arborescens*), a treelike cactus with spiny, braidlike joints, appears alongside smooth sotol and beargrass in the foothills. Nipple cactus* (*Mammillaria meiacantha*), a two-inch-high round pincushion with a delightful ring of cream-colored flowers, often grows beneath protective shrubs. Rusty hedgehog cactus* (*Echinocereus viridiflorus* var. *russanthus*), with rust-red flowers sticking out like miniature torches; cob cactus* (*Coryphantha tuberculosa* var. *tuberculosa*), named for the corncoblike tubercles at the base of its stem; and the fishhook catclaw cactus* (*Glandulicactus uncinatus* var. *wrightii*), with a barrel-shaped stem and wine-colored flowers, are scattered across the park in a variety of ecological zones, from near the Rio Grande to the Chisos foothills.

Woodland

Woodland communities begin to take over at thirty-seven hundred feet, mostly above the grasslands all the way to Emory Peak. Taller trees can be seen at forty-five hundred feet, and green corridors with a wide variety of trees fill the drainages at higher elevations. Averaging about eighteen inches of annual rainfall, and much cooler than the shrub desert below, the mountainous terrain is typical of oak-juniper-pinyon woodlands of the southwestern United States.

The Chisos woodland community includes one pine, several junipers, and many oaks as well as many other small trees, including the beautiful Texas madrone, Texas mountain laurel, Mexican buckeye, littleleaf leadtree, Gregg ash, fragrant ash, netleaf hackberry, southwestern chokecherry, and rare Big Bend hophornbeam.

Mexican pinyon pine* (*Pinus cembroides* ssp. *cembroides*), a primarily Mexican species, reaches into mountainous areas of the Trans-

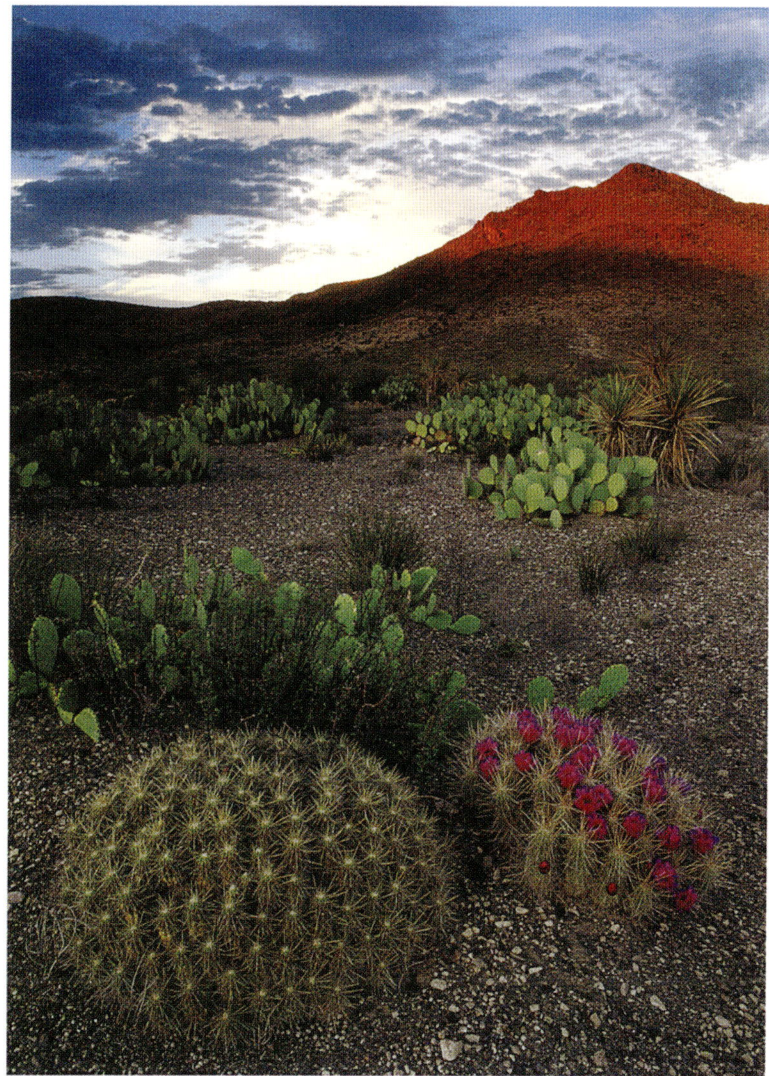

Strawberry cactus at Persimmon Gap

Pecos. Alligator juniper* (*Juniperus deppeana* var. *deppeana*) is an unforgettable tree of the woodlands, with its trademark checkered bark shaped like the back of an alligator. Drooping juniper, or weeping juniper* (*Juniperus flaccida*), is known for its graceful, drooping branches. Perhaps the least distinctive of the junipers, redberry juniper (*Juniperus pinchotii*), is identified by a dense habit, reddish-brown berries, and fissured bark. It is very common in Green Gulch and the Basin.

More than a dozen species of oaks can be found throughout the Chisos woodlands, but by far the most common are Chisos red oak (also called Graves' oak), Emory oak, and gray oak. Chisos red oak* (*Quercus gravesii*) is memorable for its broadly lobed, bristle-tipped leaves that turn a brilliant red in the fall. The semievergreen Emory oak

(*Quercus emoryi*) has dark, furrowed bark and shiny, usually toothed and spine-tipped leaves up to two and one-half inches long. Gray oak (*Quercus grisea*) is a mid-sized tree with gray-green, oblong, and smooth-edged, leathery leaves up to one and one-half inches long.

A number of rare oaks, or oaks with a restricted distribution, are also present in the Chisos Mountains. Coahuila scrub oak* (*Quercus intricata*), a thicket-forming, densely branched dwarf oak, dominates the woodlands from Laguna Meadow to Cattail Canyon. This Mexican species reaches into the United States only in the Chisos Mountains. The critically imperiled Chisos oak* (*Quercus graciliformis*), a graceful, slender oak with arching branches and dangling lancelike leaves, is endemic to the Chisos Mountains.

The pink, peeling bark and twisting branches of Texas madrone* (*Arbutus xalapensis*) make it one of the region's most spectacular trees. Although scattered at lower elevations in the grasslands, Texas mountain laurel (*Sophora secundiflora*), an evergreen tree with intoxicatingly fragrant clusters of bluish-purple flowers, often nestles into mountain canyons. Easily recognized by its fragrant pink flowers and leathery, three-chambered pods, Mexican buckeye* (*Ungnadia speciosa*) lines rocky Chisos canyons. Littleleaf leadtree* (*Leucaena retusa*), a small tree with flowers in dense, golden yellow balls, is prominent in Green Gulch and along the road to Dagger Flat.

Gregg ash (*Fraxinus greggii*) and fragrant ash (*Fraxinus cuspidata*), small trees with drooping white flower clusters and oblong, winged fruits, are very common in upper Green Gulch and the Basin and on western slopes of the Chisos. Gregg ash is distinguished by spoon-shaped leaflets with rounded tips, and fragant ash by more lancelike leaves with elongated, pointed tips.

In the woodlands, the spineless netleaf hackberry (*Celtis laevigata* var. *reticulata*), with distinctive net-veined leaves, may replace the more common desert hackberry of the grasslands. Southwestern chokecherry* (*Prunus serotina* var. *virens*), a small tree with shiny, dark green leaves and brilliant white flowers, is concentrated in the Chisos, especially in shady canyons. The critically imperiled Big Bend hophornbeam* (*Ostrya virginiana* var. *chisosensis*), known for its markedly shaggy gray bark and hop-like clusters of bladdery but flattened fruits, inhabits moist, rocky sites at higher elevations.

The variety of smaller woodland shrubs is seemingly endless. Mountain mahogany (*Cercocarpus montanus*), with distinctive fuzzy-tailed, twisting fruit, lines the road through Green Gulch into the Basin. The spectacular mountain sage* (*Salvia regla*), recognized by its unusual, two-lipped crimson flowers, crosses from Mexico into the United States only in the Chisos Mountains. Scarlet bouvardia* (*Bouvardia ternifolia*), also known as firecracker bush because its brilliant scarlet flowers vaguely resemble firecrackers, is conspicuous in Green Gulch and the Basin.

Eggleaf silktassel* (*Garrya ovata* ssp. *lindheimeri*), a dense evergreen with male flowers in catkins on prominent silky stalks, is plentiful from the Chisos Basin to the South Rim. Desert ceanothus* (*Ceanothus greggii*), a stubby shrub with short, rigid branches and profuse white flower clusters, prefers more arid, open slopes, especially along the Laguna Meadow Trail.

Havard plum and fascicled bluet are more prevalent at lower elevations in both foothills and mountains. The rounded, compact Havard plum* (*Prunus havardii*) bears small white flowers with widely spaced oval petals. Once considered rare, it occurs on gravelly soils and rocky slopes at mid elevations. Fascicled bluet* (*Hedyotis intricata*), a small shrub with intricate, stiff branches and tiny white flowers, occupies exposed ridges or sits at the base of canyon walls, especially in Lower Blue Creek Canyon.

Texas kidneywood* (*Eysenhardtia texana*), with spikelike white flower clusters, and palo prieto* (*Vauquelinia corymbosa* ssp. *angustifolia*), with bouquetlike clusters of brilliant white flowers, are also common in the woodlands at middle elevations. The desert sumac of the grasslands is replaced in the woodland by its close relatives skunkbush (or fragrant) sumac and evergreen sumac. Skunkbush sumac (*Rhus trilobata*) is distinguished by its leaves, which are divided into three distinctive leaflets that turn a brilliant red or orange in the fall. Evergreen sumac (*Rhus virens*) has shiny, dark green, leathery leaves.

Although less conspicuous here

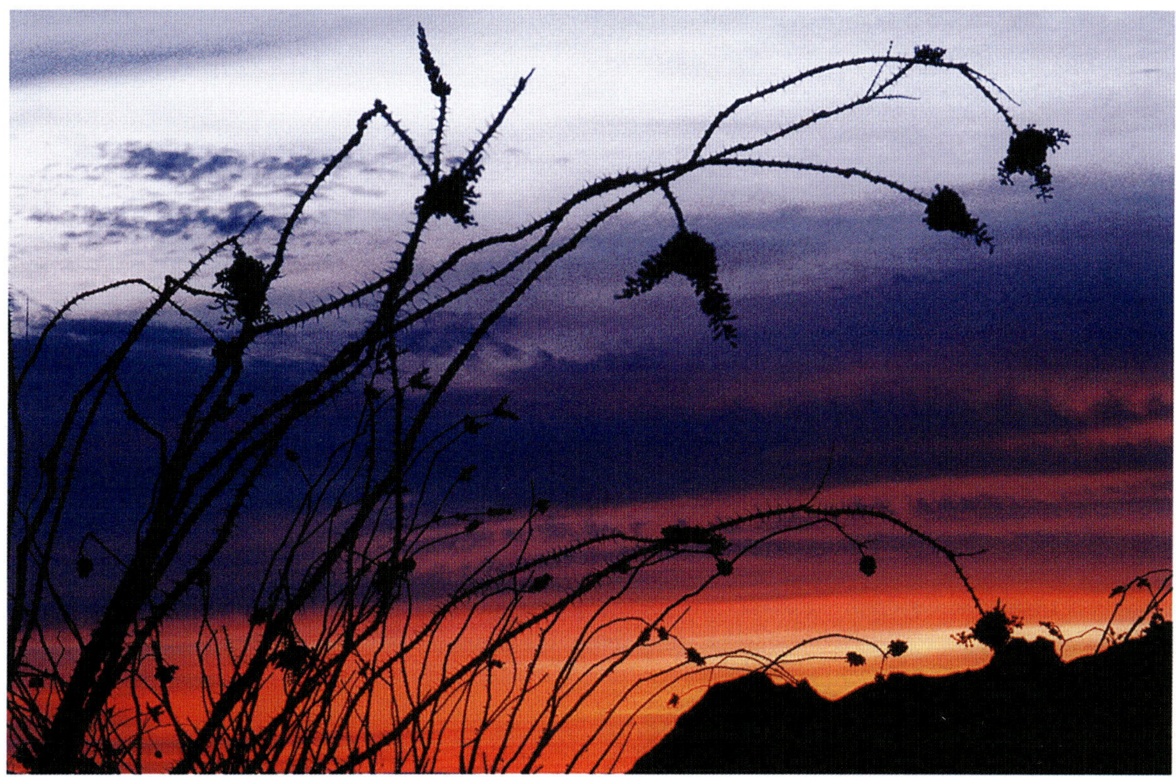

Ocotillo near Maverick Mountain

than elsewhere in the park, cacti are an important component of the woodland community, even at higher elevations. The rich wine-colored flowers and crimson fruits of the mounding Texas claret-cup cactus* (*Echinocereus coccineus* var. *paucispinus*) dot rocky volcanic soils of the woodlands. Chisos prickly pear* (*Opuntia chisosensis*), with fetching buff or peach-colored flowers, is geographically isolated, confined to the Chisos Mountains in the United States and northern Coahuila in Mexico. The globally imperiled Chaffey's pincushion cactus* (*Coryphantha chaffeyi*), less than four inches high, is a Mexican species found infrequently in the Chisos amidst rocks or decaying leaves at higher elevations.

Even within this woodland vegetation zone dramatic contrasts occur, with dense shrubbery covering cooler north-facing slopes while plants more characteristic of the desert, such as ocotillo and lechuguilla, invade drier south-facing slopes.

Moist Chisos Woodland

Comprising only 2 percent of the forested area, moist Chisos woodland covers only eight hundred acres in a few isolated locations, primarily Boot and Pine canyons and the Northeast Rim. In these special places some species reach the extreme southern limit of their ranges in the United States—the Arizona pine and the majestic Arizona cypress, the Douglas fir, bigtooth maple, and even quaking aspen. Adding to the mix in this mostly coniferous forest are drooping juniper and alligator juniper, birchleaf buckthorn, littleleaf mock-orange, and a variety of small oaks and other shrubs.

Primarily a Mexican species, the Arizona pine (*Pinus arizonica* var. *stormiae*) of the Chisos Mountains is found nowhere else in the United States. This variety, with leaves in clusters of three to five, is known only from Boot and Pine canyons and Crown Mountain. The towering Arizona cypress* (*Cupressus arizonica*) is largely confined to Boot Canyon. Douglas fir (*Pseudotsuga menziesii*), recognized by its blue-green linear leaves and reddish-brown cones with protruding bracts, is scattered through the Chisos at higher elevations.

A special attraction when its leaves turn red in the fall, bigtooth maple (*Acer grandidentatum*) frequents moist Chisos canyons such as Maple, Pine, and Pulliam. This small tree produces tiny, early blooming flowers and unusual rose-colored, winged fruits. A relic of wetter times, quaking aspen (*Populus tremuloides*) survives at two locations on the northeastern and southwestern flanks of Emory Peak. It provides an added touch of yellow in the fall.

Birchleaf buckthorn (*Rhamnus betulifolia*), a small tree with bundles of velvety leaves and black, juicy fruit, is limited to shady, moist sites in Boot, Pulliam, and Upper Pine canyons. The attractive littleleaf mockorange* (*Philadelphus microphyllus*), with bending stems and fragrant white flowers, brightens the understory in Boot Canyon and other wooded Chisos canyons at higher elevations.

Although quite limited in acreage, the moist Chisos woodland is a rich botanical resource with many unusual plant species. The rare lateleaf oak (*Quercus tardifolia*), a small evergreen tree with new leaves not appearing until summer, is known from only two locations along upper Boot Creek and one location in the Sierra del Carmen across the Rio Grande. This oak bears thick, blue-green, oblong leaves up to three and one-half inches long, with broad, spine-tipped lobes along the margins. Although it is more common in northern Mexico, the globally imperiled purple gaymallow* (*Batesimalva violacea*), a slender shrub with blue to violet flowers, is confined in the United States to a single canyon on the west slopes of the Chisos.

Human Influence

Human influence, as varied as the topography, has also shaped Big Bend and, especially in recent history, significantly affected the area's natural communities.

For the first ten thousand years of human habitation, people had little impact on the land. About 12,000 BP, before the climate changed and the earth warmed, prehistoric hunters pursued large game animals such as bison into areas north and east of the park. By 8000 BP, in a warmer climate, other nomadic hunting and gathering peoples hunted smaller desert animals with darts and used desert plants for food, medicine, and primitive goods and tools. They left behind little evidence of their presence except rock shelters, mortar holes, and fire rings. Between 1200 and 1400 AD, early farmers, whom the Spanish called Jumanos (humans), probably related to the Pueblos of the upper Rio Grande in New Mexico, began to cultivate the floodplains west of the park near Presidio.

The Spanish explorer Alvar Nuñez Cabeza de Vaca, presumably the first white man in western Texas, may have traveled through this Jumano community at Presidio about 1535. Remnants of the ill-fated Narváez expedition stranded on the Florida coast in 1527, Cabeza de Vaca and a few others made their way through the Big Bend toward Spanish colonies in Mexico (Gomez 1995).

For the next 250 years, Spaniards crisscrossed Big Bend in search of cities of gold, good farmland, Christian converts, and slaves, but they never tried to settle what they considered the desolate country they called El Despoblado (the uninhabited, or uninhabitable). They built presidios (garrisons to protect their holdings and missions) along the Rio Grande—one in 1774 at the San Vicente, Mexico, river crossing to stop Apache incursions into Mexico. In 1787 they attacked a Mescalero Apache band camped in the Chisos Mountains (Maxwell 1985). The Spanish finally achieved a limited measure of success against the Apaches, and short-lived peace, in 1791.

As early as 1720 Mescalero Apaches dominated the Big Bend, replacing the Chisos, an apparently fierce but little-known tribe whose name and origin remain a mystery. Perhaps from areas just south of the Rio Grande, the Chisos may have summered in the Chisos Mountains and made it a bastion against invaders. After 1750 Comanche hunters and raiders, pushed south by Anglo settlers, began to threaten the Apache domain. Using the Big Bend as a staging ground for raids deeper into Mexico, the Comanches forded the Rio Grande at Lajitas and just west of Mariscal Canyon to attack isolated ranching and farming communities (National Park Service 1983). They eventually extended their raids as far south as Zacatecas. When Mexico won its independence from Spain in 1821, the Spaniards left El Despoblado to the Comanches and Apaches.

Neither the Mexicans nor the Texans, who won independence from Mexico in 1836, could contend with these Native Americans, who lived off the land and traveled long distances on annual raids into northern Mexico. Passing through the park, the Great Comanche War Trail remained an active trail of destruction until about 1845. But in spite of their long dominance, these Native American tribes had little lasting impact on the environment, leaving only artifacts behind and the area largely unspoiled.

In 1848, after the war between the United States and Mexico had settled the boundary dispute, a war veteran, Ben Leaton, established the first American community in the Big Bend at Presidio. Leaton and others set up mercantile businesses to take advantage of a new Chihuahua Trail through the Big Bend. By 1851 a route was established from the Texas Gulf Coast port of Indianola through San Antonio, then west to Fort Davis, Presidio, and Chihuahua (Tyler 1975). Trail activity increased when the Gold Rush brought Forty-Niners on their way to search for gold in California and a handful of permanent settlers.

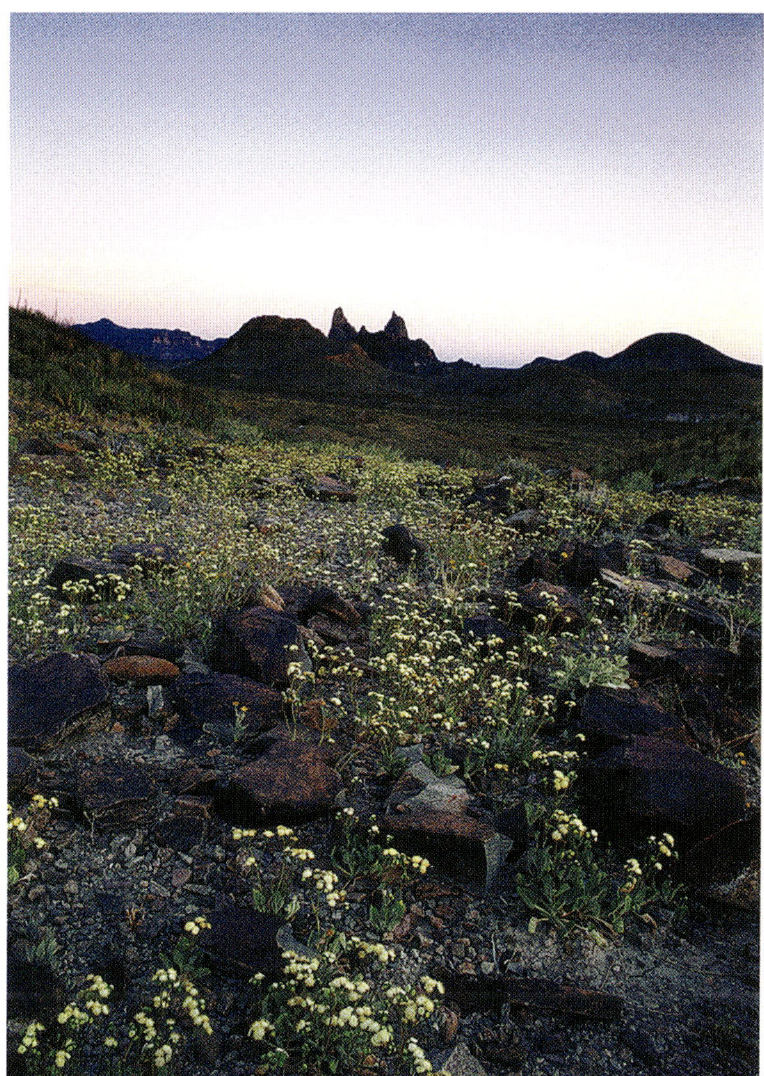

Naked brittlestem with Mule Ears Peaks in distance

Only after the Comanches and Apaches were gradually defeated, first by the destruction of bison herds and then by the construction of U.S. military forts, did settlers come in significant numbers. When the Mescalero Apache chief Victorio was killed by Mexican soldiers in 1880, and the railroad was extended to Alpine in 1882, a new wave of would-be ranchers, miners, merchants, and farmers came to the Big Bend (Gomez 1995). By 1900 Mexican settlers farmed the floodplain on both sides of the river, and after 1920, when boundary troubles and raids by Mexican bandits ended, Anglo farmers joined in.

In 1894 quicksilver mining began at Terlingua, just west of the present-day national park, and it boomed during World War I. Mariscal Mine, at the north end of Mariscal Mountain in the park, also produced some mercury. Lead, zinc, and silver were mined in the Sierra del Carmen, and a six-mile-long aerial tramway carried ore across the Rio Grande toward railroad transport in Marathon, Texas. Irrigated farming communities at Boquillas, below Castolon, and at Terlingua Abaja developed to feed the miners. Timber was cut to build housing and fuel mine furnaces, and many native cottonwoods and other trees along the Rio Grande and Terlingua Creek quickly fell. When the mining activity ended, the farming communities vanished, but the land was scarred.

The arrival of the railroad stimulated a population, land speculation, and ranching boom in the region, and by 1900 large cattle, sheep, and goat ranches dominated the Big Bend landscape. But the days of large ranches and open range did not last, as many new small ranchers began to fence their land. Soon, only the remnants of open range remained. Ranching became an industry, with a new breed of ranchers raising new cattle breeds and cattle for feedlots. Activity peaked during World War I, when beef prices spiked.

The boom had a cost. Throughout large parts of the Big Bend, native vegetation was removed by farming, ranching, and some commercial activities. Guayule, the rubber plant, was harvested for rubber intermittently between 1909 and 1926, and only scattered shrubs were left in harvested areas (Maxwell 1985). Candelilla, the wax plant, was harvested for wax, and large pure stands were removed from much of the Trans-Pecos.

In the 1940s, shortly before the national park was established, park lands were disastrously overgrazed, and grasslands were depleted. Tornillo Flat, once prime tobosa grassland where pronghorns grazed, is still shrub desert even today, sixty years later. At higher elevations, chino grama grass was stripped from large parts of the Chisos foothills.

Park protection has brought back much of the native vegetation, but threats to the environment continue. Big Bend natural biodiversity is reduced by the spread of nonnative, invasive species, such as tamarisk, that choke out more beneficial native plants such as mesquite. Illegal trade in cacti endangers some species. The flow of the Rio Grande, as well as creeks and springs, has been greatly diminished by groundwater use, and the potential mining of West Texas water for export looms menacingly on the horizon.

By far the biggest hazard now is air pollution, which could inescapably alter the essential character of Big Bend National Park. Once famous for 115-mile panoramic views, the park is now moderately hazy on most days, and clear views are rare. As far as the eye can see is becoming a shorter and shorter distance. Summer visibility has been cut in half, from about 85 miles to 43 miles (Novovitch 2000). Much of the worsening pollution is caused by the petrochemical industry and by coal-burning power plants, many still operating without pollution-control devices in north-central Texas, the Gulf Coast, the eastern United States, and Mexico (National Park Service 2006). In addition to the visibility problem, researchers question whether nitrogen deposition from polluted air is altering vegetation types.

Park History

The creation of Big Bend National Park required a difficult battle waged over two decades, but those who saw the area as a last frontier to be preserved for future generations finally came out victorious. To the public and reluctant legislators, the park was promoted as a geological laboratory and an outstanding scenic spectacle containing the only "biological island" (the Chisos Mountains isolated in the surrounding Chihuahuan Desert) in the National Park system.

Everett Townsend—state representative from Alpine, a former Texas Ranger, and U.S. customs inspector inspired by the beauty of the Big Bend country—was one of the leaders of the park movement. In 1933 he and fellow Texas state representative Robert M. Wagstaff of Abilene introduced legislation creating a state park (Maxwell 1985). Park supporters were influenced by the writings and photographs of those who had seen the Big Bend country firsthand and had come to understand the "Big Bend mystique." They were encouraged by a 1930 magazine article by J. Frank Dobie, the renowned University of Texas folklorist, and by an old 1901 magazine article by Robert T. Hill describing his geological survey trip through the canyons of the Rio Grande. A collection of photos by W. D. Smithers, the foremost photographic historian of the Big Bend Country, was used to argue the park's case (Jameson 1996).

Texas Canyons State Park, consisting of the three great river canyons, was authorized in mid-1933; it was enlarged to include the Chisos Mountains and renamed Big Bend State Park later that same year. With 225,000 acres, it became the largest state park in Texas. Soon after approval of the state facility, a national park campaign began. President Franklin Roosevelt, a strong park supporter, signed enabling legislation in 1935, but establishment of the park was contingent on the state's providing the federal government with title to park lands. Only after nine years of intense struggle did the national park become a reality. Some ranchers saw the park as a haven for mountain lions and other large predators, and some politicians opposed granting mineral rights to the federal government. The Depression and the onset of World War II made fundraising problematic.

Park advocates eventually prevailed. Walter Prescott Webb, the famous historian and colleague of J. Frank Dobie at the University of Texas, completed a highly publicized trip through Santa Elena Canyon in 1937. Dobie visited the area with representatives of *National Geographic* magazine in 1938 (Jameson 1996).

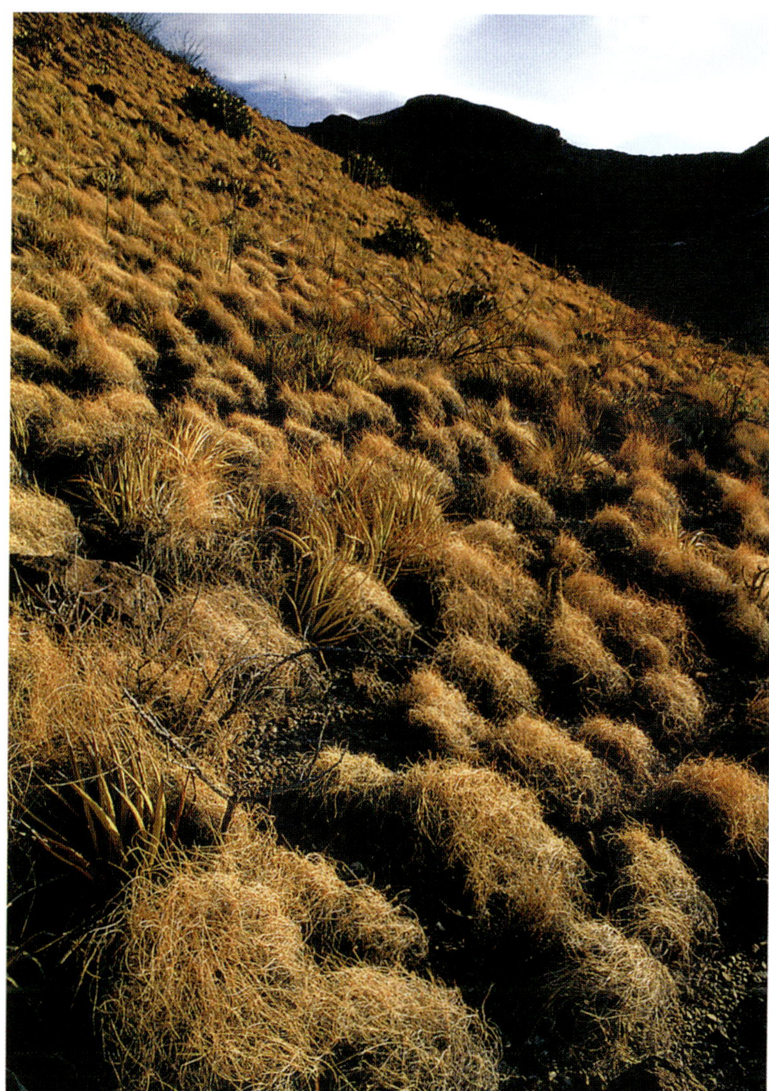

Chino grama and other grasses near Burro Mesa

The lands—more than 700,000 acres—were finally deeded to the federal government, and Big Bend National Park was established on June 12, 1944. By 2001, with the addition of the Pitcock Rosillos Mountain Ranch and the Harte Ranch in the northern Rosillos Mountains, the park encompassed 801,163 acres, the largest protected expanse of Chihuahuan Desert in the United States. In spite of its size, the park averages only three hundred thousand visitors annually.

With a long and convoluted hundred-mile-plus border with Mexico, Big Bend National Park also boasts an important bicultural and international dimension. The park was granted International Biosphere Reserve status by the United Nations Educational, Scientific and Cultural Organization (UNESCO) in 1976, and in 1978 a nearly two-hundred-mile stretch of the Rio Grande, including sixty-nine miles of the park's southern boundary, was designated a Wild and Scenic River (National Park Service 2006).

The Big Bend State of Mind

Human perceptions of the Big Bend environment over time have been sharply divergent and polarized. Yet these dichotomous perceptions are a key to what has become known as the Big Bend mystique and explain why some believe the Big Bend to be a "state of mind" (Madison and Stillwell 1958).

El Despoblado and Heightened Senses

The first Spanish maps showed Big Bend in the heart of a vast territory labeled "Terra Incognita" (unknown land). Considering it unknown, forsaken, and desolate, the Spanish referred to Big Bend as "El Despoblado." Others echoed these sentiments. Lt. William Echols of the U.S. Army Corps of Topographical Engineers, who crossed the area in an experimental camel caravan in 1860, described Big Bend as "a picture of barrenness and desolation" (Tyler 1975).

The isolation, of course, is very real for all to see. The park headquarters, Panther Junction, is 70 miles from the nearest small town, Marathon, and 125 miles from the nearest interstate highway at Fort Stockton. San Antonio is 420 miles away, and El Paso lies 323 miles to the northwest. The entire park, about 45 miles from west to east and 60 miles from north to south, sits at the southern tip of Brewster County, a county the size of Maryland with a population of less than nine thousand people. The park alone is about the size of Rhode Island. Although viewed with dismay by some, this isolation is prized by others. In his 1976 book *Quicksilver: Terlingua and the Chisos Mining Company,* Kenneth Ragsdale summarized the contradictory views of Big Bend as "a heaven or hell, a land of serene beauty or barren ugliness, a place of soothing solitude or haunting loneliness." Park visitors will probably feel and must sort out similar paradoxical impressions. At some point, a Big Bend state of mind may "kick in." Visitors may be fed up and ready to leave, or they may feel closer to a natural world far removed from human influence. For some, the stillness brings a heightened awareness—a comforting feeling that the senses are ready to function, a bit neglected but happy to perform. Sparse sounds, like the hooting of an owl, pierce the silence like musical chords, and these selective sounds become an integral part of the desert experience. But the desert also speaks in another way, through the absence of sounds of the unnatural environment that we experience every day in cities and towns.

Inhospitality and the Last Frontier

Native Americans were not being complimentary of the region when they said that after making the earth, sea, and sky, the Great Spirit dumped the leftover rocks in a heap on the Big Bend. Over the decades many have spoken of its inhospitable nature, exemplified by impassable terrain, devastating heat, drought, and flash floods. In 1937 Walter Prescott Webb seemed to echo the Native American view: "The physiographic order of the Big Bend is somewhat like the order of a great city built of stone and brick—wrecked by an earthquake. Perhaps order once prevailed there, but some mighty force wrecked the place, shook it down, turned it over, blew it up, and set it afire" (Nelson 2002).

In the 1960s U.S. astronauts practiced driving the moon rover over this wrecked landscape and its moonlike badlands. Some, however, view this same inhospitable landscape as a rugged "Last Frontier" or "God's Country." A Mexican cowboy did not think ill of the Big Bend when he described this land more than a century ago as a place "where rainbows wait for rain, and the big river is kept in a stone box, and water runs uphill and mountains float in the air, except at night when they go away to play with other mountains" (National Park Service 1983).

Writing in the *San Antonio Express* in 1896, William Ferguson spoke of the Big Bend as a region of unspoiled beauty and grandeur: "Nowhere have I found such a wildly weird country. A man . . . becomes awe-struck by nature in her lofty moods. Emotions are stirred by the grandeur and beauty of the scenery and the ever-changing play of light and shadows. . . . No painter could mix colors to justly portray it. No words can describe its splendour" (Maxwell 1968).

One man's "earthquake" was another man's "nature in her lofty moods."

Life on the Edge

Many look at the Big Bend desert and see dry, bare soil mostly devoid of life, occupied only by armed and menacing plants. Jules Leclerc, a Frenchman traveling to Veracruz in 1885, observed, "Each plant in this land is a porcupine. It is nature armed to the teeth" (Tyler 1975).

In 1916 a newsreel cameraman working for *Collier's* magazine described the country from Marathon to the Rio Grande: "The country isn't bad. It's just worse. Worse the moment you set foot from the train, and then, after that, just worser and worser" (Madison and Stillwell 1958).

Yet others are surprised at the abundance of life and paradoxical beauty of the desert. They are attracted by the resilience of plants—their ability to recover from drought and even thrive against seemingly impossible odds. Many see this amazing process of squeezing life out of so little, the perception of life on the edge, as the desert's ultimate drama.

This drama plays out on the desert floor. The prickly pear, adapted to life on the edge with scant and unpredictable moisture, produces a glorious flower in the spring, one to rival the floral displays of a tropical rainforest. The desert lantana shrub, mejorana, boasts a jewel-like bloom that gathers and reflects light like a brooch. The grass chino grama curls into playful natural sculptures untouched by human hands.

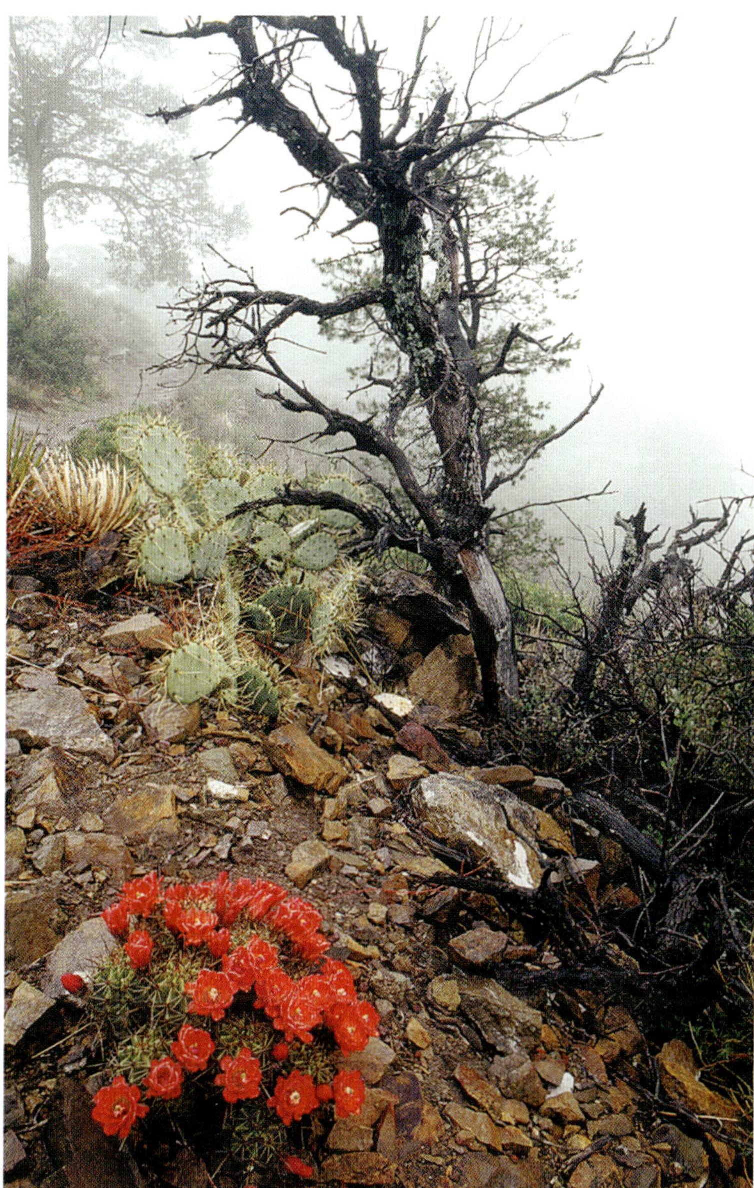

Texas claret-cup cactus on the Lost Mine Trail

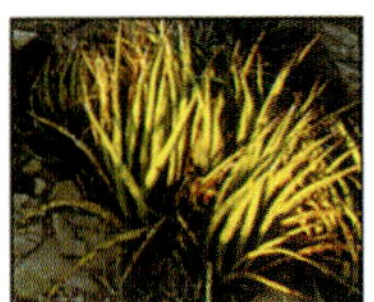

HOW PLANTS ARE NAMED

In this guide, each plant is designated by two names, a common name and a two-word scientific (or botanical) name. In the description of each plant, alternate common names are frequently mentioned, and in the note section the derivation of the scientific name is discussed. Learning about these names and how they originated may help the reader understand and appreciate the plants.

Often rooted in myth or folklore, many common names were handed down by word of mouth through generations and offer a special insight into a plant's history, or habit. A common (or scientific) name may indicate a feature of the species, its place of origin or natural habitat, the name of the botanical collector, or its medicinal use. Occasionally, these names are remarkably imaginative and apt descriptions that enrich our language. A common name may capture an essential plant characteristic so well that it becomes a valuable aid in plant identification. For example, "living rock," "sea urchin," and "whiskerbrush pincushion" are insightful and memorable names of the cacti they identify, and each in its own way acts as a trigger for plant recognition in the field.

However, the use of even the most ingenious common names for plants is problematic. A plant may be known by many different names, different from one region or country to another, and from one language to another. Plants may share the same overused colloquial name, or they may have no common name at all. During the Renaissance, when expeditions brought back new species from throughout the world, scientists recognized that each species needed a unique name.

In 1753 Swedish botanist Carl von Linné, better known by his own Latinized name, Carolus Linnaeus, proposed a new system of naming plants, the binomial system, which uniquely identifies a plant with two Latin words. Linnaeus also proposed a system of classifying plants according to their sexual characteristics. Although later botanists developed an improved system for plant classification, plants are still named today using Linnaeus's binomial system.

The unique two-word scientific name of a plant consists of the genus name (a noun beginning with a capital letter), and a second, descriptive word (beginning with a small letter) known as the "specific epithet." The two words together uniquely name the species. Occasionally, a third word is added that identifies a subspecies (abbreviated ssp. or subsp.) or a variety (abbreviated var.). A genus is a group of related species that share at least one common feature (for example, flower structure). Genera are grouped into families according to a wider range of shared characteristics.

Plants in this guide are organized in alphabetical order by plant family, genus, and species, except that members of the agave (Agavaceae) and cactus (Cactaceae) families appear first. Except for a small number of old names still considered valid, all family names carry an -aceae ending (meaning "belonging to") and are derived from the name of an included genus (for example, Agavaceae from the genus *Agave* with the -aceae ending). Classifying plants into related groups such as genera and families makes it easier to compare species and examine their similarities and differences.

Laws of botanical nomenclature, based on early work by Swiss botanist Augustin de Candolle and his son, Alphonse, were adopted in 1867, and in 1952 the International Code of Botanical Nomenclature set forth more comprehensive rules for naming plants. For a new species name to be valid, the author of the name must prepare a plant description for a "type"—a set of features for the "typical" plant. The type is usually a "type specimen," housed in a herbarium, or less often a collection of botanical drawings. In this guide, a note accompanying each plant description usually mentions the collector of the type specimen and the botanist who first validly published the plant name.

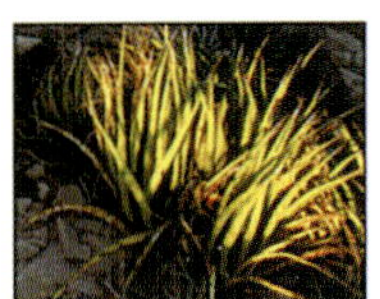

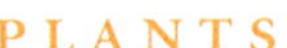

CHECKLIST OF PLANTS

ALL THE PLANTS in this checklist are organized under their family name. The scientific name is in the left column and the common name in the right. The plants appear in alphabetical order by family name, except that the agaves (Agavaceae) and cacti (Cactaceae) appear first. The family name is followed by the page numbers in which that family appears and the plant's scientific and common names.

AGAVACEAE—AGAVE 36–42

Agave havardiana	Havard agave
Agave lechuguilla	Lechuguilla
Dasylirion leiophyllum	Smooth sotol
Nolina erumpens	Beargrass
Yucca faxoniana	Giant dagger
Yucca thompsoniana	Thompson yucca
Yucca torreyi	Spanish dagger

CACTACEAE—CACTUS 43–69

Ariocarpus fissuratus var. *fissuratus*	Living rock cactus
Coryphantha chaffeyi	Chaffey's pincushion cactus
Coryphantha duncanii	Duncan's pincushion cactus
Coryphantha echinus	Sea urchin cactus
Coryphantha macromeris var. *macromeris*	Big-needle pincushion cactus
Coryphantha ramillosa	Whiskerbrush pincushion cactus
Coryphantha sneedii var. *albicolumnaria*	Silverlace cactus
Coryphantha tuberculosa var. *tuberculosa*	Cob cactus
Echinocactus texensis	Horse-crippler cactus
Echinocereus chisoensis var. *chisoensis*	Chisos hedgehog cactus
Echinocereus coccineus var. *paucispinus*	Texas claret-cup cactus
Echinocereus dasyacanthus	Texas rainbow cactus

Echinocereus enneacanthus var. *enneacanthus*	Strawberry hedgehog cactus
Echinocereus stramineus var. *stramineus*	Strawberry cactus
Echinocereus viridiflorus var. *russanthus*	Rusty hedgehog cactus
Echinomastus intertextus	Woven-spine pineapple cactus
Echinomastus mariposensis	Mariposa cactus
Echinomastus warnockii	Warnock's cactus
Epithelantha bokei	Boke's button cactus
Ferocactus hamatacanthus var. *hamatacanthus*	Giant fishhook cactus
Glandulicactus uncinatus var. *wrightii*	Catclaw cactus
Mammillaria heyderi var. *heyderi*	Heyder's pincushion cactus
Mammillaria lasiacantha	Golf ball cactus
Mammillaria meiacantha	Nipple cactus
Mammillaria pottsii	Potts' mammillaria cactus
Neolloydia conoidea var. *conoidea*	Texas cone cactus
Thelocactus bicolor var. *bicolor*	Glory of Texas cactus

Cactaceae—Cylindropuntia 70–72

Opuntia imbricata var. *arborescens*	Tree cholla
Opuntia imbricata var. *argentea*	Big Bend cholla
Opuntia kleiniae	Candle cholla

Cactaceae—Grusonia 73

Opuntia aggeria, densispina, schottii vars. *schottii* & *grahamii*	Club cholla

Cactaceae—Opuntia 74–78

Opuntia azurea var. *aureispina*	Golden-spined prickly pear
Opuntia azurea var. *parva*	Big Bend purplish prickly pear
Opuntia chisosensis	Chisos prickly pear
Opuntia rufida	Blind prickly pear
Opuntia spinosibacca	Spiny-fruited prickly pear

Acanthaceae—Acanthus 79–86

Anisacanthus linearis	Dwarf anisacanth
Carlowrightia arizonica	Arizona carlowrightia
Carlowrightia serpyllifolia	Trans-Pecos carlowrightia
Dyschoriste schiedeana var. *decumbens*	Spreading snakeherb
Justicia pilosella	Hairy tubetongue
Justicia warnockii	Warnock justicia
Ruellia parryi	Parry ruellia
Stenandrium barbatum	Shaggy stenandrium

Amaranthaceae—Amaranth 87–89

Froelichia arizonica	Arizona snakecotton
Tidestromia carnosa	Fleshy tidestromia
Tidestromia suffruticosa	Shrubby tidestromia

Anacardiaceae—Sumac 90

Rhus microphylla	Desert sumac

Apiaceae—Parsley 91

Aletes acaulis	Stemless aletes

Apocynaceae—Dogbane 92–95

Amsonia longiflora var. *longiflora*	Tubular slimpod
Haplophyton crooksii	Arizona cockroach plant
Telosiphonia hypoleuca	Bottomwhite rocktrumpet
Telosiphonia macrosiphon	Plateau rocktrumpet

Aristolochiaceae—Birthwort 96

Aristolochia coryi	Cory Dutchman's pipe

Asclepiadaceae—Milkweed 97–101

Asclepias texana	Texas milkweed
Cynanchum pringlei	Pringle swallow-wort
Cynanchum racemosum var. *unifarium*	Talayote
Matelea reticulata	Pearl netleaf milkweed
Sarcostemma torreyi	Soft twinevine

Asteraceae—Sunflower 102–127

Acourtia runcinata	Stemless perezia
Ageratina wrightii	Wright ageratina
Bidens bigelovii var. *bigelovii*	Bigelow beggarticks
Brickellia laciniata	Splitleaf brickellbush
Brickellia lemmonii var. *conduplicata*	Sandlot brickellbush
Brickellia veronicifolia	Veronicaleaf brickellbush
Carphochaete bigelovii	Bigelow bristlehead
Cirsium turneri	Turner thistle
Conoclinium dissectum	Palmleaf thoroughwort
Flourensia cernua	Tarbush
Flyriella parryi	Chisos Mountain brickellbush
Hieracium schultzii	Roughstem hawkweed
Nicolletia edwardsii	Edwards nicollet
Parthenium argentatum	Guayule
Pericome caudata	Tailleaf pericome
Perityle aglossa	Rayless rockdaisy
Perityle bisetosa var. *scalaris*	Twobristle rockdaisy
Perityle dissecta	Slimlobe rockdaisy
Perityle parryi	Heartleaf rockdaisy
Perityle vaseyi	Margined rockdaisy
Psathyrotes scaposa	Naked brittlestem
Stevia ovata var. *texana*	Roundleaf stevia
Stevia viscida	Viscid stevia
Tagetes micrantha	Licorice marigold

Thymophylla micropoides	Woolly dogweed
Xylothamia triantha	Threeflower goldenweed

Betulaceae—Birch 128

Ostrya virginiana var. *chisosensis*	Big Bend hophornbeam

Bignoniaceae—Catalpa 129–130

Chilopsis linearis ssp. *linearis*	Desert willow
Tecoma stans	Trumpetflower

Boraginaceae—Borage 131–137

Cryptantha palmeri	Cory cryptantha
Heliotropium confertifolium	Leafy heliotrope
Heliotropium glabriusculum	Greeneye heliotrope
Heliotropium molle	Soft heliotrope
Lithospermum viride	Green gromwell
Omphalodes aliena	Mexican navelseed
Tiquilia greggii	Plume tiquilia

Brassicaceae—Mustard 138–145

Cardamine macrocarpa var. *texana*	Texas largeseed bittercress
Nerisyrenia camporum	Mesa greggia
Selenia dissecta	Texas selenia
Sisymbrium auriculatum	Earlobe mustard
Streptanthus carinatus ssp. *carinatus*	Lyreleaf twistflower
Streptanthus cutleri	Cutler twistflower
Synthlipsis greggii	Gregg keelpod
Thelypodium texanum	Texas thelypody

Bromeliaceae—Pineapple 146

Hechtia texensis	Texas false agave

Buddlejaceae—Buddleja 147–148

Buddleja marrubiifolia	Woolly butterflybush
Buddleja scordioides	Escobilla butterflybush

Campanulaceae—Bluebell 149

Lobelia berlandieri var. *brachypoda*	Berlandier lobelia

Caryophyllaceae—Pink 150–153

Cerastium axillare	Trans-Pecos chickweed
Drymaria pachyphylla	Thickleaf drymary
Paronychia jamesii	James nailwort
Stellaria cuspidata	Mexican starwort

Celastraceae—Bittersweet 154–155

Mortonia sempervirens ssp. *scabrella*	Rough mortonia
Schaefferia cuneifolia	Desert yaupon

Cleomaceae—Cleome 156

Polanisia uniglandulosa	Mexican clammyweed

Commelinaceae—Spiderwort 157

Tradescantia brevifolia	Trans-Pecos spiderwort

Convolvulaceae—Morning Glory 158–163

Bonamia ovalifolia	Bigpod bonamia
Bonamia repens	Creeping rockvine
Dichondra argentea	Silver ponyfoot
Dichondra brachypoda	New Mexico ponyfoot
Ipomoea costellata	Crestrib morning-glory
Ipomoea lindheimeri	Blue morning-glory

Crassulaceae—Orpine 164–167

Echeveria strictiflora	Longpetal echeveria
Sedum havardii	Havard stonecrop
Sedum wrightii	Wright stonecrop
Villadia squamulosa	Stonecrop

Cucurbitaceae—Gourd 168–169

Ibervillea tenuisecta	Slimlobe globeberry
Sicyos glaber	Smooth bur cucumber

Cupressaceae—Cypress 170–172

Cupressus arizonica	Arizona cypress
Juniperus deppeana var. *deppeana*	Alligator juniper
Juniperus flaccida	Drooping juniper

Cuscutaceae—Dodder 173

Cuscuta indecora	Pretty dodder

Cyperaceae—Sedge 174

Cyperus seslerioides	Texas flatsedge

Ephedraceae—Ephedra 175

Ephedra torreyana var. *powelliorum*	Torrey ephedra

Ericaceae—Heath 176

Arbutus xalapensis	Texas madrone

Euphorbiaceae—Spurge 177–184

Acalypha monostachya	Round copperleaf
Bernardia obovata	Desert myrtlecroton
Chamaesyce chaetocalyx var. *triligulata*	Three-tongue spurge
Croton pottsii var. *thermophilus*	Zigzag croton
Euphorbia antisyphilitica	Candelilla
Euphorbia eriantha	Woollyflower spurge
Jatropha dioica var. *graminea*	Leatherstem
Phyllanthus polygonoides	Knotweed leafflower

Fabaceae—Legume 185–201

Acacia rigidula	Blackbrush acacia
Acacia roemeriana	Roemer acacia
Brongniartia minutifolia	Littleleaf brongniart
Calliandra humilis var. *humilis*	Dwarf fairy duster
Dalea formosa	Feather dalea
Dalea frutescens	Black dalea
Dalea neomexicana	New Mexico dalea
Dalea pogonathera var. *pogonathera*	Bearded dalea
Dalea wrightii var. *wrightii*	Wright dalea
Eysenhardtia texana	Texas kidneywood
Leucaena retusa	Littleleaf leadtree
Lupinus havardii	Big Bend bluebonnet
Mimosa emoryana var. *emoryana*	Emory mimosa
Mimosa turneri	Turner mimosa
Phaseolus grayanus	Gray bean
Pomaria melanosticta	Parry caesalpinia
Senna pilosior	Trans-Pecos senna

Fagaceae—Beech 202–204

Quercus graciliformis	Chisos oak
Quercus gravesii	Chisos red oak
Quercus intricata	Coahuila scrub oak

Fouquieriaceae—Ocotillo 205

Fouquieria splendens	Ocotillo

Garryaceae—Silktassel 206

Garrya ovata ssp. *lindheimeri*	Eggleaf silktassel

Hydrangeaceae—Hydrangea 207–209

Fendlera linearis	Narrowleaf fendlerbush
Philadelphus mearnsii	Mearns mockorange
Philadelphus microphyllus	Littleleaf mockorange

Hydrophyllaceae—Waterleaf 210–211

Nama havardii	Havard nama
Nama torynophyllum	Mat nama

Koeberliniaceae—Allthorn 212

Koeberlinia spinosa	Allthorn

Lamiaceae—Mint 213–218

Agastache micrantha var. *micrantha*	White giant hyssop
Agastache pallidiflora ssp. *neomexicana* var. *havardii*	Havard giant hyssop
Hedeoma mollis	Hairy hedeoma
Salazaria mexicana	Bladder sage
Salvia regla	Mountain sage
Stachys bigelovii	Rock betony

Linaceae—Flax 219

Linum berlandieri var. *filifolium*	Berlandier flax

Loasaceae—Stickleaf 220–221

Cevallia sinuata	Stinging cevallia
Eucnide bartonioides	Yellow rocknettle

Lythraceae—Loosestrife 222

Nesaea longipes	Stalkflower nesaea

Malpighiaceae—Malpighia 223

Janusia gracilis	Propellerbush

Malvaceae—Mallow 224–225

Batesimalva violacea	Purple gaymallow
Malvella lepidota	Scurfy mallow

Nyctaginaceae—Four-O'clock 226–228

Acleisanthes angustifolia	Narrowleaf moonpod
Acleisanthes parvifolia	Littleleaf moonpod
Cyphomeris gypsophiloides	Red cyphomeris

Oleaceae—Olive 229–230

Forestiera angustifolia	Desert olive
Menodora longiflora	Showy menodora

Orchidaceae—Orchid 231–239

Deiregyne confusa	Hidalgo ladies tresses
Dichromanthus cinnabarinus	Scarlet ladies tresses
Epipactis gigantea	Giant helleborine
Hexalectris grandiflora	Giant coral root

Hexalectris revoluta	Curly coral root
Hexalectris spicata var. *spicata*	Crested coral root
Hexalectris warnockii	Texas purplespike
Malaxis wendtii	Wendt's malaxis
Stenorrhynchos michuacanum	Michoacán ladies tresses

Orobanchaceae—Broomrape 240–241

Conopholis alpina var. *mexicana*	Mexican squawroot
Orobanche ludoviciana ssp. *multiflora*	Manyflowered broomrape

Papaveraceae—Poppy 242

Argemone chisosensis	Chisos pricklypoppy

Passifloraceae—Passionflower 243

Passiflora tenuiloba	Spreading-lobe passionflower

Pinaceae—Pine 244

Pinus cembroides ssp. *cembroides*	Mexican pinyon pine

Poaceae—Grass 245

Bouteloua ramosa	Chino grama

Polemoniaceae—Phlox 246–247

Ipomopsis havardii	Havard ipomopsis
Loeselia greggii	Papercup loeselia

Polygalaceae—Milkwort 248–250

Polygala lindheimeri var. *parvifolia*	Rock milkwort
Polygala macradenia	Glandleaf milkwort
Polygala scoparioides	Broom milkwort

Polygonaceae—Knotweed 251

Eriogonum hemipterum var. *hemipterum*	Chisos Mountain buckwheat

Portulacaceae—Purslane 252

Phemeranthus aurantiacus	Orange flameflower

Primulaceae—Primrose 253

Samolus ebracteatus var. *cuneatus*	Limerock brookweed

Pteridaceae—Maidenhair Fern 254–255

Pellaea intermedia	Creeping cliff brake
Pellaea ternifolia	Trans-Pecos cliff brake

Ranunculaceae—Crow's Foot 256–257

Anemone tuberosa var. *texana*	Tuber windflower
Aquilegia longissima	Longspur columbine

Rhamnaceae—Buckthorn 258–260

Ceanothus greggii	Desert ceanothus
Condalia ericoides	Javelina bush
Condalia viridis	Green condalia

Rosaceae—Rose 261–264

Prunus havardii	Havard plum
Prunus serotina ssp. *virens* var. *virens*	Southwestern chokecherry
Purshia ericifolia	Heath cliffrose
Vauquelinia corymbosa ssp. *angustifolia*	Palo prieto

Rubiaceae—Madder 265–267

Bouvardia ternifolia	Scarlet bouvardia
Galium uncinulatum	Bristly bedstraw
Hedyotis intricata	Fascicled bluet

Rutaceae—Citrus 268

Thamnosma texana	Dutchman's breeches

Sapindaceae—Soapberry 269

Ungnadia speciosa	Mexican buckeye

Scrophulariaceae—Figwort 270–275

Castilleja mexicana	Twistleaf paintbrush
Leucophyllum candidum	Boquillas silverleaf
Leucophyllum frutescens	Purple sage
Penstemon baccharifolius	Baccharisleaf penstemon
Penstemon havardii	Havard penstemon
Seymeria scabra	Limpia seymeria

Selaginellaceae—Spikemoss 276

Selaginella lepidophylla	Flower of stone

Simaroubaceae—Simarouba 277

Holacantha stewartii	Crucifixion thorn

Solanaceae—Nightshade 278–280

Chamaesaracha villosa	Shaggy false nightshade
Lycium puberulum var. *berberioides*	Silvery wolfberry
Nectouxia formosa	Puckering nightshade

Sterculiaceae—Cacao 281

Ayenia microphylla	Dense ayenia

Verbenaceae—Vervain 282–283

Bouchea spathulata	Spoonleaf bouchea
Lantana macropoda	Mejorana

Viscaceae—Mistletoe 284

Phoradendron hawksworthii	Rough mistletoe

Vitaceae—Grape 285

Cissus trifoliata	Ivy treebine

Zygophyllaceae—Caltrop 286–287

Guaiacum angustifolium	Guayacán
Larrea tridentata	Creosote bush

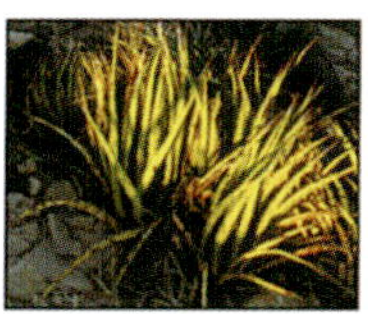

The Plants

Havard Agave

Agave havardiana

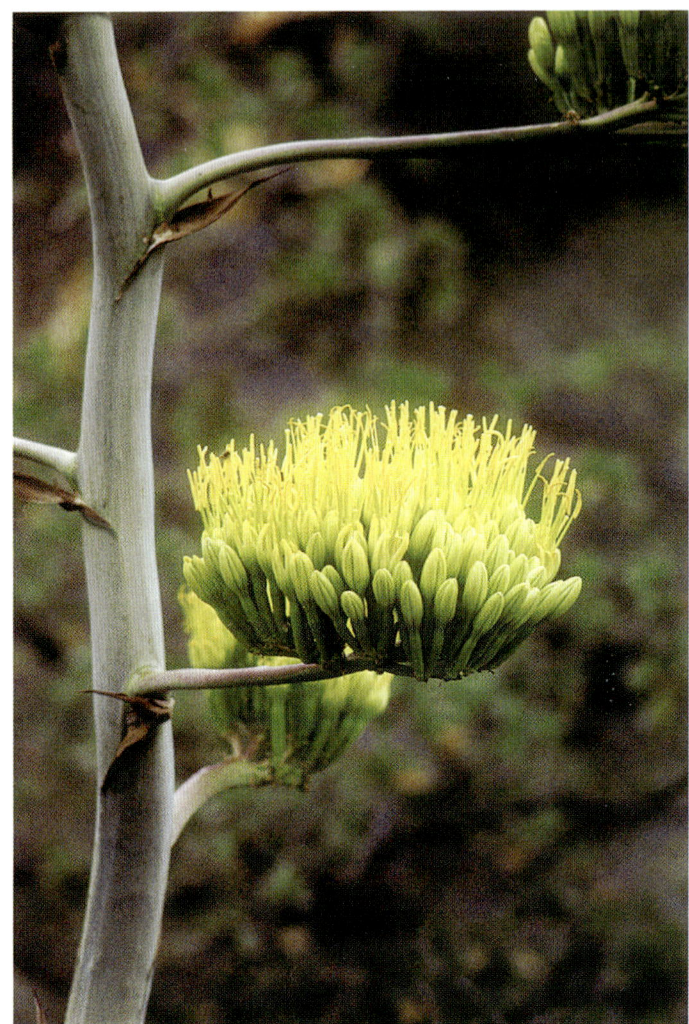

ALSO KNOWN AS Havard maguey and century plant, Havard agave is the largest native agave in Texas, with a flowering stalk up to twenty feet or more. Within ten to twenty years of age, this giant succulent puts forth a flowering stalk and blooms just once before dying. The thick, fleshy leaves are rigid, gray-green or bluish-green, with sharp, mostly reflexed teeth on the margins and vicious, dark-colored spines at the tips. Broadly lance-shaped, each leaf is one to two feet long.

Usually blooming in early summer, the flower stalk grows as much as eighteen inches a day. The stout branches extend like candelabra arms, each branch bearing a crowded cluster of tubular yellow flowers. Visited by a parade of birds and insects, Havard agave's flower clusters have been described as a banquet table, smorgasbord, candelabra, and the desert's lazy susan.

This agave is most common in the Glass and Chisos mountains of Brewster County, but it occurs in other Trans-Pecos mountains (Davis, Chinati, Madera) and extends into northern Coahuila, Mexico. In the park this majestic plant grows from grasslands up forested canyons to the highest mountain slopes. Look for it in Green Gulch and the Basin, at Boot Spring, and in the Rosillos Mountains at the park's northwest corner.

Agaves have a very long history as food, drink, and fiber. The Aztecs made pulque, an alcoholic beverage, from the sap and fed the drink to victims in religious rituals involving human sacrifice and cannibalism.

Note: The genus name *Agave* means "noble" in Greek. In 1911 William Trelease, the first director of the Missouri Botanical Garden, named this species for Valery Havard, who had collected it in the Guadalupe Mountains in 1881. Havard, a French-born physician with the U.S. Army Medical Corps, collected plants while stationed at various western outposts.

Lechuguilla

Agave lechuguilla

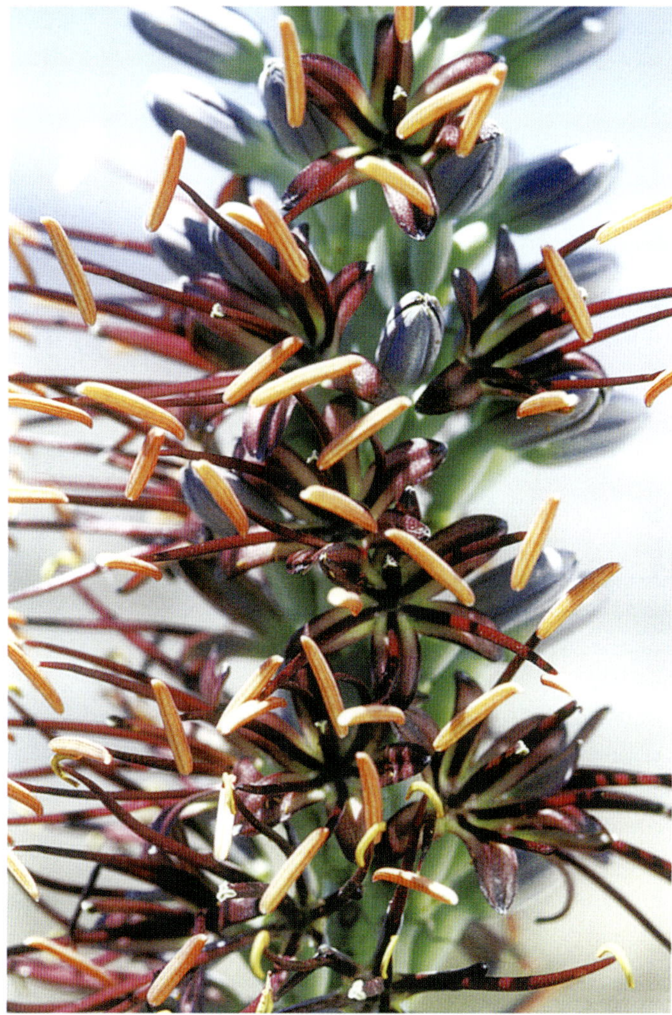

LECHUGUILLA ("little lettuce" in Spanish) is a succulent forming a rosette of basal leaves up to sixteen inches high. Each rosette consists of as many as thirty rigid, lance-shaped leaves with sharp terminal spines and widely spaced, downward-hooked teeth on the edges. It is aptly called shin dagger because the stout, yellow-green leaves can puncture shins and even automobile tires.

Lechuguilla reproduces mostly through offshoots from the underground stems of parent plants and may spread into almost impenetrable colonies blanketing entire hillsides. Geographically restricted to the Chihuahuan Desert, it is considered this desert's indicator species.

Within three to twenty years of age, lechuguilla blooms once and then dies. In late spring and summer, it produces a leafless spike up to twelve feet high with dense flower clusters near the top. The stalk may grow eight inches in a single day. In shades of purple, red, and yellow, funnel-shaped flowers with linear, petal-like lobes and prominent stamens form in clusters of two to four along the stalk.

Lechuguilla may be the dominant plant on open, arid slopes, gravelly limestone hills, and bajadas. It is often associated with other agave and yucca, sotol, creosote, and grama grasses at middle to low elevations. This succulent ranges from southeastern New Mexico through Trans-Pecos Texas south to Mexico City. In the park it is easiest to find in the Chisos foothills.

Note: Native Americans made twine, rope, clothing, and sandals from the leaf fibers. The root was used as soap. This species was collected by Charles Wright in 1849 and described by John Torrey in 1859. Wright traveled with a military expedition from San Antonio to El Paso, walking much of the way and bringing back fourteen hundred plant species. Torrey was a pioneering American botanist, mentor, and collaborator of Asa Gray, and also coauthor with Gray of *Flora of North America* (1838–43).

Smooth Sotol

Dasylirion leiophyllum

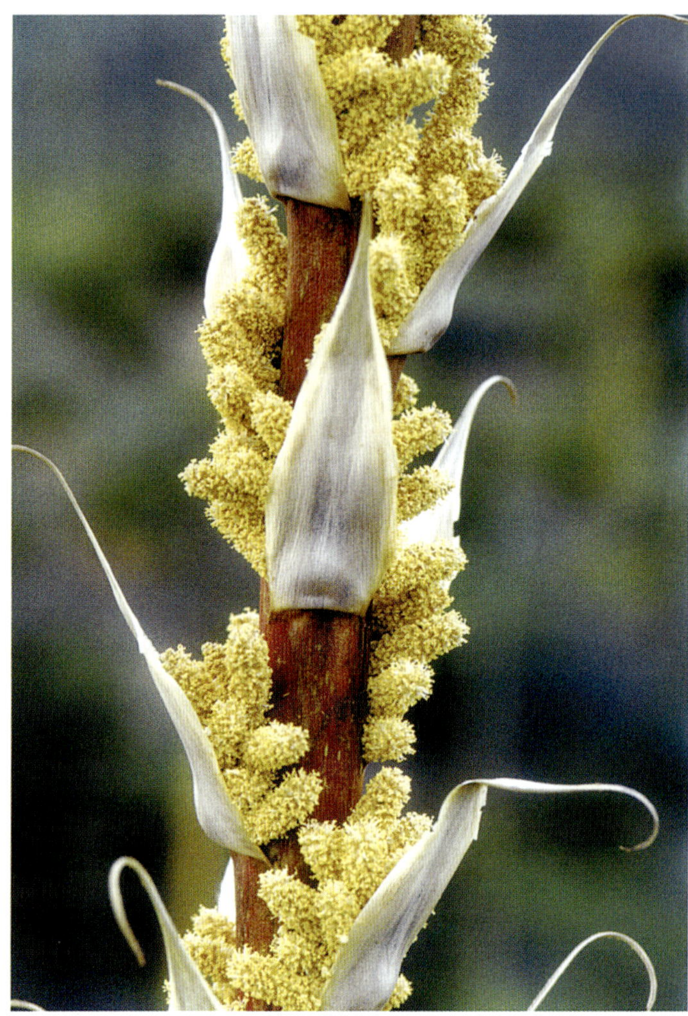

For much of the year, smooth sotol, or desert candle, consists of a thick, bulbous, mostly underground stem, which is covered with a crowded rosette of slender, ribbony leaves. The smooth green leaves, up to nearly three feet long but no more than an inch wide, have reddish-orange, backward-curving, hooked prickles on the sides and brushy, fibrous tips.

From May to August, smooth sotol produces sturdy but narrow flower stalks up to fifteen feet or more in height, with minute white flowers in a dense, elongated cluster at the upper end. Male and female flowers appear on separate plants (male flowers are shown here). The fruit is a leathery oblong capsule, one-fourth inch long, with three thin wings.

Another typical species of the Chihuahuan Desert, this shrub is prominent at mid elevations in a community known as "sotol grassland" surrounding the higher mountains. It prefers gravelly limestone hillsides and south-facing slopes, growing with lechuguilla, beargrass, yucca, and chino grama. Smooth sotol is distributed from southern New Mexico through Trans-Pecos Texas into Chihuahua and Coahuila, Mexico.

Although mostly a grassland plant, smooth sotol appears throughout the park, from Persimmon Gap in the north to Arch Camp on the Rio Grande. Sotol Vista, in the Chisos foothills, is the ideal place to observe this member of the agave family, and the sunset.

Note: The genus name *Dasylirion* is from the Greek *dasy* (thick) and *lirion* (lily), a reference to the dense flower clusters. The species name, from the Greek *leios* (smooth) and *phyllon* (leaf), refers to the smooth leaf surfaces. This species was described in 1911 by William Trelease, a specialist in oaks, agave, and cactus who, while director of the Missouri Botanical Garden, named twenty-five hundred plant species. Smooth sotol was collected at Presidio del Norte in 1880 by Valery Havard, a U.S. Army physician stationed at Fort Davis and Fort Stockton.

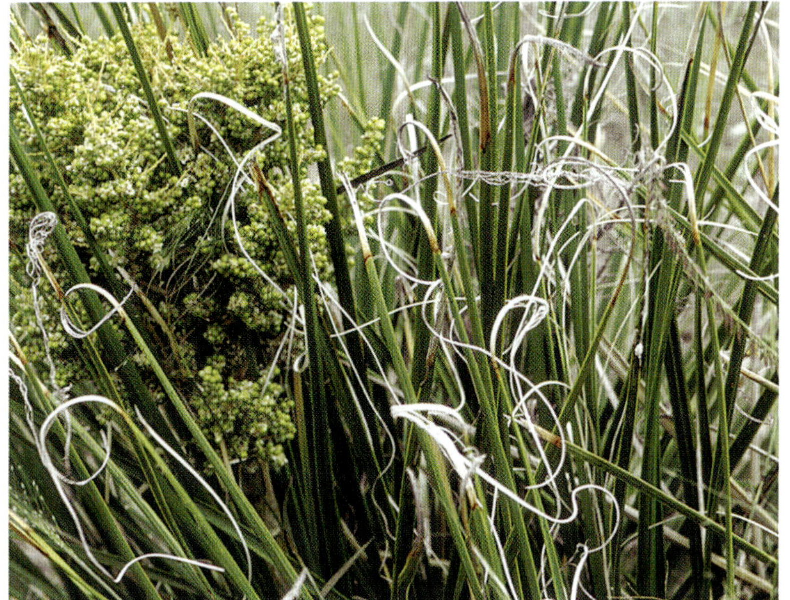

Beargrass

Nolina erumpens

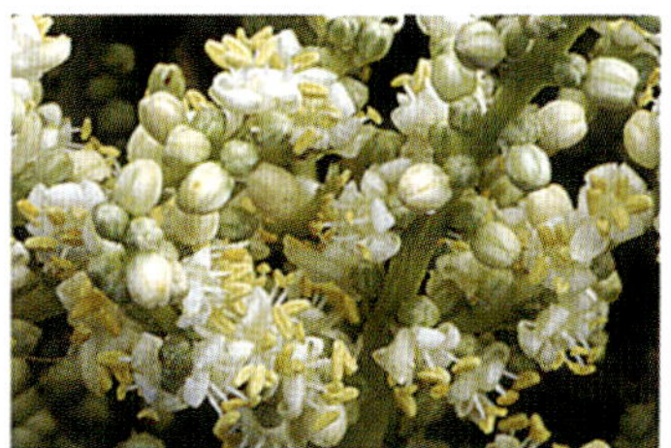

ALSO KNOWN as mesa sacahuista and foothill basketgrass, beargrass is a widely sprawling, grasslike plant as much as eight feet high and wide. From a bulbous underground base, it forms a rosette of narrow, radiating, grasslike leaves. Each stiff leaf, three to five feet long but only one-half inch wide, has finely serrated edges and thin strands of loose, curling fibers at the tips.

In late spring and early summer, beargrass produces a flowering stalk that usually is not higher than the leaves. Tiny, creamy white flowers form at the top of the stalk in a densely branched spike as much as three feet long and seven inches wide. The flowers attract scores of wasps, bees, and ants. The fruits are thin-walled, inflated capsules.

Beargrass is present on limestone and igneous soils, open rocky slopes, and grasslands, often in the foothills of higher mountains. This shrub is widespread in the Trans-Pecos, from Hudspeth County east to Terrell County, and in the Sierra del Carmen and Sierra Maderas del Carmen of Coahuila, Mexico.

In the park beargrass is easy to spot in the Chisos, from Ward Spring and foothills west of Oak Creek Canyon to Casa Grande and on the trail to the South Rim. This stem succulent has also been recorded in the Rosillos Mountains.

Note: For Native Americans, beargrass was a utility basket of goods: leaves were used to make rope, cord, mats, baskets, and basket handles, and young flowers were used as food. The genus *Nolina* is named for Abbé C. P. Nolin, eighteenth-century French director of the royal nurseries at Versailles and protégé of Madame de Pompadour. This species was collected in 1851–52 by Charles Wright, part-time botanist and surveyor on Colonel Graham's U.S.-Mexico boundary survey, and described (as *Dasylirion erumpens*) by New York botanist John Torrey in 1859.

Giant Dagger

Yucca faxoniana

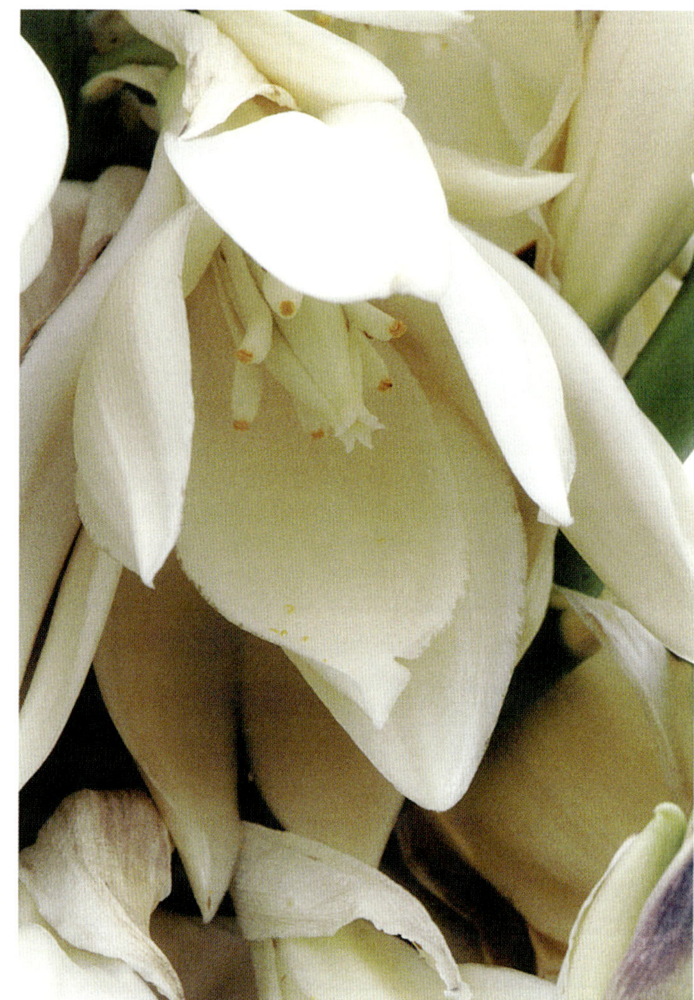

Also known as faxon yucca, Carneros yucca, and palma samandoca, giant dagger is a hulking, treelike succulent covered with curtains of defunct leaves and crowned with a symmetrical cluster of fresh dagger-shaped blades. These usually unbranched behemoths may reach thirty feet high, including the flower spike. Each stout, spine-tipped leaf is smooth and pale green or yellowish-green in color, with curling, thread-like fibers on the edges.

An unforgettable Chihuahuan Desert display occurs, typically in April, at Dagger Flat, when a forest of these "trees" bursts into bloom. This yucca produces stout, tightly branched flower spikes, which may tower above the leaves. Up to four inches long, the creamy white, thick-petaled flowers are shaped like hanging bells and seldom fully open. They are pollinated by moths (of the *Tegeticula* genus), which deposit eggs in the flowers. The oblong fruits are pulpy, sweet, and sometimes beaked.

Giant dagger is a conspicuous occupant of limestone soil on high desert plateaus, bajadas, and rocky bluffs. Its range includes Trans-Pecos Texas (from southern Hudspeth County southeast to Brewster County) and northern Mexico (from Coahuila and Chihuahua to San Luis Potosí and Zacatecas). In the park it is mostly scattered throughout the northern Dead Horse Mountains.

Giant dagger was used extensively by Native Americans: young flowering stalks as food, leaves as fiber, and roots as soap. In Mexico it has been used to make living fences and walls.

Note: The plant in the park is considered by some botanists to be *Yucca carnerosana*, a species related to but genetically distinct from *Yucca faxoniana* (based on Karen Clary's 1997 DNA study). *Yucca carnerosana* was named for Carneros Pass in Coahuila, Mexico, where Cyrus Pringle collected it in 1891. In 1905 Charles Sargent, first director of the Arnold Arboretum at Harvard, named *Yucca faxoniana* for Charles Faxon, the botanical illustrator of his work, *The Silva of North America* (1890–1902).

Thompson Yucca

Yucca thompsoniana

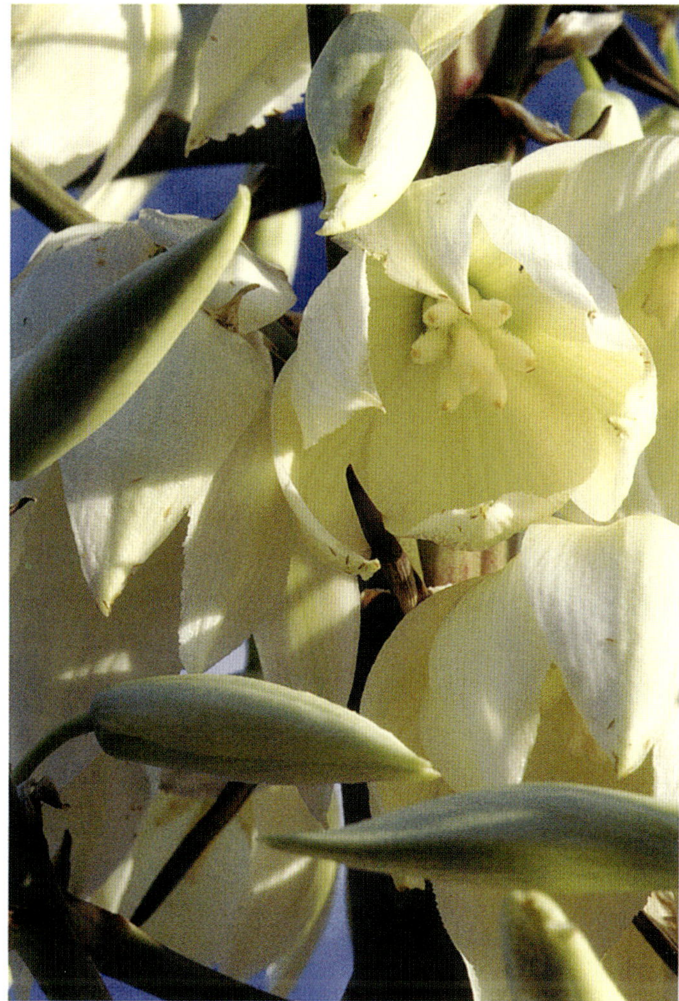

Perhaps more graceful and shapely than giant dagger, Thompson yucca is smaller, usually with a single, unbranched trunk up to ten feet high. A rosette of leaves, often symmetrical in shape, radiates from the top of the trunk, while the trunk bears a skirt of dead leaves. Sometimes slightly twisted, the thin, flexible leaves are linear, finely toothed along the edges, and spine-tipped. Up to two feet long, they are yellow-green to blue-green with yellowish margins.

From late March to early May, the flowers appear in a towering cluster within or more commonly above the leaves. With as many as forty short branches, the spike may reach three feet high. The hanging, bell-shaped flowers are creamy white, with three petal-like outer sepals and three broader inner petals. The fruit is a dry capsule, about two inches long, with a long, flaring, and twisted beak.

Also known as beaked yucca, soyate, and palmita, Thompson yucca is most prevalent on limestone soil in desert scrub habitats on the open rocky slopes and foothills of desert mountains. It is heavily concentrated in southeastern Brewster County and Coahuila, Mexico, and scattered elsewhere from Presidio to Pecos and Del Rio.

In the park Thompson yucca is conspicuous at Dagger Flat, growing with giant dagger and on the slopes above. Widespread in the Dead Horse Mountains, it is also encountered in the Santiago and Rosillos mountains.

Note: Yucca is the native Caribbean name for cassava, misapplied to these plants. This species was collected in Chihuahua in 1852 by John Bigelow, surgeon and botanist on Major Emory's U.S.-Mexico boundary survey, and described by William Trelease in 1911. Trelease named the species for Charles Thompson, agave specialist with the Missouri Botanical Garden. The species in the park was long considered a distinct species, *Yucca rostrata,* found only in Brewster County and Coahuila, Mexico, but it is now often subsumed under the name *Yucca thompsoniana.* Others consider *Yucca thompsoniana* a small northern variant of *Yucca rostrata* (Clary 1997).

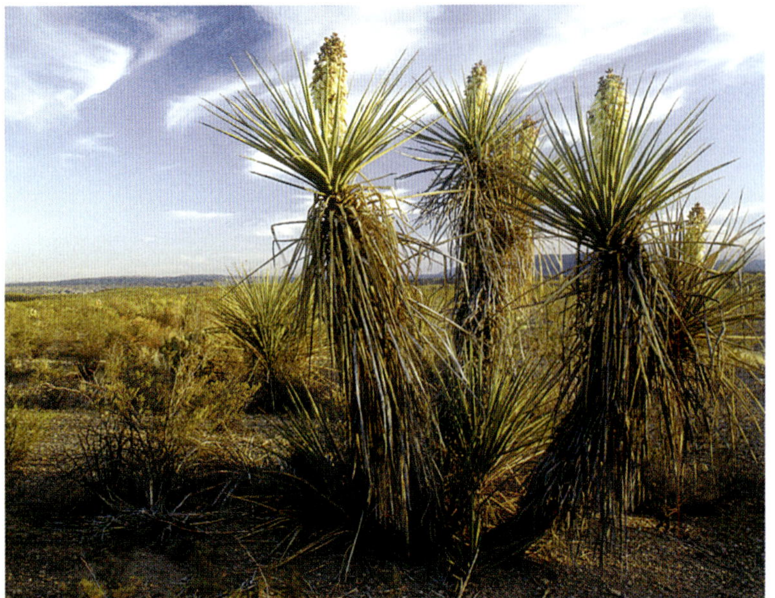

Spanish Dagger

Yucca torreyi

Spanish dagger, or Torrey yucca, is also referred to as "Old Shag" because of its shaggy, disheveled appearance. This treelike yucca is about six feet high in the park, but it reaches twenty feet or more elsewhere in its range. Spanish dagger has bayonet-shaped, yellow-green leaves that radiate from the top of one or several trunks or branches. The rigid but fleshy leaves, two to four feet long, taper to sharp spines. Loose strands of tough, coarse fiber curl along the edges. The unkempt trunk is cloaked with a skirt of withered leaves.

Despite its untidy look, Spanish dagger produces a spectacular flower spike. Creamy white, bell-shaped flowers up to three inches long are borne on a short stalk in a dense cluster within or partly above the leaves. The fragrant, waxy flowers hang from small branchlets of the flower spike and seldom fully open. Three or four inches long, the dark brown fruit is a pulpy, cylindrical capsule.

Spanish dagger fares well in desert scrub, on open arid slopes, and in grassland. It is dispersed through southwestern Texas from the upper Rio Grande Plains (where it merges with the related *Yucca treculeana*) and Edwards Plateau through the Trans-Pecos into southern New Mexico and Chihuahua and Coahuila, Mexico.

In the park Spanish dagger is wide ranging at low to mid elevations, especially in the Dead Horse Mountains. It typically blooms in spring in alternate years, but old-timers say you can find one blooming any month of the year.

Note: Native Americans ate the yucca flowers, fruits, and seeds; the roots were made into soap or a laxative; the leaves were used to make clothing and sandals. In 1859 New York botanist John Torrey defined this yucca as a variety (var. *macrocarpa*) of another species (*Yucca baccata*). John Shafer, museum custodian at the New York Botanical Garden, elevated it to species status in 1908.

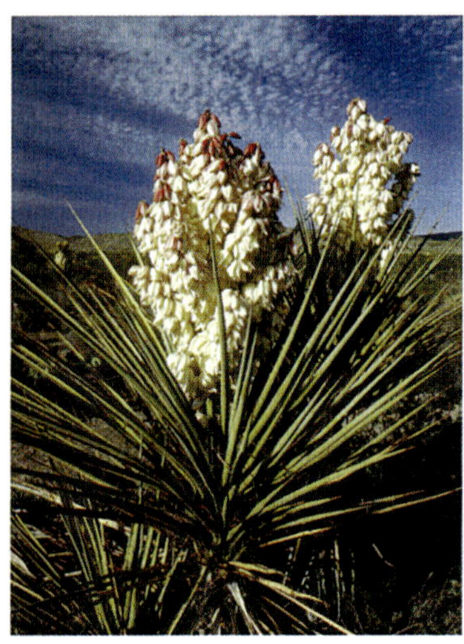

Living Rock Cactus

Ariocarpus fissuratus var. *fissuratus*

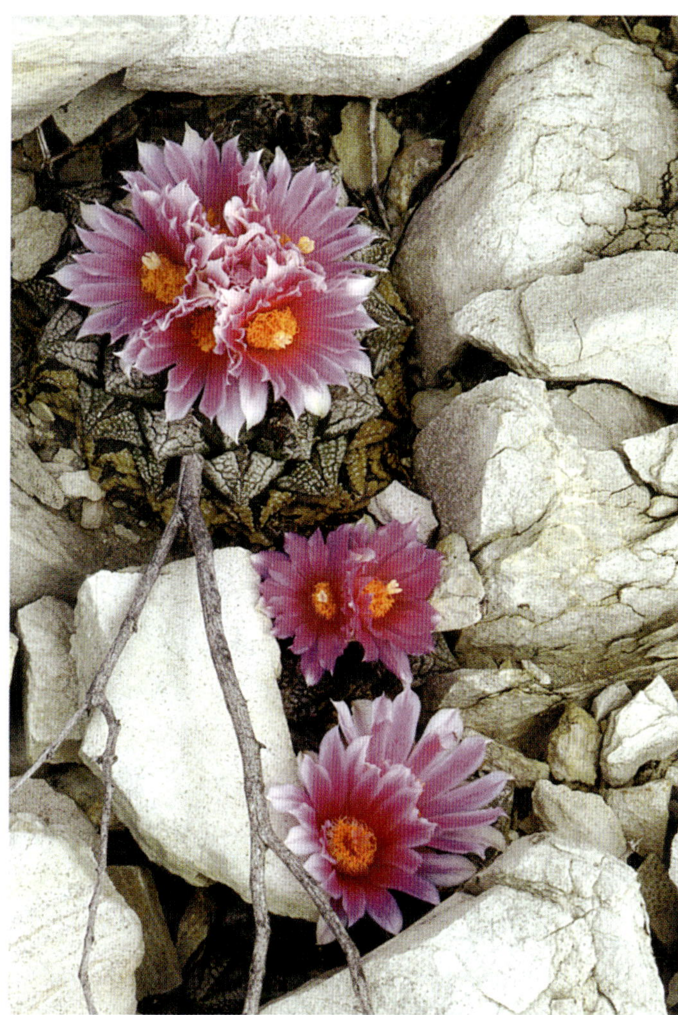

THE DESERT is home to some of the most bizarre plants on earth, and living rock cactus is one of the most peculiar. For most of the year, it blends into the rocky limestone soils of the Dead Horse Mountains, Mariscal Mountain, and the hills along the Rio Grande. You may step on one before you notice it. Spineless and flush with the ground, it has triangular, warty tubercles that overlap in a star-shaped pattern.

Two to five inches across, this cactus ekes out an existence between rocks or partially shaded by lechuguilla and creosote. In extremely dry conditions, living rock is almost invisible: it literally shrinks into the surrounding rocky soil. Moisture is stored in the taproot, and during droughts the root shrinks, dragging the stem underground. These spineless plants survive by blending into their native habitat. As added protection, they store foul-tasting, poisonous alkaloids in their fissured bodies.

A cottony tuft often grows from the center of the cactus, but in October and early November, lovely pink flowers replace the cotton. Then these cacti that have gone unnoticed suddenly seem everywhere, all around your feet. The strange conjunction of lush blooms atop tough, warty stems is unforgettable.

Other common names for this uncommon plant are star cactus, chautle, and dry whiskey, but it is known to the Tarahumara of northern Mexico as peyote cimarron, and its healing and religious powers are highly regarded. The Tarahumara have used the cactus to cast spells on enemies and build stamina in their long-distance runners.

In the United States living rock cactus is confined to Trans-Pecos Texas (from southern Hudspeth County east to Val Verde County). In northern Mexico, living rock and three other related species survive the barren landscape.

Notes: The name *Ariocarpus* is from the genus *Aria* and the Greek *karpos* (fruit), for the *Aria*-like, or by implication pear-shaped, fruit. The species name *fissuratus,* Latin for "fissured," applies to the tubercles. This species was collected by Arthur Schott in 1852 on the "Río Bravo del Norte" during the Emory boundary survey and described by George Engelmann in 1856.

Chaffey's Pincushion Cactus

Coryphantha chaffeyi

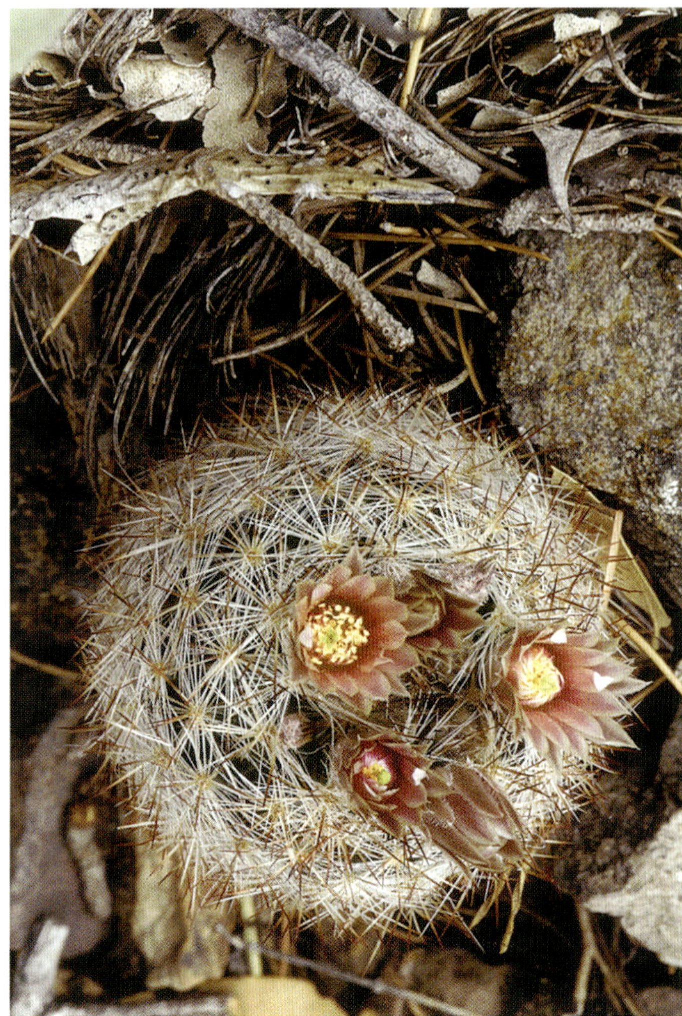

CHAFFEY'S PINCUSHION cactus, a mostly Mexican species, is found infrequently in the Chisos Basin and higher in the mountain woodlands. It grows in the igneous rock and leaves on mountain slopes, often unnoticed in the rubble.

Resembling the common cob cactus (*Coryphantha tuberculosa*) in appearance, this pincushion is almost always found at higher elevations in oak-juniper-pinyon forests. Also known as biscuit cactus, the plant is usually solitary but sometimes forms small clusters of only a few stems.

Chaffey's pincushion flowers are smaller than the common cob and seldom open widely, and the petals are more sharply pointed. Flower color is variable, but the petals are mostly cream-colored with a broad pale brown to salmon or rust midstripe. The stigma of this flower is green, while the common cob's stigma is white to yellow.

This pincushion is known from the Mexican states of Zacatecas, San Luis Potosí, Durango, and Coahuila, and the plants in the Chisos seem to be outliers. It is considered imperiled globally and critically imperiled in Texas.

You may notice Chaffey's pincushion hidden amidst rocks or leaves below Emory Peak, along the trail through Boot Canyon to the South Rim, or in Upper Blue Creek Canyon. This elusive cactus usually blooms in April and May.

Note: The genus name *Coryphantha* is from the Greek *corypha* (top) and *anthus* (flower): these cacti bear flowers on new growth at the stem tips. This species was described (as *Escobaria chaffeyi*) in 1923 by Nathaniel Britton, first director of the New York Botanical Garden, and Joseph Rose, first professional botanist with the Smithsonian Institution, in their four-volume work *The Cactaceae.* They named the species for Elswood Chaffey, a British-born physician and mining company doctor who had collected it in the Mexican state of Zacatecas in 1910.

Duncan's Pincushion Cactus

Coryphantha duncanii

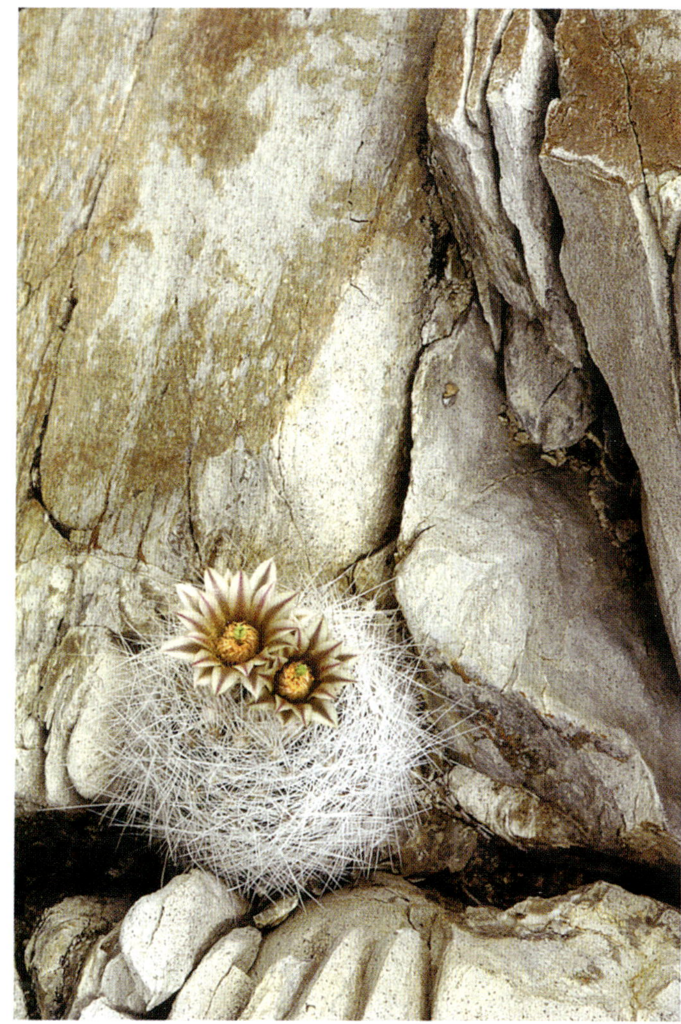

Also known as Duncan's snowball, Duncan's pincushion is one of the most unobtrusive and difficult to find cacti. It is classified as imperiled or critically imperiled both globally and in Texas and (like all plants in the park) should not be disturbed.

This small cactus is usually unbranched, often wedged between rocks on white limestone hills near or overlooking the Rio Grande. When healthy and mature, the spherical or conical stems may be densely covered with small white spines. Obscured by the weak, flexible, and bristly spines, the plants may indeed resemble tiny snowballs.

The pincushion is a difficult-to-identify species, somewhat similar to cob cactus (*Coryphantha tuberculosa*). Cob cactus is usually larger and blooms later in the spring, producing flowers with white or pale pink petals and white stigmas. The flowers of Duncan's pincushion are pale white or cream colored, delicately accented with peach or brownish midstripes and green stigmas. Perhaps one inch long, the petals seldom open widely and may be noticeably fringed.

This topflowered cactus is also sometimes confused with golf ball cactus (*Mammillaria lasiacantha*), which grows in similar, partially overlapping habitat. But the smaller golf ball cactus has a more spherical shape, much shorter spines that resemble fuzz, and relatively large flowers (compared to the stem width) with petals that fully open and even curl backward at the tips.

Duncan's pincushion is rare in southern Brewster and southeastern Presidio counties in Texas, an isolated area in southwestern New Mexico, and perhaps adjacent Coahuila and Chihuahua, Mexico. Another scarce attraction of the Big Bend country, this pincushion is one of Big Bend's earliest bloomers: I have photographed flowers in late January, but February is more typical.

Note: J. Pinckney Hester—cactologist, rare plant collector, and author of a number of articles in *Desert Plant Life*—described this species in 1941. Hester named this pincushion for Frank Duncan, the owner of a mining claim near Terlingua where the plant was first collected (Powell and Weedin 2004).

Sea Urchin Cactus

Coryphantha echinus

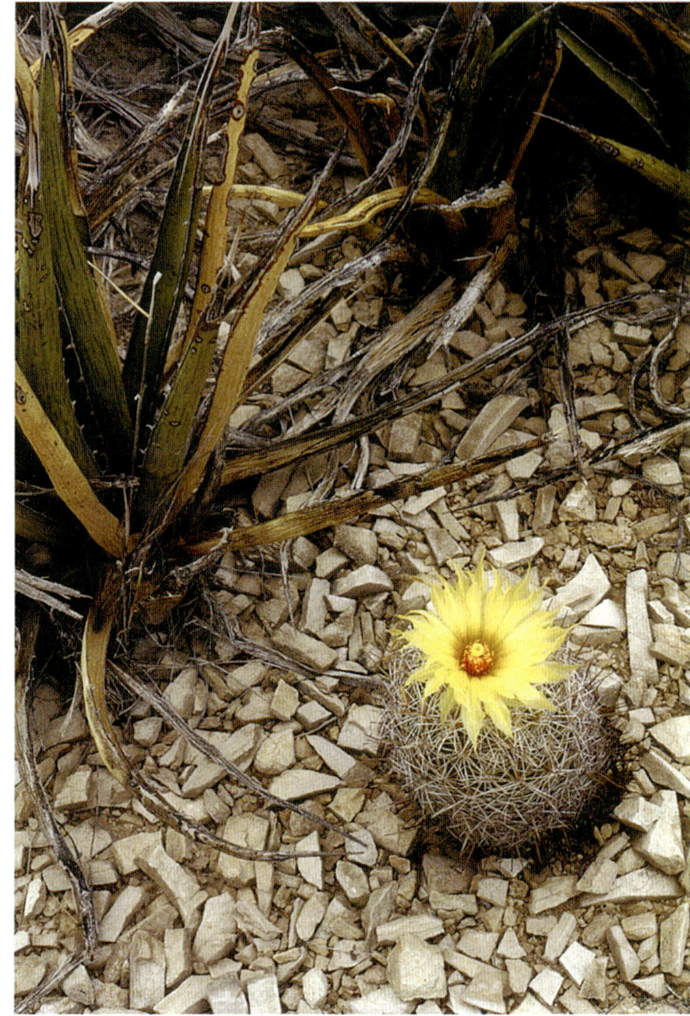

Sea urchin cactus's spherical or conical body, about four inches high, may be totally concealed by dense radial spines, and because its thick and rigid lower central spines stick straight out from the stem, this cactus of the desert floor clearly resembles a sea urchin from the ocean floor.

Also known as rhinoceros cactus and hedgehog Cory cactus, sea urchin is often solitary, but you may see clusters of the plant in its favorite habitat: limestone soils of the Dead Horse and Mariscal mountains and on Mesa de Anguila above Lajitas. I have also witnessed numerous individuals growing with rusty hedgehog cactus (*Echinocereus viridiflorus* var. *russanthus*) in the Chisos foothills.

Sea urchin is found within a limited domain—mostly the southern and eastern portion of the Trans-Pecos to Del Rio and adjacent Chihuahua and Coahuila, Mexico. It also occurs sporadically northeast of the Trans-Pecos to Howard and Coke counties (Powell and Weedin 2004).

This intriguing plant blooms all too briefly from late spring through summer, especially after summer storms. But what a bloom it is! Sea urchin produces startling sulphur-yellow flowers up to three inches across with sharp-pointed petals and striking red stamens. The intensity and purity of color and contrast when it first blooms is one of Big Bend's sweet summer surprises.

Of course the best things don't last: an individual flower may begin to open in late morning, be fully open in the bright light of noon, show unmistakable signs of wilting soon thereafter, and close completely by midafternoon. Green fleshy fruits about one inch high follow the flowers in late summer and fall.

Note: The species name *echinus* in Greek means "spine" or "sea urchin." This cactus, described (as *Mammillaria echinus*) by German-born Saint Louis botanist George Engelmann in 1856, was collected by Charles Wright in 1849. Wright, a Yale-educated but mostly self-taught botanist, worked as a schoolteacher and surveyor in East Texas and taught at Texas's first college, near La Grange, but his love of exploration led him into positions as a botanist on military survey expeditions.

Big-Needle Pincushion Cactus

Coryphantha macromeris var. *macromeris*

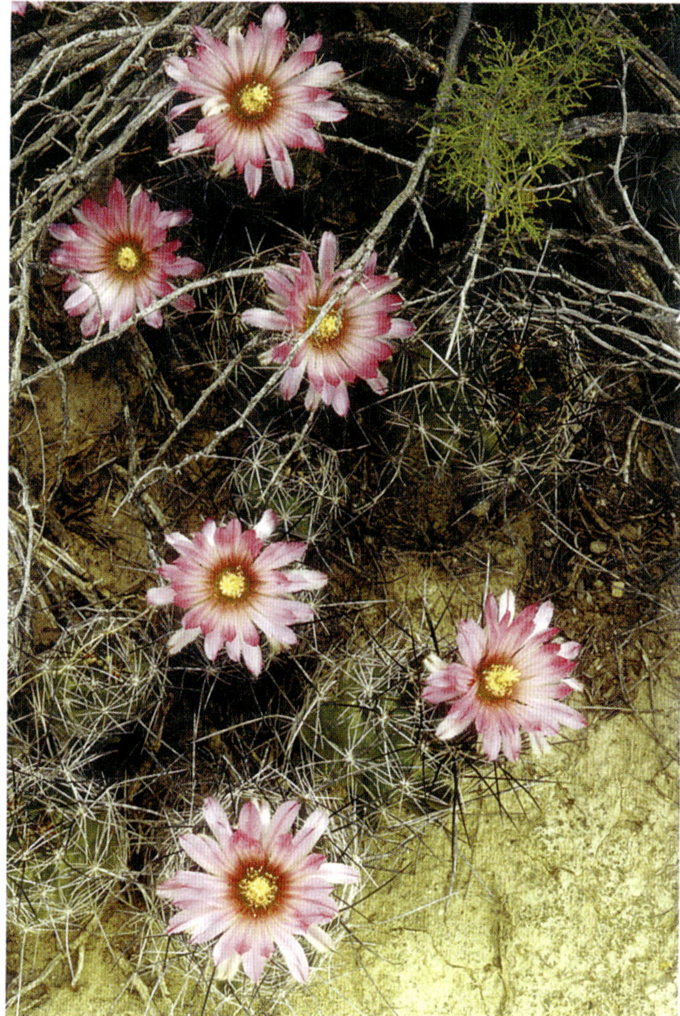

You can learn much about this cactus from its graphic common names. Big-needle pincushion—also known as big nipple Cory cactus, nipple beehive, and long mamma—has loose, globular stems with long, flabby, mammarylike tubercles.

This cactus ranges through the Trans-Pecos west into southern New Mexico and south into Mexico through Coahuila and parts of Chihuahua, Durango, and Zacatecas. In Big Bend, it is a resident of clay flats in the Rio Grande floodplain and the sides and crests of gravel hills up to thirty-five hundred feet in elevation.

Single stems are not uncommon, but usually this cactus forms many-stemmed mats as much as three feet across. Often it makes a home under the protection of creosote and mesquite.

Big-needle pincushion has white to gray radial spines and long, blackish central spines projecting haphazardly in all directions. This feature contributes to the unkempt appearance for which it is known and luckily discourages collectors. The bulky, tubercled stems, clearly visible beneath the untidy spines, gather and reflect light beautifully, like dark green, wrinkled peppers.

Quickly sprucing up in summer, this cactus produces large magenta or rose-pink blooms with dainty, fringed petals. Subsequently, it sports sweet edible fruit, like small purplish-green grapes. Seek out big-needle pincushion cactus as far north as Persimmon Gap or along the River Road, especially from the Gravel Pit to Solis on slopes above the road. Occasionally spreading mounds several feet across are nestled beneath mesquite.

Note: The species name *macromeris,* from Greek words meaning "large parts," probably refers to the broad tubercles. This species was described by George Engelmann in 1848 in the botany appendix to Adolph Wislizenus's *Memoir of a Tour of Northern Mexico.* Wislizenus, a German-born surgeon and freelance botanical collector from Saint Louis, had collected near Chihuahua in 1846, unaware of the U.S.-Mexico War. He was temporarily arrested as a spy but eventually joined Colonel Doniphan's military expedition across northern Mexico, collecting many new species along the way.

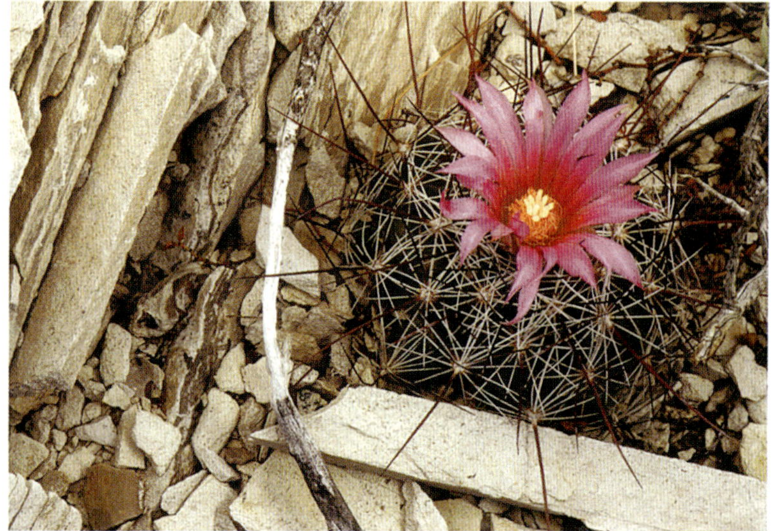

Whiskerbrush Pincushion Cactus

Coryphantha ramillosa

Also known as bunched Cory or Big Bend Cory cactus, the rare whiskerbrush pincushion cactus inhabits remote, sparsely vegetated limestone hills near the Rio Grande. It was listed in the United States as Threatened in 1979 because of its small numbers, limited range, and spotty distribution. Confirmed at about twenty-five sites in the park, whiskerbrush populations cross into Terrell County and the Lower Canyons of the Rio Grande and into northern Coahuila, Mexico.

In many ways this plant resembles big-needle pincushion cactus: it has dark green globular stems with large nipplelike tubercles. But whiskerbrush is more spherical and squatty, usually less than four inches high, with a bristly aspect. It has white radial spines and long, mostly dark, central spines, which stick out in all directions like whiskers.

Unlike the clustering and colonizing big-needle pincushion, the whiskerbrush is mostly solitary. It tends to grow in rock crevices and on ledges, not in the open or under mesquite. Inhabiting an inhospitable area with minimal rainfall and brutal temperatures, whiskerbrush may seem dead much of the time, like stubble.

This cactus produces pale pink to wine-colored flowers at the stem tips, nestled amidst the surrounding upswept spines. Whiskerbrush blooms mostly in late summer, often triggered by soaking summer rains. Juicy green to gray-green fruits mature in late fall.

Along the Rio Grande, this rarity is associated with two other cacti with very circumscribed distribution: golden-spined prickly pear and Big Bend cholla.

Note: The species name *ramillosa* means "with many small branches." This species was discovered in 1936 by A. R. Davis, a cactus collector from Marathon, Texas, and described in 1942 by Ladislaus Cutak, horticulturist with the Missouri Botanical Garden and author of *Cactus Guide* (1956). It was first collected at Reagan Canyon in Brewster County.

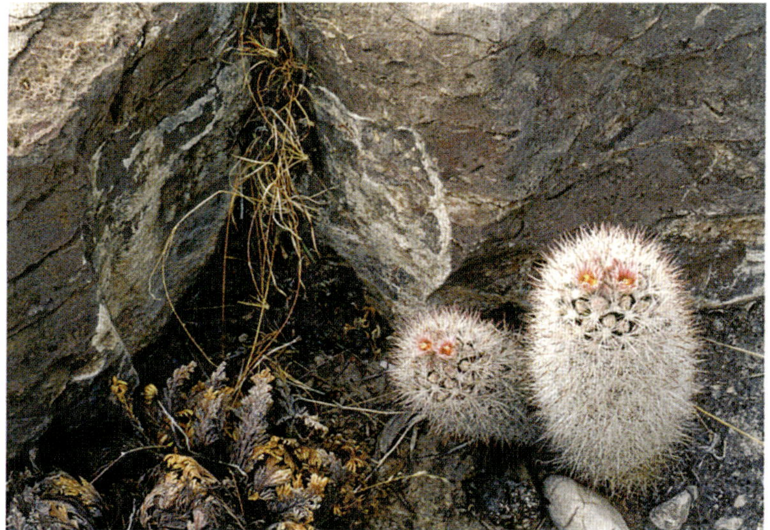

Silverlace Cactus

Coryphantha sneedii var. *albicolumnaria*

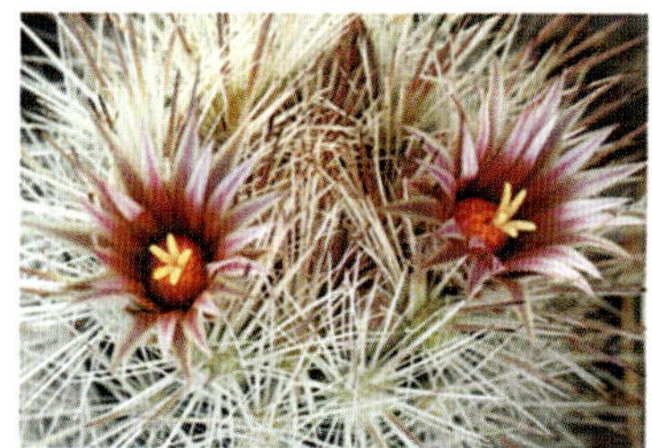

Also known as snowcone nipple and white column cactus, silverlace cactus is usually solitary or sparsely branched, reaching up to ten inches high. Often a little taller, straighter, and bulkier than the somewhat similar common cob cactus (*Coryphantha tuberculosa*), this plant acquired its name from the many bristling, brittle, snowy white spines that radiate in all directions and obscure the stem.

In spring dainty funnel-shaped flowers, not opening until early afternoon, emerge at the crest of the silverlace stems. Their slender, lance-shaped petals are pink or white with pink midstripes. Unlike cob blooms, the flowers are smaller and seldom open widely. Up to one inch long, pale green or red, club-shaped fruits appear in summer and fall.

In Texas, silverlace cactus grows in the Big Bend near the Rio Grande, especially in southern Brewster County, Presidio County, and rarely, Pecos County. Although not really known from Mexico, this cactus has been identified in Chihuahua and probably is present in Coahuila as well (Powell and Weedin 2004). It is classified both globally and in Texas as vulnerable or imperiled.

In the park silverlace cactus is restricted to limestone soils in desert scrub habitats and desert mountains and alluvial flats near the Rio Grande, from Boquillas Canyon in the east to Lajitas in the extreme southwest.

Note: The species name *sneedii* honors J. R. Sneed, who collected the first specimens (of *C. sneedii* var. *sneedii*) in the Franklin Mountains in 1921. The varietal name *albicolumnaria* means "white column" in Latin. This variety, described in 1941 as a species (*Escobaria albicolumnaria*) by J. Pinckney Hester, cactus and succulent collector and writer, is now mostly regarded as a variety of *C. sneedii.*

Cob Cactus

Coryphantha tuberculosa var. *tuberculosa*

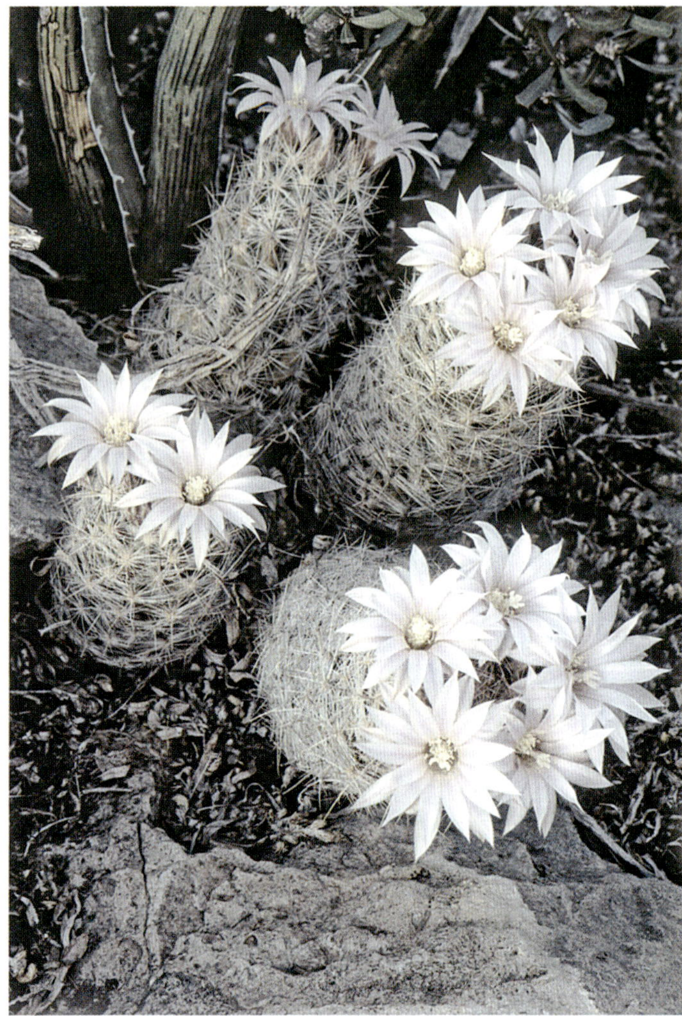

THE MOSTLY cylindrical stems of cob cactus may grow singly, but they usually form many-branched clumps of up to fifty stems. The stems stand upright or grow at strange, even ninety-degree angles, to the rest of the cluster. The older parts of the cob stem often lose their spines, exposing dry gray tubercles that resemble a decaying corncob.

Cob cactus blooms profusely in spring and sometimes again after summer rains, putting forth white to pale pink flowers at the stem apex. The flowers open widely, and the petals often curl back toward the stem. Bright red, cylindrical fruits may form when the cob is still flowering, and it is a whimsical sight: a corncoblike cactus with brilliant red fruits beside delicate pink flowers.

In a good spring, entire hillsides suddenly come alive with flowering cob, growing at low to mid elevations in rocky crevices or on ledges. In the park cob cactus is easily accessible from Persimmon Gap to the foothills of the Chisos and south and east to the Rio Grande.

Two varieties are recognized (Powell and Weedin 2004). The one pictured here (var. *tuberculosa*) forms large clumps in limestone habitats. This variety is quite common through Coahuila, eastern Chihuahua, and northeastern Durango, Mexico, and into Trans-Pecos Texas and southern New Mexico. Usually unbranched and more globular in shape, varicolor cob (var. *varicolor*) prefers igneous and novaculite habitats. It is mostly limited to the southern Trans-Pecos (Brewster, Presidio, and Jeff Davis counties) but may reach into adjacent Mexico.

Note: This species was collected in 1852 by John Bigelow in Chihuahua during the U.S.-Mexico boundary survey and described by George Engelmann in 1856 (as *Mammillaria tuberculosa*). In 1832 Engelmann had moved to the Saint Louis area, where he established a medical practice, founded the German newspaper *Das Westland,* recruited botanical collectors for expeditions throughout western North America, described more than six hundred new species from these and his own explorations, and played a critical role in the creation of the Missouri Botanical Garden.

Horse-Crippler Cactus

Echinocactus texensis

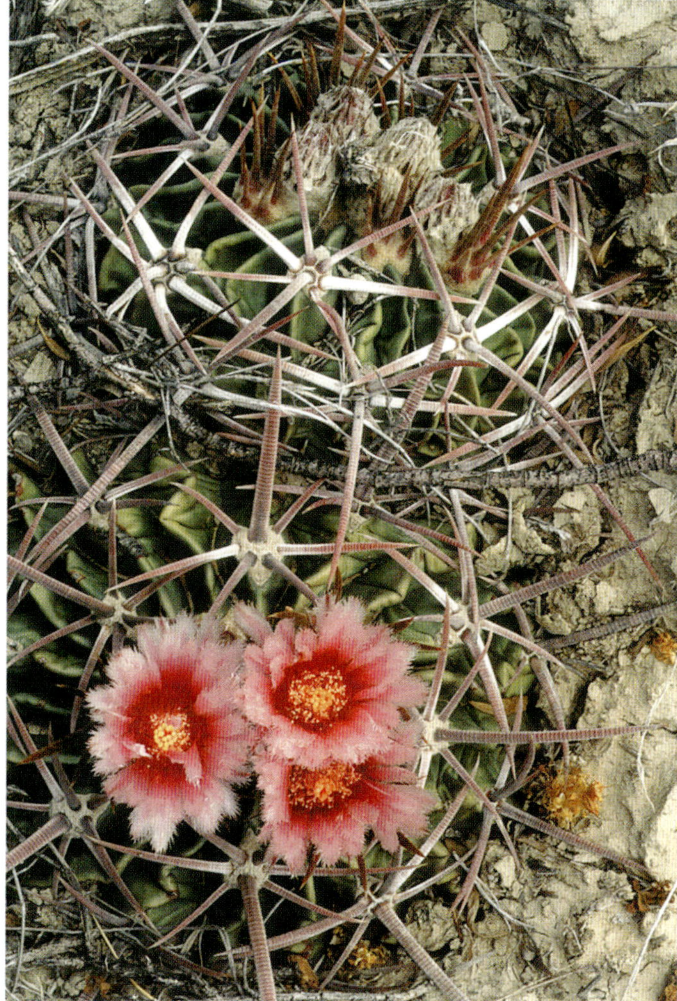

ALSO KNOWN AS devil's head, devil's pincushion, and manca caballo, horse-crippler is another cactus of the rugged Big Bend country with menacing spines. Horse-crippler's flat stem is often unnoticed by horses and cattle, and the long, arching spines can easily pierce horses' hooves and even puncture tires. Ranchers have tried to eradicate the species, dragging fields with chains to uproot the plants.

This cactus, which can be a foot wide, has a broad, pale green, circular stem with a hard surface that is flattened or domelike and only slightly elevated off the ground. Tough, pinkish spines, whiter near the base and often curving downward, protrude from prominent, irregular ribs on the stem surface.

Although rather forbidding in its spiny armor, horse-crippler produces fetching pink or salmon flowers with brilliant red throats. Frequently the petals are noticeably fringed at the tips. Not too long after the late spring blooms, red, jumbo-sized oval fruits up to two inches long emerge at the stem's woolly center.

The mostly solitary horse-crippler's wide range encompasses northeastern Mexico, much of Texas (from the Rio Grande Plains throughout the Edwards Plateau and eastern Trans-Pecos to the Panhandle), and southeastern New Mexico. In Big Bend National Park it occupies silty soils and sandy flats, especially near Dog Canyon and on Tornillo Flat.

Note: The genus name *Echinocactus* is from the Greek *echinos* (hedgehog) and *cactus* (a spiny plant, and an earlier genus name). This species was described by cactologist Carl Hopffer in 1842 in a German gardening magazine. The type specimen was grown from seed, but the plant was collected in Mexico in 1836 by Jean Louis Berlandier and in Texas in 1844 by Ferdinand Lindheimer. The naturalist Berlandier, born on the French-Swiss border, collected in northern Mexico and Texas for his teacher, Swiss botanist Augustin de Candolle, and later became a physician in Matamoros, Mexico. Lindheimer, a Frankfurt-born guide for German settlers and Texas botanical collector who discovered several hundred new species, later served as editor of the New Braunfels, Texas, German-language newspaper.

Chisos Hedgehog Cactus

Echinocereus chisoensis var. *chisoensis*

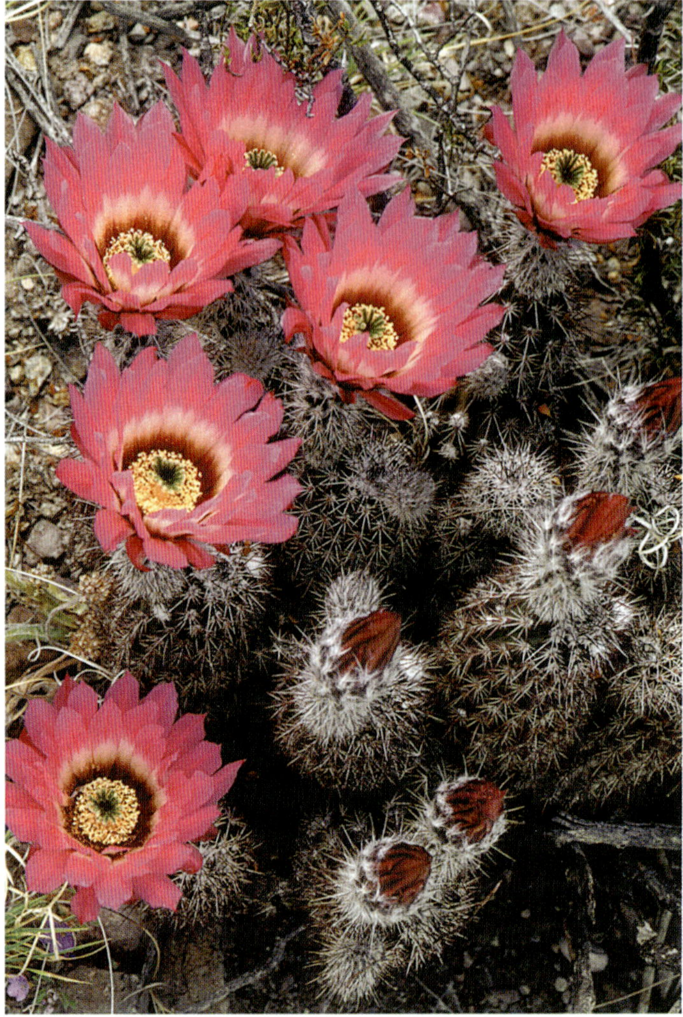

Chisos hedgehog cactus is a seldom seen species growing on desert pavement and terraces south and east of the Chisos Mountains. Isolated at about a dozen locations in a small rectangular area (roughly ten miles long and three miles wide) within Big Bend National Park, Chisos hedgehogs may number no more than one thousand plants. Their habitat apparently does not extend into Mexico.

Also known as Chisos Mountain hedgehog, this cactus is often well camouflaged under "nurse" plants, especially lechuguilla and creosote. The shade of nurse plants provides relatively more moisture and sun protection for the hedgehog seeds.

Once seen, you won't easily mistake this plant for another cactus again. Due to spine coloration, the green, slightly pointed or conical stems have a maroonish cast most of the year. Sometimes nearly a foot high, the stems are easily visible beneath the short spines. White, cottony hairs frequently grow from areoles, especially at the stem tip. Individual plants, often single, may branch just above the ground or grow near others. It is not uncommon to find these cacti growing within large clusters of sea urchin cactus, or young plants protected in dog cholla mounds.

Large tricolored flowers, bursting forth in early spring, are this cactus's most arresting feature. Although not opening widely, the blooms are easily two inches across, pink at the top with a red base and a cream-colored midriff. It is a strange juxtaposition: half-hidden plants sporting oversized, knockout flowers that could win a best-of-show trophy anywhere.

Considered a relict species surviving in a very changed environment, Chisos hedgehog was listed by the U.S. Fish and Wildlife Service as Threatened in 1988. It is classified as critically imperiled and especially vulnerable to extirpation in Texas.

Note: The genus name *Echinocereus* is from the Greek *echinos* (hedgehog or spiny) and the Latin *cereus* (wax candle, an earlier genus name) because the stems are shaped like spiny candles. William Marshall, of Arizona's Desert Botanical Garden, described this species in 1940, based on a specimen collected by Frank Radley in Big Bend National Park in 1939.

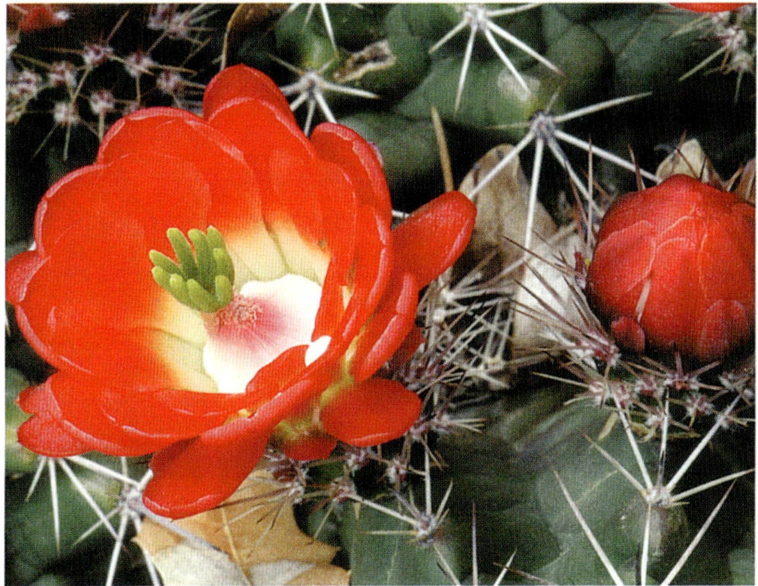

Texas Claret-Cup Cactus

Echinocereus coccineus var. *paucispinus*

TEXAS CLARET-CUP is the red jewel of the Chisos, accenting the mountain landscape with a magical scarlet color. The flower color resembles the deep hue of a fine claret wine, hence the plant's name. The thick stems, usually six to ten inches high, branch to form mounds, constructing clumps up to several feet across with eighty-five or more members. In stormy weather, these clumps with their fiery red blossoms light up the mountain trails like glowing lanterns.

This cactus thrives in rocky volcanic soils in the Chisos woodlands. Curiously the waxy flowers (with stiff rounded petals, a delicate cream-colored base, and green stigma lobes) remain open day and night for several days. A spring bloomer, this hedgehog has male and female plants that are functionally different: fruits are produced only on female plants that have been cross-pollinated by male plants (Powell and Weedin 2004).

The stems have relatively few gray or black spines and broad, sometimes wrinkled grooves that separate very obvious ribs. After the blooms have gone, the mounds dot the hills again with juicy crimson fruits that can be made into delicious preserves.

In rocky mountain hideaways and oak-juniper-pinyon woodlands, the Texas claret-cup variety in the park enlivens hills across the Trans-Pecos from Culberson County east to Del Rio. Its range includes portions of the Edwards Plateau, southeastern New Mexico, and northern Coahuila. This variety is vulnerable globally and in Texas. In the park, you will probably recognize this cactus in Green Gulch or on the Lost Mine and Pinnacles trails.

Note: The species name *coccineus,* "scarlet" in Latin, refers to the red flowers. The variety name *paucispinus* means "few-spined." This variety was described (as *Cereus paucispinus*) by Saint Louis botanist George Engelmann in 1856.

Texas Rainbow Cactus

Echinocereus dasyacanthus

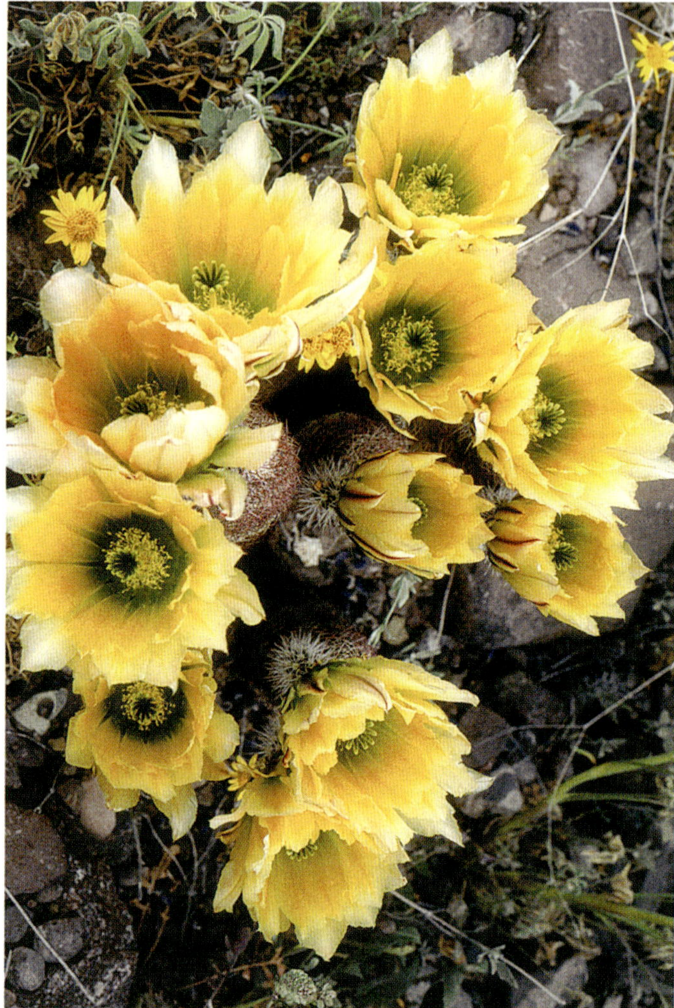

After the pineapple, golf ball, and Duncan's pincushion cacti bloom at the start of the spring cycle, the rainbows follow with their exquisite trumpet flowers heralding the full outbreak of spring. With tall cylindrical stems and waxy yellow flowers, the Texas rainbow cactus resembles a yellow wax candle.

Preferring limestone soils, rainbows are scattered throughout the park from desert flats to grasslands below four thousand feet. Single or often branching, the rainbows' robust stems may reach twelve inches high and four inches wide. The stems have narrow ribs sometimes completely covered by short, interlocking radial spines.

This cactus gets its name from the colors of its bristly spines, which may grow around the stem in a different shade each year, often forming horizontal, rainbowlike bands of colors. Spine colors may vary from white or yellow to tan, pink, purple, or brown.

In the spring this panoply of spine colors is accompanied by huge three- to four-inch-wide flowers, usually lemon yellow but occasionally almost orange. The flowers sport elongated petals with ragged edges and greenish bases, and their elegance is accentuated with green stigma lobes. In the park you may rarely locate a rainbow with rose-pink or magenta flowers. This cactus produces spiny fruits that are green to purple when mature.

Texas rainbow is widely distributed from northern Coahuila and northeastern Chihuahua through the Trans-Pecos to El Paso and into southern and central New Mexico. In the park my favorite place to find these cacti with their trademark color bands is Burro Mesa on the Apache Canyon Trail. At one time it was thought that these extremely banded rainbows were a distinct variation of the species.

Note: The species name, from the Greek *dasy* (shaggy, dense) and *acanthus* (spine), means "shaggy" or "densely spined." This species was described by George Engelmann in 1848 and collected by Adolf Wislizenus during his travels in northern Mexico. Wislizenus had previously traveled the Oregon Trail and authored *A Journey to the Rocky Moutains in the Year 1839.*

Strawberry Hedgehog Cactus

Echinocereus enneacanthus var. *enneacanthus*

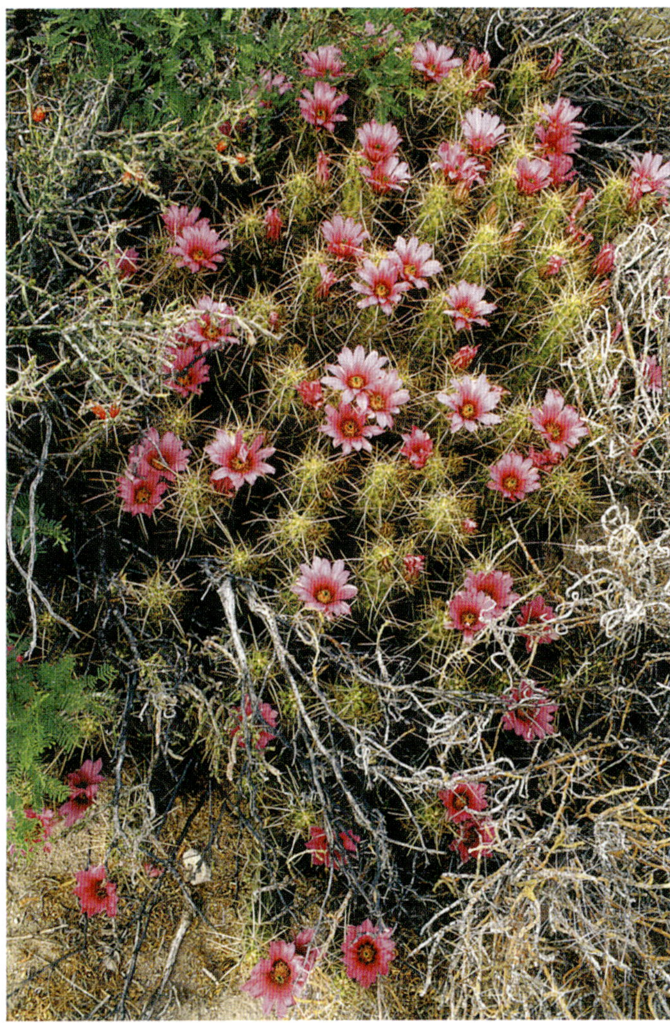

TWO SPECIES of "strawberry" cacti appear in the park. Strawberry hedgehog is clump-forming like its close relative strawberry cactus (*Echinocereus stramineus* var. *stramineus*), but it usually has fewer stems arranged in looser, flatter clusters or lower and not typically dome-shaped mounds. In clusters of up to one hundred, the plant's green stems are less obscured by spines. Up to six inches or more in length, the tapered, pickle-shaped stems may seem flabby, loose, and hollow to the touch.

Also known as pitaya and alicoche, this hedgehog frequently sprawls under the shade of mesquites or other shrubs in mostly alluvial habitats in silt, clay, and gravel from the Rio Grande to three thousand feet of elevation. The related strawberry cactus is more often encountered at higher elevations, on rocky slopes and mesas in full sun.

Compared to its close relative, strawberry hedgehog's spring-blooming pink to magenta flowers are not quite so large or translucent, and its fruits are not considered as juicy, flavorful, or aromatic.

This variety of strawberry hedgehog (var. *enneacanthus*) spills across northern Mexico (from Durango and San Luis Potosí) north into Brewster County and west mostly in a band near the Rio Grande to El Paso. A distinct variety (var. *brevispinus*) is dispersed from eastern Brewster County to the Texas Hill Country near Kerrville and south to northeastern Mexico.

Although not as attractive as its showy cousin, this cactus is, nevertheless, quite special. In spring, in sandy areas by the Rio Grande, these "pickles" seem to bloom profusely under almost every mesquite, not so much a surprise as a welcome certainty. Look for this hedgehog at Boquillas Flats on the way to Boquillas Canyon.

Note: The species name *enneacanthus* is from Greek words meaning "nine-spined," a reference to the number of radial spines. This species was collected by Adolf Wislizenus south of Chihuahua, Mexico, in 1847. A naturalist, mapmaker, artist, historian, and plant collector, Wislizenus traveled with a gun-running caravan into northern Mexico in 1846.

Strawberry Cactus

Echinocereus stramineus var. *stramineus*

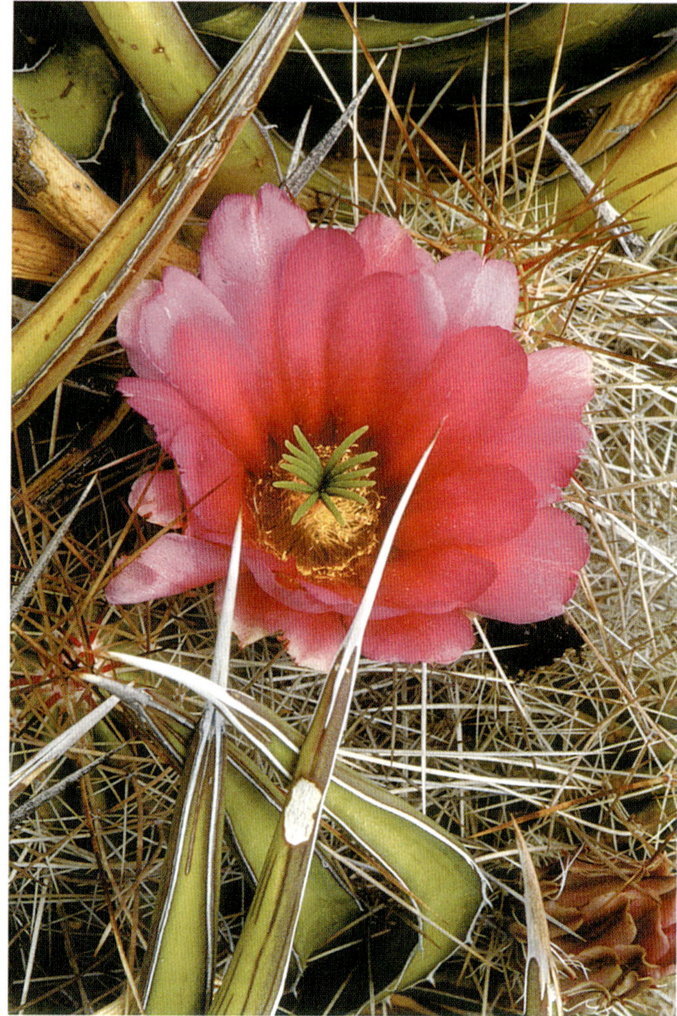

Also known as strawpile hedgehog cactus and spine mound cactus, strawberry cactus blankets open rocky slopes at many park locations but is especially beautiful at Persimmon Gap, the park's northern entrance. This cactus forms immense domelike, compact mounds, sometimes more than three feet across, with as many as several hundred stems. Alongside lechuguilla and chino grama in rocky limestone outcrops encircling the Chisos, it can take over hillsides.

With long, straw-colored translucent spines obscuring the stems, the glistening strawberry cactus mounds brighten the hills well before the onset of spring. Then in April and May a lovely transformation begins as three- to five-inch-wide magenta flowers cover the golden mounds with as many as forty to a single dome.

Long green stigma lobes protrude from satiny flowers that are deep magenta at the base but lighter near the tips. The flowers open about 11 a.m. and for a few brief days dominate the landscape. This is a glorious time in the desert: you learn firsthand the role of color in nature and the emotions evoked by natural color. After the brilliant flower displays are gone, those who know the desert well sample the juicy, reddish-brown, strawberrylike fruits or eat them as a delicacy with sugar and cream.

Populations of this cactus stretch across northern Mexico into the southern Trans-Pecos and west to El Paso and southern New Mexico.

Note: The species name *stramineus,* Latin for "made of straw," refers to the spines. This species (also known as *E. enneacanthus* var. *stramineus*) was collected in 1851 by Charles Wright on Colonel Graham's U.S.-Mexico boundary survey and described by George Engelmann in 1856. In 1852 Wright had taken extensive cacti collections to Engelmann in St. Louis and other specimens to Asa Gray at Harvard. Never in the Southwest again, he joined the Ringgold expedition to the North Pacific the following year.

Rusty Hedgehog Cactus

Echinocereus viridiflorus var. *russanthus*

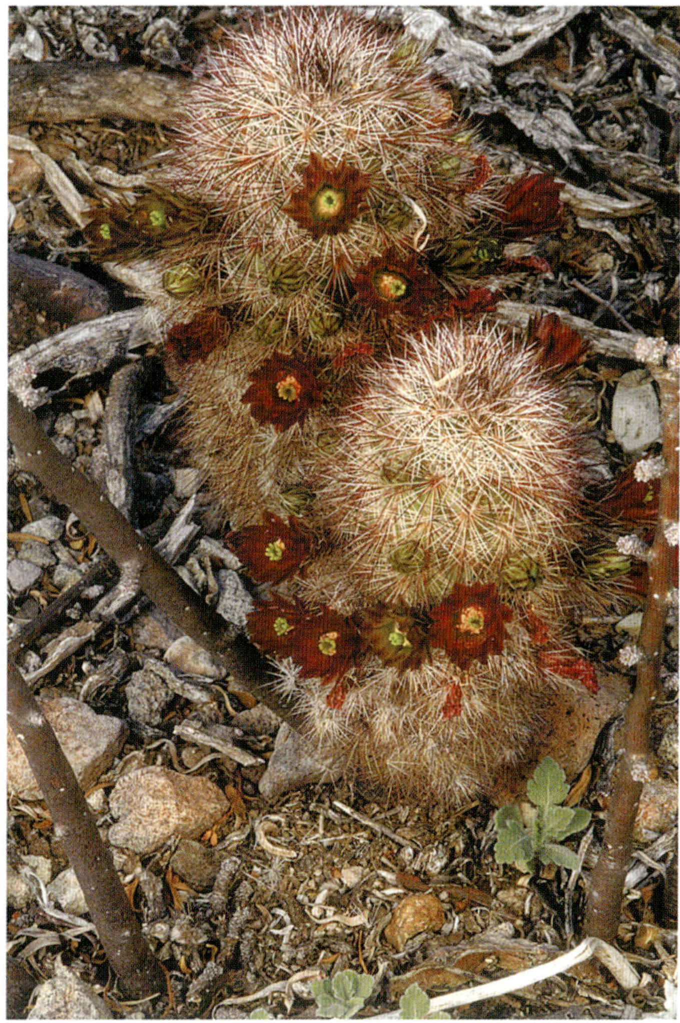

Also known as brown-flowered cactus and torch cactus, rusty hedgehog is an engaging member of the Chihuahuan Desert landscape. Flowering in March and April, this plant may have a somewhat clumsy or awkward appearance: it is seemingly weighted down with funnel-shaped, russet flowers which ring the stem at many different levels.

Notably, flowers arise from old growth of previous years rather than new growth at the stem tips. The rust-colored petals do not open widely but stick out from the stem's ribs like miniature torches. Green to reddish-brown globular fruits one-half inch in length follow the flowers about two months later.

Rusty hedgehog is covered with slender and interlocking bristlelike spines, mostly reddish or white with gray to purple tips. The spines are occasionally so thick that they completely conceal the stem, giving the cactus a shaggy, bottlebrush appearance. Often growing in the open, this cactus also does well in the partial shade of bushes, especially creosote.

This cactus species is endemic to central and especially southern Brewster County, although it may occur in adjacent Coahuila, Mexico (Powell and Weedin 2004).

In the park rusty hedgehog is widespread from Chihuahuan Desert scrub habitats to grasslands, mountain foothills, and woodlands, mostly on igneous soils. This hedgehog can be observed from Persimmon Gap in the north to Castolon in the south and from the Dead Horse Mountains in the east to the Maverick Badlands in the west. It prospers at all but the highest elevations of the Chisos.

Note: The species name *viridiflorus,* from Latin, means "green-flowered," one of the flower colors of the species. The varietal name, *russanthus,* refers to the "russet-colored" flowers of this variety. Rusty hedgehog was described in 1969 by Del Weniger, longtime professor and chairman of the biology department at Our Lady of the Lake College in San Antonio and author of *Cacti of the Southwest* (1969).

Woven-Spine Pineapple Cactus

Echinomastus intertextus

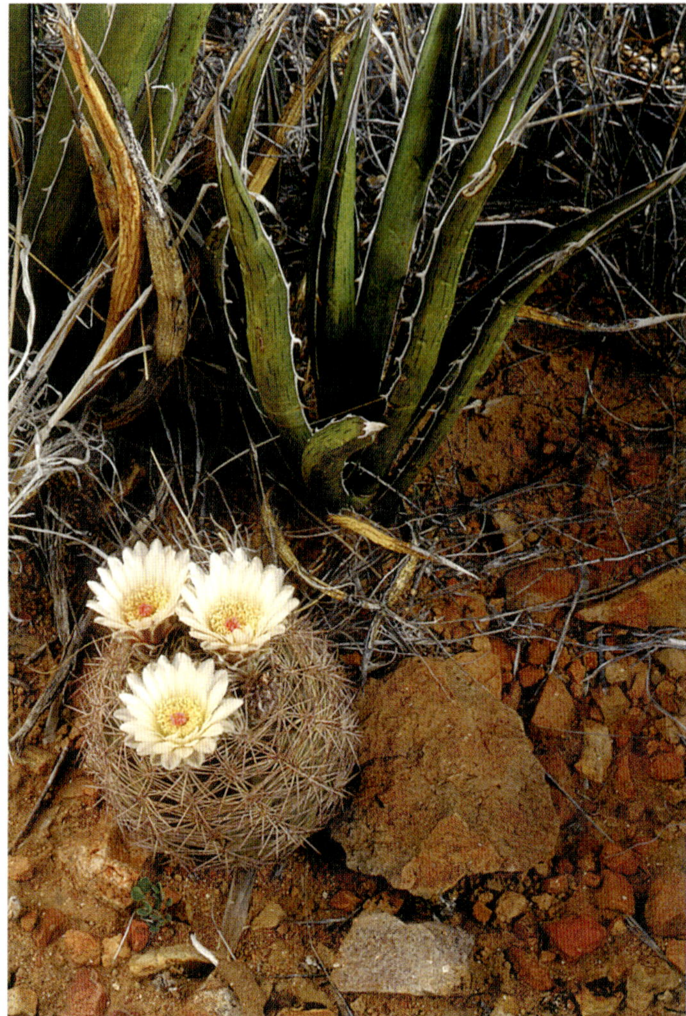

ALONG WITH Mariposa (*Echinomastus mariposensis*) and Warnock's cactus (*Echinomastus warnockii*), woven-spine pineapple is one of three "pineapple" cacti in the park. It is distinguished by a solitary green stem, egg- or pineapple-shaped, which is up to six inches high and readily visible beneath the spines. The spines are straw-colored with pink to red tips or red throughout. By comparison, Warnock's cactus has a blue-green, glaucous stem that is usually evident beneath chalky, blue-tipped spines, while the solitary Mariposa stem is typically obscured by overlapping ash-white spines.

Also known as blanco viznagita and early bloomer, woven-spine pineapple is often perched on a narrow pedestal-like stalk slightly above the ground. It is a comical sight to find this pineapple resting precariously atop a partially exposed, flimsy root.

The white flowers of Warnock's cactus and Mariposa have green stigmas, but this cactus produces white flowers with very prominent red stigmas and somewhat ragged petals. These pineapple cacti are harbingers of spring, flowering as early as February.

While Warnock's cactus and Mariposa flourish in desert habitats, woven-spine pineapple frequents higher elevations in open grassland. It is an uncommon resident of the west slopes of the Chisos near Lower Oak Creek Canyon and is separated from larger populations in the central Trans-Pecos, especially the Davis Mountains. This cactus is also present in the Franklin Mountains near El Paso, southwestern New Mexico, southeastern Arizona, and Chihuahua and Sonora, Mexico (Powell and Weedin 2004).

Note: The genus name *Echinomastus*, from Greek words meaning "spiny-breasted," alludes to the spiny tubercles. The species name *intertextus* ("interwoven" in Latin) refers to the intertwined spines. This species was described by George Engelmann in 1856. Two varieties have been identified: var. *intertextus* and var. *dasyacanthus* (which is restricted to New Mexico and the Franklin Mountains near El Paso). The plants in the park, although resembling variety *dasyacanthus*, may be hybrids of variety *intertextus* and Warnock's cactus (*E. warnockii*) (Powell and Weedin 2004).

Mariposa Cactus

Echinomastus mariposensis

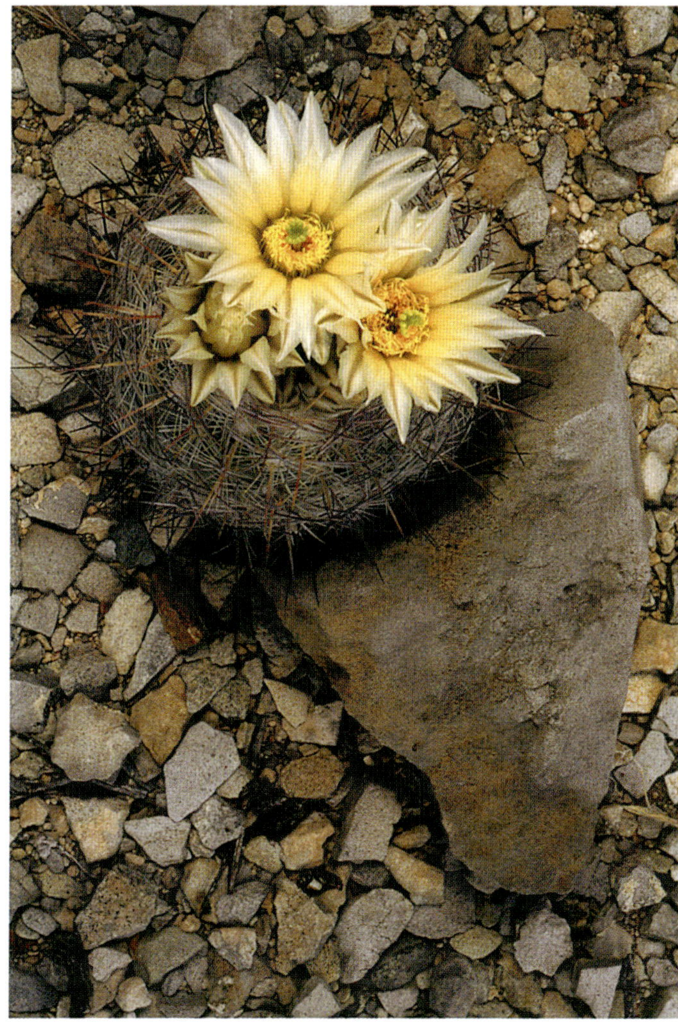

Listed as Threatened under the Endangered Species Act in 1979, Mariposa cactus was once quite common around Terlingua. It was discovered by the legendary J. Pinckney Hester, a cactus lover and determined explorer of the Big Bend country, and described by him in 1945 in the magazine *Desert Plant Life.*

Mariposa means butterfly in Spanish, but the cactus is named for the Terlingua quicksilver mining company—Mariposa Mines—that once operated there. It has also been known as Lloyd's Mariposa and golf ball cactus, although many plants are the size of tennis balls or even larger.

With white flowers and solitary, blue-green stems, Mariposa is somewhat similar to the other pineapple cacti in appearance, but the stems, no more than four inches high, tend to be more globe-shaped and are usually hidden beneath numerous overlapping, ash-white radial spines. This cactus is also distinguished by a tuft of larger upswept, bluish-gray to black central spines near the stem tip and by lower, shorter central spines pointing slightly downward.

The white flowers, appearing as early as late February, have rather prominent green stigmas and petals with pale midstripes, which are flesh-colored to pink or tan. You may notice that many of the early spring flowers are partially eaten: the petals are a favorite delicacy of ants, beetles, rodents, and birds. The green, sometimes yellow-green, fruits are globe-shaped or oblong, less than one-half inch long and not quite as wide.

Nearly wiped out at many locations by collectors, this pineapple cactus still survives within the protective confines of Big Bend National Park. Mariposa is known from about thirty sites in Brewster County, many in the park, from southern Presidio County and Coahuila, Mexico.

Unlike the other desert-dwelling pineapple, Warnock's cactus, Mariposa is largely restricted to Boquillas limestone habitats, in crumbling gravelly soil or more stable rocky outcrops. More common than previously believed, it can be abundant in small areas, especially near exposed hilltops in the Dead Horse Mountains in the open or near small shrubs, such as creosote and lechuguilla.

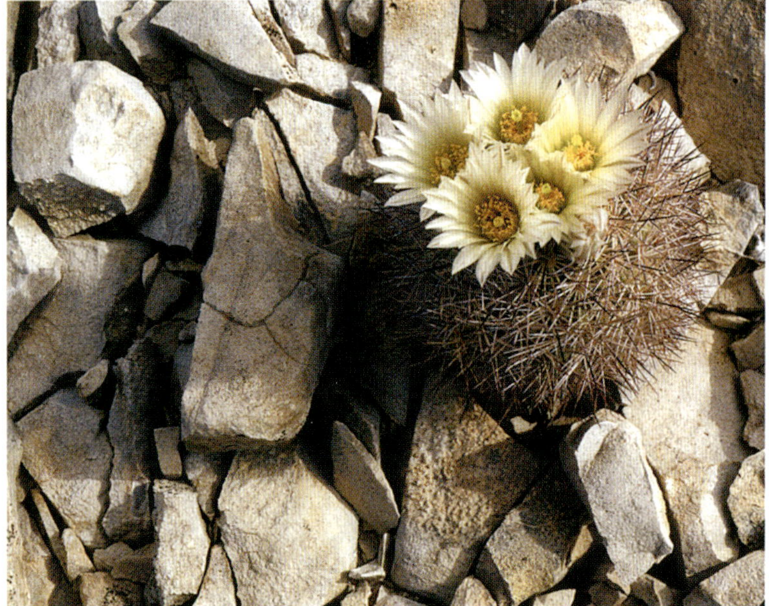

Warnock's Cactus

Echinomastus warnockii

THIS PINEAPPLE cactus is named for the late Barton Warnock, well-known botanist of the Big Bend country and former professor at Sul Ross State University. Warnock's cactus has a blue-green, mostly solitary, waxy stem, two to as much as eight inches high, which is nearly but not fully obscured beneath stiff, dense spines. Spreading rather haphazardly in all directions, the spines are either chalky blue near the tips or blue-gray throughout, sometimes giving the entire cactus a bluish-gray cast.

Early flowering like other pineapple cacti, this species can produce white flowers in February. Opening shortly after noon, the blooms have light green stigmas and pale green to tan midstripes on the petals. Pale green globular fruits less than one-half inch wide follow the flowers.

Warnock's cactus is a frequent inhabitant of a wide variety of desert habitats, often low limestone hills, gravelly mesas, and alluvial flats, with creosote and lechuguilla. In the United States, this pineapple is confined to the Trans-Pecos, from southern Hudspeth to southern Presidio and Brewster counties, where it is most plentiful. Considered vulnerable globally and in Texas, it also reaches into adjacent Chihuahua and probably Coahuila, Mexico.

In the park Warnock's cactus is common below forty-five hundred feet, especially in the Dead Horse Mountains. Good examples can be seen at Dagger Flat, where it grows near Texas rainbow cactus, catclaw cactus (*Glandulicactus uncinatus* var. *wrightii*), and giant dagger yucca.

Note: This species was described in 1969 by cactologist and prolific botanical author Lyman Benson, who named it for his friend, Warnock. Benson, chairman of the botany department at Pomona College in California, authored many works on cacti and other plants, including a comprehensive 1982 study of *The Cacti of the United States and Canada.*

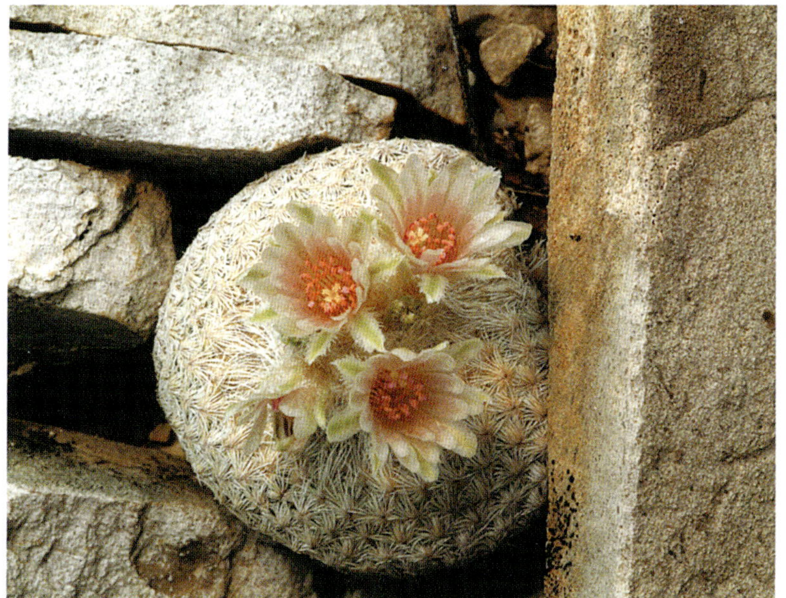

Boke's Button Cactus

Epithelantha bokei

At first glance resembling a tiny white mushroom, Boke's button cactus is an extraordinary plant to stumble across in the desert landscape. And stumble across it you may because it blends so well with its native white limestone habitat in the Dead Horse Mountains.

Also known as Boquillas button cactus, these miniatures are usually solitary, less than two inches high and wide, with minute, dense, closely appressed, smooth-to-the-touch white spines covering the globose stems. A distinctive tuft of spines also forms at the woolly, slightly sunken stem apex, and delicate white to pale pink flowers emerge from the depressed center in late spring. These flowers can easily be missed: they may open only on bright hot days and often well after noon. When the bloom is gone, the button puts forth slender, red, club-like fruits.

Boke's button is limited to southern Brewster and Presidio counties and adjacent Mexico. Globally and in Texas, it is classified as vulnerable to extinction. Much rougher to the touch, not as white, and with smaller flowers, the similar common button cactus (*Epithelantha micromeris* var. *micromeris*) may go unnoticed. More wide ranging in the Trans-Pecos, it is less extensive in the park, primarily in the northern Dead Horse Mountains.

Like the living rock and peyote cacti, Boke's button contains poisonous alkaloids that keep it from becoming a meal for desert denizens. The Tarahumara of Chihuahua use button cacti medicinally, assigning them magical powers. For some reason, they call them hikuli mulato (dark-skinned peyote).

Rarely, monstrous or freak forms (cristates) of this cactus are noted in the park. Some plants seem genetically predisposed to form cristates under adverse conditions (injury, disease, over- or under-nourishment, sudden change in soil conditions, extreme weather).

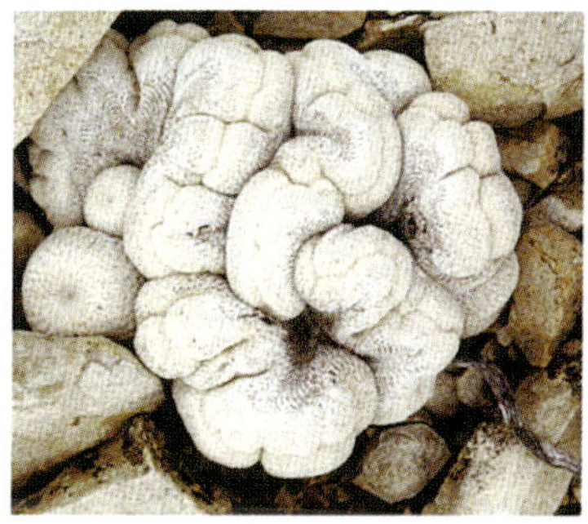

Note: The genus name *Epithelantha* is from Greek words meaning "flower upon nipple," referring to the formation of flowers at the tubercle tips. This species was described by cactologist Lyman Benson in 1969 and named for University of Oklahoma botanist Norman Boke, the former editor of the *American Journal of Botany,* who studied the genus *Epithelantha* and had collected this plant in 1955.

Giant Fishhook Cactus

Ferocactus hamatacanthus var. *hamatacanthus*

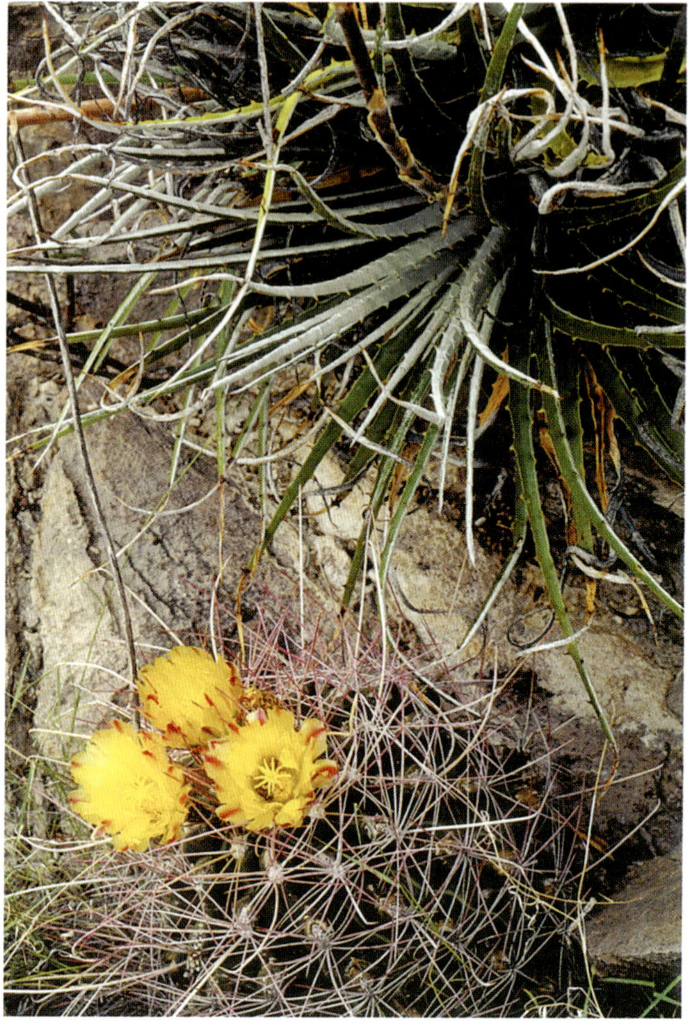

Giant fishhook cactus has many aliases, but some of the most used are Turk's head, biznaga de limilla, and whiskered barrel. These barrels hang off the edges of bluffs, especially on slopes of the Dead Horse Mountains or limestone hills overlooking the Rio Grande, but they can be viewed from the Rio Grande to the Chisos.

Since their pinkish or straw-colored, hooked, and wiry spines twist in all directions, gathering and reflecting light, it is easy to spot these cacti from a great distance. The main central spine, hooked at the end, can reach a length of six inches. But this is actually one of the smaller barrel cacti of the desert Southwest, mostly less than twelve inches high and about six inches wide.

Giant fishhook cactus is usually single but may cluster into a dramatic color display of small barrels six feet across. These cacti are most beautiful in early summer, when glossy and fragrant lemon-colored flowers are centered at the barrel's top amidst glistening spines. When mature, the green, oblong fruits turn maroon and become soft and juicy. The broad-ribbed, globular green stems may turn purple in winter, adding even more color.

This variety of giant fishhook is most prominent west of the Pecos River to the Davis Mountains and also in south-central New Mexico and northern Mexico south to San Luis Potosí. A close relative (var. *sinuatus*) is more prevalent east of the Pecos River through Tamaulipan thorn scrub to the southern tip of Texas (Powell and Weedin 2004).

Giant fishhook has been documented at Castolon, in the Rosillos Mountains in the northwestern corner of the park, at Oak Spring below the Window, and even on the South Rim.

Note: The genus name *Ferocactus,* from the Latin *ferus* (fierce), refers to the stout spines. The species name *hamatacanthus,* from the Greek *hamata* (hooked) and *akantha* (thorn), describes the hooked central spine. This species was described in 1846 by F. Muehlenpfordt, a Hanover cactus cultivator, in the German gardening magazine *Allgemeine Gartenzeitung.*

Catclaw Cactus

Glandulicactus uncinatus var. *wrightii*

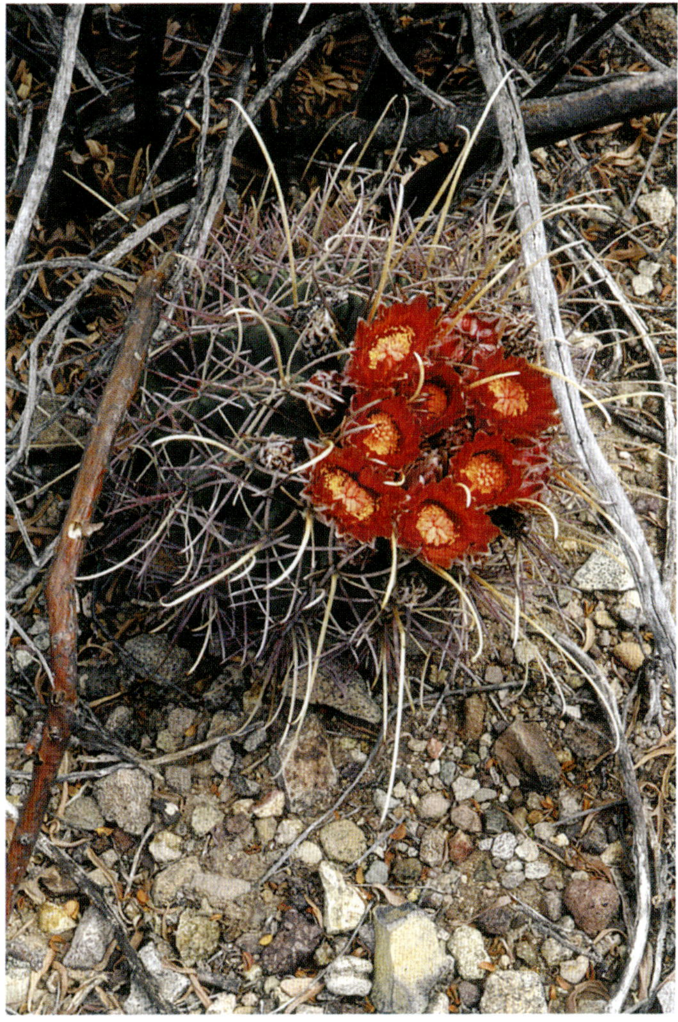

Also known as eagle-claw cactus, this fishhook cactus usually forms single, barrel-like stems about six inches tall and three inches wide. These stems have a bluish-green to gray cast with widely separated and obvious protruding ribs.

Because long, bristly spines may extend four inches and hook at the ends, catclaw cactus often resembles a thick clump of grass. The slender spines are multicolored—red, tan, straw, or gray—and this mixture of bristling colors adds to the overall impression of the plant. Often this cactus may be hidden by grasses or concealed beneath desert shrubs.

Although catclaw cactus usually seems in disarray, from March to May it redeems itself, producing funnel-shaped flowers in delightful shades of garnet, maroon, deep wine red, or russet. Forming a circle at the stem tip, these blooms are accented with pale yellow or orange stigmas and creamy yellow anthers. They seldom fully open. Clustered at the top of the fishhook in a cage of spines, shiny red oval fruits eventually supplant the flowers.

This fishhook ranges from Zacatecas through northern Mexico into most Trans-Pecos counties and southern New Mexico. It has been recorded, rarely, in southern Texas (Starr County).

In the park catclaw cactus is far-reaching but infrequent, from limestone outcrops above the Rio Grande to the Chisos foothills and from desert scrub habitats in the Dead Horse Mountains on the east of the park to Mesa de Anguila at the far southwest tip near Lajitas. Superb examples often line the Mule Ears Spring Trail, sometimes near nipple cactus (*Mammillaria meiacantha*). Both bloom in midspring.

Note: The genus name *Glandulicactus,* from the Latin *glandula* (small gland), refers to the distinctive areolar glands of these cacti, and the species name, *uncinatus,* means "hooked." This fishhook variety was described by George Engelmann in 1856 and named for Charles Wright, who collected it in 1852 on the Graham survey of the U.S.-Mexico boundary.

Heyder's Pincushion Cactus

Mammillaria heyderi var. *heyderi*

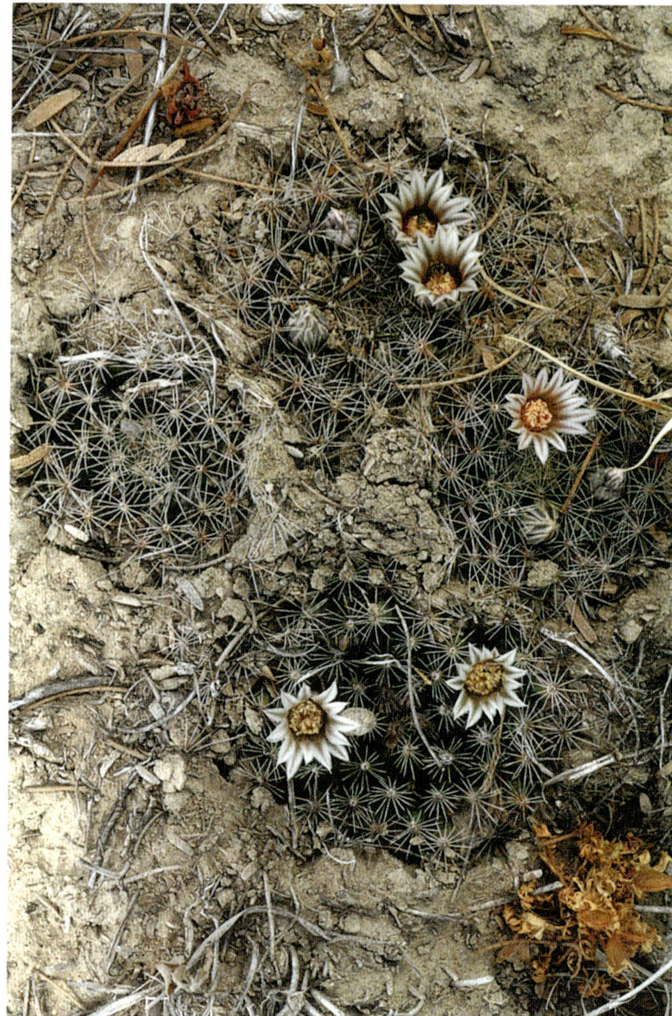

Two very similar species of pincushion cacti are recognized in the park: Heyder's pincushion cactus, also known as pancake pincushion and flattened mammillaria, at lower elevations, often in limestone or silty soils, and nipple cactus (*Mammillaria meiacantha*) in woodlands at elevations above four thousand feet.

Less round and flatter than nipple cactus, Heyder's pincushion forms a small, spiny mat flush with the ground. Comparatively small, almost needlelike, radial spines cluster at the tips of readily visible, cone-shaped tubercles. About three to six inches wide, this cactus recedes in drought, becoming covered with sand or gravel. It may look half dead or just on the edge of survival.

When Heyder's pincushion puts forth flowers, the white or cream-colored blossoms with pale pink to tan or greenish-brown midstripes seldom form a neat, tidy ring around the tip of the stem. The stems are often so covered with dust that the spring flowers appear to be growing directly out of the soil. This pincushion sports crimson fruit, compared to the purplish-red color of nipple cactus fruit. Birds feast on the sweet and sour fruit. Settlers of this desert put them in pies.

Heyder's pincushion is distributed throughout northeastern Mexico (Coahuila to Tamaulipas) and along the Rio Grande in the United States from Brownsville into parts of southeastern New Mexico. It also extends from the South Texas Plains and Edwards Plateau north to southwestern Oklahoma.

In the park the silty soils of Dog Flats near Dog Canyon and Tornillo Flat are the preferred habitats of these pincushions. The tubercled stems contain sticky white latex that Native Americans used ceremonially and also medicinally to treat earache.

Note: The genus name *Mammillaria,* from the Latin *mamilla* (breast), refers to the nipple-shaped tubercles on the stem. This species was described in 1848 by German horticulturist F. Muehlenpfordt in the gardening magazine *Allgemeine Gartenzeitung* and was named in honor of Edward Heyder, a nineteenth-century German cactus grower.

Golf Ball Cactus

Mammillaria lasiacantha

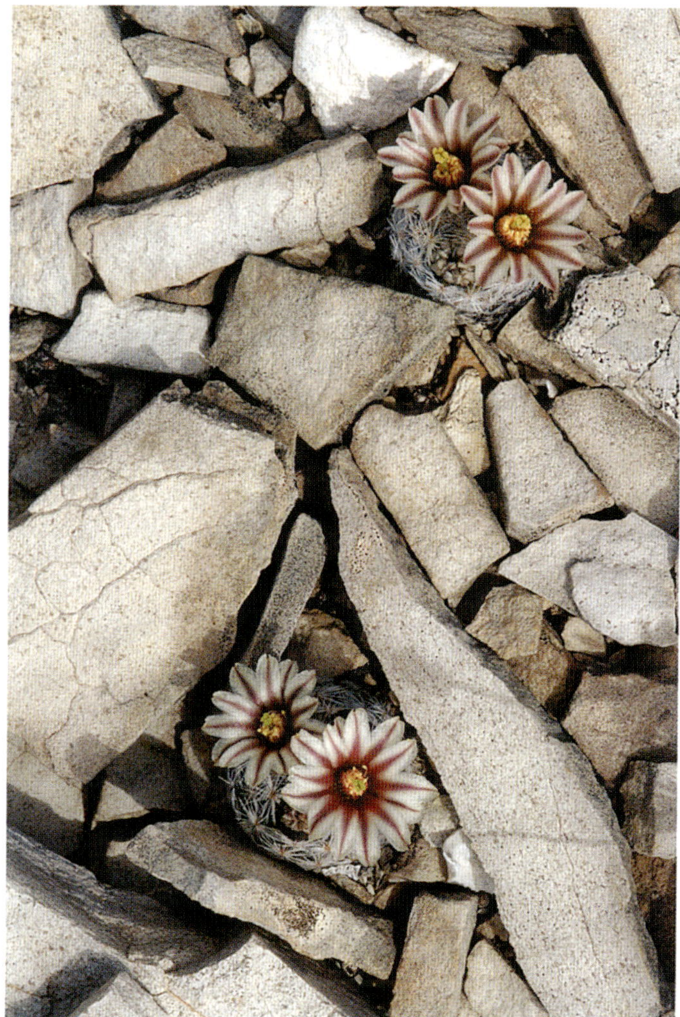

Golf ball cactus, also known as lacespine pincushion and fuzzy mammillaria, shares much the same habitat as the button cactus, mostly the limestone hills of the Dead Horse Mountains. Tiny, bright white, bristly spines cover the oval stem, which averages no more than one inch high and one inch wide. The spines are so minute and interlaced, and lie so flat against the stem, that the cactus can be easily handled.

Often hidden between rocks and protected by fine but densely interlocking spines, this diminutive plant is usually solitary on open limestone slopes and gravelly flats in lechuguilla scrub habitats.

These dwarfs begin blooming in very early spring, producing relatively large (up to three-fourths-inch-wide) white to cream flowers that open widely and may completely cover the small stems. The flower petals are highlighted with a distinctive red-brown midstripe. This pincushion may be even more engaging in summer, when it sprouts bright scarlet fruit, like miniature clubs, atop the white "golf balls."

Golf ball cactus dots the landscape from northern San Luis Potosí and Zacatecas across northern Mexico into large portions of Trans-Pecos Texas and southern New Mexico. It has also been confirmed as far west as Sonora, Mexico.

These cacti are quite difficult to discern because they blend so well into their white limestone habitat. If you search carefully in or near the Dead Horse Mountains, you may discover golf balls nearly everywhere, especially from Hot Springs to La Clocha, in low hills overlooking the river or along the Old Ore Road.

Note: The species name *lasiacantha* is derived from Greek words *lasios* (hairy or shaggy) and *akantha* (spine), referring to the harmless fuzzy spines. This species was described by George Engelmann in 1856, having been collected in 1855 by Arthur Schott at the Pecos River in Texas. Schott was a jack-of-all-trades on Major Emory's U.S.-Mexico boundary survey: a botanical, geological, and zoological collector; a surveyor and topographer; and an artist of the landscape and the people. He also surveyed the Rio Grande from Eagle Pass to the mouth of the Pecos River.

Nipple Cactus

Mammillaria meiacantha

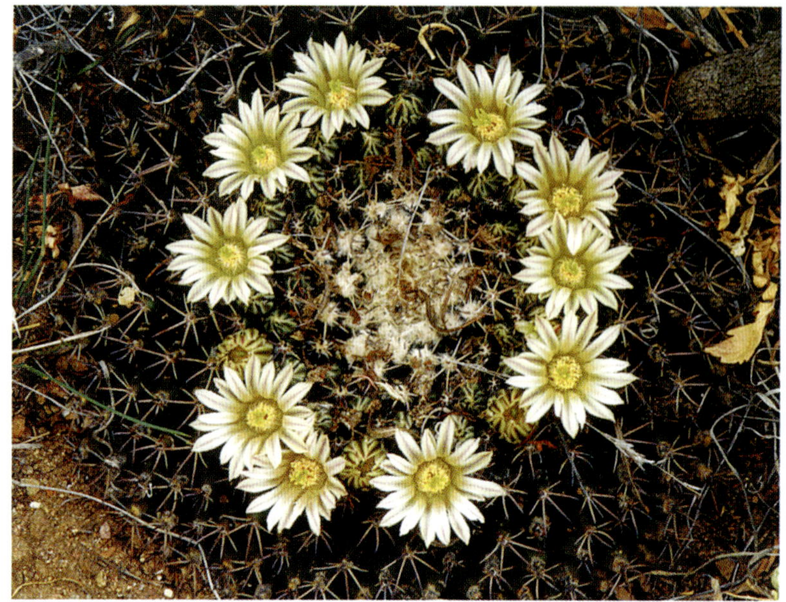

ALSO KNOWN AS little chilis or biznaga de chilitos, this round pincushion barely rises above the ground, usually no more than two inches high and six inches wide. Nipple is an apt name because nipplelike tubercles with spines at the tips cover the plant.

Similar to Heyder's pincushion cactus, nipple cactus is, however, a species of grasslands, mountain foothills, and woodlands. Difficult to detect, it is normally located in partial shade, hidden beneath trees, other plants, or amidst fallen leaves.

In spring, cream or pinkish-white flowers form a crown encircling the center of the pincushion's flat stem. On rare occasions, red to purple, acornlike fruit from a previous bloom form a brilliant red ring accompanying the lovely flower crown.

The broad range of this pincushion includes central and southern New Mexico, Trans-Pecos Texas, and part of northern Mexico (primarily Coahuila). In the Trans-Pecos it is most plentiful from the Guadalupe Mountains south through the Davis to the Chinati and Chisos mountains.

You can find superb examples of nipple cactus between Panther Junction and the Basin turnoff. Walk cross-country toward the Chisos, and you will be rewarded.

Note: The botanical name *meiacantha* is from the Greek *meion* (less) and *akantha* (thorn), a reference to the fewer number of spines compared to other pincushion cacti. This species was described by George Engelmann in 1856. A specimen that had been collected by John Bigelow in 1853, on "cedar plains east of the Pecos," was later chosen to represent the species. After completing his work on the U.S.-Mexico boundary survey, in the fall of 1853 Bigelow joined Lieutenant Whipple's Pacific Railroad Survey, which traveled from Fort Smith, Arkansas, to Los Angeles in 1853–54.

Potts' Mammillaria Cactus

Mammillaria pottsii

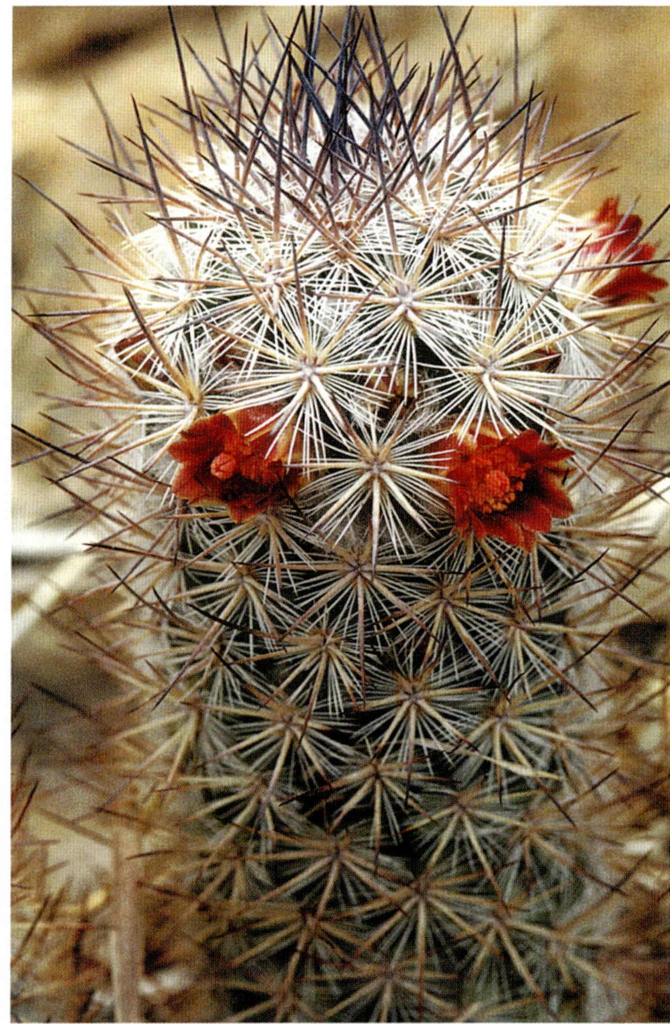

Potts' mammillaria cactus has a tall and slightly skinny cylindrical shape (up to about eight inches high and two inches wide) and white spines so dense that they hide the stem. In addition, white woolly hair or fuzz often grows between the tubercles. These features give this nipple cactus a fuzzy, grayish-white or chalky appearance, like the tail of some furry animal. For this reason, Potts' mammillaria is also aptly known as foxtail cactus.

A more pleasing quality is the tiny, ruddy red or maroon, bell-shaped flowers no more than one-half inch across. They form a circle below the stem tip and seldom fully open. Bright red and smooth club-shaped fruits appear in April, not long after the blooms.

Potts' cacti are often situated behind or within the protection of larger plants. Older stems may branch profusely and awkwardly, sometimes almost horizontally before turning upward. When not in bloom, this plant may be confused with cob cactus or silverlace cactus.

With their curious flowers, the oddball Potts' cacti have no counterparts in the United States. This species is widely dispersed across northern Mexico, from eastern Chihuahua to Nuevo León and south to Zacatecas, but enters the United States only in the Big Bend in southern Presidio and southwestern Brewster counties.

Potts' mammillaria is concentrated at a few locations in the south and west of the park, on limestone mesas and slopes and low gravelly hills. I have photographed these plants on or near Mesa de Anguila, finding them in bloom as early as February. They are most accessible near Terlingua and Lajitas but may occur on Mariscal Mountain as well.

Note: This species was described in 1844 by German prince Joseph Salm-Dyck, an amateur botanist and horticulturist who maintained an extensive collection of cacti, agave, and succulents at his Düsseldorf estate. The species is named for either John or Frederick Potts. John, manager of the Chihuahua mint in northern Mexico, sent cacti to Friedrich Scheer, cactus specialist at the Kew Botanic Gardens in London. John's brother, Frederick, a mining engineer in the Sierra Madre, apparently collected many of the plants.

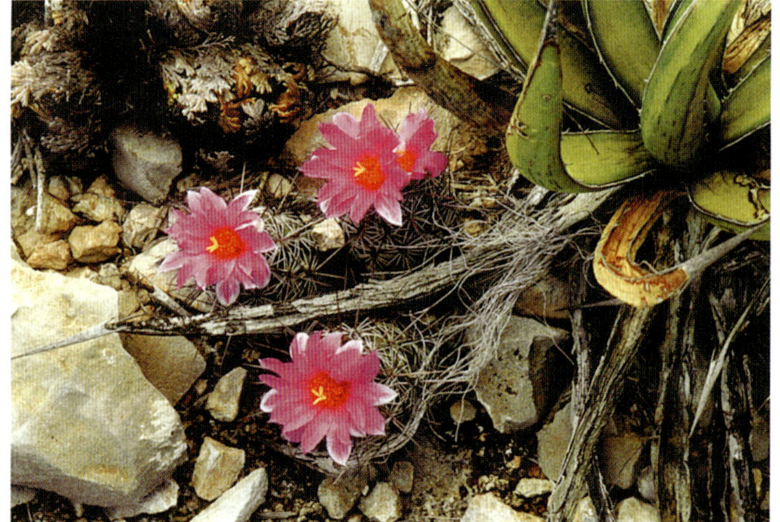

Texas Cone Cactus

Neolloydia conoidea var. *conoidea*

A SELDOM SEEN attraction of the park's northeastern corner, Texas cone cactus, usually no more than five inches high, grows in the limestone hills of the Dead Horse Mountains. Mostly cylindrical, or perhaps slightly cone-shaped, in the United States it is encountered only in Texas, hence the name.

Texas cone stems may be solitary, but they often branch into small clumps. These stems have stout blackish to grayish central spines and shorter white radial spines, giving the appearance of a white cactus with black and gray accents. White woolly hairs may protrude conspicuously at the stem tip, adding to the whitish aura.

Usually in later spring or early summer, vivid pink to magenta flowers perhaps two inches wide emerge from the tip of this top-flowering cactus. Small, dry fruits, greenish-brown to olive in color, are produced in late fall.

In the United States, Texas cone cactus is a component of desert scrub communities in the Trans-Pecos, primarily in Brewster, Pecos, and Terrell counties. It is also present in northern Mexico (Coahuila to Tamaulipas and south to Aguascalientes and Hidalgo).

Sometimes blooming profusely, Texas cone cactus is locally common on rocky hills above Dagger Flat. It is often situated in broken rock on hilltops and cliff edges in association with Warnock's cactus, strawberry cactus, and guayule.

Note: The genus *Neolloydia* was described in 1922 by Britton and Rose and named for McGill University botanist Francis Lloyd (1868–1947). The species name *conoidea* in Greek means "conelike," probably referring to the stem shape. This species was first described in 1828 by Swiss botanist Augustin de Candolle, a pioneer in plant classification who spent his last twenty-five years trying to describe all seed plants.

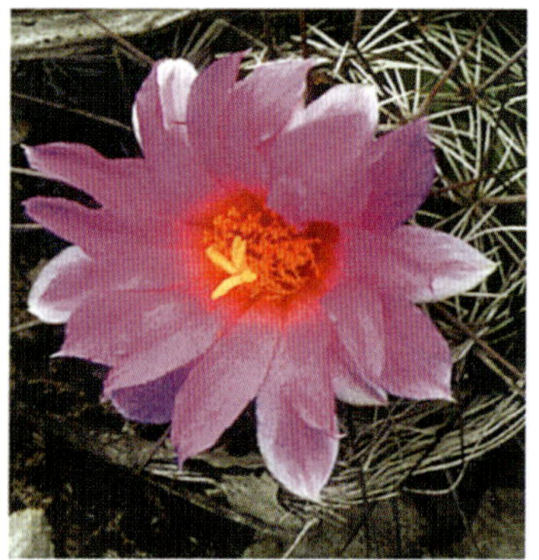

Glory of Texas Cactus

Thelocactus bicolor var. *bicolor*

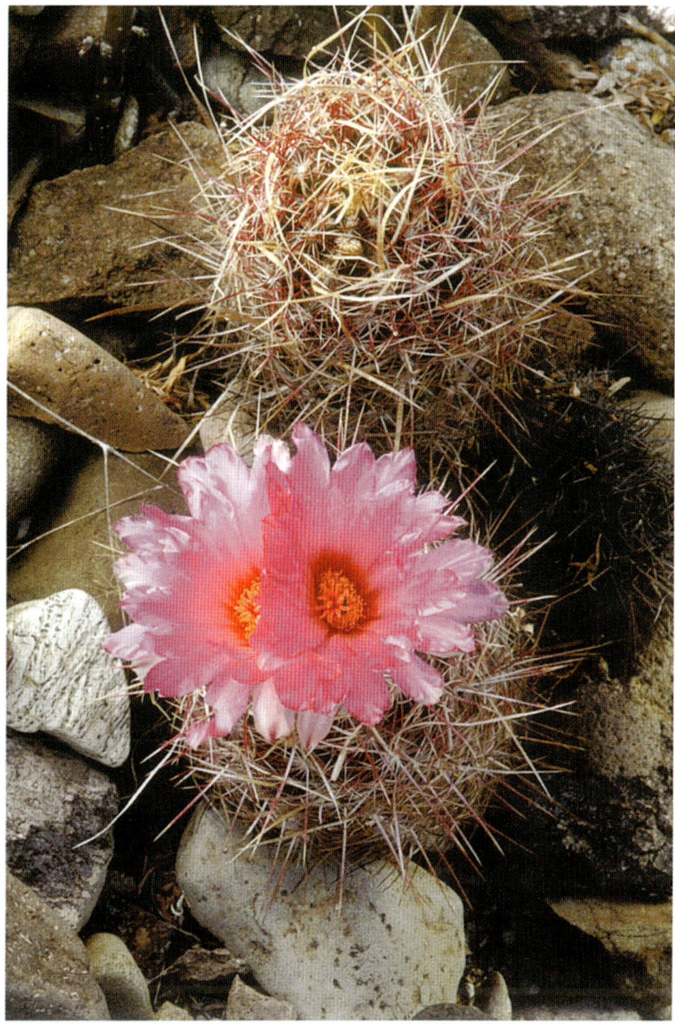

Also known as Texas pride, the glory of Texas is arguably the most beautiful of all Texas cacti. Growing from the apex of the stem, exotic magenta or fuchsia flowers with a darker scarlet throat may open three inches wide. The satiny petals frequently curl backward at the tip. This exquisite flower may also be one of the most finicky: it tends to bloom for only a few hours on a single day and even then only under the brightest light of the desert sun.

The stems, easily reaching six to eight inches high and half as wide, are egg-shaped to cylindrical with rounded ribs and often spiraling tubercles. Many of the spines seem to stick out haphazardly and are bicolored—pink or red in the center and straw-colored or tan on the end. Some spines are so thin and papery that they twist, bend, or curl, adding to the aura of the plant.

The glory of Texas might be better described as the glory of Mexico, where it is more common (eastern Chihuahua to Tamaulipas and south to Zacatecas and San Luis Potosí). This cactus spills over into Texas only in the Big Bend, in southern Brewster and Presidio counties and in the Lower Rio Grande Valley (Starr County).

Large specimens of this cactus flourish at isolated sites just north of the Rio Grande, far from anywhere, in silty washes and on gravelly hillsides. You might find examples on the hills above Paso Lajitas on the Mexican side of the border and at Lajitas on the Texas side. I have even found a few strays beside Rough Run Creek near Maverick Mountain.

Note: The genus name *Thelocactus,* from the Greek word *thele* (nipple), refers to the nipplelike tubercles. This variety was described by George Engelmann in 1856 (as *Echinocactus bicolor* var. *schottii*). The name *schottii* honored Arthur Schott, German immigrant and first assistant surveyor for the U.S.-Mexico boundary survey in 1851–53. Schott, who found this variety in 1853 "near Mier" in Mexico, was an accomplished artist whose drawings were made into tinted lithographs to illustrate Major Emory's survey report.

Tree Cholla

Opuntia imbricata var. *arborescens*

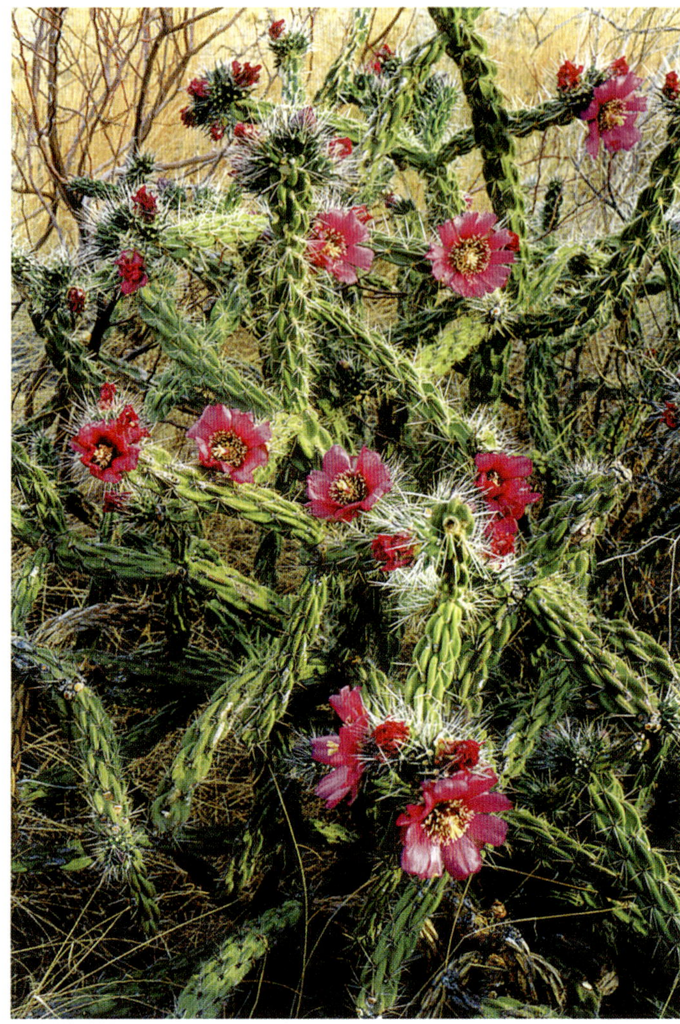

Also known as cane cholla, walking-stick cholla, coyonostyle, velas de coyote (coyote candles), and devil's rope, tree cholla is a treelike, jointed cactus that branches into strange shapes. Usually forming a short solid trunk at the base, the "tree" can exceed six feet in height. Bearing tan, yellow, or white spines, the joints have distinctive overlapping, braided tubercles.

In late spring and early summer, large magenta flowers up to three inches wide emerge from the stems. Knobby, tuberculate yellow fruits ripen in August but often remain through the winter.

Common in the park, this cactus thrives at between four thousand and six thousand feet, especially on the Chisos foothills and Burro Mesa and in Green Gulch. A hardy cactus, tree cholla even survives the winter temperatures as far north as Colorado. New plants can form from fallen joints as well as from seeds.

Conspicuous throughout the Trans-Pecos, this cholla also reaches into the Edwards Plateau and Texas Panhandle, north to southern Colorado, west to southeastern Arizona, and south in Mexico to San Luis Potosí and Zacatecas.

In cold weather, tree cholla joints may turn purplish. When the winter sun is low on the horizon, the colors of the yellow fruits and purple cholla joints are richest. Then a mountain slope covered with these contorted "trees" makes a bizarre forest indeed. Tree cholla is strange in death as well. The stems form white, braided skeletons that sculptors must surely envy. Walking sticks and lamp stands are made from the dead stems.

Note: The species name *imbricata* (overlapping) refers to the cholla joints. The variety name *arborescens* means "treelike." This variety (also named var. *imbricata*) has been known at least since 1848, when it was described by George Engelmann (as a species, *Opuntia arborescens*). The type specimen had been collected by Augustus Fendler in 1847, near Santa Fe, New Mexico. Fendler, a German immigrant who trained under George Engelmann, was sent by Asa Gray to collect in New Mexico.

Big Bend Cholla

Opuntia imbricata var. *argentea*

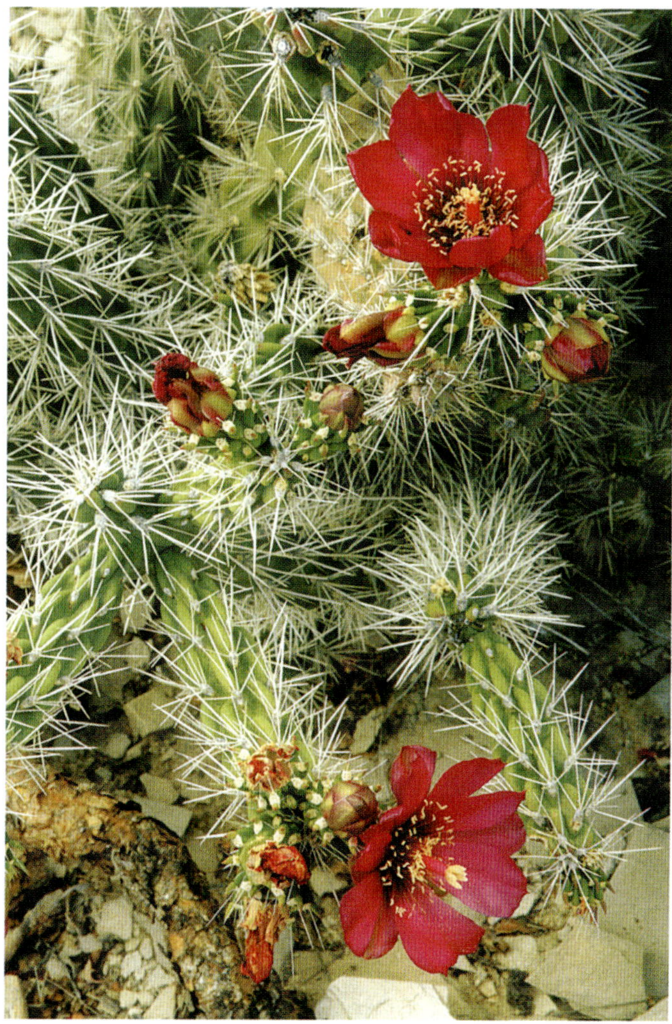

Also known as Mariscal cholla or silverspine cholla, Big Bend cholla is a rare local variety of the common tree cholla. It is native to Mariscal Mountain and the surrounding low hills, creosote flats, mesquite thickets, and desert washes, although it crosses into adjacent Coahuila and Chihuahua, Mexico. Because of its restricted geographic range and small populations, this variety is classified as critically imperiled globally and in Texas.

Compared to tree cholla, this erect shrub is shorter (less than four feet high) and more densely branched with shorter but stouter stem joints and smaller, more compact tubercles. The joints of Big Bend cholla are more noticeably glaucous (covered with a whitish, waxy coating) with dense spines encased in silvery, highly reflective sheaths.

This thicket-forming cholla blooms in April, producing two-inch-wide, deep magenta flowers, which are quite waxy and darker than tree cholla blooms. Yellow tubercled fruits, one inch or more long and toplike in shape, follow in the fall.

In the park good examples of this stubby cholla sit atop white reeflike hills above the Rio Grande. Clusters of Big Bend cholla are also ensconced on and near Mariscal Mountain, their glistening spines lighting the way in the early morning or late afternoon sun. They appear with silverlace cactus, lechuguilla, dog cholla, and creosote in deep limestone soils.

Note: Reputedly the genus name *Opuntia* was first used by the ancient Greek botanist Theophrastus to describe an entirely different plant near the old Greek town of Opus. However, some suggest that the word is from the Papago Indian name *opun.* Whatever the case, the genus was described in 1754 by an English gardener, Philip Miller, in the fourth edition of his *Gardeners Dictionary.* The varietal name *argentea* is "silver" in Latin. A relatively recent discovery, this cholla variety was first described in 1956 by Margery Anthony, an ecologist at Chico State College in California and specialist in the *Opuntia* of the Big Bend region.

Candle Cholla

Opuntia kleiniae

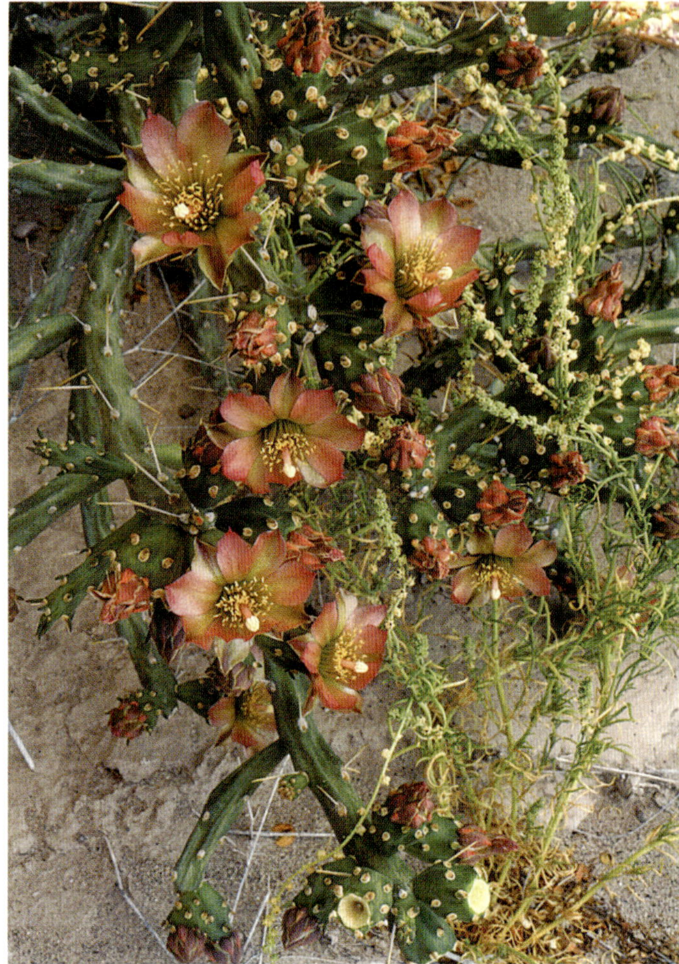

ALSO KNOWN AS Klein cholla and tasajillo, this scraggly shrub that reaches six feet high or more, may form impenetrable thickets, trapping the unwary.

The cholla's green, purple-tinged stems sometimes branch profusely and erratically, like sprawling candelabras. Its joints are no more than one-half inch wide, smaller than tree cholla but broader than desert Christmas cholla (*Opuntia leptocaulis*). The needlelike spines, one to four in each cluster, are covered with thin, papery sheaths.

Candle cholla flowers may open up to one and one-half inches wide. The pinkish-bronze flowers with flat rounded petals are extremely waxy in appearance and translucent in the sun. Blooming in late spring and summer, they seem to take on a wide range of shades and hues: individual blossoms may be tinged with purple, green, maroon, or cream. Pink filaments and creamy white stigma lobes are added touches of color.

This cholla produces fleshy, red-orange mature fruit, somewhat tubercled but spineless. It apparently prospers in sandy soils near the Rio Grande, often perched on low limestone hills above the river or on desert flats and alluvial washes.

Candle cholla is wide ranging but spotty in distribution from northern Mexico through Trans-Pecos Texas (especially along the Rio Grande and in the Davis Mountains) into southern and central New Mexico. It is very scattered elsewhere (southern panhandle, north-central Texas, southwestern Oklahoma). In the park candle cholla dominates a large stretch of the floodplain near Black Dike, where Smoky Creek drains into the Rio Grande.

Note: This species was described in 1828 by Augustin de Candolle. It was one of sixty-three cacti collected in Mexico from 1824 to 1834 by Irish naturalist Thomas Coulter. Coulter, who managed silver mines in Mexico, maintained a small cactus garden and sent specimens to de Candolle in Geneva. He also collected in Arizona and California, and Coulter pine is named in his honor. The species name *kleiniae* may refer to the resemblance of candle cholla to plants of the genus *Kleinia,* succulent species in the aster (Asteraceae) family endemic to Africa (Anthony 1956).

Club Cholla

Clumped Dog Cholla
Opuntia aggeria

Schott Dog Cholla
O. schottii var. *schottii*

Graham Dog Cholla
O. schottii var. *grahamii*

Big Bend Devil Cholla
O. densispina

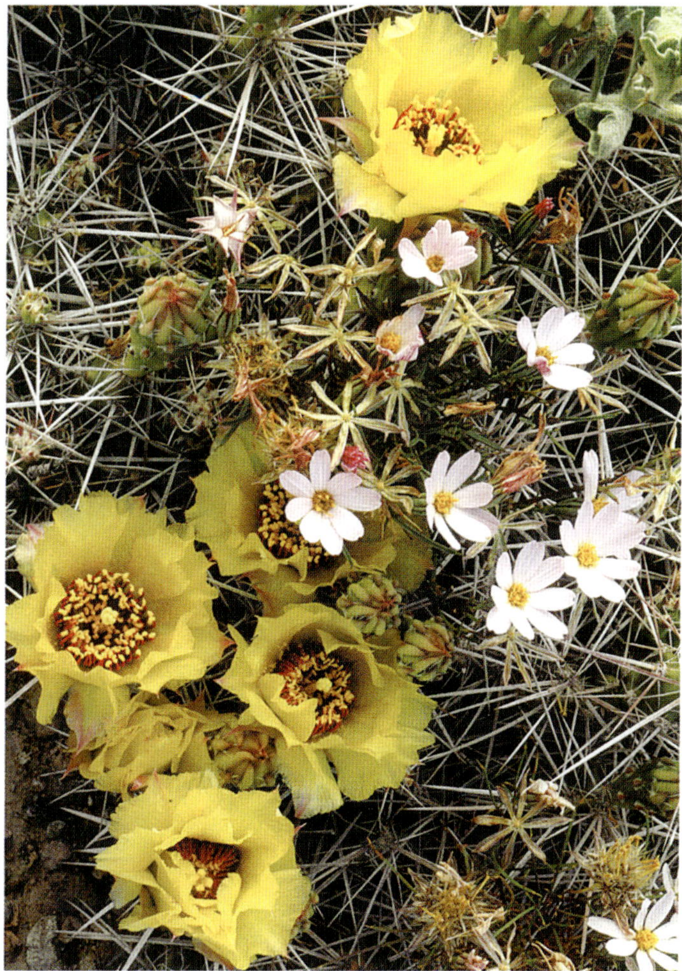

CLUB CHOLLA forms low mats several feet wide of club-shaped, warty joints often prostrate on the edges and upright near the center. Opening around noon, the lemon-yellow flowers are two inches across and showy.

Three similar species, one with two varieties, inhabit the park. The most common, clumped dog cholla, has firm joints and commonly chalk-white spines, the longest often twisted or curved. Blooming in late March and April, it spreads across southern Brewster and southeastern Presidio counties and adjacent Coahuila.

Schott dog cholla, or dog turd cholla (so named because it clings to shoes), has easily separated joints and frequently flat spines, some sharp-edged. It may form chains or loose mats. Blooming in early summer, it is distributed over the Rio Grande Plains, adjacent Coahuila, and the eastern Trans-Pecos. Graham dog cholla has smaller, slightly firmer stems and more, often cylindrical, spines. Blooming in late spring, it ranges from northern Mexico to Brewster County and west to El Paso and southern New Mexico.

Big Bend devil cholla resembles clumped dog cholla but forms more compact, rounded mounds. In firm clumps up to ten feet across, it has notably more spines and denser spine clusters. Blooming in late spring and early summer, it is restricted to clay soil near the Rio Grande, from Mariscal Mountain and Solis to the Dead Horse Mountains. Rarely you may encounter cholla with off-color blooms; I have found salmon-pink flowers on plants off the Old Ore Road.

Note: The species name *aggeria* is from the Latin *aggerare* (to mound). This plant was collected by Margery Anthony in 1948 on Tornillo Flat. The name *schottii* honors Arthur Schott, who collected these plants by the Pecos River in 1853. Variety *grahamii*, collected by Charles Wright near El Paso in 1851, was named for Col. James Graham, chief of the U.S.-Mexico boundary survey scientific corps. The species name *densispina* connotes "densely spined." This plant was described by Sul Ross State University botanists Barbara Ralston and Richard Hilsenbeck in 1992 and collected by them near Solis in 1989.

Golden-Spined Prickly Pear

Opuntia azurea var. *aureispina*

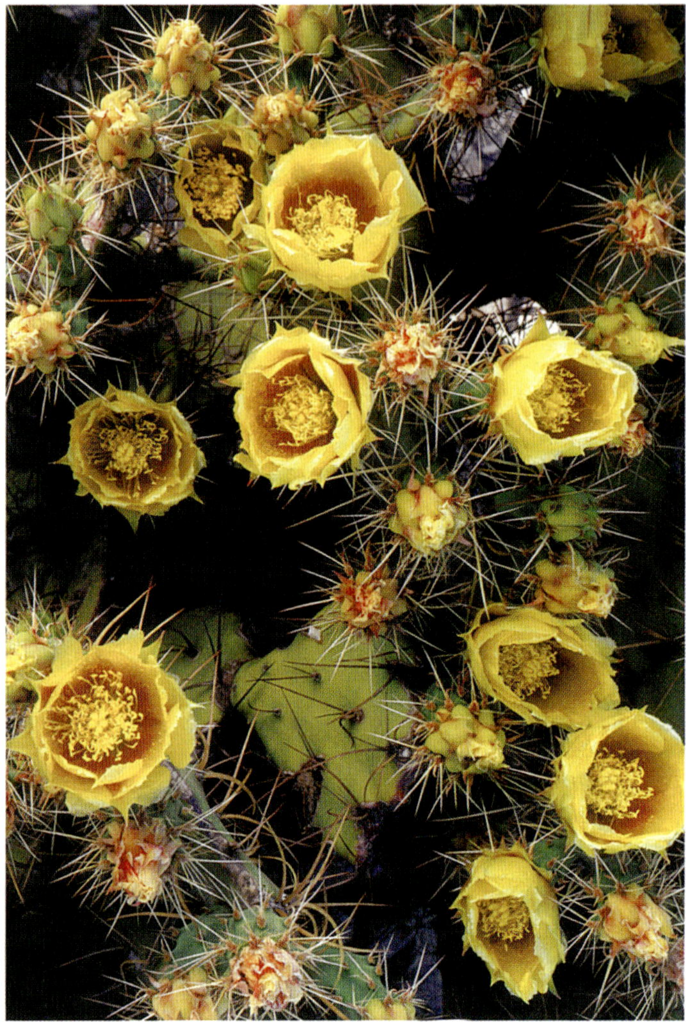

A BULKY SHRUB with an obvious, short, and notably spiny trunk, the golden-spined prickly pear is seldom much more than three feet high. It is noted for its golden-yellow spines, often curving or twisting, that may extend two inches, covering the yellow-green or sometimes bluish-green pads. Highly variable, these spines may be red at the base, or a darker color overall, mostly brownish to almost black.

Considered rare in Big Bend National Park and known from only a few scattered locations, golden-spined prickly pear occupies rocky limestone slopes and canyons and desert flats usually below two thousand feet in elevation. Localized along or near the river at the park's southeastern limit, this cactus probably also occurs in adjacent Coahuila. It is classified as critically imperiled globally and in Texas.

In spring this impressive shrub produces large yellow flowers, perhaps two and one-half inches wide, often with orange to red centers. Frequently, the blossoms do not open widely. The burlike, felty fruits bear a few stiff yellow spines and sometimes yellowish bristles and glochids. At first green and somewhat fleshy, they become tan, dry, and shrunken at maturity.

Golden-spined prickly pear was first discovered in 1985 by A. M. Powell of Sul Ross State University, coauthor of *Cacti of the Trans-Pecos and Adjacent Areas* (Powell and Weedin 2004). The cactus hybridizes with other prickly pear species in its limited habitat.

Look for isolated examples of this plant, also known as Rio Grande prickly pear, near the Rio Grande, especially on Boquillas limestone flats.

Note: The species name *azurea* ("blue" in Latin) refers to the bluish cast of the prickly pear pads. The variety name *aureispina,* from Latin, means "golden-spined." This prickly pear was described (as a variety of *Opuntia macrocentra*) by Kenneth Heil and Steven Brack in 1988 and reclassified by Powell and Weedin (as a variety of *Opuntia azurea*) in 2004.

Big Bend Purplish Prickly Pear

Opuntia azurea var. *parva*

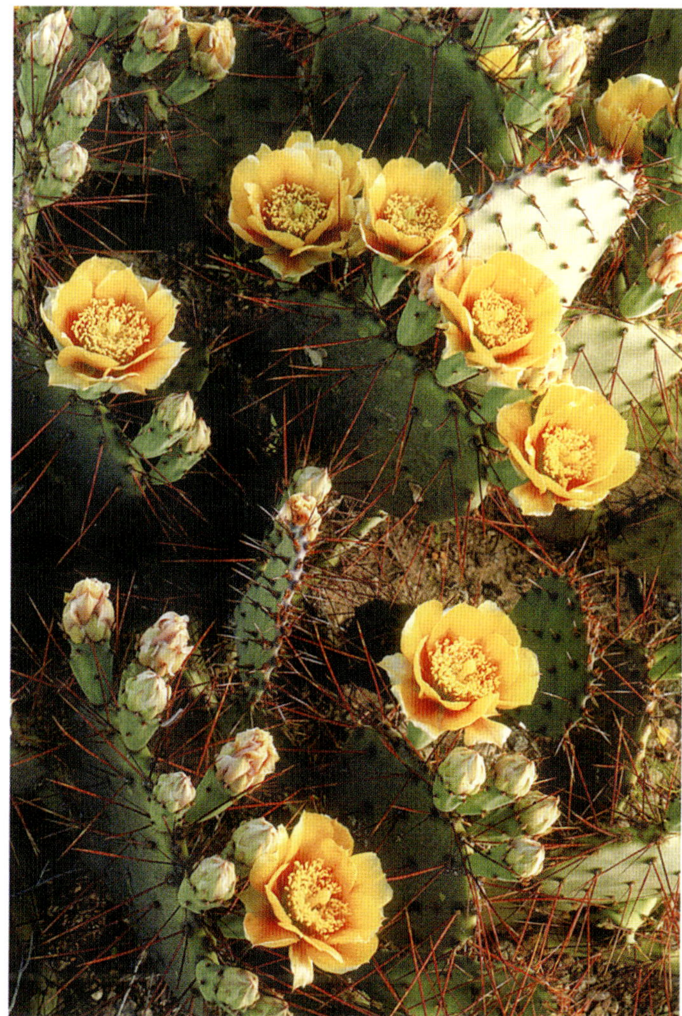

The Big Bend purplish prickly pear is rightly famous for the color of its pads, which varies widely from blue-green with a light purple tinge on the edges to a deep reddish-purple covering the entire pad. The purple hue is more pronounced under extreme weather conditions, such as very cold winters or drought.

Either widely spreading or strictly erect with a short trunk, this cactus is usually no more than three feet high, with many crowded branches. From about three to more than five inches wide, the pads are rounded at the top and narrowly wedge-shaped at the base. Long spines in varying shades, frequently black, reddish-brown, or maroon, with whitish tips, grow from each pad. Four or even five inches long, these spines tend to protect the plant from deer and javelina.

A spring bloomer, Big Bend purplish prickly pear produces lush yellow flowers with rich red centers. On occasion, you may see as many as sixty to seventy blossoms, each as much as three inches wide, on a single plant. Seldom blooming much more than a single day, an individual flower quickly ages from yellow-red to orange-red as the day progresses. Red to purple fruits follow the blooms. These fruits are spineless and somewhat fleshy, but they dry with age.

As described by A. M. Powell and J. F. Weedin (2004), this is one of two *O. azurea* varieties with purple coloring in Big Bend National Park. It is limited primarily to the Chihuahuan Desert scrub habitat around the Chisos Mountains in Brewster County and, presumably, adjacent Coahuila, Mexico.

In low light, early in the morning or late in the day, the purple color of this prickly pear may dominate the desert landscape, blending with the colors of purple sage and the hues of a rising or setting sun.

Note: The varietal name *parva* is Latin for "small," a reference to the relatively small size of the prickly pear pads compared to other species. A. M. Powell and Shirley Powell collected the type specimen in 1993 near Study Butte in Brewster County.

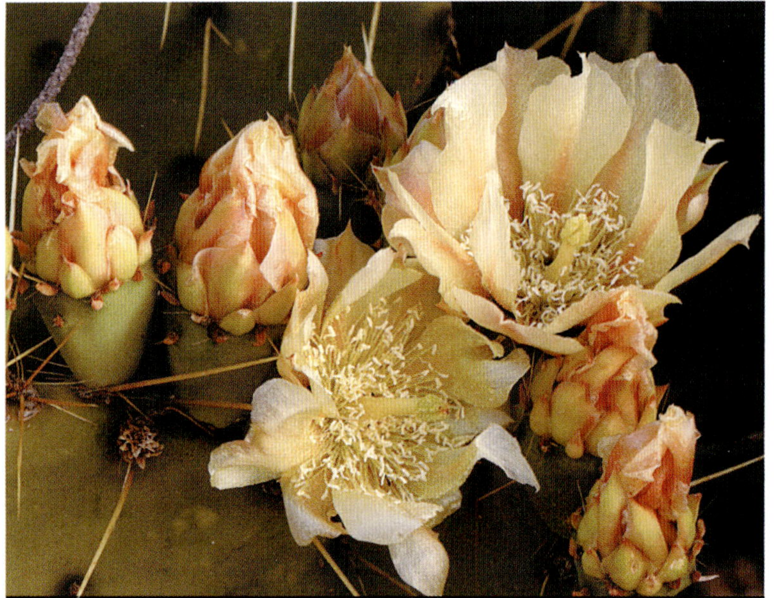

Chisos Prickly Pear

Opuntia chisosensis

Chisos prickly pear is an upright, sometimes spreading shrub, seldom over three feet high, with rounded or broadly egg-shaped, blue-green or gray-green pads. It grows from a large base but lacks a distinct trunk. The mostly yellow spines, sometimes more than two inches long, may age to darker shades. Spines along the upper edges of the pads may be conspicuously curved.

This cactus blooms slowly over the course of several weeks. It produces captivating pastel flowers, occasionally two and one-half inches across, in shades of pale yellow to buff or peach. The red or purplish fruits that follow are about one inch long, shaped like a globe or a barrel. They are noticeably fleshy and spineless.

In the United States, Chisos prickly pear is apparently confined to the Chisos Mountains, although it is present in the Sierra Maderas del Carmen across the Rio Grande in northern Coahuila, Mexico. It frequents pinyon-oak-juniper forests and mountain grasslands at elevations from about five thousand to over seven thousand feet.

This prickly pear is readily observed throughout the Chisos, from Upper Green Gulch and the Basin to much higher elevations. The delicate flowers bloom primarily in May at Panther Pass, Pummel Peak, Boot Spring, and many other sites. Seek them out along the Window, Lost Mine, Pinnacles, and Emory Peak trails.

Note: Margery Anthony, a student of Trans-Pecos cacti in the 1950s but now retired from Chico State College in California, described this species in 1956 as a variety (*Opuntia lindheimeri* var. *chisosensis*). The variety was elevated to species status by cactologist David J. Ferguson in 1986. Chisos prickly pear is now considered to be more closely related to golden-spined prickly pear and Big Bend purplish prickly pear (*Opuntia azurea* varieties) than to the species with which it was originally associated.

Blind Prickly Pear

Opuntia rufida

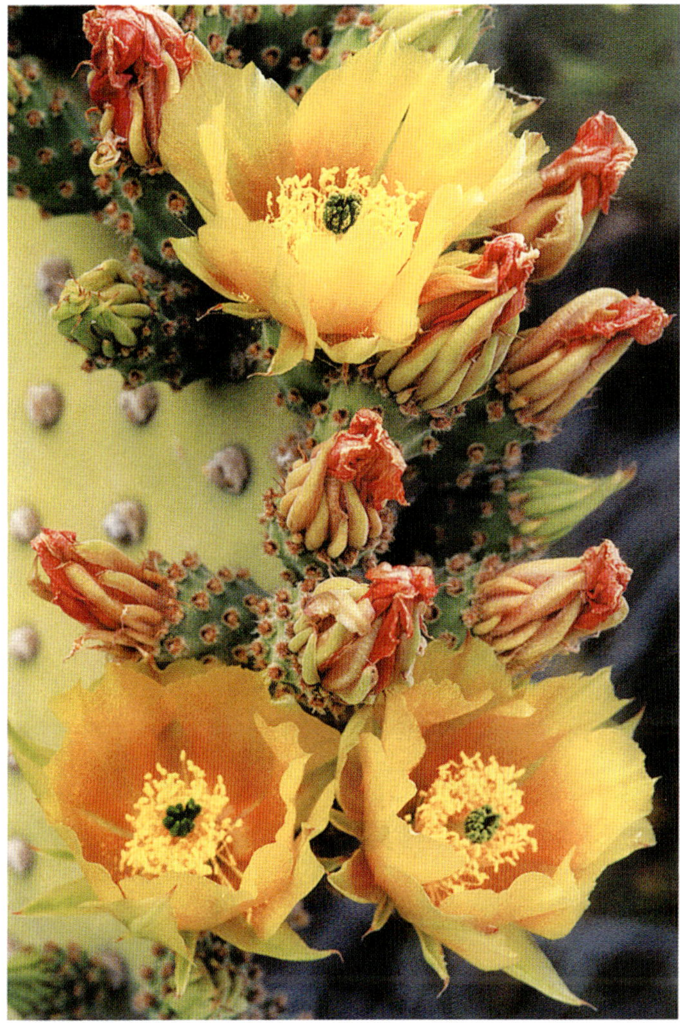

Blind prickly pear, a stout, erect, much-branched shrub up to five feet high, has no prickly spines. Instead, it is covered with clusters of bristly reddish-brown glochids (minute barbed hairs). The glochids emerge from areoles (cushionlike areas where spines usually grow) on the round, woolly pads, replacing spines. But the blind pear is far from harmless: when brushed against or dislodged, these bristles can blind cattle, hence the name.

Blind prickly pear is a Mexican species that enters the United States only in the Trans-Pecos. Its range stretches from southeastern Hudspeth County to Presidio County and across Big Bend National Park into the Lower Canyons of the Rio Grande. It is also extensive in northern Mexico (in Coahuila and Chihuahua south to Durango). A frequent sight on desert flats, rocky slopes, and gravelly and sandy soils below four thousand feet, this cactus commands vast expanses near Boquillas and along the road from Panther Junction to Boquillas Canyon.

In April and May this cactus produces yellow flowers aging to orange up to three inches across. The exquisite flowers are a stark contrast to the cold, gray-green pads. Ripening in summer, the fleshy red fruits are spineless but often dotted with glochids.

In changing desert light—the dwindling light of dusk or in a stormy landscape—this prickly pear is an imposing, even eerie presence.

Note: The species name *rufida,* from the Latin *rufus* (red), refers to the ruddy red glochids. This species was collected by John Bigelow in 1852 in Chihuahua at Presidio del Norte and described by George Engelmann in 1856. After joining the Pacific Railroad Survey and collecting in northern California, Bigelow in 1856 wrote *Memoir of the Life and Public Services of John C. Frémont,* about the explorer, governor, and U.S. senator from California. The memoir was used in Frémont's unsuccessful campaign for the U.S. presidency.

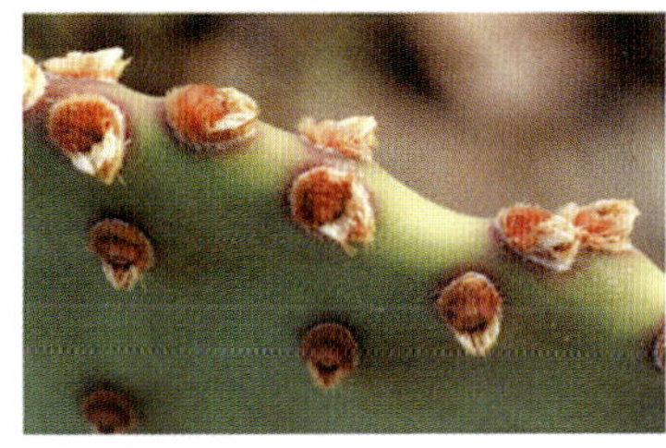

Spiny-Fruited Prickly Pear

Opuntia spinosibacca

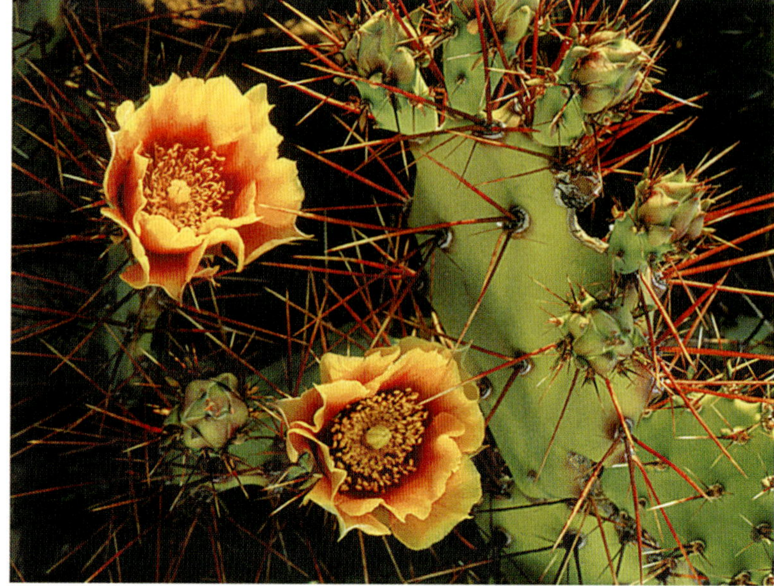

Spiny-fruited prickly pear is a compact shrub as much as four feet tall or more, growing from a short trunk or extensively branching at the base. The pale green or yellow-green pads, usually about eight inches long and six inches wide, are covered with spreading and highly reflective, reddish-orange spines. This species is distinctive for the elevated or mounded areoles (cushionlike areas) from which the thick spines protrude and for the purple blotches that often appear near these dark areoles.

Spiny-fruited prickly pear is most attractive in spring, when it sports golden yellow flowers sometimes nearly three inches across with deep red centers. But this prickly pear is known for its spiny fruits, an uncommon trait for fleshy-fruited prickly pears. The oval fruits are one and one-half inches long with a distinct rim around the top where the spines are concentrated.

This cactus is extremely circumscribed in distribution, known only from southern Brewster County on limestone hills near the Rio Grande. It can be found above Hot Springs and the tunnel and on fractured limestone rocks near Boquillas Canyon. Although undocumented in Mexico, this prickly pear almost certainly occurs across the river near Boquillas.

Spiny-fruited prickly pear is presumed to be of hybrid origin, a cross between golden-spined prickly pear (*Opuntia azurea* var. *aureispina*) and another cactus. A. M. Powell and J. F. Weedin (2004) have speculated that the other parent is a red-spine form of sweet prickly pear (*Opuntia dulcis*). Take the time to scrutinize this plant when it is backlit by the evening sun: it reflects the sun's glow like a burning bush.

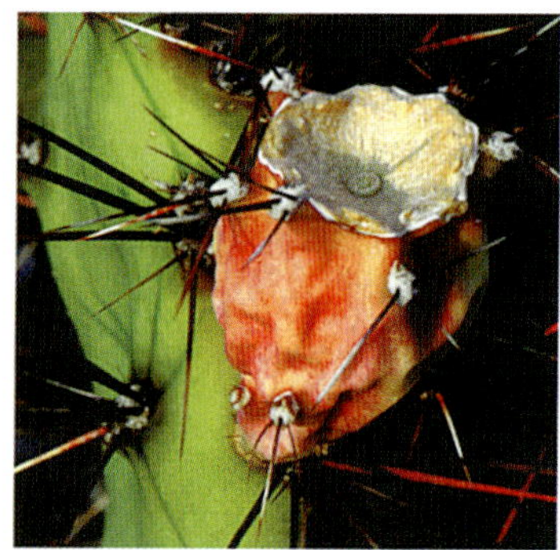

Note: The species name *spinosibacca* is from Latin words meaning "spiny-fruited." This cactus was first described in 1956 by Margery Anthony. The type specimen had been identified in the park in 1948 near Rio Grande Village.

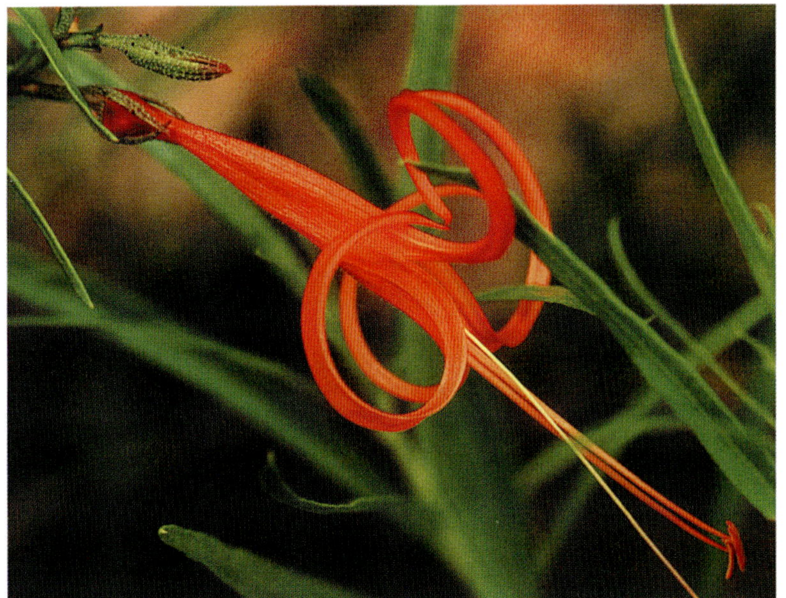

Dwarf Anisacanth

Anisacanthus linearis

Until it blooms, dwarf anisacanth is an easily missed, nondescript shrub. About three to six feet high, it has slender, arching branches with grayish white, shredding bark and sparse foliage. The leaves are linear or very narrowly lance-shaped, usually less than one-eighth inch wide. Its upper leaves may be entirely stemless.

In summer and fall, orange-red trumpetlike flowers, single or paired, bloom with a flourish and hang like ornaments from the weak branches. On rare occasions, you may find a bush with yellow flowers. Each funnel-shaped flower is a superb hummingbird attraction; they are two inches long with curling straplike lobes split almost to the base.

Also known as muicle, dwarf anisacanth grows in gravelly arroyos, almost always in dense brush. In the United States, this shrub is primarily centered in Brewster County (around Marathon and in Big Bend National Park, mostly in the Chisos foothills), although it has been reported in Presidio and Hudspeth counties. It is also a resident of northern Mexico from southeastern Chihuahua to southwestern Nuevo León.

In the park dwarf anisacanth is easy to locate in desert washes winding through the Grapevine Hills, or in Blue Creek Canyon. It is also plentiful on the Oak Spring and Window trails.

Note: The genus name, derived from the Greek *anisa* (unequal) and *acanthus* (plants of the Acanthaceae family), emphasizes differences between the first species (*A. quadrifidus*) and other family members. The species name *linearis* refers to the linear leaves. This shrub was collected by John Moore and Julian Steyermark in 1931 in Blue Creek Canyon and described by Stanley Hagen in 1941 as a variety of *Anisacanthus insignis.* The *insignis* species was collected in 1847 by Josiah Gregg in Chihuahua along the Río Conchos. A related species (*Anisacanthus puberulus*) with pink flowers blooms near Oak Spring.

Arizona Carlowrightia

Carlowrightia arizonica

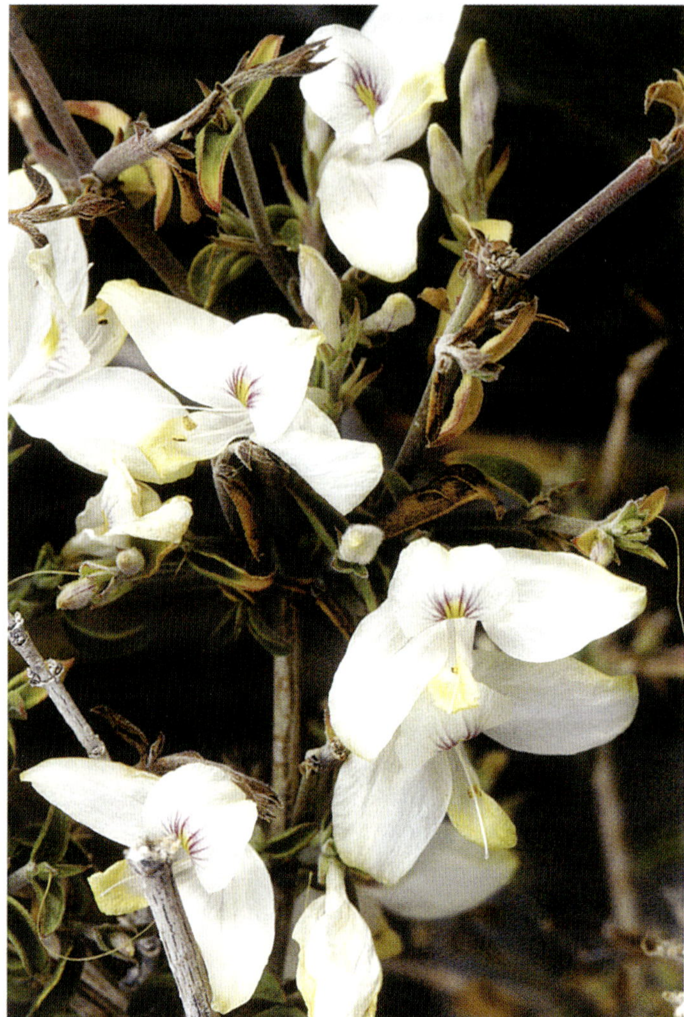

ARIZONA carlowrightia is seldom over two feet high, with many slender, ashy branches. About one-fourth inch wide and less than one inch long, the paired leaves are lance- or egg-shaped and smooth on the edges.

When not in bloom, this unassuming plant is a jumble of drab branches. Even when in bloom, the shabby branches may seem to have snared pieces of white paper blown by the wind. On closer inspection, you realize the white papers are bird-shaped, orchidlike flowers as delicate as any desert landscape might conceal.

Appearing from spring to early fall, each blossom consists of a short, slender tube opening into four unequal lobes. The bloom's upper lip is etched with thin maroon streaks across a yellow eye, and the lower lobe is narrow and folded. The overall appearance of the white flowers suggests birds in flight.

This often shrubby perennial is occasional on dry, rocky soil in the deserts of far west Texas, southern Arizona, and California. It is concentrated in southern Brewster and Presidio counties in Texas but is more common south of the border, from northern Mexico to Oaxaca and Sinaloa and south to Costa Rica.

In the park Arizona carlowrightia is sparse but perhaps easiest to spot along the Rio Grande near Boquillas Canyon and in Santa Elena Canyon. It has also been recorded several thousand feet higher in the Chisos foothills, especially near Wilson Ranch and Blue Creek Canyon and in the Sierra Quemada near Dominguez Spring.

Note: In 1878 Harvard botanist Asa Gray named the genus *Carlowrightia* for botanical collector Charles Wright. Considered the father of American botany, Gray edited the *American Journal of Science,* wrote a series of botany textbooks and extensive studies of North American flora, and recruited and trained many botanical collectors. He described this species from a specimen collected by Edward Palmer in 1867. Palmer found the plant near Camp Grant in Pinal County, Arizona, hence the species name *arizonica.*

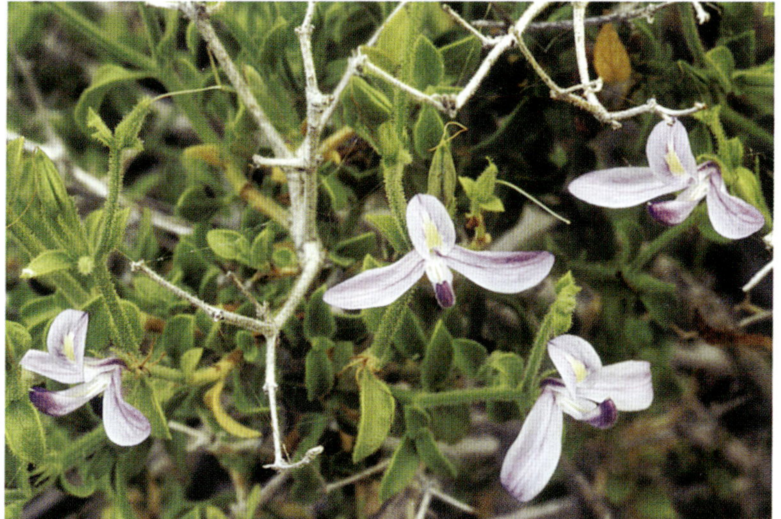

Trans-Pecos Carlowrightia

Carlowrightia serpyllifolia

Trans-Pecos carlowrightia is a low herbaceous shrub or shrubby herb up to about one foot high with nearly white, thin branchlets. The light green leaves, one-half inch long, are oblong to egg-shaped, pointed at the tip, and tapering at the base.

Blooming in summer and fall, the flowers are pale purple or cream with purple streaks and splashes. Each flower consists of a very slender, short tube deeply divided into four petal-like lobes. Look closely: the upper lobe has a distinctive flame-shaped yellow eye, and the lower lobe is folded lengthwise to enclose the stamens. The lateral lobes spread out like wings and may curl backward.

Trans-Pecos carlowrightia is often situated in rocky, brushy canyons and drainages or on open slopes and sometimes near springs and periodic creeks. This subshrub is prevalent in northern Mexico and the southern Trans-Pecos in Brewster and Presidio counties. In Mexico, populations reach throughout parts of Chihuahua and Coahuila south to northeastern Durango and northern Zacatecas.

This member of the acanthus (Acanthaceae) family appears at many locations in the Chisos, especially in the foothills at middle elevations. It has been identified at Ward Spring, in Oak Creek Canyon, at the base of Bailey Peak, in the Basin, and on moist west slopes of the Chisos. Trans-Pecos carlowrightia can also be seen in the Grapevine Hills and in the Pinnacles and Rosillos mountains.

Note: This species was described by Asa Gray in 1886 and had been collected by Vermont Quaker Cyrus Pringle near Jimulco, Coahuila, in 1885. Pringle, a horticulturist and rare plant enthusiast, spent twenty-six years collecting in Mexico for the Gray herbarium. A conscientious objector during the Civil War, Pringle wrote *The Record of a Quaker Conscience,* published posthumously in 1918. The species name *serpyllifolia* refers to the "thymelike leaves."

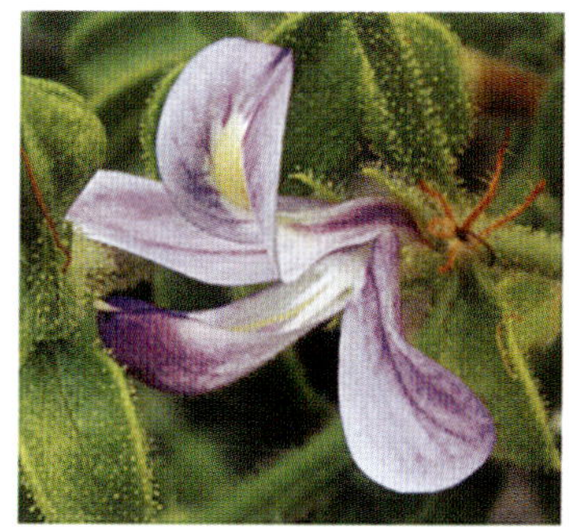

Spreading Snakeherb

Dyschoriste schiedeana var. *decumbens*

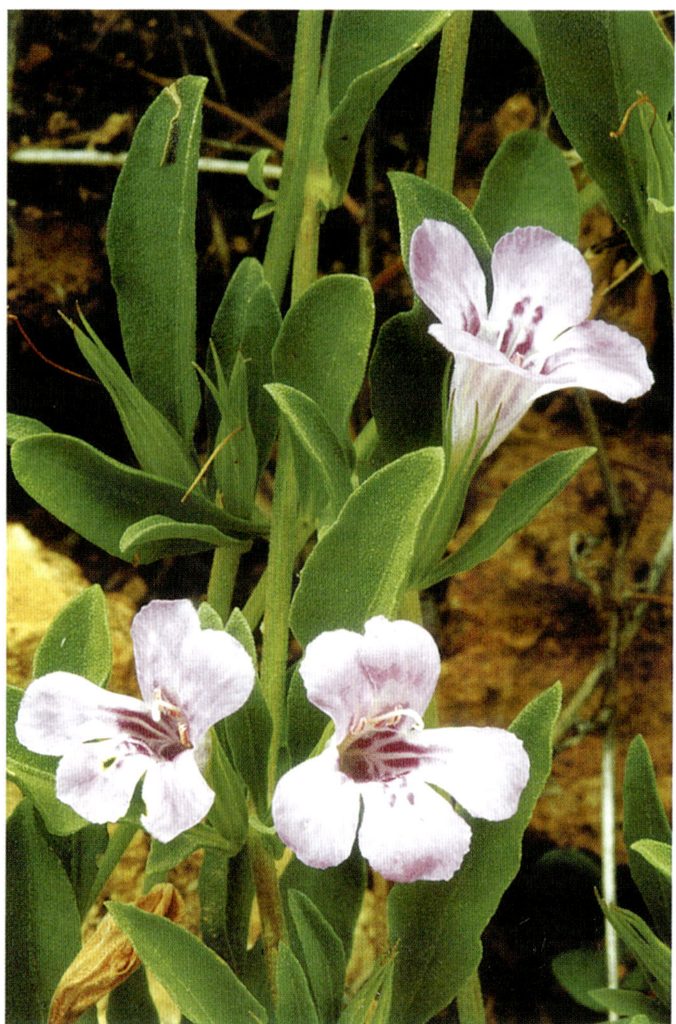

Spreading snakeherb is a low perennial, usually less than one foot high, that may spread across or trail over the ground. Although the plant is mostly prostrate, with hairy, four-angled stems, the leaves are often distinctly erect and bunched at the stem nodes. The thick, grayish-green blades, about one inch long and one-fourth inch wide, are spatula-shaped or broadly linear, occasionally slightly wavy or curly along the edges, and covered with a coat of curved or crinkled hairs.

Perhaps three-fourths inch long, the striking lavender flowers are stalkless and borne at the base of leaves, mostly in groups of one to three. On a short, funnel-shaped tube, each blossom is two-lipped, with an erect, partly divided upper lip and a broad, deeply three-lobed lower lip. The throat of the flower tube may be streaked with dark purple or red.

Spreading snakeherb fares well on both limestone and igneous soils in oak-juniper-pinyon woodlands and grasslands, mostly at elevations above four thousand feet. In Texas this variety is encountered only in the Trans-Pecos, from the Davis Mountains south to the Chisos Mountains. It is more far-reaching elsewhere: southeastern Arizona, New Mexico, and especially in northern Mexico from eastern Sonora to Chihuahua and Durango.

In the park spreading snakeherb blooms from May to September. This attractive herb is most abundant in the Basin of the Chisos and on the Laguna Meadow and Pinnacles trails, but it has also been found on Crown Mountain and in Upper Blue Creek Canyon.

Note: The genus name *Dyschoriste,* from Greek words *dys* (with difficulty) and *choristos* (separated), refers to the tightly closed valves of the fruit pods. Described by Asa Gray in 1878 (as *Calophanes decumbens*), this plant was renamed *Dyschoriste decumbens* by German botanist Otto Kuntze in 1891. In 1999 James Henrickson and Richard Hilsenbeck treated it as a variety of *D. schiedeana,* another species that Kuntze had named for German naturalist Christian Schiede. Schiede had collected in Mexico during the 1820s. The variety name *decumbens* means "lying down."

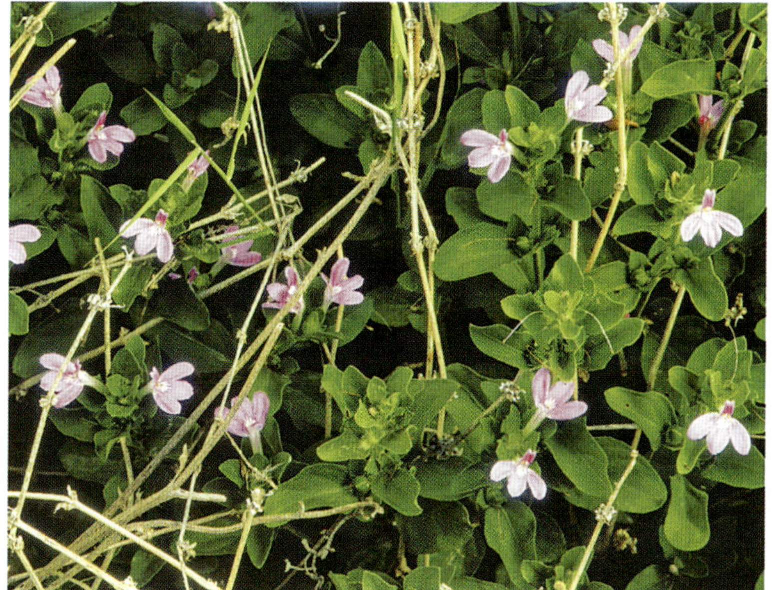

Hairy Tubetongue

Justicia pilosella

Hairy tubetongue is a low perennial herb four to twelve inches high, with many weak, hairy stems branching from the base. Mostly upright but sometimes sprawling, it may form colonies in hospitable habitats. The leaves are oblong to egg-shaped, from one-half to one and one-half inches long, often with blunt or rounded tips. These paired leaves form tight bundles along the stems.

Blooming from spring to fall, pinkish-violet or pale magenta flowers appear at the base of upper leaf clusters. They often bloom profusely, but individual blossoms quickly fall. Each flower, up to one inch long, has two lips at the end of a very slender, long white tube. The narrow upper lip is small and notched at the tip; the lower lip is much larger and divided into three distinct lobes with a white spot at the base.

Dry rocky and gravelly soils, desert scrub, and brush are characteristic habitats of this low herb. It also colonizes alluvial flats, stream banks and springs, and roadside depressions. Hairy tubetongue is dispersed across the southern and eastern Trans-Pecos into central and north-central Texas and south Texas. It crosses into southern New Mexico and Mexico (mostly in parts of Chihuahua to Tamaulipas but also south in Hidalgo and Puebla).

In the park this perennial pops up at widely scattered locations: along the road to Persimmon Gap, on the Hot Springs Trail beside the Rio Grande, north of Pulliam Bluff, and in the Paint Gap Hills.

Note: The species name *pilosella* means "with small hairs." This species was described in 1847 (as *Monechma pilosella*) by Christian Nees, botanist in Breslau, Germany, and prolific botanical writer who described seven thousand species. It was collected in Mexico by both Josiah Gregg and Adolph Wislizenus that same year. This species was reclassified as *Justicia pilosella* in 1990 by Sul Ross State University botanist Richard Hilsenbeck.

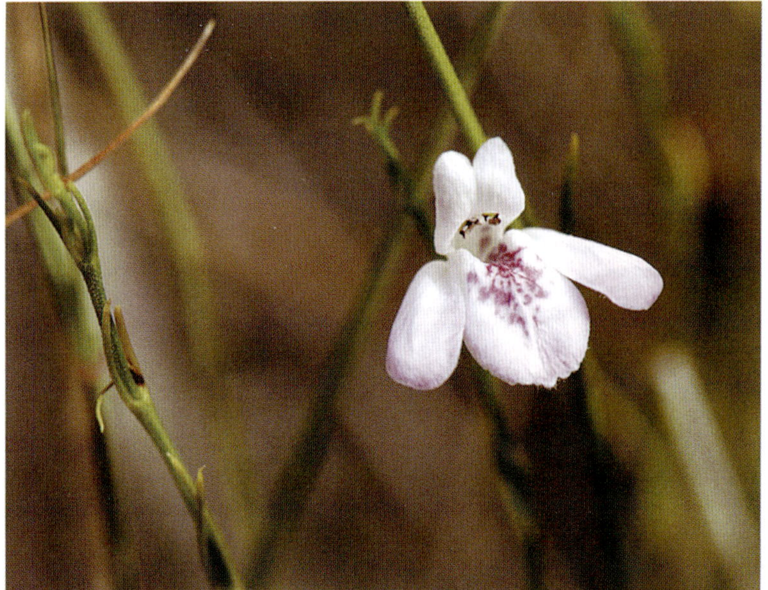

Warnock Justicia

Justicia warnockii

WARNOCK JUSTICIA is a low shrubby perennial about one foot high, almost grasslike in appearance, with a sturdy, grayish stem and many thin green branches. About one-half inch long, the stemless leaves are smooth and needlelike.

The ephemeral, solitary flowers open in early morning but begin to wilt and fall as the sun rises. White with purple markings in the throat, these two-lipped flowers have a distinctly notched upper lip and a broad, three-lobed lower lip. Sparse flowers appear from late spring to fall.

Warnock justicia sprawls over dry rocky outcrops, limestone slopes, and canyons in desert mountains. In the United States, this perennial is an unusual occupant of the southern Trans-Pecos, especially near the Rio Grande, from southern Hudspeth County east to Crockett County. It is also native to eastern Chihuahua and western Coahuila, Mexico.

In the park Warnock justicia is easily overlooked in or near the Dead Horse Mountains. It has been noted at Dead Man's Cut near the tunnel, along the Old Ore Road, and at McKinney Spring. It has also been seen on the west slopes of the Chisos, south of Ward Spring. The photo was taken in Telephone Canyon in August.

Globally and in Texas Warnock justicia is classified as vulnerable throughout its range.

Note: *Justicia* is an old genus, named in 1753 for the Scottish horticulturist James Justice (1698–1763). Justice, the first person in England to grow a pineapple to fruit, lost his fortune in the tulip trade, or "Tulip Madness," that swept Europe. B. L. Turner, University of Texas botanist and former student of Barton Warnock at Sul Ross State University in Alpine, described Warnock justicia in 1951, naming it for his mentor, who had collected the plant in 1941 in the Glass Mountains of Brewster County.

Parry Ruellia

Ruellia parryi

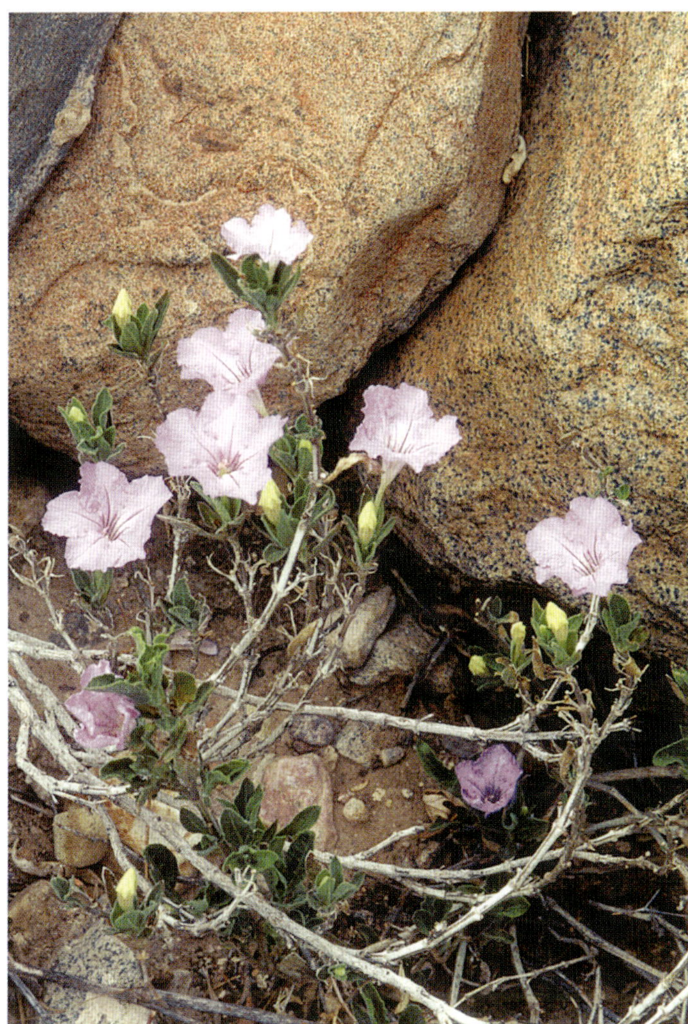

ALSO KNOWN AS wild petunia, Parry ruellia is a widely spreading low subshrub that grows to about sixteen inches high, with many pale gray, brittle branches that become conspicuously white with age. These branches may spread almost horizontally near the ground, and roots may be exposed. Gradually tapering to a short stem, the narrow leaves are mostly oblong or elliptical, about three-fourths inch long and noticeably hairy.

Blooming in spring and summer, the funnel-shaped, solitary flowers are usually very pale lavender. As much as one inch across, the flowers have a slender tube that opens abruptly into five broad, flaring lobes.

Primarily at elevations below five thousand feet, this subshrub occupies rocky slopes and canyons and arroyos leading to higher mountains in both limestone and igneous soils. Parry ruellia is pervasive throughout Trans-Pecos Texas, from Hudspeth County to Del Rio, and also reaches into southeastern New Mexico. This member of the acanthus (Acanthaceae) family is also a typical species in parts of northern Mexico from Chihuahua to Nuevo León and south to Durango.

In the park Parry ruellia is most prominent in the rocky foothills of the Chisos. Likely sites include Glenn Spring by Chilicotal Mountain in the east; Government Spring and Rough Spring to the north; Blue Creek and Oak Creek canyons and Burro Spring to the west; and along Smoky Creek to the south.

Note: Described in 1753 by Swedish botanist Carolus Linnaeus, the genus *Ruellia* is named for Jean Ruelle (1474–1537), a Parisian physician and dean of the School of Medicine in Paris. Ruelle translated Dioscorides's *Materia Medica* from Greek into Latin (1516) and authored a comprehensive botanical treatise in 1536, *De Natura Stirpium Libri Tres.* This species was described (as *Dipteracanthus suffruticosus*) by John Torrey in 1859 and later named for Charles Parry, botanist on the Emory boundary survey who had collected it near Presidio in 1852. A student of Torrey at Columbia, Parry later explored the Rocky Mountains of Colorado and became known as the "King of Colorado Botany."

Shaggy Stenandrium

Stenandrium barbatum

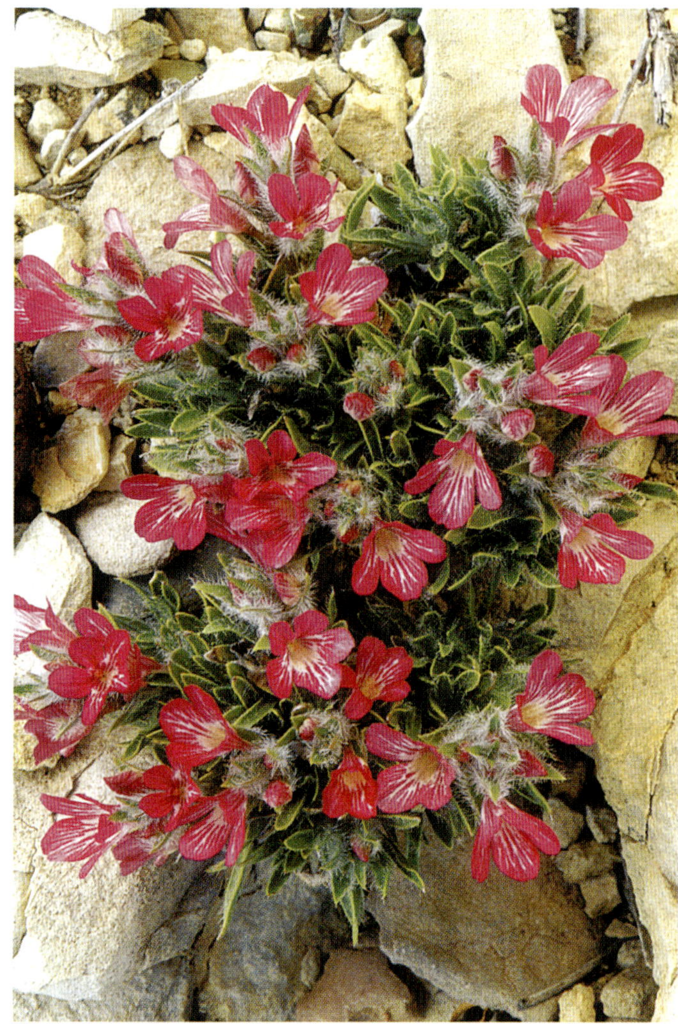

SHAGGY STENANDRIUM is a perennial herb only a few inches high with several stems forming a gray, tufted mound. The short-stemmed or stemless leaves, about one inch or more in length, are densely crowded and overlapping. They are usually oblanceolate (lance-shaped but broadest above the middle and tapering at the base) or narrowly spatula-shaped. Both stems and leaves of this dwarf herb are conspicuously shaggy with white hairs.

Reddish-pink flowers, about one-half inch wide, appear on short spikes in small, leafy clusters. This member of the acanthus (Acanthaceae) family has funnel-shaped flowers consisting of a slender tube that flares into five petal-like lobes. Each two-lipped blossom has five rounded lobes with veinlike white streaks on the lower lip and a white, bearded throat.

Shaggy stenandrium is frequently associated with lechuguilla on dry limestone soils and especially on rocky, gravelly, and crumbly desert slopes and barren flats. This low herb inhabits probably every county of the Trans-Pecos as well as southern New Mexico and northern Mexico (northeastern Chihuahua and northwestern Coahuila).

In the park the best place to view this spring bloomer is Dog Flats, where it is often embedded in low gravelly hills off the Dog Canyon Trail. You might also check near Persimmon Gap, the north entrance to the park, or at the north end of Mariscal Canyon near the Rio Grande. Blooming mostly from March to May, this perennial adds delicate pastel hues to the desert's palette.

Note: The genus name *Stenandrium* is derived from the Greek *stenos* (narrow) and *andros* (male, stamen) for "narrow stamen." The species name *barbatum* (bearded) may refer to the conspicuously hairy stems and leaves or the flower's bearded throat. This species was described by John Torrey and Asa Gray in the 1855 Pacific Railroad Survey report, but the type specimen had been collected in 1851 by Charles Wright along the Pecos River in Texas. Torrey, New York state botanist, compiled *Flora of New York* (1843) and in 1858 founded what later became the Torrey Botanical Club.

Arizona Snakecotton

Froelichia arizonica

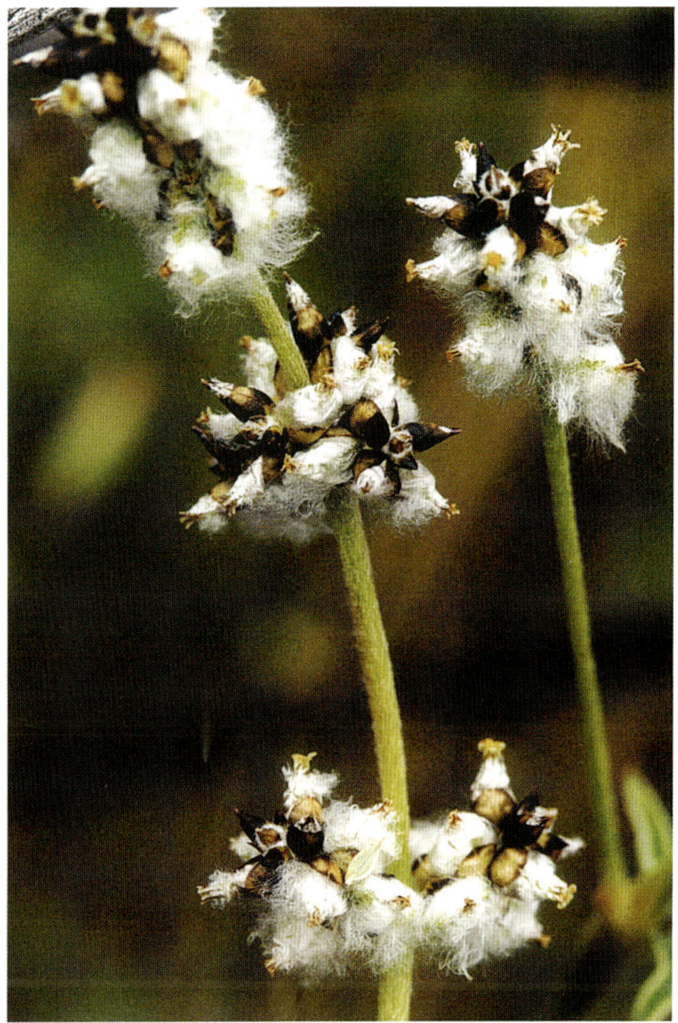

ARIZONA SNAKECOTTON is a slender perennial herb, seldom much over two feet high, with several sturdy, silky-haired stems growing from a thick, woody taproot. These stems are rarely or sparingly branched in the upper reaches. Up to two or occasionally three inches long, the paired leaves are mostly crowded near the base of the stem, with fewer, much smaller leaves along the lower stems. The blades are narrowly lance-shaped, short-stemmed or stemless, with white, woolly hairs.

Mostly in summer and fall, tiny pale yellow flowers, only one-eighth inch long, are spirally arranged in sturdy, cottonlike spikes. Lacking petals, the minuscule blooms consist of two-lipped, five-lobed, tubular sepals encased in dense, bright white hairs. Flowering throughout the year but especially in summer and fall, they resemble little jewels mounted in the woolly spikes.

Rather weedy in appearance and well known for invading disturbed habitats, Arizona snakecotton seems somewhat out of place on dry, open, rocky slopes and grassy flats and along gravelly arroyos, on both limestone and igneous soils. This herb ranges throughout southeastern Arizona, southwestern New Mexico, and Trans-Pecos Texas from El Paso to Brewster County. Its distribution extends into northern Mexico from Chihuahua to Nuevo León.

In the park Arizona snakecotton is scattered throughout the Chisos foothills and higher mountains and less frequently throughout the Dead Horse Mountains. Look for this perennial at Onion Spring, in Green Gulch, and on slopes above Lower Oak Creek Canyon or at a much higher elevation along Boot Creek above Boot Spring.

Note: In 1794, Marburg University botanist Conrad Moench named the genus *Froelichia* in honor of the German physician and botanist at Erlangen, Joseph von Froelich (1766–1841). The species *arizonica* was described in 1917 by Paul Standley, curator of the U.S. National Herbarium. John Thornber, dean of the College of Agriculture at the University of Arizona, and David Griffiths, U.S. Department of Agriculture horticulturist, had collected the species in 1902 in Arizona's Santa Rita Mountains.

Fleshy Tidestromia

Tidestromia carnosa

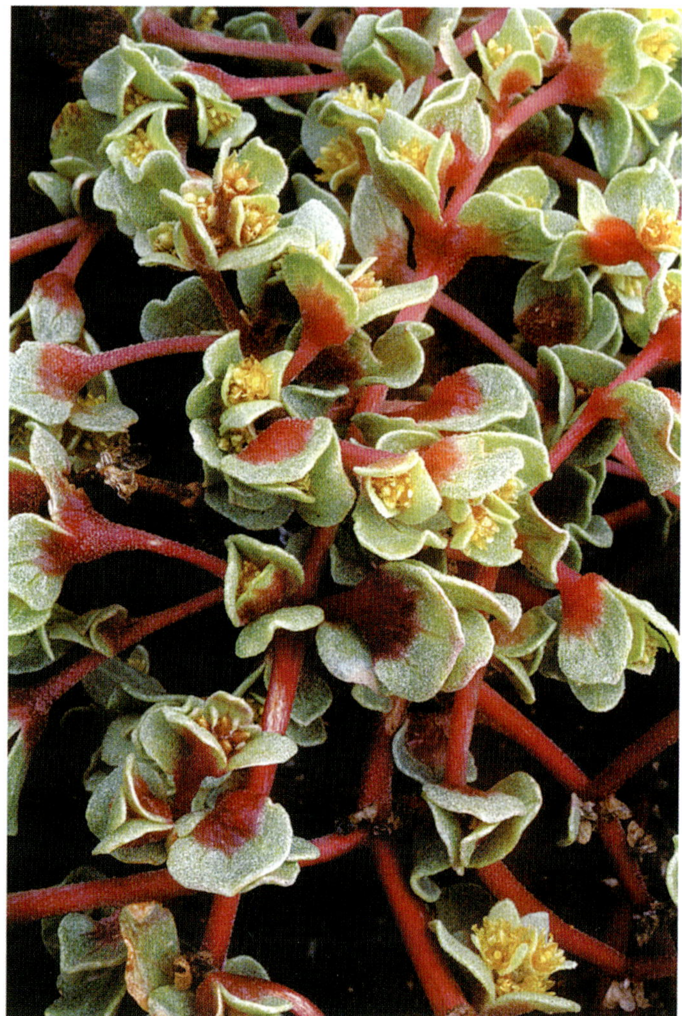

An often prostrate, coarse annual with widely branching stems, fleshy tidestromia spreads to form large mats. The stems and leaves are yellowish-green and very succulent but markedly membranous when dry. Up to one inch long and almost as wide, the spoon-shaped blades are usually nearly hairless. They may grow in clusters of three with one larger than the other two.

Mostly in late summer and fall, minute yellow flowers form in groups of two or three at the base of leaves. They are distinctly woolly and a fragrant attraction for insects. When the fall weather turns wintry, the yellow-green mats turn a deep, dark red, a perfect accent to the yellow flowers.

This herb is endemic to the Chihuahuan Desert of the Trans-Pecos and adjacent Mexico. It occurs in Brewster and Presidio counties, especially in gypseous or saline clay soils; northern Chihuahua and northern Coahuila; and, reportedly, El Paso County in similar habitat.

Fleshy tidestromia blankets clay dunes and barren slopes on the west side of the park, especially the Maverick Badlands and Terlingua Abaja off the Old Maverick Road. It can be prolific in sandy soils but also in gravel and along roadsides. This annual spreads over gravelly slopes between Castolon and Santa Elena Canyon and may bloom profusely after summer monsoon rains.

The conservation of fleshy tidestromia is a matter of concern. Globally, it is vulnerable throughout its range and in Texas imperiled.

Note: In 1916 this genus was named for Ivar Tidestrom, a Swedish-born botanist who collected for the U.S. Department of Agriculture in the western states and published two key works on western plants, *Flora of Utah and Nevada* (1925) and *Flora of Arizona and New Mexico* (1941). The species name *carnosa* connotes "fleshy." This herb was identified as a variety by Julian Steyermark in 1932 (*Cladothrix lanuginosa* var. *carnosa*) from a specimen that he and John Moore collected in Brewster County east of Study Butte in 1931. In 1943, Harvard botanist Ivan Johnston elevated it to species status.

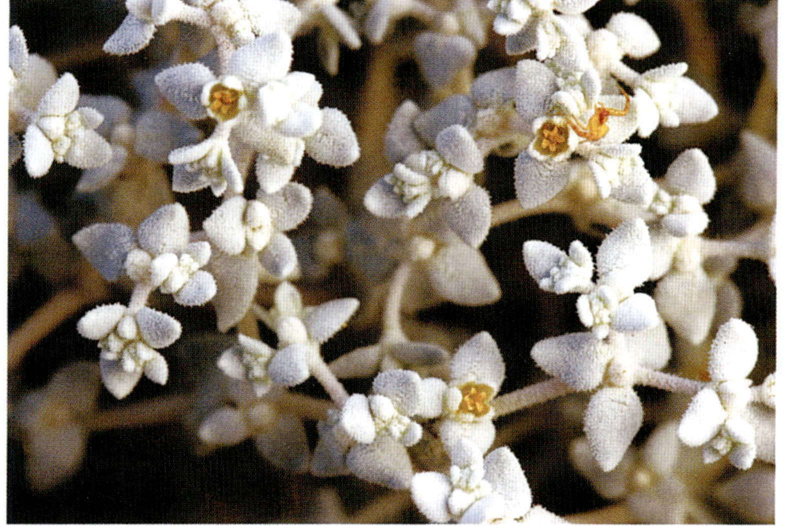

Shrubby Tidestromia

Tidestromia suffruticosa

Shrubby tidestromia is an erect subshrub, woody at the base, and usually not over one foot high but occasionally reaching two feet. Covered with very fine hairs, this member of the amaranth (Amaranthaceae) family is pale gray or almost white. It grows in compact mounds or rounded domes with many forking branches.

The slender branches are covered with small, short-stemmed oval leaves that often form in groups of three along the stems. They are coated with grayish, branching hairs.

The delicate flowers are difficult to see without a magnifying glass. No more than one-eighth inch wide, they are half hidden in densely hairy, tight clusters (of one to a few) at the base of leaves. Blooming primarily from May to August, each minute flower has as many as five yellow sepals with white, woolly hairs outside, but it lacks traditional petals.

This compact shrub grows in desert scrub, on open rocky and gravelly hillsides, in disturbed areas and along roads, in gypseous clay and limestone soils. Its distribution is disjunct and spotty: from southern New Mexico to El Paso and, in a separate part of the Trans-Pecos, largely from southeastern Presidio County to Pecos and Terrell counties. It is also found in Coahuila, Mexico.

Shrubby tidestromia frequents lower elevations of the park, especially the Dead Horse Mountains and limestone soils near the Rio Grande. It has been documented at Hot Springs and Boquillas, at Muskhog Spring, and on Mesa de Anguila.

Note: The species name *suffruticosa* is from the Latin *suffrutescens* (literally "subshrubby"). This species was collected by Charles Wright on Colonel Graham's U.S.-Mexico boundary survey in 1851–52 and described (as *Alternanthera suffruticosa*) by John Torrey in 1859.

Desert Sumac

Rhus microphylla

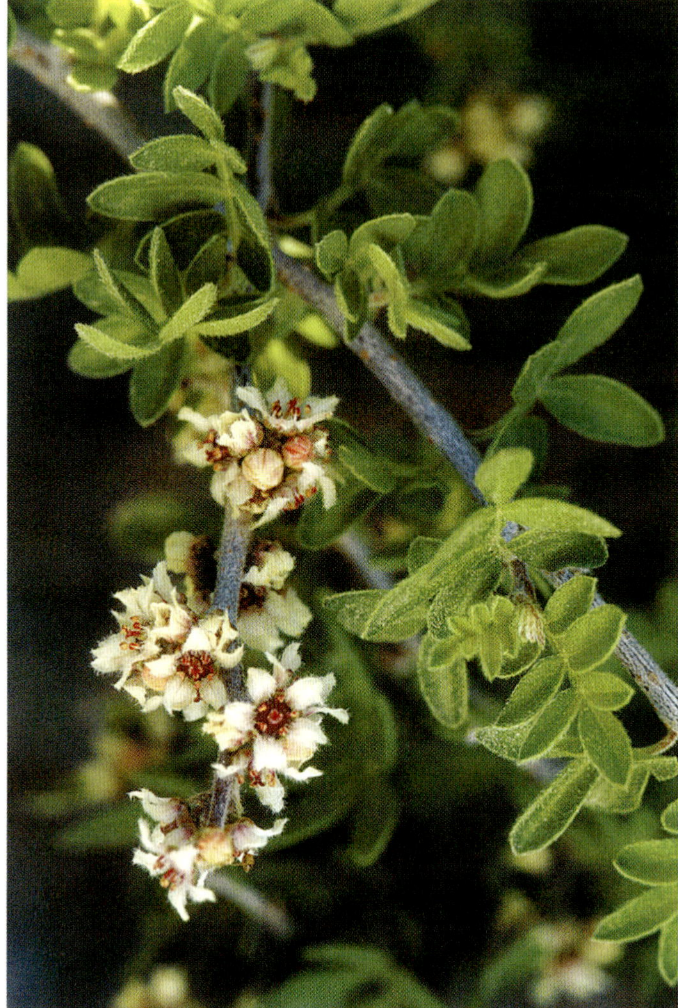

ALSO KNOWN AS littleleaf sumac (pronounced "shumac"), correosa, and agrillo, desert sumac is a densely branched shrub with many rigid and somewhat spiny short, crooked branchlets. Perhaps eight feet high and just as wide, these shrubs sometimes form extensive thickets. The compound leaves, up to one and one-fourth inches long, are divided into five to nine small leaflets, which are variable in shape: elliptic, oblong, or egg-shaped.

Often described as inconspicuous, the tiny flowers may go unnoticed because they appear in early spring before the new leaves form. In tight clusters at the ends of spur branches or at the base of leaves, the flowers are less than one-fourth inch long. Each bloom has five creamy white, spatula-shaped petals that are noticeably fringed with hairs. The flowers are followed by clusters of red-orange, sticky-hairy fruits about one-fourth inch wide. This sumac was once called sodapop bush because the bittersweet fruits were used by settlers as a lemon substitute in "sumac-ade."

This shrub is most often observed on dry, rocky or gravelly flats, slopes, and mesas and also in desert washes and scrub habitats and brushy thickets and canyons. Populations stretch across the Trans-Pecos into the Edwards Plateau, Upper Rio Grande Plains, and large areas of the Panhandle. They reach west as far as southeastern Arizona and south into northern Mexico from Nuevo León to San Luis Potosí and Zacatecas.

In the park desert sumac is perhaps most common in foothills around the Chisos Mountains. Examples can easily be located in Green Gulch and the Basin, in Oak Creek Canyon, and at Dugout Wells, but you might come across this sumac in desert habitats almost anywhere at elevations up to six thousand feet.

Note: *Rhus* is an old Latin name for sumac, from the Greek name *rhous* used by Theophrastus to describe one species, *Rhus coriaria.* The term *microphylla* refers to the small leaves. This species was described in 1852 by George Engelmann and probably first collected by Ferdinand Lindheimer in 1850 at Comanche Springs near New Braunfels in the Texas Hill Country.

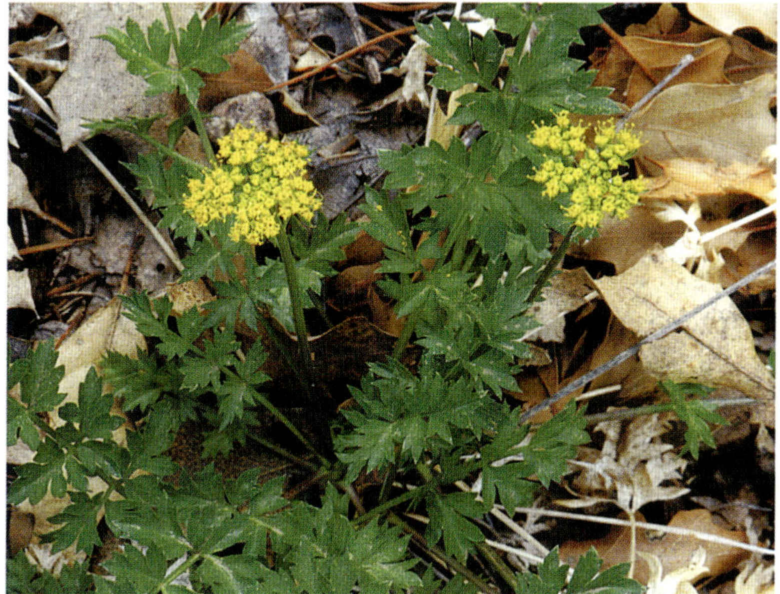

Stemless Aletes

Aletes acaulis

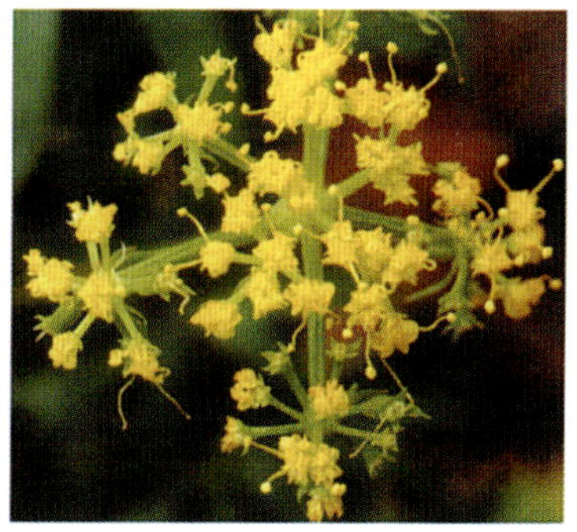

ALSO KNOWN AS stemless Indian parsley, this member of the parsley (Apiaceae) family is a clump-forming perennial herb that reaches about fourteen inches in height and is indeed stemless, with highly dissected, almost fernlike leaves and naked flowering stalks that tower above the leaves. The compound, oblong blades, up to four inches long, are once or twice divided into small, wedge-shaped leaflets, which are in turn deeply lobed and cleft and pointed at the tip.

In spring and summer, dainty yellow flowers appear atop a leafless stalk as much as ten inches long. The bulging flower cluster (umbel) consists of eight to fifteen subclusters, each on a stem radiating from the same point like the spokes of an umbrella. Each tiny blossom has five rounded petals—broadest at the tip and conspicuously curling backward—and five protruding stamens.

This herb is seen on rocky mountain slopes and in wooded canyons at elevations above five thousand feet in igneous soil, often in the shade of trees and boulders. Its wide range encompasses the Rocky Mountains of Colorado and New Mexico, the mountains of Trans-Pecos Texas, and the states of Chihuahua and Coahuila, Mexico.

In the park this perennial has been recorded at higher elevations of the Chisos, on top of Casa Grande, on the north slopes of Mount Emory, and along the Boot Spring Trail. You might walk by stemless aletes on the Pinnacles Trail not far below the Pinnacles.

Note: The genus name *Aletes,* Greek for "wanderer," probably refers to the many genera in which the type species *A. acaulis* had previously been placed. The species name *acaulis* means "stemless." This species was collected in 1853 by survey botanist John Bigelow at San Antonita, New Mexico, and described (as *Deweya acaulis*) by John Torrey in 1857 in Whipple's Pacific Railroad Survey report. John Coulter and Joseph Rose moved this species into the genus *Aletes* in 1888.

Tubular Slimpod

Amsonia longiflora var. *longiflora*

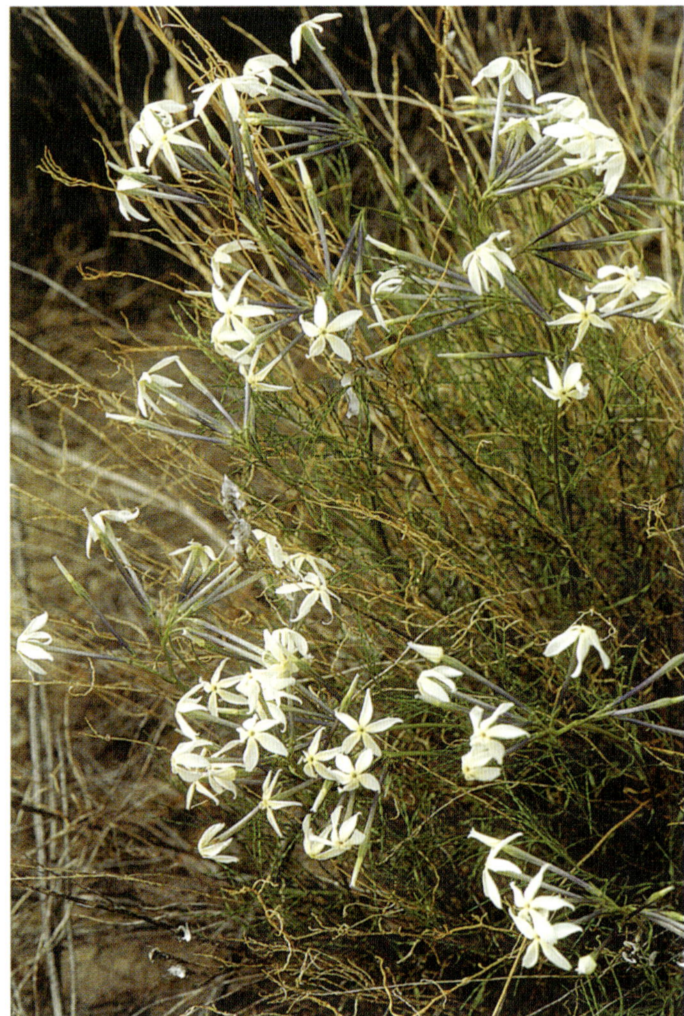

Tubular slimpod or tubular bluestar is a grasslike perennial herb up to two feet high with thick woody roots but rather weak, slender stems. The stems are usually upright but often sprawling. Thin, almost threadlike leaves up to two inches long are sometimes alternate but often form whorls (small groups of leaves borne at the same point) along the branches.

In loose clusters, white, trumpet-shaped flowers, sometimes with a distinctive bluish cast, radiate from the tips of the stems. Blooming in spring, each fragrant flower consists of a very slender tube, up to one and one-half inches or more, that flares abruptly into five narrow, petal-like lobes. If you look closely, you may notice that the elliptic lobes are twisted to the left, or backward. The yellowish fruit is a very slender, smooth follicle (a dry, single-chambered pod that splits along one seam) sometimes more than five inches long.

Tubular slimpod prefers limestone hills and flats, open grassy slopes, and sandy arroyos, and is often found on clay shales and caliche soils. Widespread but uncommon, it appears sporadically in most counties of the Trans-Pecos, south-central New Mexico, and eastern Coahuila, Mexico.

In the park this dogbane is known primarily from McKinney Spring in the Dead Horse Mountains. I have photographed it there many times and in the runoff area below, especially in March. It also occurs on Mesa de Anguila.

Note: The genus name *Amsonia* was first used in 1757 by John Clayton, county clerk and early plant collector of colonial Virginia, and the genus was described in 1788 by Thomas Walter. Clayton probably used the name to honor his contemporary John Amson, a physician and mayor of Williamsburg, Virginia. The species name *longiflora* (long-flowered) refers to the flower tube. The species was collected by Charles Wright in 1851 near El Paso and described by John Torrey in 1859. Torrey, the nation's first professional botanist, was often asked by the government to describe plant collections from the western United States, Mexico, and Central America.

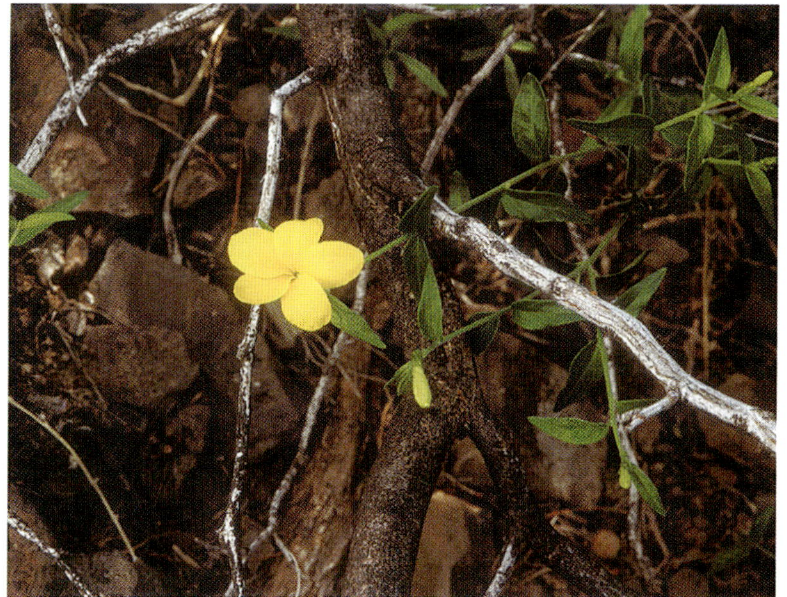

Arizona Cockroach Plant

Haplophyton crooksii

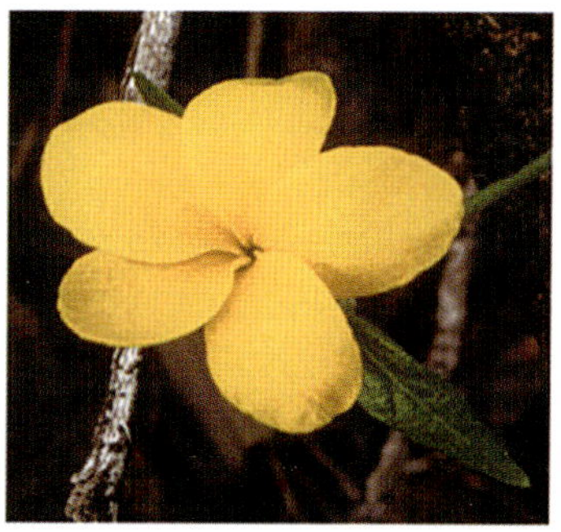

ALSO KNOWN AS hierba de la cucaracha and atempatli, Arizona cockroach plant is a weak subshrub no more than two feet high with many slender green branches. It may be partially concealed in or under larger shrubs. On very short stalks, the lance-shaped leaves are alternate or opposite, never much more than one inch long.

Arizona cockroach plant produces splashy yellow flowers one inch across. Each flower consists of a short slender tube that opens into five broad, rounded lobes. The flowers are easily recognizable because the petal-like lobes often twist to the left, forming a spiral.

Like some other members of the dogbane (Apocynaceae) family, this shrub is poisonous and has been used in Mexico as an insecticide. Supposedly, a mixture of dried leaves and molasses makes a good roach poison. The milky sap is also used as an insect repellent.

Although secure globally, this species is considered critically imperiled in Texas (NatureServe 2005). Rather scarce in the Trans-Pecos—in mountains and canyons of El Paso, Hudspeth, Presidio, and Brewster counties—it is more wide-ranging in southwestern New Mexico, southeastern Arizona, and northern Mexico from Sonora east to Coahuila.

Arizona cockroach plant had not been reported in the park until 2003, when I spotted a shrub blooming in June near the Blue Creek trailhead; two years later I discovered several bushes near the west end of the Dodson Trail.

Note: The genus name *Haplophyton* is from Greek words meaning "simple, or single, plant." This species was described by Lyman Benson in 1943 and named in honor of D. M. Crooks, a specialist in drug plants for the U.S. Department of Agriculture. Benson coauthored *The Trees and Shrubs of the Southwestern Deserts* (1954, revised 1981).

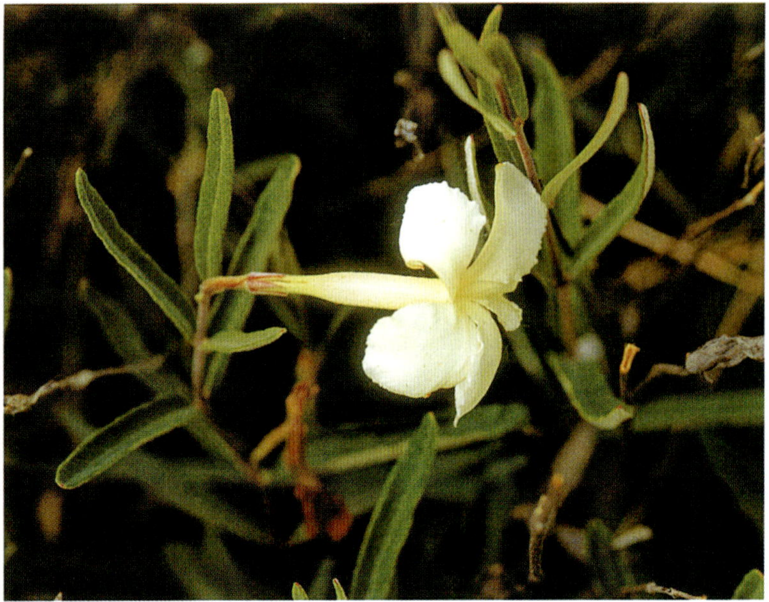

Bottomwhite Rocktrumpet

Telosiphonia hypoleuca

Bottomwhite rocktrumpet, also known as guirambo, is a mostly erect, low subshrub never much over a foot high with herbaceous stems that contain a white, sticky sap. Perhaps two inches long, the narrow oblong leaves are dark green on the upper surface but covered with white wool below. The margins of the short-stalked blades may be noticeably curled.

White trumpet-shaped flowers, with slender tubes and spreading petal-like lobes, appear in small clusters of one or two. Blooming in summer and early fall, the fragrant flowers usually open in late afternoon and close before noon the next day.

Bottomwhite rocktrumpet is usually situated on open rocky slopes and ledges above canyons and intermittent streams at higher elevations in igneous soil. In the United States it is restricted to Trans-Pecos mountains: the Davis, Chisos, and Chinati mountains. This low shrub is much more extensive in Mexico from Coahuila as far south as Michoacán.

Although more frequent decades ago, *T. hypoleuca* now survives in the park at a single location high in the Chisos Mountains near Laguna Meadow. It is easily distinguished from the other species in the park, plateau rocktrumpet, by its slender leaves.

Note: The genus name *Telosiphonia,* from the Greek *siphon* (tube) and *telos* (end), describes the flowers, which end in a long tube. The species name *hypoleuca,* from the Greek *hypo* (under) and *leucos* (white), refers to the white undersides of the leaves. This species was named by English botanist George Bentham (as *Echites hypoleuca*) in 1839 in *Plantas Hartwegianas,* which describes plants collected in Mexico by the German explorer Carl Theodore Hartweg. Searching for English garden plants, Hartweg traveled throughout Mexico, Central America, Ecuador, and Columbia from 1836 to 1843, sending crates of plants to the Royal Horticultural Society in London.

Plateau Rocktrumpet

Telosiphonia macrosiphon

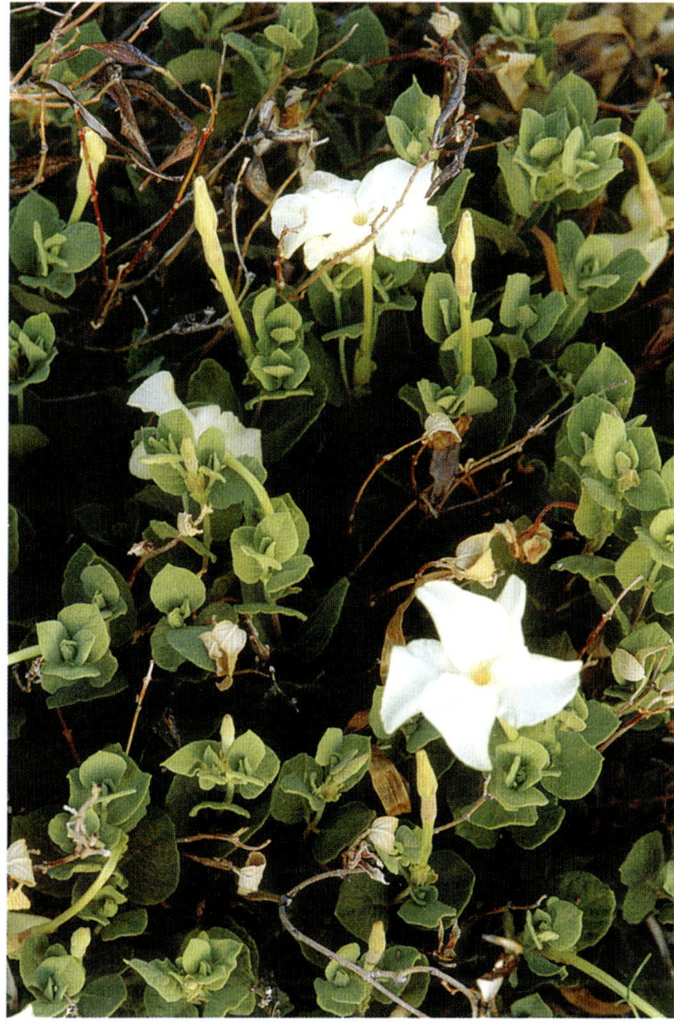

PLATEAU ROCKTRUMPET, or flor de San Juan, is a low subshrub less than one foot high, with many weak, silky-haired stems. Mostly erect but occasionally widely spreading or nearly prostrate, this rocktrumpet may form dense colonies many feet across. The stems contain a sticky, milky juice.

Egg-shaped or nearly oval, the woolly leaves are often clustered together in small bunches along the stem. Distinctly fleshy, and sometimes curly or wavy along the edges, these leaves can approach two and one-half inches long and two inches wide but are normally much smaller.

Usually solitary at the base of upper leaves, the stunning trumpet-shaped flowers have slender tubes up to three and one-half inches long. The white tubes expand into five petal-like lobes, which are twisted at the base so that they resemble propeller blades or whirligigs. Blooming in summer and early fall, the flowers open in late afternoon and close by noon the next day, and their distinctive fragrance attracts moths from long distances. The fruit is a slender, many-seeded follicle (pod) about five inches long.

Plateau rocktrumpet is most prevalent on open rocky slopes and wooded hillsides and also in chaparral in mountains and foothills. Widely distributed throughout most of Trans-Pecos Texas from Hudspeth County to Del Rio and into the Edwards Plateau, this subshrub also enters northern Mexico (in Coahuila and eastern parts of Chihuahua and Durango).

In the park, plateau rocktrumpet puts on showy displays after summer rains. Seek them out in Oak Creek and Pine canyons or along the Pinnacles and Laguna Meadow trails.

Note: The species name *macrosiphon,* from the Greek *siphon* (tube) and *macro* (large), refers to the large tubular flowers. This species was described by New York botanist John Torrey (as *Echites macrosiphon*) in 1859 and had been collected by Charles Wright on his first West Texas trip in 1849. Wright collected plateau rocktrumpet "along the Rio Grande in Texas and Chihuahua."

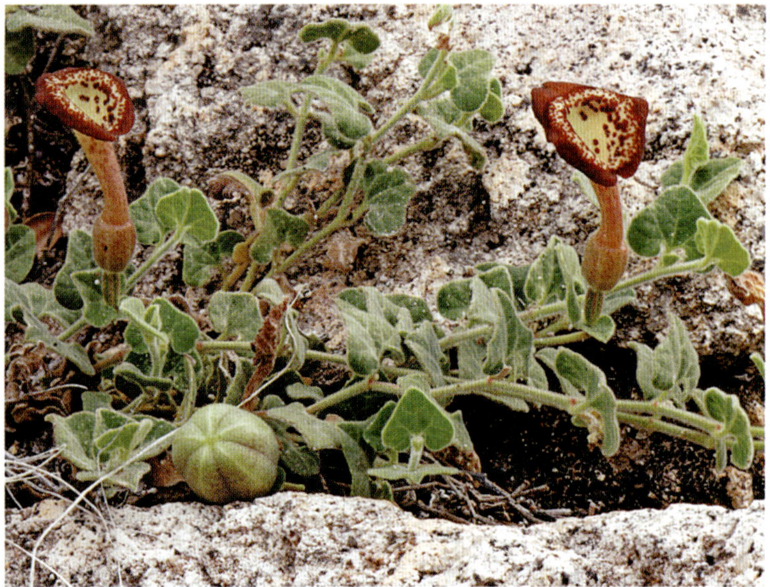

Cory Dutchman's Pipe

Aristolochia coryi

Cory Dutchman's pipe, a member of the birthwort or pipevine (Aristolochiaceae) family, is a prostrate perennial herb with sparsely hairy, vinelike stems. The two-inch-long leaves are mostly triangular: heart-shaped at the base and pointed at the tips. Noticeably veined and hairy on the margins, the long-stalked leaves are often so cupped at the base that they appear three-lobed.

More like dark-colored leaves in appearance, the bizarre blooms are scarcely recognizable as flowers. The blossoms lack petals and consist largely of a tubular calyx (usually the external green part of a flower). Funnel-like in shape, they resemble an upturned pipe. On stalks up to three inches long, these "pipes" often have a chocolate-brown upper edge and yellow markings. They are irresistible to swallowtail butterflies.

This pipevine thrives on rocky slopes, in rock crevices, among boulders, and at the base of rock slides, especially at cooler, wetter sites in canyons and near springs. It is dispersed from the Edwards Plateau near Fredericksburg to Brewster and Presidio counties in the southern Trans-Pecos and into northern Mexico.

In the park this vinelike herb blooms from April to October at widely separated locations. Look for it at the Window, Burro Spring, and the Pinnacles Mountains in the Chisos; Ernst Tinaja and Boquillas Canyon in the Dead Horse Mountains; and Santa Elena Canyon and Mesa de Anguila in the extreme southwest.

Note: The genus name *Aristolochia* is from Greek words meaning "best parturition" because a European species was used to aid childbirth. Other species have been used medicinally but are illegal in some countries because they contain toxic aristolochic acid. In 1940 Harvard botanist Ivan Johnston named this species for Victor Cory, Southern Methodist University botanist who collected it in Edwards County, Texas, in 1934.

Texas Milkweed

Asclepias texana

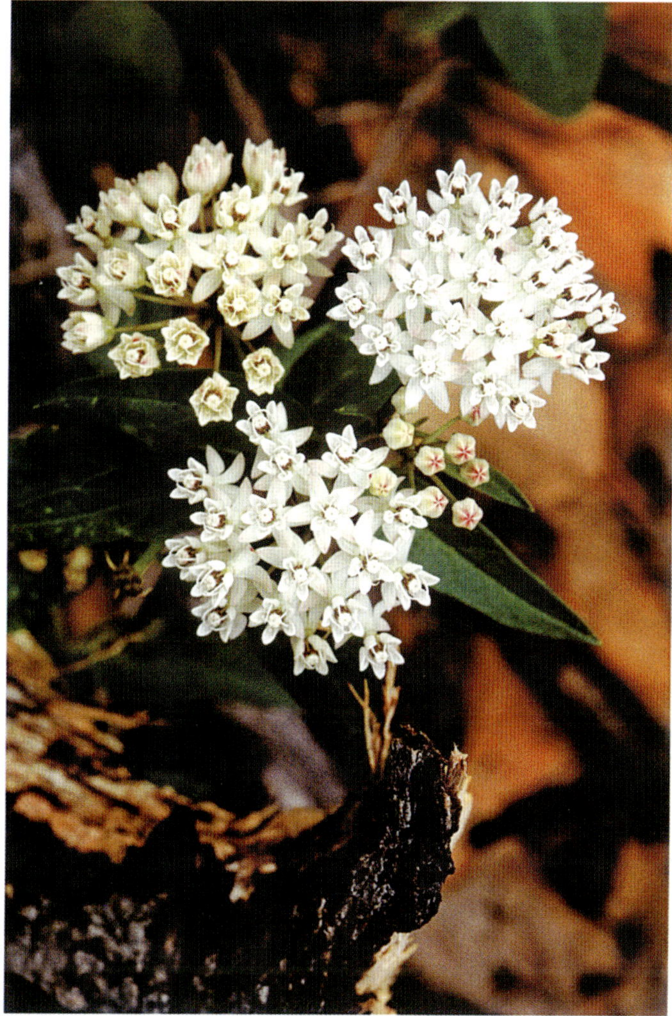

Texas milkweed is an attractive, slightly shrubby perennial herb, up to two and one-half feet high but usually much less, with slender stems branching mostly from the base and turning upward. The thin, relatively smooth opposite leaves, on stalks, are highly variable, from one to two and one-half inches long, and narrowly oblong or lance-shaped, or even egg-shaped.

Blooming in summer and fall, white flowers appear in large flat or convex bouquets, on pedestals (pedicels) radiating from a common point like umbrella spokes. Tinged with purple or rose, each blossom, one-fourth inch wide, has five petals turned sharply backward and a prominent crown (corona) typical of the milkweed family. The crown consists of five slender horns arching over a central column (a gynostegium, or fused stamens and pistil) and five surrounding outer "hoods" resembling miniature petals. The fruit is a smooth, slender pod about four inches long, tapered at both ends, and notably erect on an erect stalk.

This perennial prospers on rocky and gravelly slopes and in wooded canyons at elevations above forty-five hundred feet. It is present in the Hill Country (Austin southwest to Leakey) and Big Bend mountains (in Jeff Davis, Presidio, and Brewster counties) of Texas and in Coahuila and Nuevo León in northern Mexico.

In the park, you might find Texas milkweed at moderate elevations (in Pine Canyon and below Casa Grande) or on higher slopes (on Lost Mine and Emory peaks and above Boot Spring), but it is more readily accessible in Upper Green Gulch and the Basin.

Note: The genus *Asclepias* is named for the Greek god of medicine, Asklepios, because of the medicinal properties of some species. Species *texana* was collected by Philadelphia botanist A. Arthur Heller in 1894 near Kerrville in the Texas Hill Country and described by him in 1895 in his *Botanical Explorations of Southern Texas.* Heller, who collected extensively in the West, especially in Nevada and California, developed a large personal herbarium and described many new species in his botanical journal *Muhlenbergia.*

Pringle Swallow-wort

Cynanchum pringlei

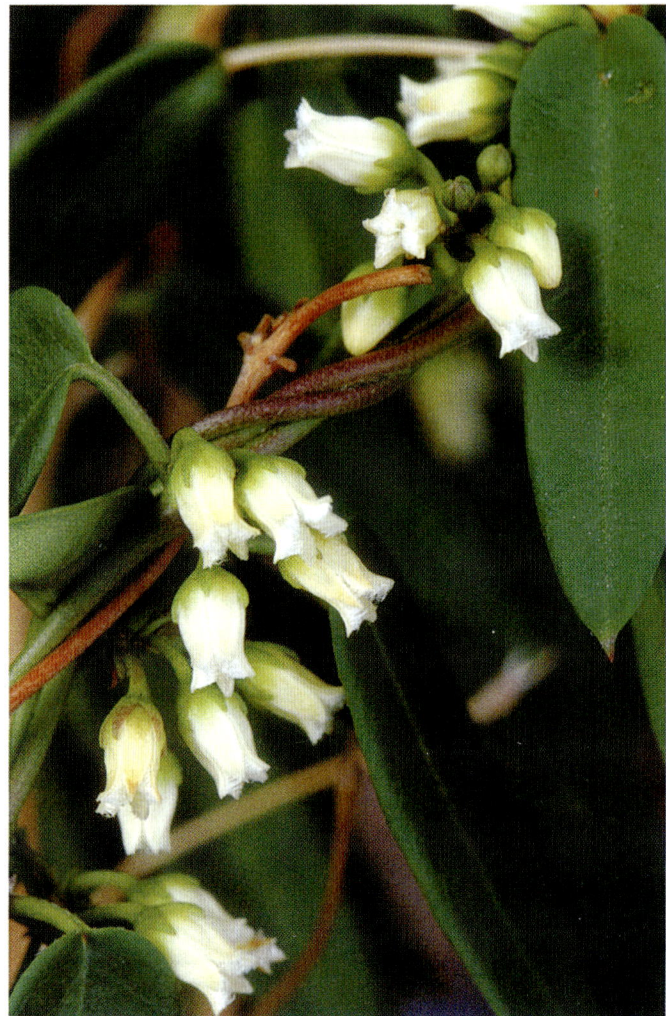

THIS MILKWEED is a woody-based perennial vine with many slender and intricately branching stems that twine through shrubs and cacti and trail across rocks. Like those of most milkweeds, the smooth stems contain white, milky latex. On slender stalks, the dark green leaves are up to two inches long and narrowly oblong. Slightly leathery and thick, the paired blades usually have a short, abruptly pointed tip.

Pringle swallow-wort is similar to its close relative, bearded swallow-wort, except its tiny, urn-shaped flowers are even smaller and seldom fully open. The minuscule creamy white flowers, less than one-eighth inch long, form small clusters at the base of leaves along the stem. Blooming in late spring and summer, these "urns" can be quite hairy at the tips and inside.

This vine flourishes in open woodlands and on open, arid mountain slopes as well as in chaparral and thorn scrub habitats. In the United States it is limited to Trans-Pecos Texas from southern Hudspeth County southeast to Brewster County, and in northern Mexico to Coahuila and eastern parts of Chihuahua and Durango.

Widely scattered throughout the park, Pringle swallow-wort has been encountered from Harte Ranch and the Rosillos Mountains to near McKinney Spring in the Dead Horse Mountains, to the Chisos. Occasionally, it can be plentiful in the Chisos foothills, especially near springs. This swallow-wort has been identified at the rock dike south of Ward Spring, at Lower Juniper Spring to the east of the Chisos, and on the small peak in the center of the Basin. Sotol Vista is an ideal place to witness it and the sunset.

Note: The genus name *Cynanchum* is a product of the Greek words *kyon* (dog) and *ancho* (to strangle), presumably because of the plant's toxicity and its use as a dog poison. Described by Asa Gray as *Metastelma pringlei* in 1886, this species is named for Cyrus Pringle, who collected it in Chihuahua in 1885. Pringle, who collected in Mexico from 1885 until his death in 1911, founded the Pringle Herbarium in Vermont and donated five hundred thousand specimens to the University of Vermont and other institutions.

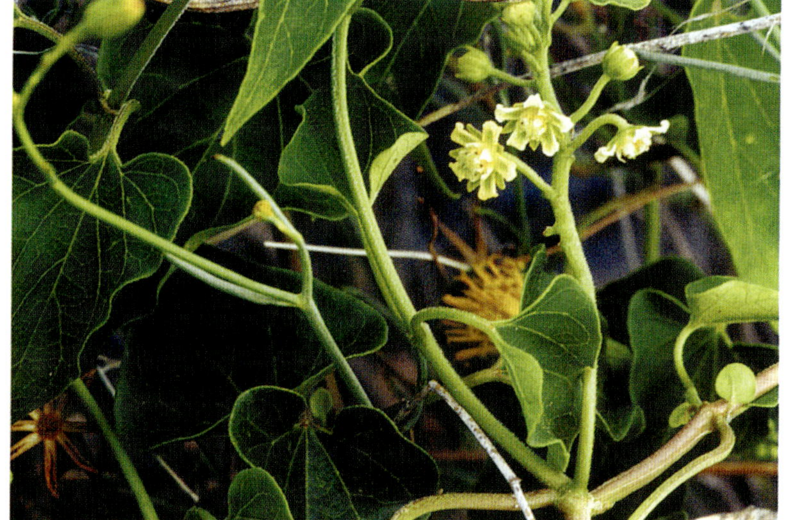

Talayote

Cynanchum racemosum var. *unifarium*

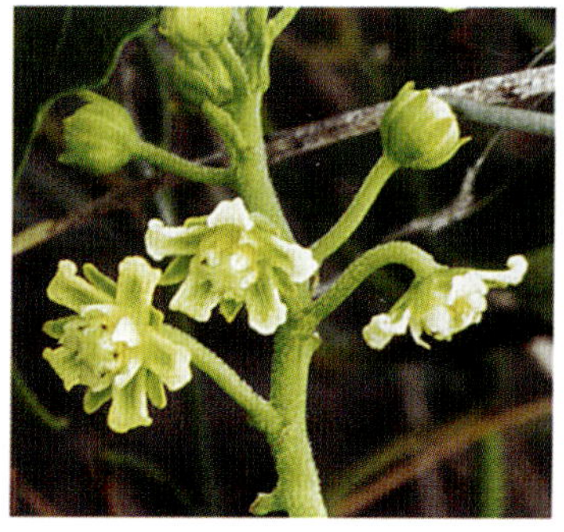

TALAYOTE is a robust, twining vine up to ten feet long with stems that clamber through and hang down from trees and large shrubs. On long curving stems, the paired leaves are heart-shaped or triangular and tapered at the tip. Dark green on the surface and veined, the thick, slightly leathery leaves may approach four inches in length.

Blooming primarily in summer, small succulent flowers one-fourth inch wide form clusters at the base of leaves. On long stems, they are creamy white to pale yellow, with touches of light green. Like other milkweed flowers, they have a prominent central crown. The smooth, rather fat dry pods, about four inches long and one inch wide, may appear while the vine is still blooming.

Unknown elsewhere in the United States, talayote does well in chaparral and brushy, wooded canyons, frequently near creeks and springs, in parts of mostly southern, central, and western Texas. The vine can be locally common in the southern Trans-Pecos, from Fort Davis to Del Rio, and is widespread in northeastern Mexico (from Coahuila to Nuevo León and south to Veracruz).

In the park talayote is resident at springs and creeks, and in wooded canyons of the Chisos: at Ward Spring, in Pine and Panther canyons, along Oak Creek, and in the Basin. In good years it can form thick stands in Blue Creek Canyon, where this photograph was taken at the end of June.

Note: This variety was collected in 1846 by Ferdinand Lindheimer on the Guadalupe River in Comal County near New Braunfels, Texas. Adolf Scheele, botanist and priest in Heersum, Germany, described this variety as a species (*Gonolobus unifarius*) in 1849. The species name *racemosum* refers to the shape of the flower cluster (a raceme); the variety name *unifarium* implies "in a single series."

Pearl Netleaf Milkweed

Matelea reticulata

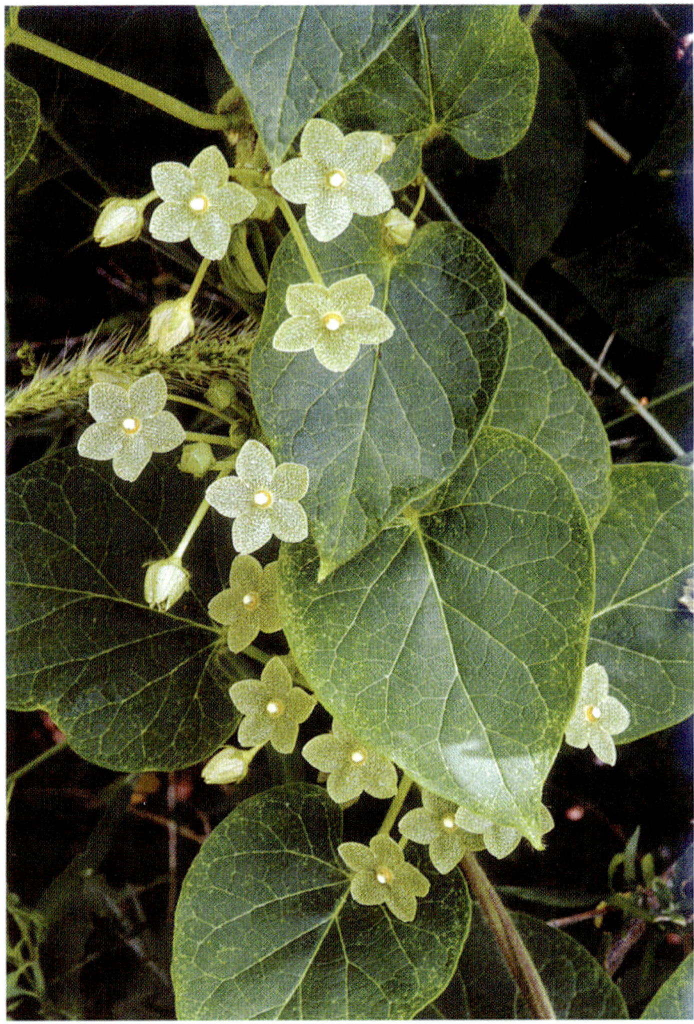

PEARL NETLEAF milkweed, or netted milkvine, is a twining vine with slender, hairy stems. The large yellow-green leaves, four inches or more in length, are heart-shaped with pointed tips and sturdy stalks.

The most distinctive feature of this perennial herb is its attractive greenish flowers. The blossoms are wheel-shaped with prominent net-like veins and a small pearl-colored central crown. In full bloom, sprawled on a boulder or hanging from a small tree, this vine may look more like an out-of-place bejeweled necklace than a common plant.

At first green but drying to pale brown, narrow fruit pods up to five inches long and three-fourth inch wide follow the blooms. They are prickly and tapered at both ends.

Far reaching in northeastern Mexico (from eastern Chihuahua to Tamaulipas and south to San Luis Potosí), pearl netleaf milkweed also ranges throughout the South Texas Plains and Edwards Plateau into north-central Texas and the south-eastern Trans-Pecos (in Brewster, Terrell, and Val Verde counties). It blooms from April to October on both rocky and sandy soils in diverse habitats: mostly shady mountain woodlands but also open thickets and sandy washes.

Look for this wiry vine in the Chisos Basin, especially along the Window Trail in July. Elsewhere in the Chisos, pearl netleaf milkweed has been documented in Pulliam Canyon, on the lower east slopes of Crown Mountain, and below Bailey Peak. Outside the Chisos, it has been found on the north side of Talley Mountain and at Harte Ranch.

Note: In 1775, Jean Baptiste Aublet, a French botanical explorer of South America, described the genus *Matelea* in a famous flora of French Guiana, which named more than four hundred new species. Aublet was sent to Martinique and later French Guiana in search of medicinal plants for the Royal Apothecary. The species name *reticulata* (netted) refers to the network of veins on the green flower. This species was collected in 1845 by Ferdinand Lindheimer near New Braunfels, Texas, and described by Asa Gray in 1877 (as *Gonolobus reticulatus*).

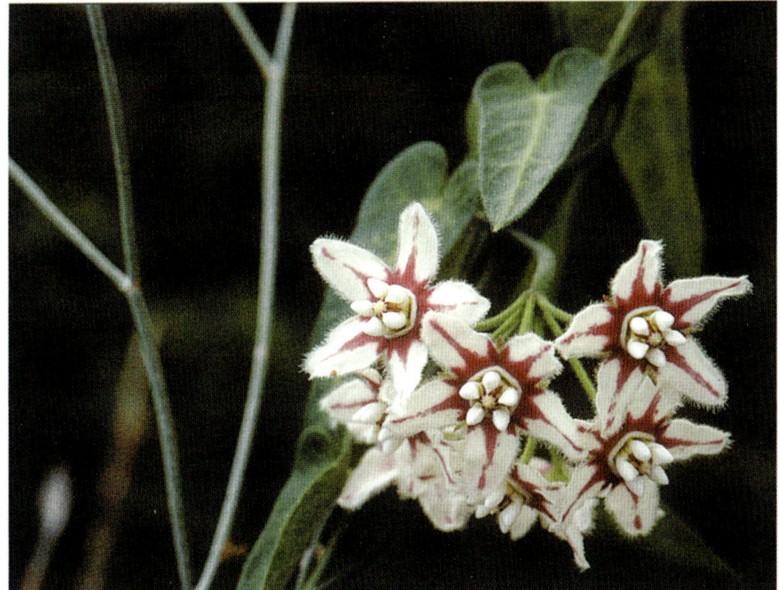

Soft Twinevine

Sarcostemma torreyi

SOFT TWINEVINE, or Torrey twinevine, is a distinctly hairy perennial vine with wiry, twining stems. The large green leaves, up to two inches long, are triangular: pointed at the tip and heart-shaped at the base. Somewhat leathery, the drooping blades are covered with spreading hairs.

Flower clusters (umbels) with as many as ten flowers form on long stalks, each blossom with a separate small stem. The wheel-shaped flowers have five pointed petals and a conspicuous central crown consisting of five white, inflated sacs. Each white petal is deep red at the base with distinct red midstripes. Attractive to butterflies and bees, the fragrant flowers bloom in late spring and summer. About three inches long, the hairy, dry fruit pods are thickest near the middle and tapered at the ends.

Soft twinevine inhabits arid rocky slopes and brushy woodlands in foothills and mountains and also gravelly soils at lower elevations. In the United States, this vine is confined to Brewster and Presidio counties of the Big Bend, although it has a broader distribution in northern Mexico (from eastern Chihuahua to Tamaulipas).

In the park soft twinevine is concentrated in the Chisos at middle elevations: to the west at Oak and Ward Spring, to the east at Lower Juniper Spring and Crown Mountain, to the north at Green Gulch and Pulliam Bluff, and in the Basin. It also occurs outside the Chisos at Persimmon Gap, in the Rosillos Mountains, and at Boquillas Canyon on the Rio Grande.

Note: The genus name *Sarcostemma,* from the Greek *sarkos* (flesh) and *stemma* (crown or wreath), refers to the fleshy flower crown. This species was collected by John Bigelow (with Parry, Wright, and Schott) in "S.W. Texas" in 1852 and described by Asa Gray (as *Philibertia torreyi*) in 1876. Gray named the species for John Torrey, his former teacher and collaborator.

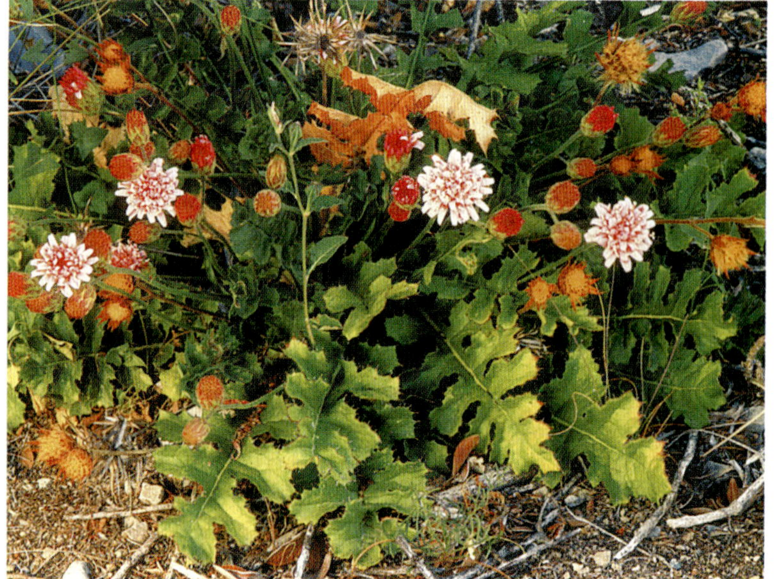

Stemless Perezia

Acourtia runcinata

Stemless perezia, peonia, or featherleaf desert peony is a stemless perennial herb forming a rosette of basal leaves. Instead of stems, it produces several naked flower stalks up to one foot high. Crowded leaves, radiating from the center, are as much as six inches long or more, with the edges deeply cut, almost to the midrib, into rounded, wavy lobes. The lobes are edged with tiny bristly teeth.

Fragrant flower heads one inch wide form at the top of elongated scapes or sometimes just above the leaves. Each flower head contains forty to fifty minute florets, mostly pinkish but also in varying shades from lavender to rose to purple to nearly white. Blooming from spring to fall, each minuscule floret is a thin tube, two-lipped at the rim.

Stemless perezia sprawls over rocky limestone and volcanic soils, often hidden among boulders or under shrubs and in brush. It grows in southern and south-central Texas and farther west throughout the southeastern Trans-Pecos to Presidio. Populations also stretch across northern Mexico (from Chihuahua to Tamaulipas and south to Hidalgo).

In the park this perennial is most often seen in the Dead Horse Mountains: near Hot Springs and McKinney Spring and at Devil's Den and Dog Canyon in the northeast. It has also been recorded on Fresno Creek, at Onion Spring, and in the Chisos Basin.

Note: The species name *runcinata* means "incised" in Latin. Described in 1830 (as *Clarionea runcinata*) by David Don, botanist at King's College and librarian at the Linnaean Society of London, this species was collected in Mexico during the Sessé and Mociño expedition (1788–1803). Asa Gray placed the species in the *Perezia* genus in 1849. *Perezia* was named for Lorenzo Pérez, a sixteenth-century Spanish apothecary and author of a history of drugs. In 1978 the plant was placed in the genus *Acourtia* by B. L. Turner, University of Texas botanist. This genus may be named for Mrs. Mary Gibbes A'Court, a British amateur botanist.

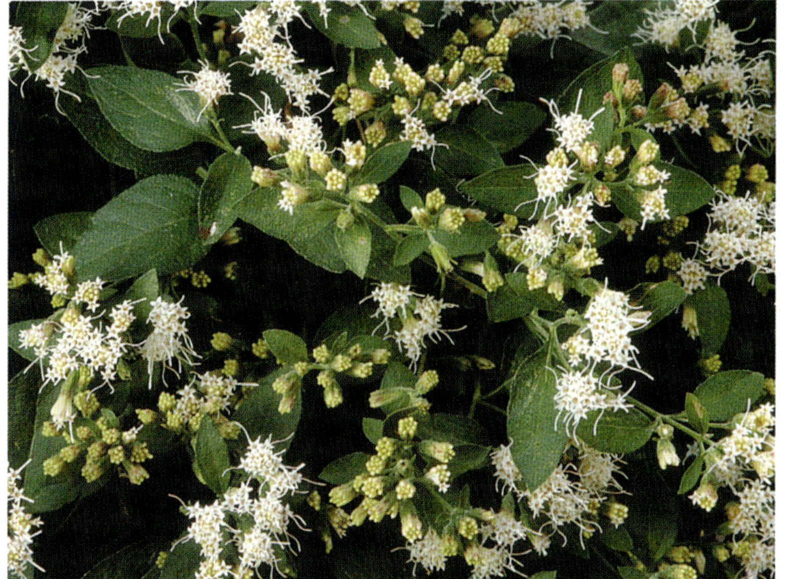

Wright Ageratina

Ageratina wrightii

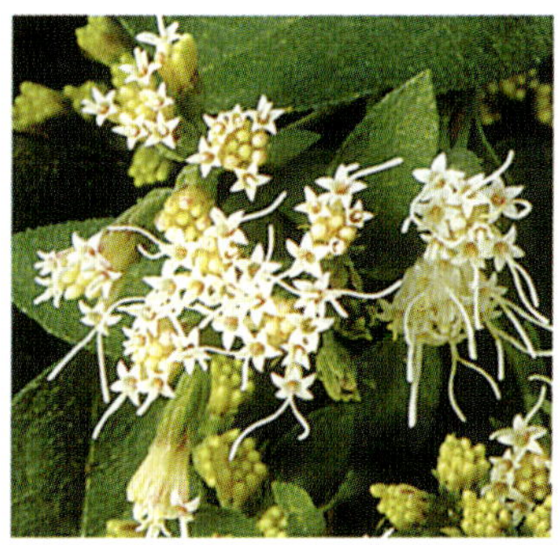

ALSO KNOWN AS Wright thoroughwort and Wright snakeroot, Wright ageratina is a widely spreading but often rounded low shrub under two feet high with many short and very leafy branches. The opposite leaves are egg-shaped or triangular, mostly less than one inch long, on short and sometimes slightly winged stems. They may have a rough, even grainy surface.

In fall, these low shrubs are bedecked with clusters of white or slightly pinkish flower heads near the tips of branches. Each small head consists of up to a dozen tiny flowers crowded together in a disk-like cup. The minute flowers are tubular with five protruding teeth.

Wright ageratina spills across rocky, open and wooded slopes in foothills and higher mountains in limestone and igneous soils. A well-established fixture throughout much of the Trans-Pecos, from El Paso to Brewster County, the shrub is most prominent in the Chisos, Davis, and Guadalupe mountains. It also crosses into the Chihuahuan Desert portions of southern New Mexico and southeastern Arizona and deep into Mexico from Chihuahua and Coahuila to Aguascalientes and Jalisco.

In the park Wright ageratina is probably most abundant at mid elevations of the Chisos from the Basin to Laguna Meadow and along the Pinnacles Trail to higher mountains. Yet it can be observed high on Emory Peak and outside the Chisos in the Dead Horse Mountains.

Note: The genus *Ageratina* was described in 1841 by Édouard Spach, an Alsatian botanist and professor of natural sciences in Paris. The name is a diminutive reference to the genus *Ageratum*, implying "like a little ageratum." The species was first described by Asa Gray in 1852 (as *Eupatorium wrightii*) and named for Charles Wright, who had collected it in the Guadalupe Mountains east of El Paso in 1849.

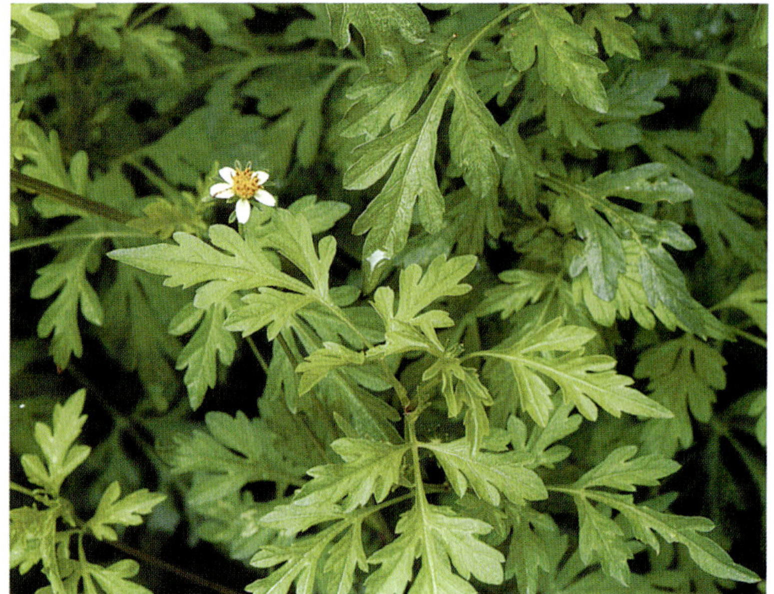

Bigelow Beggarticks

Bidens bigelovii var. *bigelovii*

Bigelow beggarticks is a tall annual up to three feet high, with much-branched and leafy, four-angled stems. The highly dissected leaves, up to three inches long, are usually divided into three to five stalked leaflets, with each of these again deeply divided or lobed.

The charming flowers are mostly solitary at the top of long stalks, sometimes as much as six inches above the leaves. Really collections of flowers, they consist of up to five white ray florets and many tiny yellow, five-toothed disk florets in a central disk.

Beggarticks have acquired amusing names (bur marigold, sticktights, harvest lice, and devil's pitchforks) referring to the unusual barbed fruits (achenes). The plant has two kinds of four-sided fruits: slender inner fruits one-half inch long and a few shorter, club-shaped exterior fruits. Crowned with a few barbed bristles, the dry fruits cling to clothes.

On occasion, this annual can quickly become pervasive, spreading like a weed after good summer rains. Restricted in Texas to mountainous Trans-Pecos terrain (from the Guadalupe Mountains south to the Chisos), this variety of Bigelow beggarticks reaches into New Mexico and southeastern Arizona, as far north as southern Colorado, and into Coahuila, Mexico.

In the park the herb blooms from June to September, often along Oak Creek and the South Rim Trail and in Upper Cattail Canyon. It frequently lines the road into the Basin.

Note: The genus name *Bidens,* from Latin meaning "two-toothed," refers to the fruits' barbed bristles. In the U.S.-Mexico boundary survey report of 1859, Asa Gray named this species for survey botanist John Bigelow. The species was apparently first collected in 1841–42 by the Bavarian explorer Baron Karwinski, on a Russian botanical expedition to Mexico.

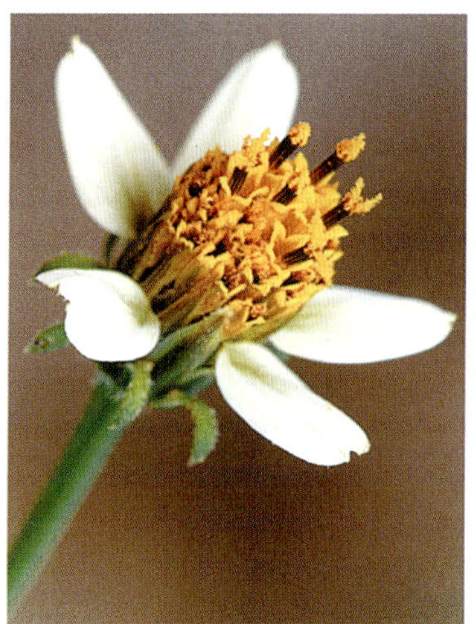

Splitleaf Brickellbush

Brickellia laciniata

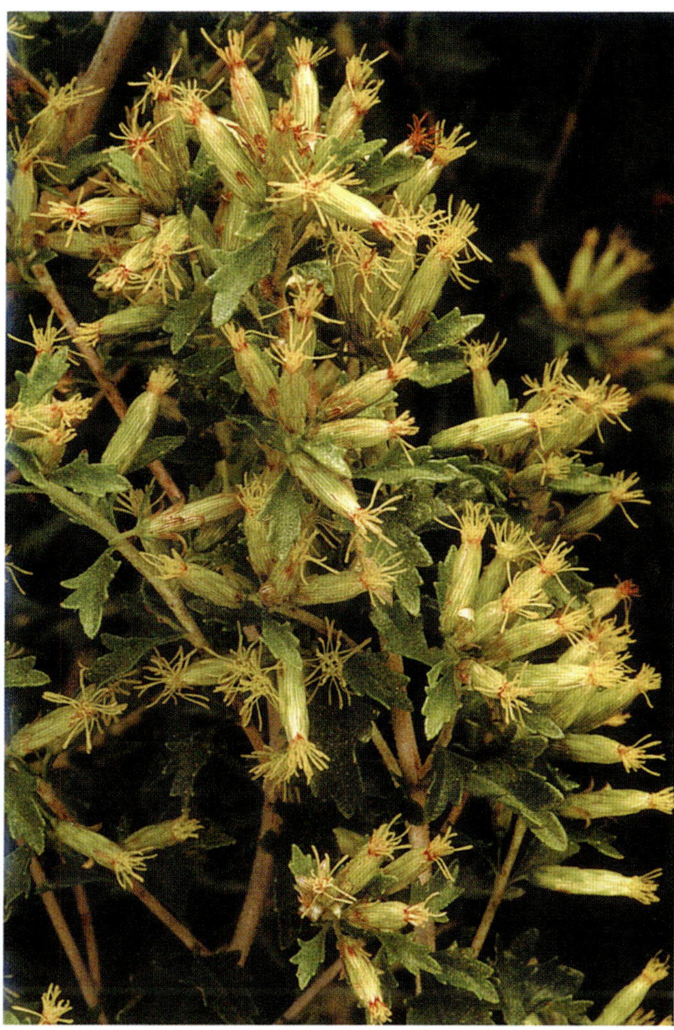

SPLITLEAF BRICKELLBUSH, or cutleaf brickellbush, is a compact, densely branched shrub three to five feet high and often just as broad. The smooth, light brown stems are thick with leaves. This plant is named for its crowded, oblong to egg-shaped leaves, on very short stalks, which are deeply toothed or lobed. The thin leaves, rarely more than three-fourths inch long, are gland-dotted and often sticky to the touch.

The greenish-yellow flower heads emerge in tight leafy clusters. The heads lack ray florets and consist exclusively of florets growing from a central disk. Each floret is a narrow tube with five tiny threadlike lobes at the tip. The fruits are ten-ribbed cylindrical achenes (dry, one-seeded, with the ovary wall free from the seed) with a crown of rough or feathery bristles at the top.

Blooming in late summer and fall, splitleaf brickellbush is probably the most wide-ranging *Brickellia* in the Trans-Pecos, extending east to Del Rio, west to southwestern New Mexico, and south into Mexico as far as Zacatecas and San Luis Potosí. It frequents washes and arroyos, gravelly sites near creeks and springs, roadside depressions, and other disturbed locations.

In the park this dense shrub is easy to locate in Lower and Upper Green Gulch. I have photographed splitleaf brickellbush in the Basin, on the slopes of Casa Grande, on the Window Trail, and along Oak Creek, usually in September.

Note: In 1824 South Carolina botanist Stephen Elliott named the genus *Brickellia* for John Brickell, an Irish-born Georgia physician and amateur botanist who, until his death in 1809, collected plants in South Carolina and near his Savannah, Georgia, home. Elliott was one of the founders of the *Southern Review* quarterly and author of the widely respected botanical work *Sketch of the Botany of South Carolina and Georgia.* The species name *laciniata* refers to the deeply incised leaves. The species was collected by Charles Wright on his 1849 West Texas expedition and described by Asa Gray in 1852.

Sandlot Brickellbush

Brickellia lemmonii var. *conduplicata*

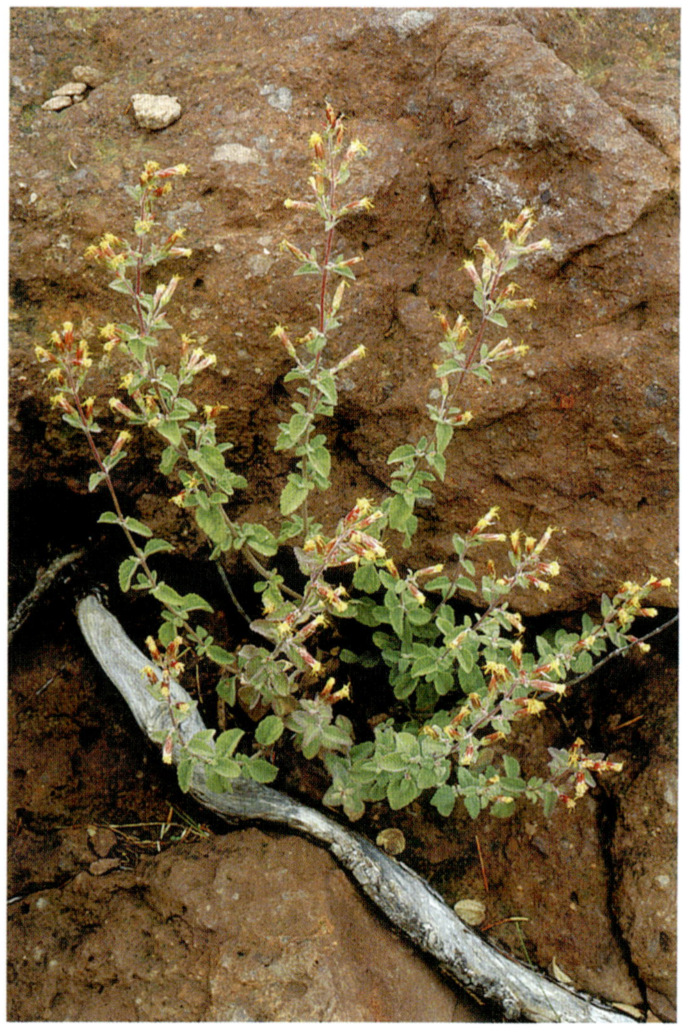

Also known as southwestern brickellbush, sandlot brickellbush is a perennial herb up to two and one-half feet high with one or several frequently unbranched, sometimes reddish or purplish stems. The nearly stemless leaves are broadly egg-shaped to triangular with blunt teeth along the edges. Rarely exceeding two inches, the mostly paired leaves are slightly leathery, noticeably veined underneath, and dotted with yellowish glands. In open sun, the blades have a tendency to fold lengthwise at the middle.

Often in leafy, elongated clusters, the cylindrical flower heads, frequently tinged with red, contain ten to twelve minute yellow disk flowers. They bloom in late summer and fall. The fruits are brownish-black, ten-ribbed, oblong achenes (dry and one-seeded) covered with bristles.

Sandlot brickellbush appears sparingly on igneous soil at higher elevations in shady canyons and rocky woodlands. It is a rather inconspicuous inhabitant of mountains in the southern Trans-Pecos and the Mexican states of Coahuila, Tamaulipas, and San Luis Potosí.

In the park, you might notice this perennial in Upper Green Gulch and the Basin, in Pine Canyon, on the Window and Lost Mine trails, and on the Pinnacles and Laguna Meadow trails leading to the higher Chisos.

Note: In 1882 Asa Gray named the species *lemmonii* for schoolteacher and California botanist John Lemmon, who collected it at Apache Pass in the Chiricahua Mountains of southeastern Arizona in 1881. Variety *conduplicata* (meaning "folded") was collected by Cyrus Pringle in the Mexican state of San Luis Potosí in 1890 and described in 1907 (as a variety of *Brickellia betonicifolia*) by Benjamin Robinson, curator of the Gray Herbarium at Harvard University. This species is also known as *Brickellia conduplicata.*

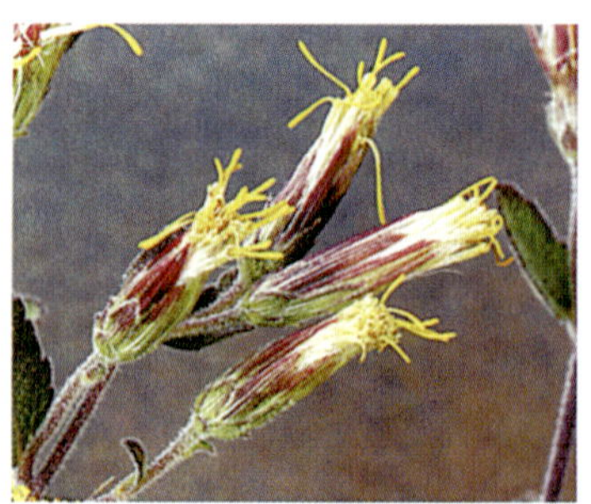

Veronicaleaf Brickellbush

Brickellia veronicifolia

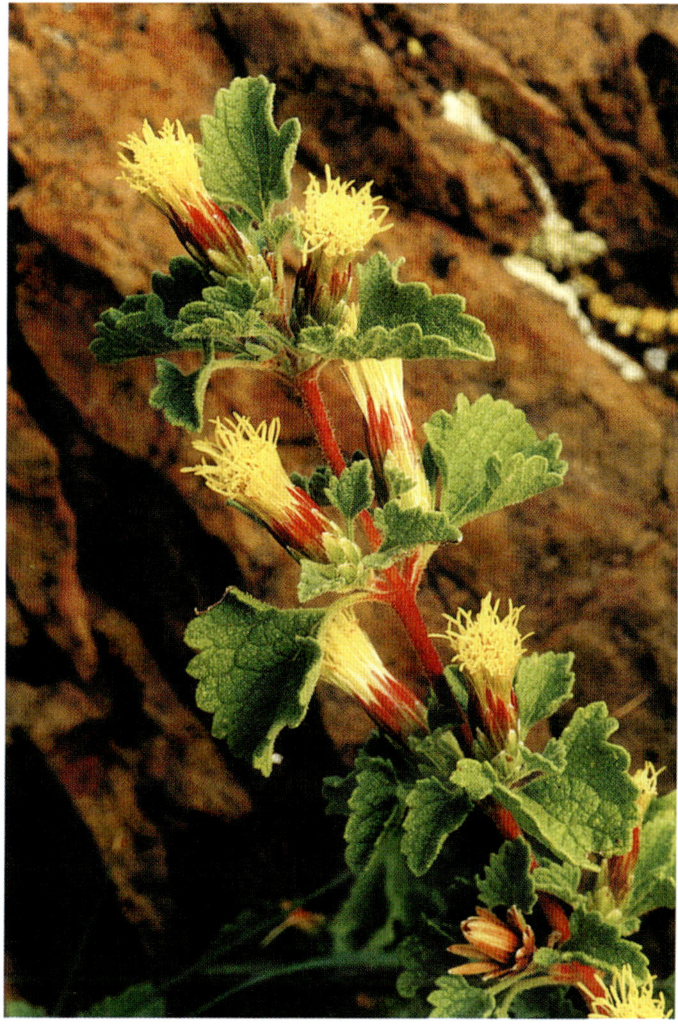

VERONICALEAF BRICKELLBUSH is a small shrub, or subshrub, up to twenty inches high, often intricately branched from near the base. The younger stems especially are conspicuously red or maroon. No more than one-half inch long, the leaves are broadly egg- or kidney-shaped, with a veiny and wrinkled surface. The paired blades, with rounded teeth along the margins and a blunt tip, may be noticeably hairy and dotted with glands.

On short stalks, the flower heads form in small clusters on the leafy branchlets. One-fourth inch across, the narrow flower heads are pale yellow but often tinged with red. Blooming in fall, the rayless, disklike heads contain tiny tubular florets with five teethlike lobes. The cylindrical fruit is an achene (dry and one-seeded) with ten prominent ridges and a crown of about twenty-five rough bristles.

Veronicaleaf brickellbush prefers rocky, mountainous sites at mid to high elevations. In the United States this brickellbush is isolated in the Chisos Mountains. Although critically imperiled in Texas (NatureServe 2005), it is much more common in Mexico (from Chihuahua east to Nuevo León and south to northern Durango, Veracruz, and Puebla).

In the park the most likely spot to view this low shrub is on the Chinese Wall Trail leading to the summit of Casa Grande. Veronicaleaf brickellbush has also been reported on the Lost Mine Trail, in Green Gulch, and along the Laguna Meadow Trail. I have photographed good examples in Boot Canyon and Upper Cattail Canyon in October.

Note: The species name *veronicifolia* connotes "with leaves like those of the genus *Veronica.*" This species was collected by Prussian explorer Baron Alexander von Humboldt and his French colleague Aimé Bonpland near Mexico City in 1803 and described (as *Eupatorium veronicifolium*) by German botanist Carl Kunth in 1820. From 1799 to 1804, Humboldt and Bonpland explored parts of South America, Cuba, and Mexico. Asa Gray moved this species into the genus *Brickellia* in 1852. The plant in the park has often been treated as a distinct variety (var. *petrophila,* meaning "rock-loving"). It was collected by Cyrus Pringle near Chihuahua in 1885.

Bigelow Bristlehead

Carphochaete bigelovii

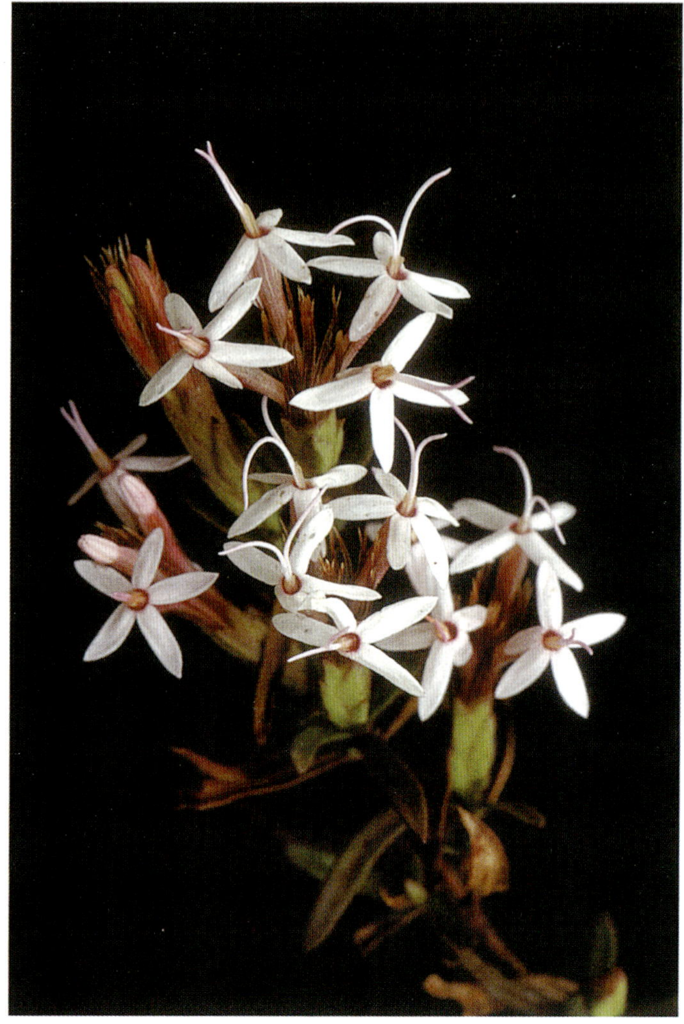

Bigelow bristlehead is considered a small shrub or subshrub, although it is rather spindly, rarely more than one and one-half feet high, with slender, reddish-brown branchlets. Paired or sometimes clustered and overlapping, most leaves are narrowly oblong and usually less than one inch long. They are stemless and often densely covered with dots of resin.

This plant's pure white to pink cylindrical flowers with strikingly narrow, radiating lobes are the main attraction. Often low to the ground at the ends of short branchlets, the blossoms resemble tiny white stars draped over the stems and leaves. Each flower head lacks ray flowers and is comprised of four to six small disk flowers. They bloom mostly in spring.

The columnar fruits are also distinctive: although less than one-half inch long, they are topped with about fifteen stiff and sharp-pointed, protruding bristles, hence the name bristlehead.

In Texas this bristlehead is centered in the mountains of Brewster, Presidio, and Jeff Davis counties, although it also occurs in the Franklin Mountains near El Paso. Occupying rocky and grassy mountain hillsides, often in the shade of oaks and juniper, this subshrub is also present in southern New Mexico, southern Arizona, eastern Chihuahua, and northern Coahuila.

Bigelow bristlehead has been identified at the top of Lost Mine Trail and near Casa Grande, and it is sometimes scattered throughout upper Green Gulch and along trails to the higher mountains. You might hike by it along the Pinnacles Trail, near Boulder Meadow or the Pinnacles backcountry campsite.

Note: The genus name *Carphochaete,* from the Greek *karpho* (to dry, wither) and *chaite* (bristle), refers to the bristly dry fruit. This species was described by Asa Gray in 1852. He named the plant for John Bigelow, one of the botanists attached to the U.S.-Mexico boundary survey who, together with Charles Parry, Charles Wright, and Arthur Schott, first collected the plant along the Rio Grande in 1852.

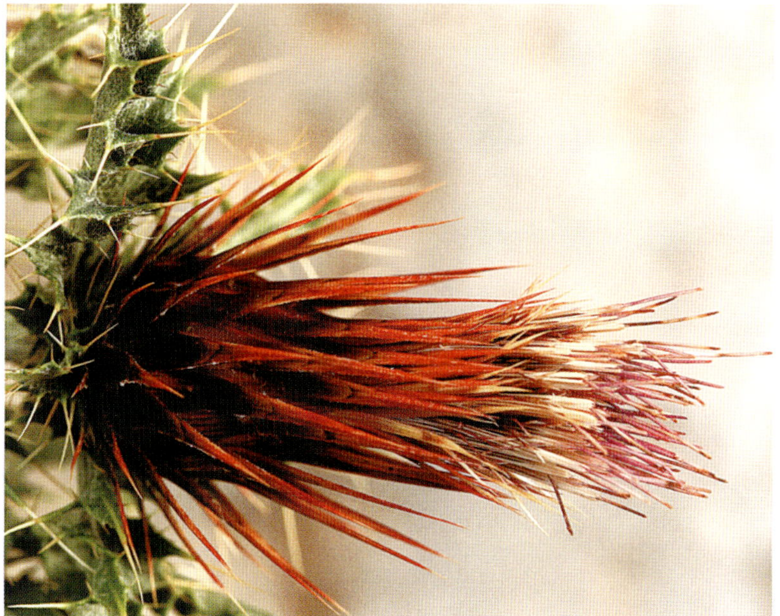

Turner Thistle

Cirsium turneri

Also known as cliff thistle, Turner thistle is a perennial herb growing in clumps up to eighteen inches high. This unusual herb hangs, more or less exclusively, from the walls of steep cliffs or limestone bluffs, or protrudes from rock fissures and crevices. Up to thirty pendulous, unbranched stems grow from the thistle's woody crown. The dense leaves, perhaps four to six inches long, are deeply lobed and spiny at the tips and along the margins. At the base, they tend to clasp the stem.

Primarily in very late spring and summer, flower heads supported by very spiny bracts appear at the stem tips. Each rayless head is a collection of sixty to eighty tiny disk flowers crowded into a bell-shaped floral cup. Each deep reddish-purple, slender blossom is divided near the tips into five linear lobes.

Turner thistle is dispersed across Brewster, Terrell, and Val Verde counties, often in relatively isolated canyons near the Rio Grande, and throughout adjacent Chihuahua and Coahuila, Mexico. It is considered vulnerable globally and in Texas.

Barton Warnock discovered this thistle in Upper Pine Canyon in 1950, but it was not rediscovered in the park until recently. I have encountered this rarity high in Maple Canyon in the Chisos Mountains and in Passionflower Canyon in the Dead Horse Mountains. It has also been confirmed at a few other locations (the Rosillos Mountains, Boquillas Canyon, and Mesa de Anguila).

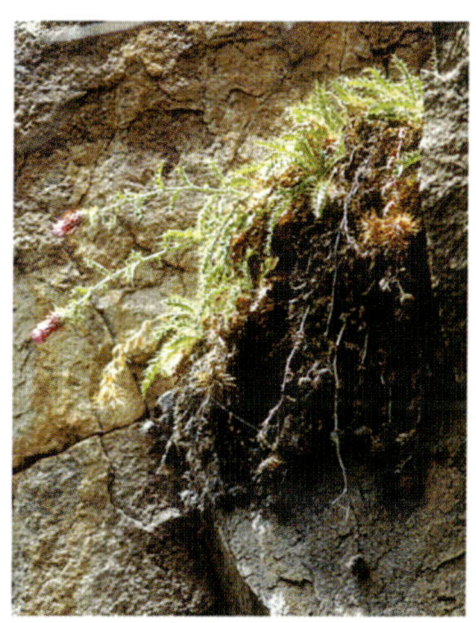

Note: The genus name *Cirsium* is from the Greek *kirsion,* meaning "thistle," which is from *kirsos,* referring to "swollen vein," which the thistle was supposed to cure. This species was collected by Barton Warnock in 1948 in Doubtful Canyon of the Del Norte Mountains southeast of Alpine, Texas, and described by him in 1960. Warnock named the thistle for his former botany student and friend, B. L. Turner, an Asteraceae and Fabaceae specialist, professor of systematic botany, and director of the Plant Resources Center at the University of Texas at Austin.

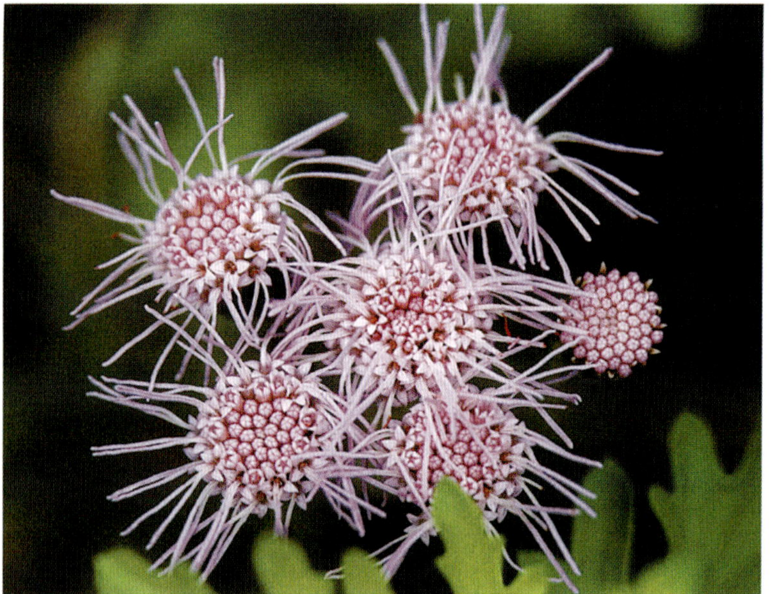

Palmleaf Thoroughwort

Conoclinium dissectum

PALMLEAF THOROUGHWORT, or mistflower, is a low, spreading perennial herb about two feet high with many erect but often spindly stems. Under favorable conditions, this herb can form large lavender patches many feet across. The short-stemmed leaves are broadly egg-shaped, often more than ten inches long and just as wide, and deeply cleft into three large lobes, which are themselves highly dissected and toothed.

Blooming from spring to fall, the flower heads appear in hemispheric clusters up to two inches wide atop long, leafless stalks. A single head contains up to eighty tiny disk florets, each with five purple or lavender toothlike lobes. Pale purple, thread-like styles, much longer than the flowers, protrude from the tiny blooms. This herb is also known as mistflower because expanses of it in bloom are reminiscent of the ethereal color of mist in early morning light.

Palmleaf thoroughwort thrives in moist mountain sites and wooded canyons and on open grassy slopes. In Texas it is most prevalent in the Trans-Pecos (from Hudspeth County east to Pecos and south to the park), although it occurs south and east to Medina County and in part of the south Texas brush country. Its range includes southern New Mexico, southeastern Arizona, and northern Mexico (from Coahuila to Nuevo León and south to San Luis Potosí).

In the park this perennial is easiest to find in the Basin of the Chisos and on the Pinnacles and Laguna Meadow trails.

Note: The genus name *Conoclinium,* from the Greek *konos* (cone) and *klinion* (little bed), refers to the cone-shaped flower heads (receptacles) bearing the florets. The species name *dissectum* refers to the dissected leaves. Asa Gray described this species in 1852, from a specimen collected by Charles Wright in 1849. *Eupatorium greggii,* described by Gray in 1884 and named for Josiah Gregg, is another botanical name for this plant.

Tarbush

Flourensia cernua

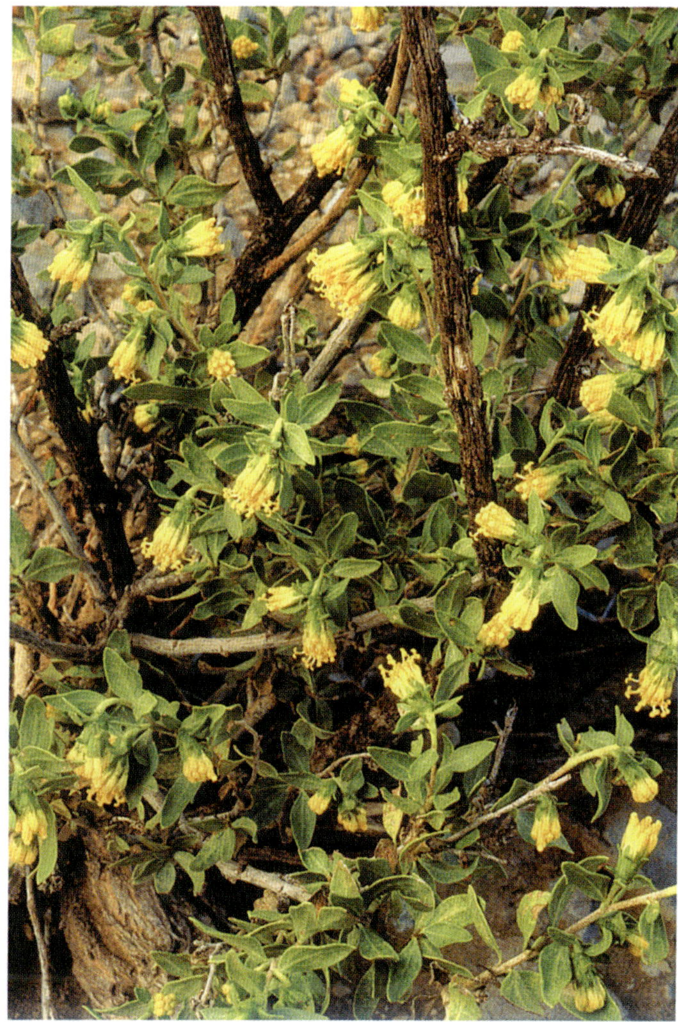

Also known as hojasen or hojase, blackbrush, and varnish bush, tarbush is a leafy, much-branched shrub three to six feet tall. Covered with grayish-black bark, the stems often seem burned. The yellowish-green, somewhat shiny leaves exude a gluey, resinous substance that smells like tar. One-half to one inch long, on short stems, these elliptic or egg-shaped leaves have a bitter taste unpalatable to livestock.

Nodding flower heads are clustered among the leaves, stalkless or on very short stalks. They lack ray flowers but contain numerous yellow and rather sticky, disk flowers. This shrub produces small, flattened, one-seeded fruits with soft shaggy hairs. In Mexico, the leaves and dried flowers are sold as hojase, a treatment for indigestion.

Tarbush, a major component of the desert scrub community, is an indicator plant of the Chihuahuan Desert. Interspersed with creosote, it dominates large stretches of low desert throughout the Trans-Pecos. Flowering from September to December, it is also distributed throughout southern New Mexico, southeastern Arizona, and northern Mexico (from Chihuahua to Nuevo León south to Zacatecas, San Luis Potosí, and Hidalgo).

In the park tarbush is a frequent sight from low desert flats to the Chisos. Large stands of this dense shrub are easily accessible at Dagger Flat, Dog Flats near the Rosillos Mountains, along Oak Creek in the Chisos Basin, and especially from Sam Nail Ranch to the west slopes of the Chisos.

Note: The genus *Flourensia* is named for Marie Jean Pierre Flourens, a French physiologist and pioneer in anesthesiology. The species name *cernua* means "nodding," referring to the rayless flower heads. Both the genus and species were described by Augustin de Candolle in 1836 from a specimen collected near Monterrey, Mexico, by naturalist Jean Louis Berlandier. Sent by de Candolle to collect plants for the Mexican Boundary Commission, Berlandier from 1827 to 1831 shipped fifty-two thousand specimens back to Switzerland. Berlandier's manuscript describing his travels was translated in 1980 as *Journey to Mexico during the Years 1826–1834.*

Chisos Mountain Brickellbush

Flyriella parryi

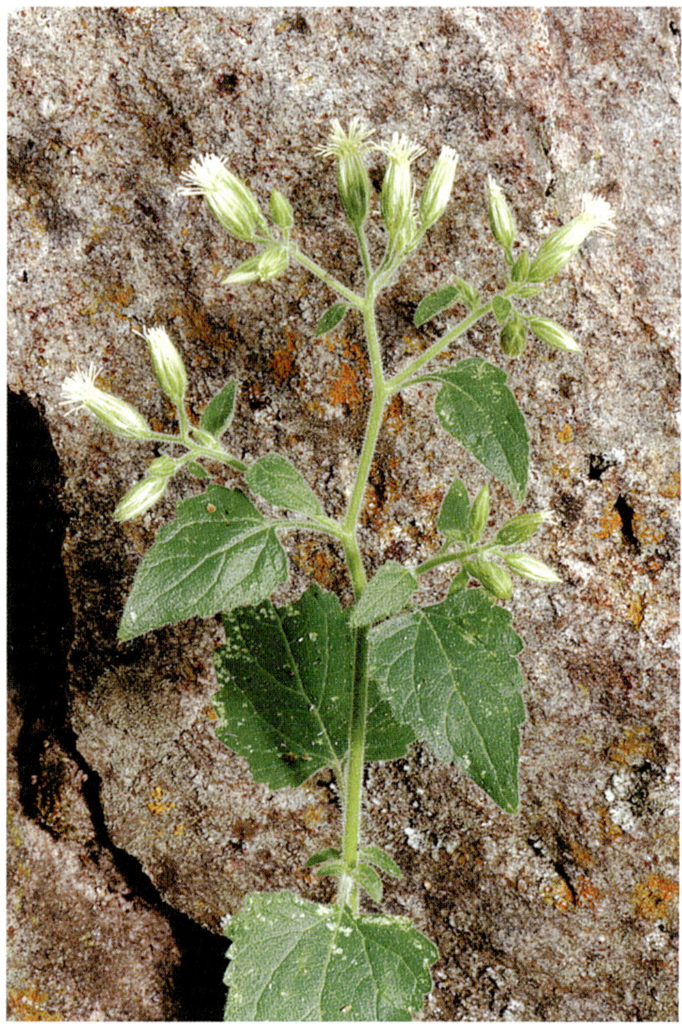

ALSO KNOWN AS Shiner's brickellbush, this brickellbush is a leafy herb less than two feet high, usually with several hairy-sticky upright stems branching at the base. Triangular to broadly egg-shaped, the coarsely serrated leaves are thin and pointed at the tips, on stalks (petioles) sometimes as long as the blades. Up to two and one-half inches long and rarely longer, the lower blades are paired, the upper often alternate along the stems.

The leaves are far more noticeable than the small, pale yellow flower heads, which grow in sparse clusters above the leaves and in smaller groups along the stems. Blooming from spring to fall, this plant lacks ray flowers: each flower head is a collection of tiny, tubular disk flowers that may look more like bristles or tiny brushes than blooms.

In the United States, this rare member of the sunflower (Asteraceae) family is known only from rocky slopes and shady canyons of the Chisos and various locations in Val Verde County (for example, Eagle Cave and Seminole Canyon State Historical Park). It is also found in northern Mexico, in the mountains of eastern Chihuahua east to Nuevo León. Globally, the herb is classified as vulnerable throughout its range, and in Texas it is imperiled.

In the Chisos, Chisos Mountain brickellbush has been documented in Upper Green Gulch; in Pine, Maple, and Pulliam canyons; on the slopes of Lost Mine Peak; and on the east side of Casa Grande. The photograph was taken in Maple Canyon in late August.

Note: This species was described by Asa Gray in 1859 (as *Eupatorium parryi*) from specimens collected on the U.S.-Mexico boundary survey. Gray named the plant for Charles Parry, a survey surgeon and naturalist. Parry did the bulk of his collecting for the survey in California and is best known for discovering the Torrey pine near San Diego. Robert King and Harold Robinson, Asteraceae specialists with the U.S. National Herbarium, placed this plant in their newly defined genus *Flyriella* in 1972. The genus is named in honor of the late University of Texas *Brickellia* specialist David Flyr, who had described this plant in 1968 as *Brickellia shineri.*

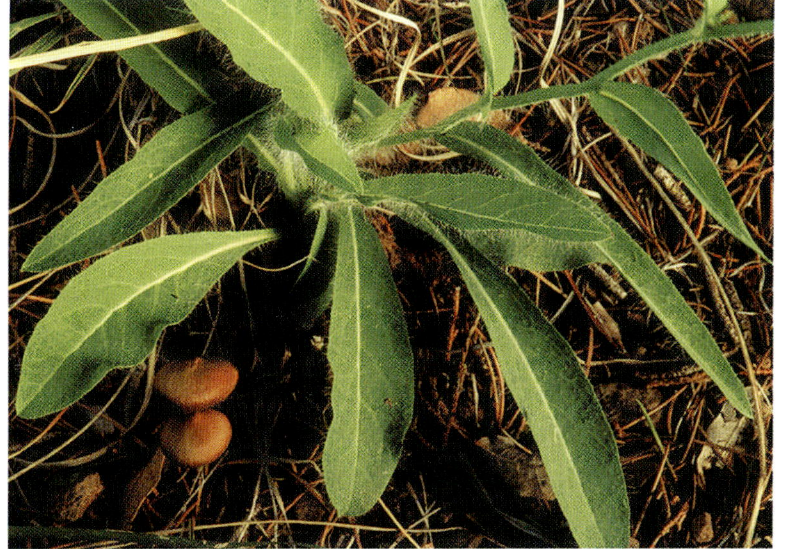

Roughstem Hawkweed

Hieracium schultzii

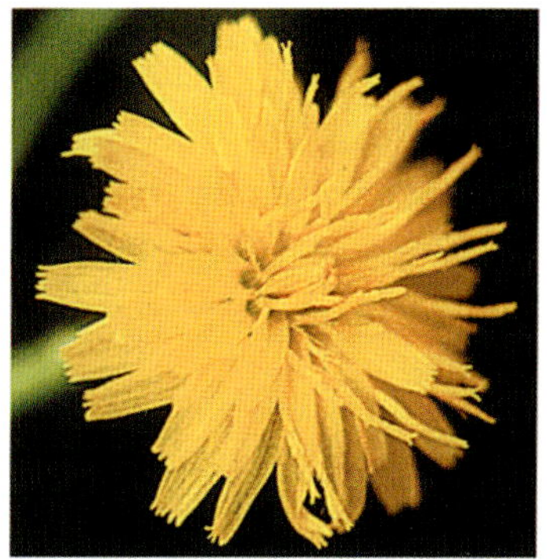

ROUGHSTEM HAWKWEED is a perennial herb one to three feet high, usually with a single sturdy but hollow stem. Sometimes branching in the upper reaches, the stem is cloaked with hairs and sticky bristles. A rosette of half a dozen or more long, oblong leaves, typically broader in the upper half and tapering near the base, hugs the ground at the base of the stem. Smaller oblong or lance-shaped leaves line the stem, and all leaves are covered with shaggy hairs.

The large yellow flower heads grow at the tips of upper branches. They consist of many narrow, overlapping yellow rays, each tipped with five tiny teeth. This robust perennial usually blooms in late summer and early fall. The fruit is a brownish-black, ten-ribbed, columnar achene.

Roughstem hawkweed inhabits mountain woodlands and canyons, often in shaded, more protected areas and in igneous soils. In the United States, this hawkweed is limited to the higher mountains of Trans-Pecos Texas (in Brewster and Jeff Davis counties), but it is much more widespread in Mexico (Baja California, Chihuahua, and Coahuila and as far south as Oaxaca, Chiapas, and Guatemala).

In the park roughstem hawkweed prospers at higher reaches of the Chisos Mountains. Recorded in Upper Green Gulch and along the Lost Mine Trail, it can also be plentiful at Laguna Meadow and around Boot Spring, along the Emory Peak Trail, and in Upper Cattail Canyon.

Note: The genus name *Hieracium* is probably derived from the Greek *hierax* (hawk), presumably because of the ancient belief that hawks ate the sap to improve their eyesight. This species was described in 1862 by Swedish botanist and hawkweed specialist Elias Fries and named for German Asteraceae specialist Carl Heinrich Schultz-Bipontinus. Also known as *Hieracium wrightii,* the species had been collected near Mexico City by Johann Schaffner, a German pharmacist who emigrated to Mexico in 1856; he became a dentist and botanist in Mexico City.

Edwards Nicollet

Nicolletia edwardsii

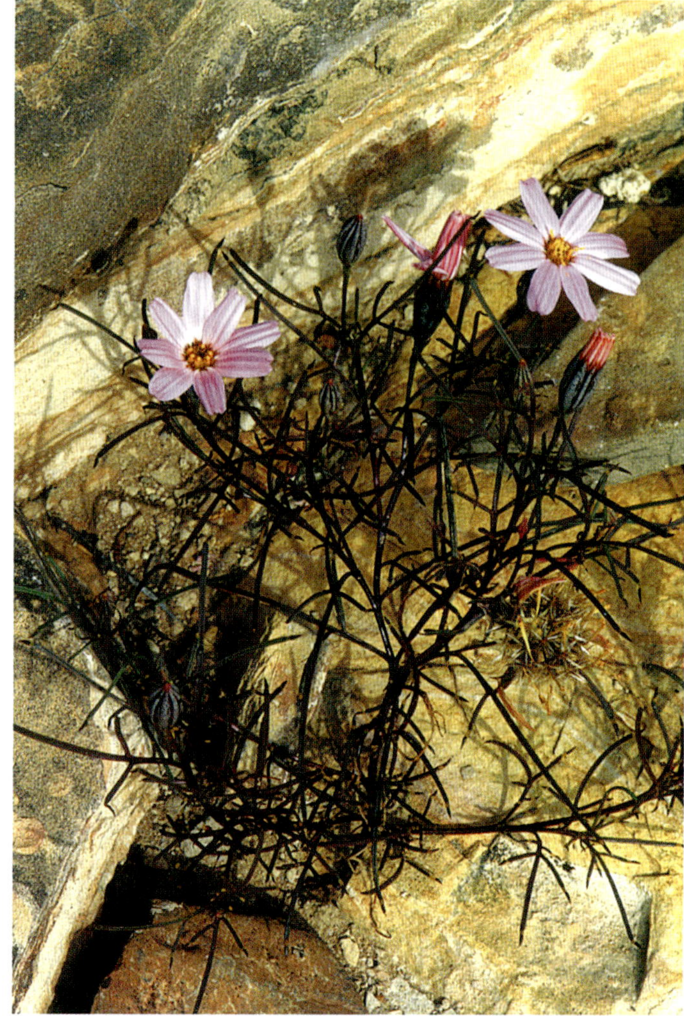

Edwards Nicollet, or Edwards' hole-in-the-sand plant, is a small annual seldom over six inches high with wiry, branching, often rust-colored stems and dark green leaves. When not in bloom, it may resemble a jumble of dark wires. Up to two inches or more in length, the highly dissected leaves have threadlike lobes dotted with transparent oil glands. These glands secrete a strongly scented, oily substance that some consider unpleasant.

For such a flimsy annual, this member of the sunflower (Asteraceae) family produces disproportionately large and attractive flower heads with fifteen to twenty-five yellow disk florets and usually eight white to pink ray flowers. Tinged with lavender or purple, toothed at the tips, and sometimes blanketing the stems, the flowers can be quite endearing.

Edwards Nicollet ranges across the Trans-Pecos near the Rio Grande from southern Hudspeth to Brewster County and south through northern Mexico to Durango and Zacatecas.

Although considered rare in Trans-Pecos deserts, this herb is not rare in the park. Edwards Nicollet has been observed at Dog Flats, and it often brightens the desert on the east side of the park from Hot Springs to Glenn Spring and on the west along Old Maverick Road. Usually described as flowering in late summer and fall, it may just as easily pop up in spring, in glorious bloom after a good rain. I photographed this wiry annual blooming near Hot Springs and above Glenn Draw in March.

Note: This genus is named for Joseph Nicollet (1786–1843), the French astronomer and geologist who immigrated to the United States in 1832 after losing his fortune in the stock market. Nicollet is best known for leading a U.S. Army survey of lands between the Mississippi and Missouri rivers. The species was described by Asa Gray in 1852 in *Plantae Wrightiana.* It had been collected by Dr. L. A. Edwards in Mexico in 1846, near the site of a famous U.S.-Mexico War battle, the Battle of Monterrey. Edwards botanized in Arkansas, Texas, and Mexico and in 1854 joined the Pacific Railroad surveys.

Guayule

Parthenium argentatum

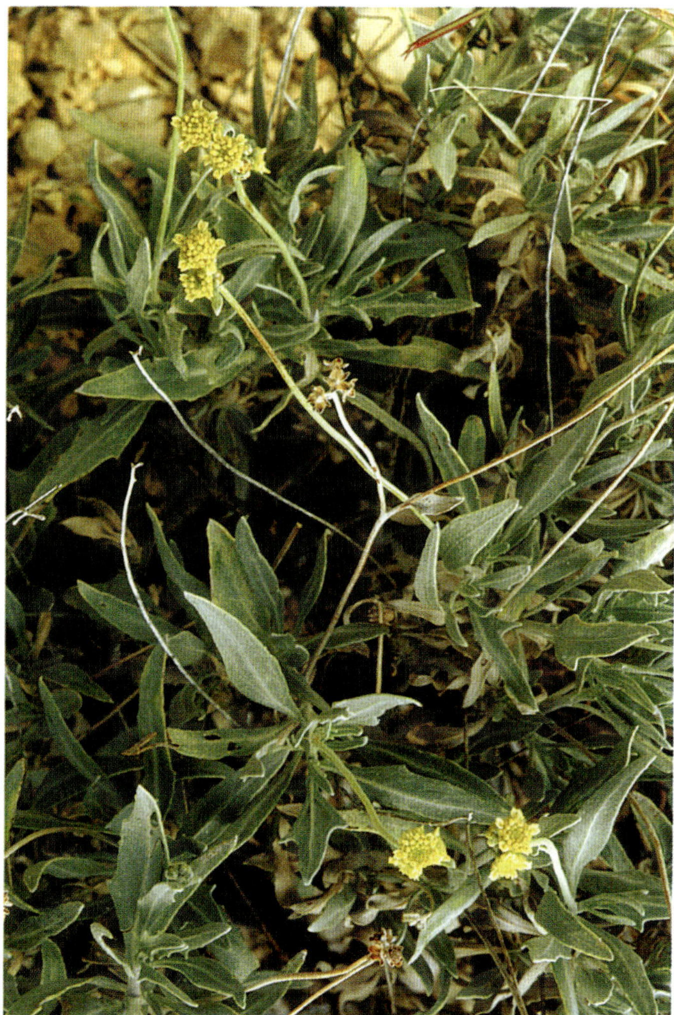

The most notable feature of this densely branched shrub, which stands about two feet high, is its silvery appearance. The color is produced by fine, hoary hairs that blanket the narrow, pointed, and often toothed leaves.

Clusters of small, yellowish-cream flower heads, with short ray and many disk flowers, rise above the silvery leaves at the tips of long, bare stalks.

Also known as rubber plant, guayule is a source of high-quality rubber, found mostly in the bark. Efforts to extract rubber from guayule have continued intermittently since 1892. A Marathon, Texas, factory produced guayule rubber during World War I. During World War II, more than four hundred tons of guayule rubber were produced in southern California. More recently, the resinous by-products of guayule rubber extraction have been used as wood preservatives. Apparently the role of the resinous latex in nature is to protect the guayule plant from insects and other living things out for a meal.

At one time a characteristic Big Bend species on low limestone hills, this shrub was severely overharvested for rubber. Guayule is now locally common at disjunct locations south of Alpine, near Fort Stockton, and in the Dead Horse Mountains. Its habitat crosses the Trans-Pecos (Pecos to Presidio) into northern Mexico, and some large populations still exist there.

In the park guayule plants remain in hills above Dagger Flat and in foothills of the Rosillos Mountains. You might see this silvery shrub off Old Ore Road near Muskhog Spring. It blooms almost anytime, especially after good summer rains.

Note: The genus name *Parthenium,* an old name for Asteraceae with white ray florets, may be from the Greek *parthenos* (girl, virgin) possibly because only the female ray florets are fertile. The species name *argentatum* means "silvery." This species was described by Asa Gray in 1859 from a specimen taken during the Emory boundary survey by John Bigelow.

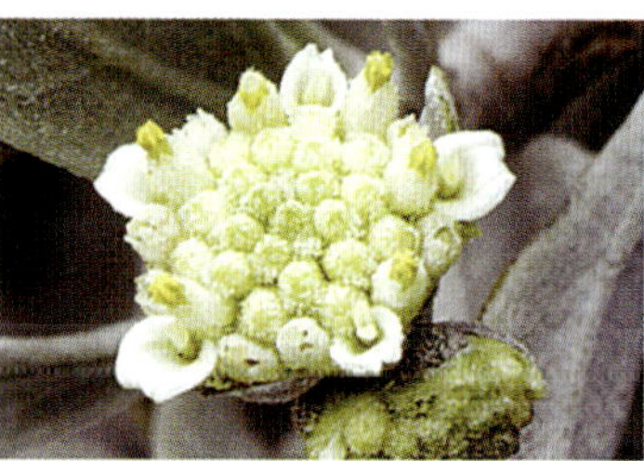

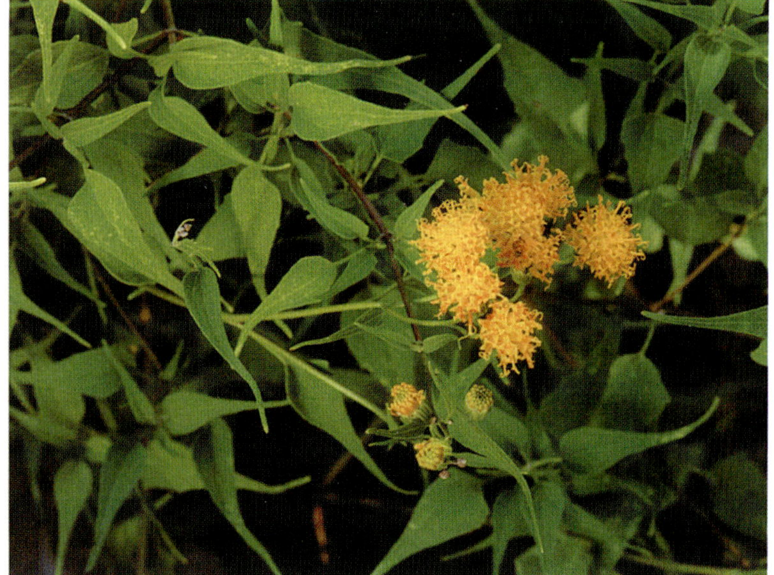

Tailleaf Pericome

Pericome caudata

ALSO KNOWN AS taperleaf or mountain tailleaf, tailleaf pericome is a perennial aromatic herb with an acrid or sharp odor and many arching or freely spreading branches. Reaching three feet or more, the stems are covered with tiny, soft hairs. This shrubby herb is named for the long tail-like tips of its tapering, triangular leaves.

Arranged in terminal clusters, the flower heads lack ray flowers but contain twenty to forty yellow disk florets. With a magnifying glass, you can see that each minute disk floret consists of a short tube with an abruptly expanded, five-lobed border.

Tailleaf pericome is sometimes known as rockslide daisy because it grows on steep talus slopes among rocks and boulders. This leafy herb has also acquired the nickname yerba de chivato (herb of the male goat) because of its goatlike scent. It blooms from late summer to fall in the Chisos, Davis, and Guadalupe mountains of the Trans-Pecos.

This perennial is spread across the southwestern United States from western Texas to southeastern California, north to parts of southern Nevada and Colorado and the Oklahoma Panhandle, and south into Chihuahua, Coahuila, and Durango, Mexico.

In the park tailleaf pericome is sparse, largely restricted to Upper Green Gulch, the east slope of Casa Grande, and the igneous rock slide at the top of Mount Emory. Search for it in late September and early October.

Note: The leaves and roots have been used to treat rheumatism. The genus name, from the Greek *peri* (around) and *kome* (hair), refers to the hairs around the edges of the fruit. The species name *caudata* means "tailed." This species was collected by Charles Wright during the Graham boundary survey in 1851 and described by Asa Gray in 1853.

Rayless Rockdaisy

Perityle aglossa

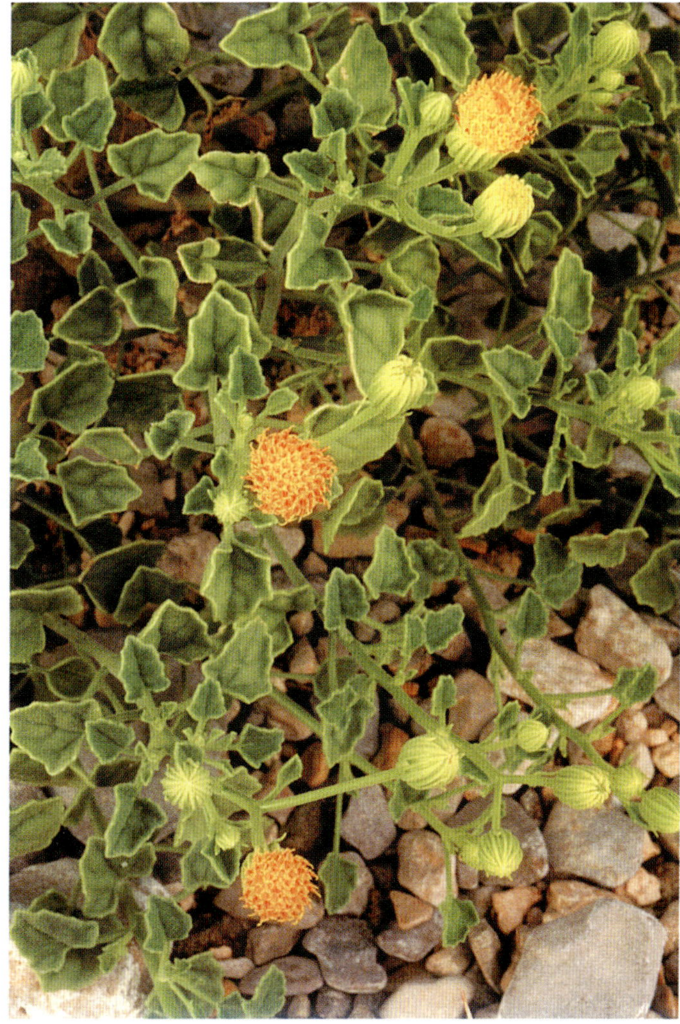

ALSO KNOWN AS rayless perityle, this small shrub is seldom much over a foot tall. The leafy, somewhat rounded plant has triangular to heart-shaped leaves up to one and one-half inches long. Both opposite and alternate, the blades are toothed and often shallowly and irregularly lobed.

Rayless rockdaisy is named for its rayless flower heads consisting exclusively of disk flowers. Usually solitary, each bell-shaped head contains many tiny yellow, tubular florets, which may have a pink or purple blush. The flattened, oblong fruit is an achene (small, dry, one-seeded, with the ovary wall free from the seed) with thick, hairy edges, crowned with scales and a single bristle.

This scarce rockdaisy is confined to Brewster and Terrell counties in the United States and the adjacent state of Coahuila in Mexico. In Terrell County, it hangs from rock cliffs in the Lower Canyons of the Rio Grande.

Blooming in summer and fall, rayless rockdaisy, like many other rockdaisies, fills pockets and fissures in barren limestone bluffs. In the park this rockdaisy has been located at the tunnel near Rio Grande Village and at a site about four miles west of Hot Springs. I chanced upon it in October in Passionflower Canyon of the Dead Horse Mountains. There the plants line the wash running through the canyon and also droop from the rocky canyon walls.

Globally and in Texas, rayless rockdaisy is classified as vulnerable to extinction throughout its range.

Note: The genus name *Perityle*, from the Greek *peri* (around) and *tyle* (knob, callus), describes the calloused edges of the fruit (achene). The first *Perityle* plant was collected on the around-the-world voyage of the HMS *Sulphur* and described by the English botanist and famous taxonomist George Bentham in 1844. The species *aglossa*, meaning "tongue-less," or by implication "rayless," was collected by Charles Parry at Arroyo San Carlos, Mexico, in 1852 during the U.S.-Mexico boundary survey and described by Asa Gray in 1853 in *Plantae Wrightiana*.

Twobristle Rockdaisy

Perityle bisetosa var. *scalaris*

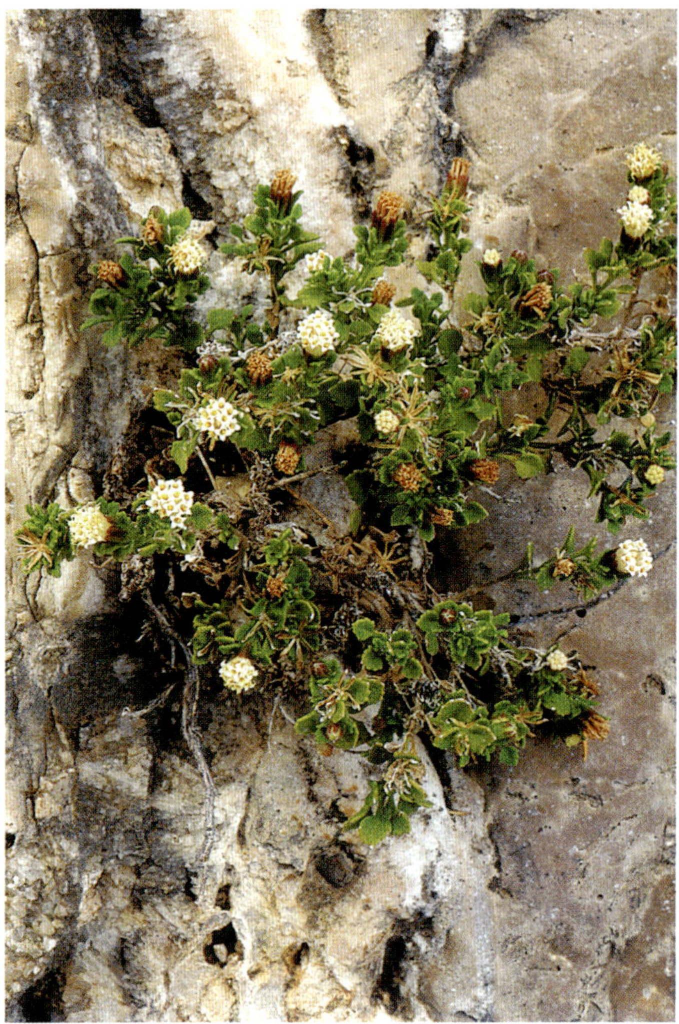

Twobristle rockdaisy is a dwarf shrub no more than five inches high and often much less. You might call it the Big Bend version of a bonsai. It forms tight, sometimes rounded clumps with many stems and tiny oval leaves one-third inch long. On distinct short stalks, the leaves are noticeably rough-hairy and mostly alternate with lobed or serrated edges.

The solitary flower heads are white and lack ray flowers: they are a collection of miniature disk flowers shaped something like funnels. Try looking at them with a magnifying glass. The fruit is small, dry, and one-seeded with a crown of usually two flat, linear bristles at the top.

Four varieties of twobristle rockdaisy have been recognized: three from Trans-Pecos Texas and one from Coahuila, Mexico. According to A. M. Powell (2004), an authority on *Perityle,* these varieties have gained a foothold at widely separated locations and have only minor differences. All four almost exclusively are rooted in small cracks and crevices in limestone rock. They bloom from spring to fall.

Also known as stairstep rockdaisy, this variety was originally discovered by Barton Warnock and described in 1967 by A. M. Powell of Sul Ross State University in Alpine. It has been known exclusively from the Black Gap Game Refuge in Brewster County (Stairstep Mountain and Big Brushy Canyon north of the park).

Twobristle rockdaisy was not identified in the park until 2004. I came across these miniature rockdaisies in early October in Passionflower Canyon of the Dead Horse Mountains, east of the Ernst Basin backcountry campsite.

Twobristle rockdaisy is ranked as critically imperiled globally and in Texas.

Note: The species name *bisetosa* implies "two-bristled." The variety name *scalaris* (stairs) refers to Stairstep Mountain at Black Gap, where this variety was first collected. Although the variety was not collected until 1965, the *bisetosa* species was collected by Charles Parry in 1852, during the U.S.-Mexico boundary survey, and described by Asa Gray the next year as part of the genus *Laphamia.*

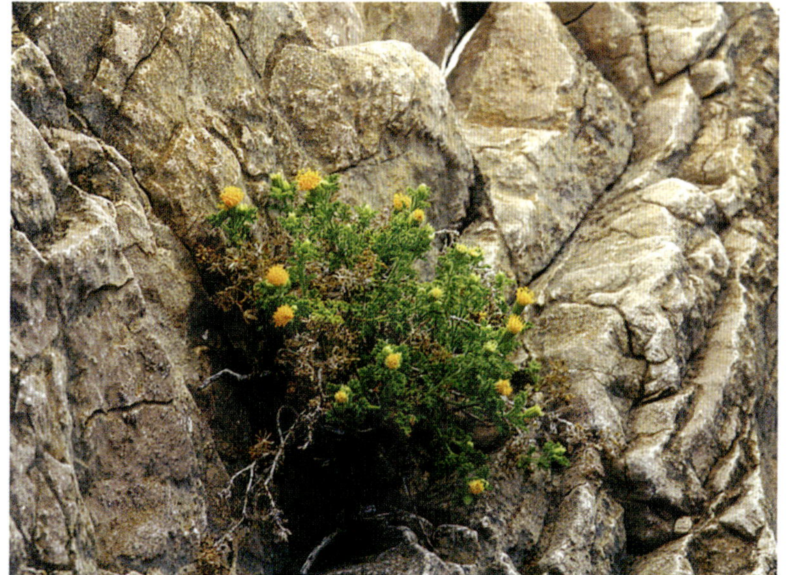

Slimlobe Rockdaisy

Perityle dissecta

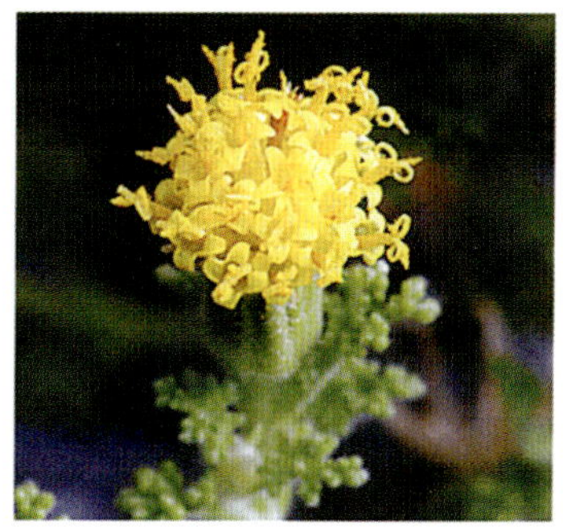

SLIMLOBE ROCKDAISY is a perennial herb or dwarf shrub up to eight inches high with short woody stems. Densely hairy, the stems form compact leafy mounds that hug the sheer rock walls out of which they grow. The leaves are no more than one inch long and half as wide, broadly egg-shaped in outline, and highly and irregularly dissected, divided (pinnately) one to three times into smaller segments. Noticeably curled along the edges, covered with fine hairs and dotted with glands, the divided blades sometimes grow in pairs or alternately along the stems.

From spring to fall, yellow flowers appear in disklike heads up to three-eighths inch long. Solitary or in small clusters at the tips of branches on their own short sturdy stalks, the heads lack ray flowers but contain twenty to thirty tiny, four-lobed disk florets. Black at maturity, the fruit is an oblong, flattened achene (dry and one-seeded) with thickened edges, capped with usually one sturdy bristle and a rudimentary crown of scales.

This woody miniature sits atop large boulders and fills the crevices of limestone rocks on steep canyon walls. It is isolated in a small portion of the Big Bend (in southern Brewster and southern Presidio counties) in the United States and the neighboring state of Chihuahua in Mexico.

In the park slimlobe rockdaisy has been confirmed in Santa Elena Canyon, the Rosillos foothills, and the Chisos, on Pulliam Bluff, and in the Basin. With fewer than twenty known occurrences globally and in Texas, it is classified as imperiled and very vulnerable to extinction throughout its range.

Note: The species name *dissecta* refers to the dissected leaves. New York botanist John Torrey described this species (as *Laphamia dissecta*) in 1853 from a specimen collected by John Bigelow the previous year. In 1884 Asa Gray placed this species in the genus *Perityle.*

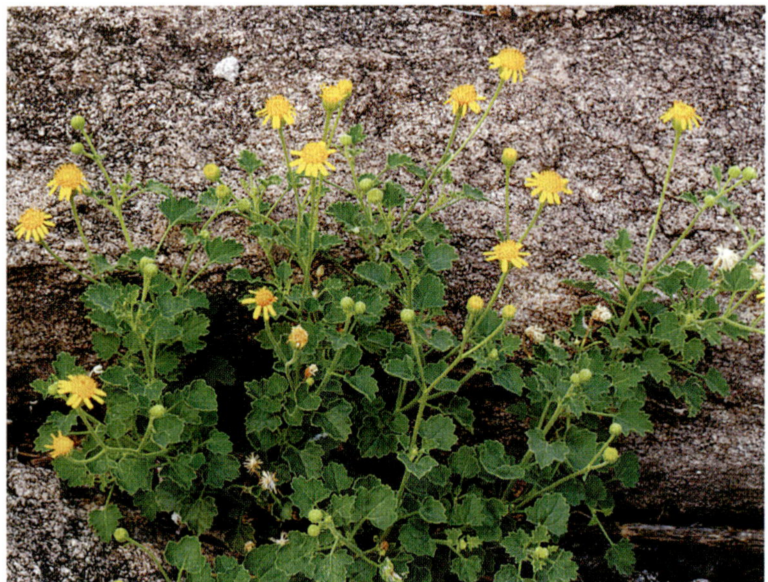

Heartleaf Rockdaisy

Perityle parryi

Heartleaf rockdaisy, or Parry's rockdaisy, is an extremely variable subshrub as much as two and one-half feet high but often growing low to the ground from narrow rock crevices or fissures in boulders or widely spreading in gravelly soils. The heart-shaped leaves are also variable, but most have very rounded teeth and are shallowly three-lobed. The long-stemmed blades may be opposite or alternate, usually about one to two inches long and more than half as wide.

The yellow flower heads form singly or in small clusters on stalks that may reach three inches or more. Each hemispheric head consists of twelve to sixteen narrowly oblong ray flowers and many tiny, tubular disk florets. They bloom from spring to fall. The small, blackish fruits (achenes) are crowned with scales and usually a single bristle.

Very circumscribed in distribution, this rockdaisy is an uncommon resident of the Big Bend country and is classified as vulnerable in Texas (NatureServe 2005). It is present only in Brewster and Presidio counties and the adjacent Mexican state of Chihuahua.

In the park you might notice heartleaf rockdaisy in Santa Elena Canyon, lodged in rock pockets at the base of canyon walls, or on top of Burro Mesa, sprawled across rocky and gravelly arroyos. It has also been reported in the Chisos Mountains, Green Gulch, the Basin, and along the Window Trail. This subshrub may hybridize with margined rockdaisy (*Perityle vaseyi*).

Note: Like rayless and twobristle rockdaisy, this species was collected by Charles Parry along the Rio Grande in Mexico in 1852, during the U.S.-Mexico boundary survey. It was described by Asa Gray the following year, in his *Plantae Wrightiana,* which was mainly an account of the plants collected by Charles Wright.

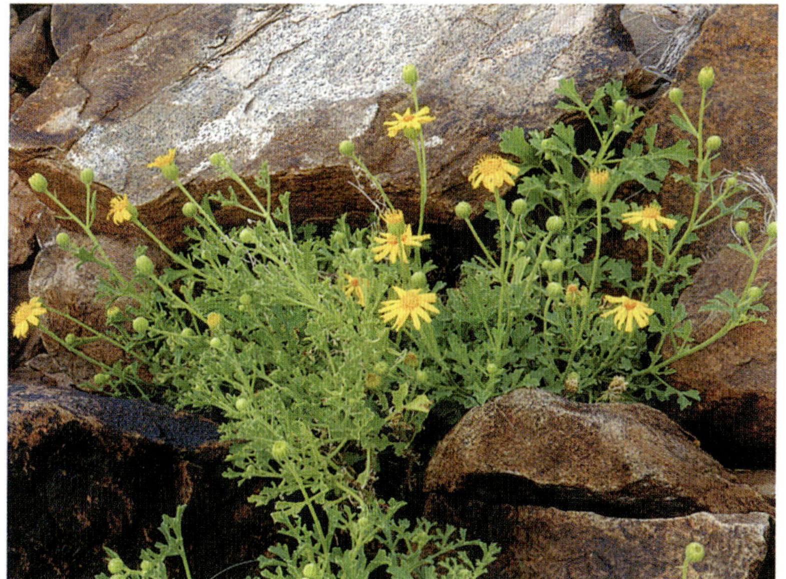

Margined Rockdaisy

Perityle vaseyi

MARGINED ROCKDAISY, or Vasey's rockdaisy, is a perennial herb or low subshrub, usually no more than one foot high, with ascending, often very leafy green stems. On conspicuous long stalks, the leaves are somewhat cross-shaped, deeply divided into three wedgelike segments, with each segment usually cut again into three distinct lobes. They may be covered with sticky hairs and exude a "clammy" odor.

The yellow flower heads form singly or in small clusters on long stalks. Each bell-shaped head may have fourteen to sixteen narrowly oblong ray flowers and very many tiny, funnel-shaped disk florets. Up to one inch wide, they bloom from spring to fall.

This perennial is well suited to sparse and eroded desert habitats, especially on gypseous clay soils, but it also flourishes in disturbed areas and along roadsides and reaches to mountain foothills and rocky slopes. Although quite extensive locally, margined rockdaisy occupies only a small part of the Big Bend (in Brewster and Presidio counties) and adjacent Mexican states (Chihuahua and Coahuila).

This rockdaisy is by far the most common and widely scattered *Perityle* in the park. It is prominent at numerous locations, especially to the west and south of the Chisos (Gano Spring, Chisos Pens, Maverick Badlands, Castolon) and in and near the Dead Horse Mountains to the southeast (Hot Springs, the tunnel, Tornillo Creek).

Note: This species was described by John Coulter in 1890 from a specimen collected in the Chisos Mountains in 1889 by Texas botanist Greenleaf Nealley. Coulter, University of Chicago botanist and founder of the *Botanical Gazette,* described a collection of plants by Nealley "from Brazos Santiago to El Paso." Coulter named the species *vaseyi* for George Vasey, the first botanist for the U.S. Department of Agriculture and curator of the U.S. National Herbarium until his death in 1893.

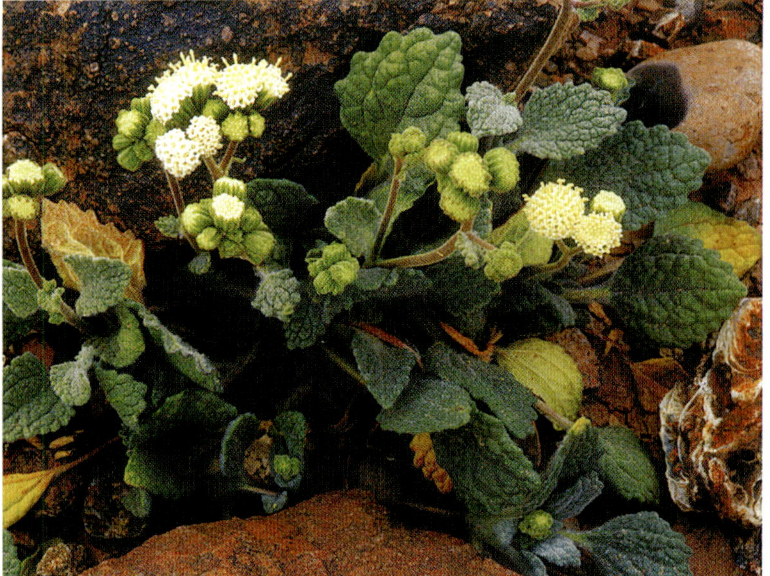

Naked Brittlestem

Psathyrotes scaposa

Naked brittlestem, or naked turtleback, is a low, weak annual about eight inches high, with mostly basal leaves and a few stout, leafless stems slightly branched above. Egg-shaped or triangular, the crowded leaves have long stems, scalloped edges, and wrinkled, bumpy surfaces. They are covered with soft, woolly hairs and often coated with a scaly crust that attracts sand and dirt.

Each of the "brittlestems" supports several flower heads. The pincushionlike heads lack ray florets and consist of tiny five-toothed disk flowers crowded on the "pincushion." Blooming from spring to fall after good rains, the slender disk florets are pale yellow to white.

This herb occurs throughout the Chihuahuan Desert portions of Trans-Pecos Texas and southern New Mexico and in adjacent Chihuahua, Mexico. It often colonizes large stretches near the Rio Grande from southern Hudspeth to Brewster County. A similar plant, desert turtleback or desert velvet, frequents the Mojave and Sonoran deserts. All these plants have bumpy leaves that resemble a turtle's shell.

Naked brittlestem can take over yellow clay soils on the park's west side, from Santa Elena Canyon to the Rattlesnake Mountains to the Maverick Badlands. It can also be abundant on clay soils at Dog Flats south of Persimmon Gap, near the Rio Grande from San Vicente to Glenn Spring, and in the Chisos foothills.

Note: The genus name *Psathyrotes* is derived from the Greek *psathyros* (brittle), referring to the stems. The species name *scaposa* (with a scape) refers to the leafless stalks. This species was described by Asa Gray in 1853 from a specimen collected by Charles Wright in 1852 on "stony hills above El Paso."

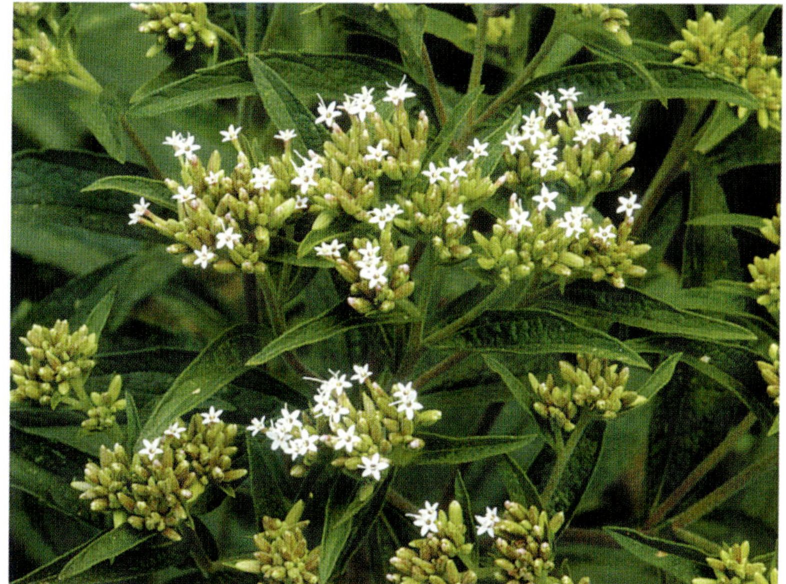

Roundleaf Stevia

Stevia ovata var. *texana*

ROUNDLEAF STEVIA, or Texan candyleaf, is an erect, often shrubby perennial herb two to four feet high that grows from a single stout stem sparsely branched in its upper half. It has large hairy leaves up to two and one-half inches long. Nearly stemless, the elongated, lance-shaped leaves are tapered near the tip and slightly scalloped on the upper edges.

Small white flowers form in compact clusters above the leaves. Ray flowers are lacking, and each flower head consists of only a handful of tiny florets crowded into a disklike floral cup. The fruit is an achene (a hard, dry, one-seeded fruit).

Roundleaf stevia is situated in mountain woodlands and on open rocky slopes, often in protected canyons or near springs and intermittent streams. Variety *texana* is limited to higher elevations in the Chisos and Chinati mountains in the Big Bend and to northeastern Coahuila in Mexico. It is classified as critically imperiled in Texas (NatureServe 2005). South of the border, four varieties of *Stevia ovata* have a much broader range throughout Coahuila and the Mexican Highlands into Central America.

In the park look for roundleaf stevia in Pine Canyon; on the Chinese Wall, Laguna Meadow, and Boot Spring trails; and in Upper Cattail Canyon.

Note: In 1797, Antonio José Cavanilles, director of the Madrid botanical gardens, named the genus *Stevia* for Pedro Jaime Esteve (1500–1556), author of a work on herbs and medicinal plants and a leader of the humanist movement at the University of Valencia. This species was collected by Baron von Humboldt and Aimé Bonpland near Mexico City in 1803 and described by German botanist Carl Kunth in 1820 (as *Stevia rhombifolia*). Variety *texana,* described by Jerold L. Grashoff in 1974, was collected in 1961 by Donovan Correll and Marshall Johnston on Lost Mine Peak in the Chisos.

Viscid Stevia

Stevia viscida

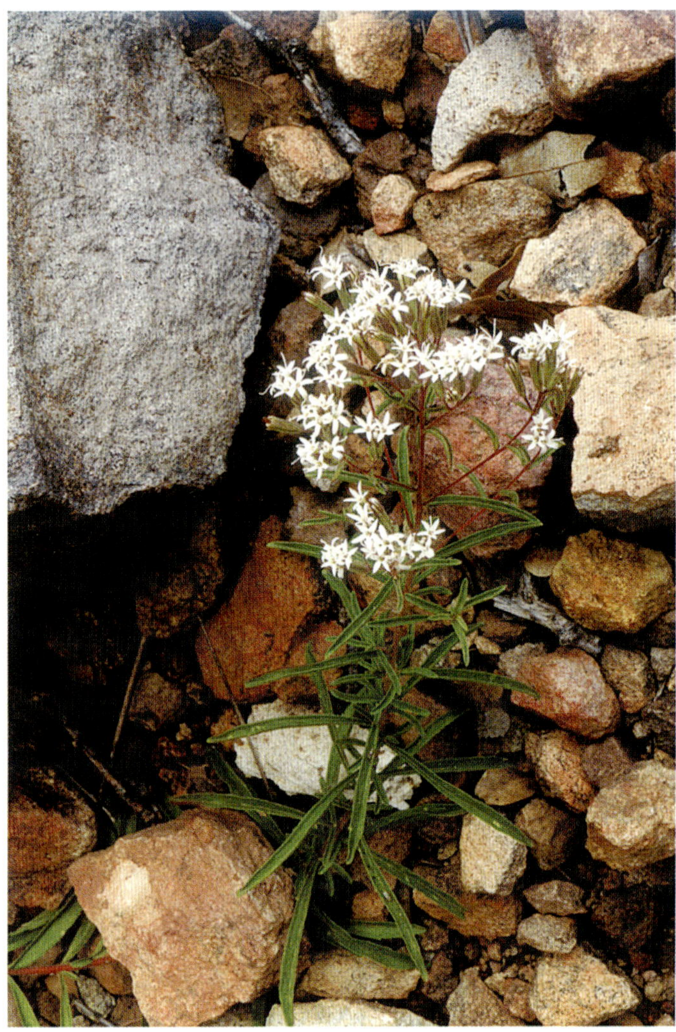

VISCID STEVIA, or viscid candyleaf, is a perennial herb usually two to three feet high with a single slender maroon stem (or occasionally several stems). Although thin, the rounded, hairy stems are mostly straight and erect. Narrow linear leaves, one to three inches long, alternate up the stem or crowd together in small bundles. The stalkless blades are blunt at the tips, relatively smooth along the edges, and dotted with oily glands.

The small flowers are bunched in brilliant white bouquets as much as eight inches across. Individual flower heads forming the bouquet lack ray flowers: one small head contains about five funnel-shaped flowers, each five-toothed at the tip and crowded into a cylindrical disk three-eighths inch long. Threadlike white styles protrude conspicuously from the hairy flower tubes.

In Texas this herb is rare, known exclusively from the Chisos Mountains, but it is more wide ranging elsewhere, in southern Arizona and primarily Mexico (Sonora east to northeastern Coahuila, and south to Jalisco and Guatemala). Viscid stevia fares well from grasslands to woodlands, on open rocky slopes, and in forested canyons near intermittent streams.

In the park this attractive perennial appears at higher elevations of the Chisos. It usually blooms in September and October near the entrance to Laguna Meadow and in Upper Cattail Canyon. This photograph was taken on the Emory Peak Trail.

Note: A related species (*Stevia rebaudiana*) was used before the Spanish Conquest by the Guarani Indians of Paraguay to sweeten maté. First used commercially in 1908, stevia is now popular worldwide as a sugar substitute. The species name *viscida* (viscid) refers to the glutinous leaves and stems. This species was described in 1820 by German systematist Carl Kunth and collected by Baron von Humboldt and Aimé Bonpland in the Mexican state of Michoacán. Kunth spent eight years (1813–20) describing Humboldt and Bonpland specimens, and then in 1829 he sailed to America on his own three-year expedition to South and Central America and the West Indies.

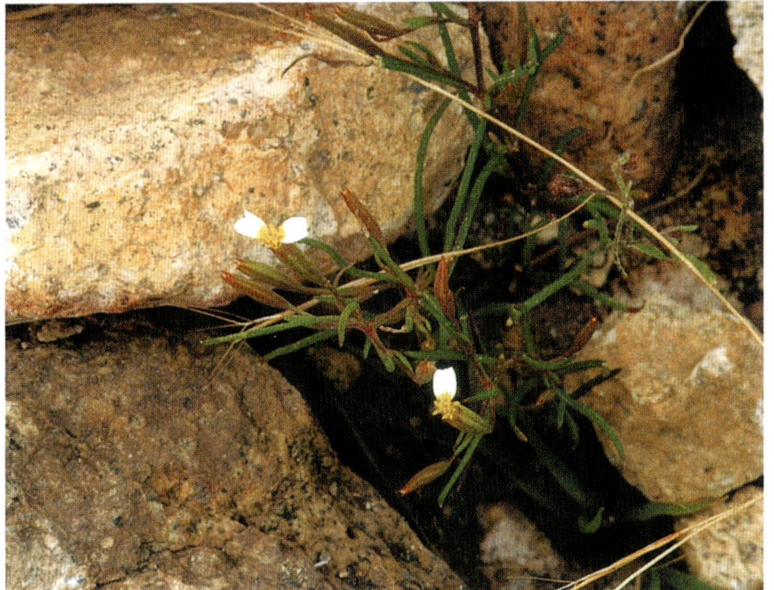

Licorice Marigold

Tagetes micrantha

SELDOM OVER six inches high, licorice marigold is one of those intriguing dwarf annuals that make the Chisos Mountains such a botanical treasure trove. Clear, microscopic oil glands cover the slender linear leaves as well as the flowers and exude a distinct aroma of licorice, which attracts attention to this otherwise easily overlooked herb.

The tiny flower heads sit atop slender but erect branches. They usually consist of one or two short white ray flowers protruding horizontally from a cluster of about six yellow disk flowers. They sometimes resemble a yellow baseball cap with a large white brim.

In Texas this delicate annual is mainly concentrated in the Chisos and Davis mountains, but its wide range extends into New Mexico and southern Arizona and deep into Mexico (as far south as Guerrero).

Licorice marigold blooms high in the Chisos from July to October in wetter but open areas. It is typically found near Boot Spring, on the South Rim Trail, and in Upper Cattail Canyon. I photographed this charming dwarf on Mount Emory in September.

A related species with much larger yellow flowers, Mexican tarragon (*Tagetes lucida*), is well known in the southern United States as a spice and tarragon substitute. Licorice marigold has a similar flavor and is sometimes mentioned in Mexican cookbooks.

Note: In 1753 the Swedish botanist Linnaeus named this genus for the Etruscan god of wisdom, Tages. The species name *micrantha* means "small-flowered." Licorice marigold was described in 1797 by Antonio Cavanilles, director of the Royal Botanic Gardens in Madrid. The specimen was cultivated from the seed of a plant discovered by Luis Née in Querétaro, Mexico. Née was a botanist on the Malaspina Expedition (1789–94) to Spanish colonies around the world.

Woolly Dogweed

Thymophylla micropoides

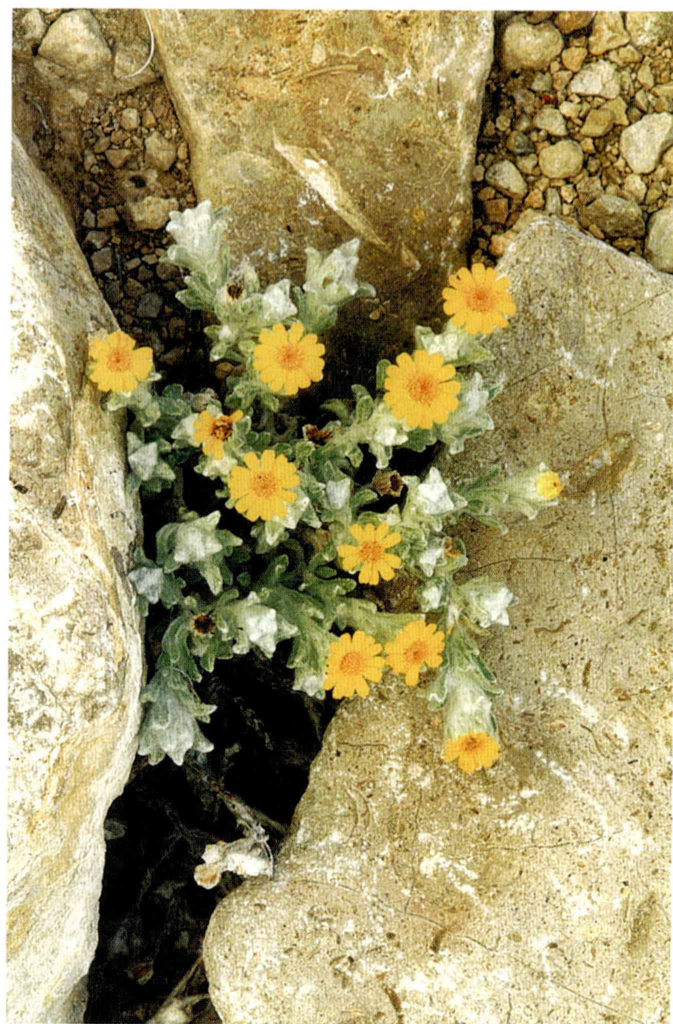

Woolly dogweed is a weak annual only two to six inches high with intensely woolly stems. The spatula-shaped, alternate leaves, only about one-half inch long, are bunched along the stems and almost hidden in the soft, cobwebby hairs. The undersides of these leaves are dotted with shiny, sometimes strongly scented oil glands that are covered by the wool and largely invisible.

Yellow flower heads one-half inch wide form at the tips of the stems. The "flowers" are actually collections of flowers: each top-shaped head has perhaps a dozen bright yellow ray flowers and about sixty tiny, darker yellow to orange disk florets. Like many dogweeds, woolly dogweed may bloom intermittently throughout the year.

Low limestone hills of desert mountains are the preferred habitat of this ephemeral annual. Woolly dogweed may approach its western limit in the park. It is most prevalent in the southeastern Trans-Pecos, from Brewster County to Del Rio, but it is also encountered farther east near Leakey and south to Eagle Pass on the Rio Grande. This annual also grows in parts of northeastern Mexico (from Coahuila east to Tamaulipas).

In the park woolly dogweed has been documented at Harte Ranch by the Rosillos Mountains, but it is much more frequent east of the park in the Black Gap Wildlife Management Area. This photograph was taken in the hills above Ernst Tinaja in early September.

Note: The genus name *Thymophylla* is derived from the Greek *thymos* (thyme) and *phyllon* (leaf), referring to the thymelike leaves. The species name *micropoides* means "like the genus *Micropus*." This species was described in 1838 (as *Gnaphalopsis micropoides*) by Augustin de Candolle and collected by Jean Louis Berlandier near Monterrey, Mexico. Berlandier served from 1827 to 1829 as botanist for the Mexican Boundary Commission and then collected on his own in northern Mexico and southern Texas. He also studied Native American tribes and authored *The Indians of Texas* in 1830.

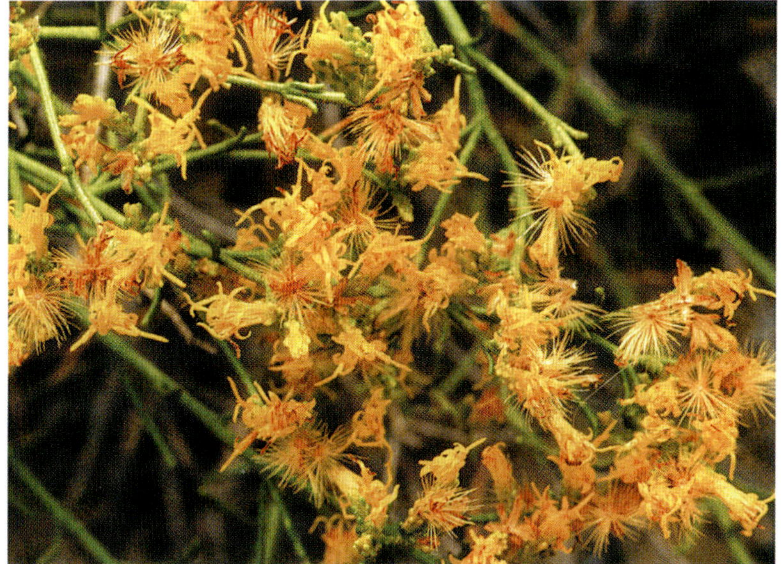

Threeflower Goldenweed

Xylothamia triantha

THREEFLOWER GOLDENWEED, or Terlingua stickbush, is a sticky, branched shrub no more than three feet high with faintly aromatic stems. Up to about three-fourths inch long, the leaves are linear or very narrowly cylindrical in shape and resin-coated.

The shrub is known as threeflower goldenweed because of its golden yellow (also commonly pale yellow) flowers. The flower heads lack rays and usually contain three disk flowers, but the number may vary (to as many as seven). Since even the bracts supporting the flower heads are sticky, Terlingua stickbush may be the more apt name.

The fruits are light brown, five-ribbed, hairy achenes (dry, one-seeded, and indehiscent) with stiff brownish bristles that, together with the pasty blooms, give the plant a distinctly disheveled appearance. The flowers and fruits seem to form a gooey mass, like yellow cotton candy.

This bush is centered in northern Mexico (in parts of Chihuahua to Nuevo León and also Durango), entering Texas only in desertic southwestern Brewster County. It is classified as critically imperiled in Texas (NatureServe 2005). Blooming from August to October, it spills across gyp-clay deposits near Study Butte, Terlingua, and Lajitas.

In the park threeflower goldenweed inhabits the floodplain between Santa Elena Canyon and Castolon. It has also been observed at Tornillo Flat and near Hot Springs. I photographed these shrubs in September near Terlingua Abaja.

Note: The genus name *Xylothamia,* from the Greek *xylon* (wood) and the genus name *Euthamia,* emphasizes its woody nature and resemblance to *Euthamia* species. The species name *triantha* means "three-flowered." This species was collected by Barton Warnock in the Chisos area in 1937 and described (as *Haplopappus trianthus*) in 1938 by U.S. Department of Agriculture botanist Sidney Blake. Blake documented many new species and coauthored the *Geographical Guide to Floras of the World* (1942, revised 1961). This shrub was placed in the new genus *Xylothamia* by botanist Guy Nesom (and others) in 1990. For many years it was known as *Ericameria triantha.*

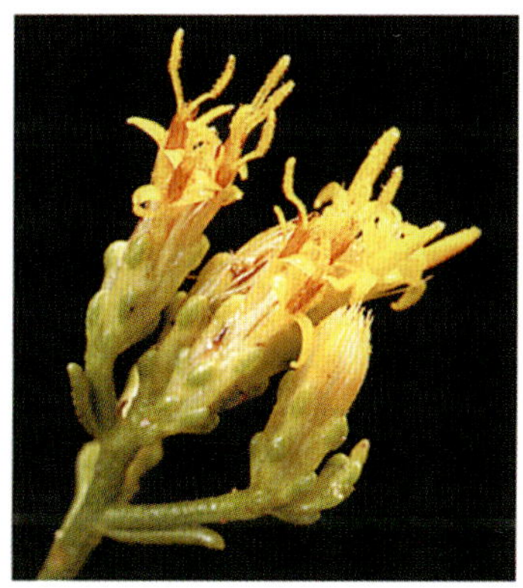

Big Bend Hophornbeam

Ostrya virginiana var. *chisosensis*

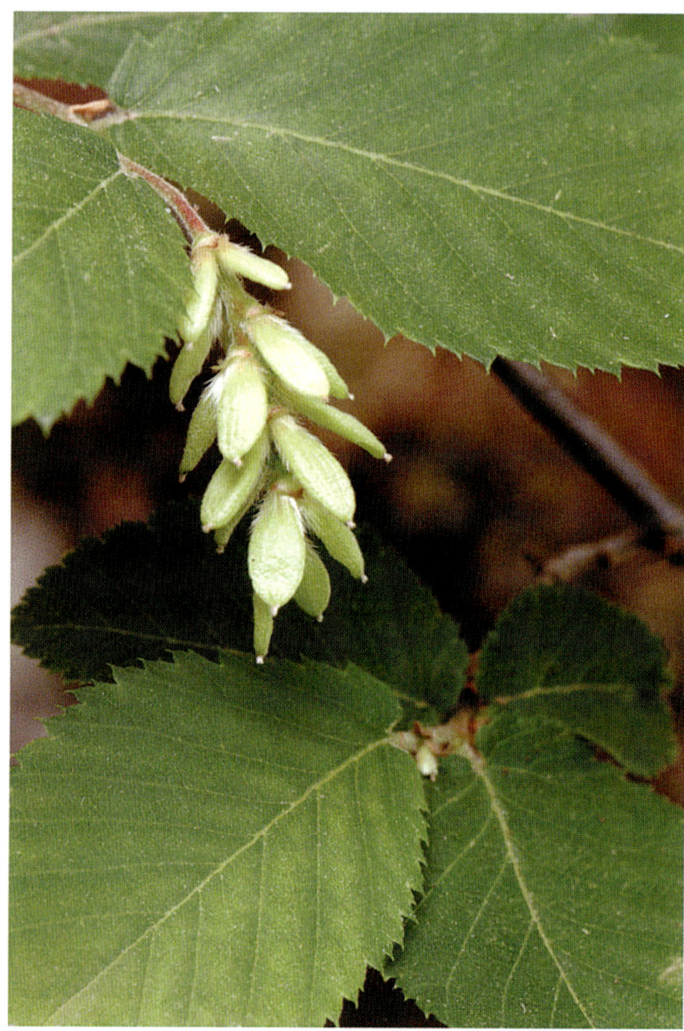

Big Bend hophornbeam is a small to medium-sized tree twenty to thirty feet high with round, slender branches and very hard wood. This member of the birch (Betulaceae) family is easily distinguished by the shaggy appearance of its light gray or brownish bark, which is often very scaly and flaking. The alternate leaves are easy to spot as well, mostly broadly elliptic in shape on short stems and very finely toothed along the edges. They may be slightly hairy and marked with conspicuous, straight veins.

Each tree has separate male and female flower clusters, which often appear before the new leaves. The male flower clusters (catkins), about one and one-half inches long, are thin and cylindrical, drooping in groups of one to three from the branch tips. Female flowers appear in solitary, spike-shaped clusters at the ends of new branchlets. The fruit is a tiny nutlet enclosed in a much larger bladdery sac. About three-fourths inch long, the densely hairy, flattened sacs form cone-shaped or hoplike clusters at the tips of new shoots.

Scarce in shady, relatively moist rocky habitats, this hophornbeam is restricted to higher elevations of the Chisos Mountains in Big Bend National Park and the Sierra Maderas del Carmen in adjacent Coahuila, Mexico. It is ranked as imperiled globally and critically imperiled in Texas, with only one to five known occurrences.

In the park Big Bend hophornbeam has been located on Emory Peak, Toll Mountain, and Crown Mountain and high in Pine Canyon, but it may be most accessible along the Pinnacles Trail.

Note: The genus name *Ostrya* is "hophornbeam" in Latin, perhaps from the Greek *ostryos* (scale), referring to the scaly fruits. Variety *chisosensis* was originally described by Donovan Correll in 1965 as a separate species. More recently, it has been treated by Chihuahuan Desert flora specialist James Henrickson as a variety of *O. virginiana.* Correll, coauthor with Marshall Johnston of the *Manual of the Vascular Plants of Texas,* collected this plant in 1964 high in the Chisos Mountains, at the base of north-facing ledges on Emory Peak.

Desert Willow

Chilopsis linearis ssp. *linearis*

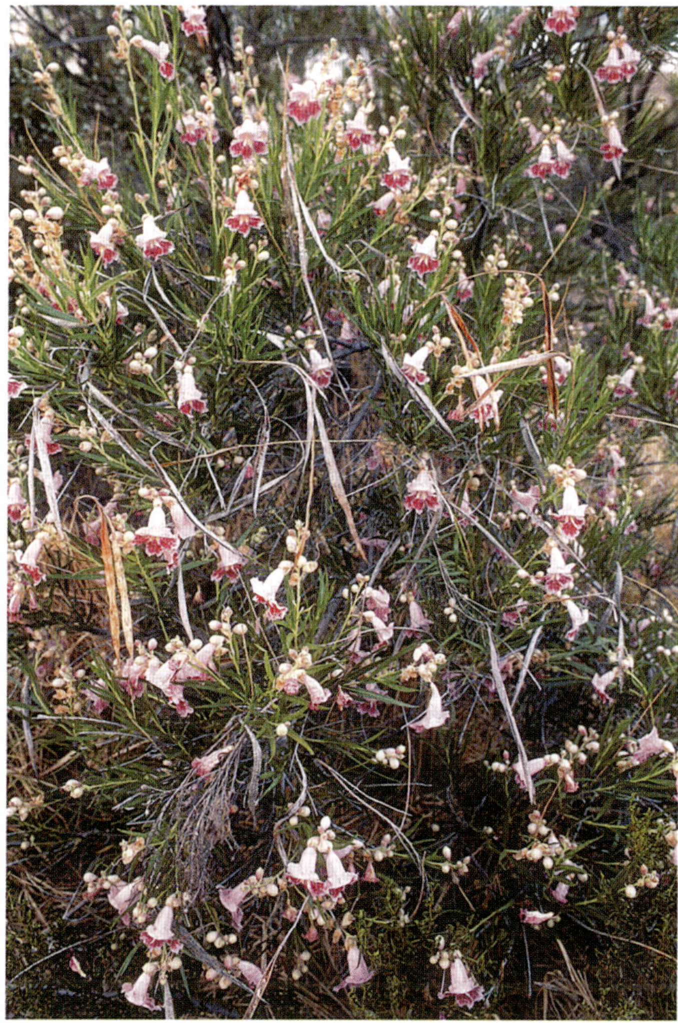

Also known as mimbre, flor de mimbre, and desert catalpa, desert willow is an upright shrub up to twenty-five feet high with dark gray, scaly bark and leaning trunks. About four inches long, the linear leaves are willowlike with smooth edges and elongated, pointed tips. Often covered with a sticky coating that seals in moisture, they may grow in small bundles radiating from a common point.

From May to July these trees produce dramatic displays of fragrant, funnel-shaped, multicolored flowers. Each two-lipped flower has a pink to white, two-lobed upper lip and a three-lobed lower lip in shades of pink, white, maroon-red, and yellow. The lower lobes have conspicuous maroon-red blotches and streaks, and the broad throat has two yellow, hairy ridges at the base. The fruits are slender, two-valved cylindrical pods up to one foot long with many winged seeds.

Desert willow prefers desert washes, sandy arroyos, streambeds, and springs. This species is conspicuous in the southwestern United States. In Texas it is plentiful in the Trans-Pecos and Hill Country but spottier elsewhere, and in Mexico it ranges from Baja California east to Tamaulipas and south to central Mexico. The subspecies in the park, with straight leaves, is limited to southern New Mexico, Trans-Pecos Texas, and northern Mexico.

In the park this attractive shrub is widespread, especially at lower elevations and near the Rio Grande. It has also been recorded near many park springs (Glenn, McKinney, Grapevine, Gano, and Hot Springs) and even in the Basin of the Chisos.

Note: The genus name *Chilopsis,* from Greek words *cheilos* (lip) and *opsis* (likeness), is a reference to the two-lipped flowers. The species name *linearis* refers to the linear leaves. Antonio Cavanilles described this species (as *Bignonia linearis*) in about 1796 from a specimen cultivated in Madrid from seeds brought from Mexico.

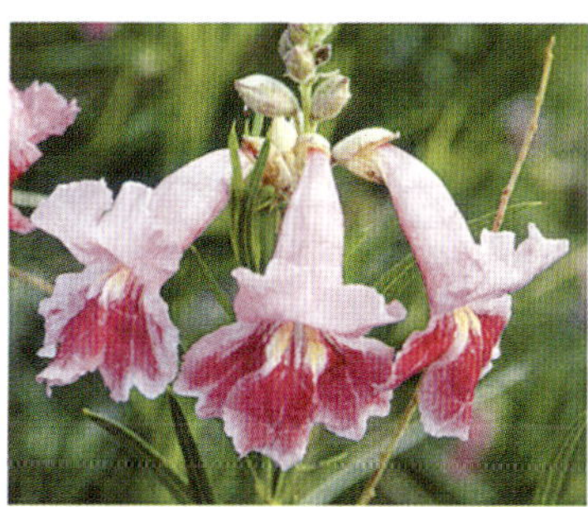

Trumpetflower

Tecoma stans

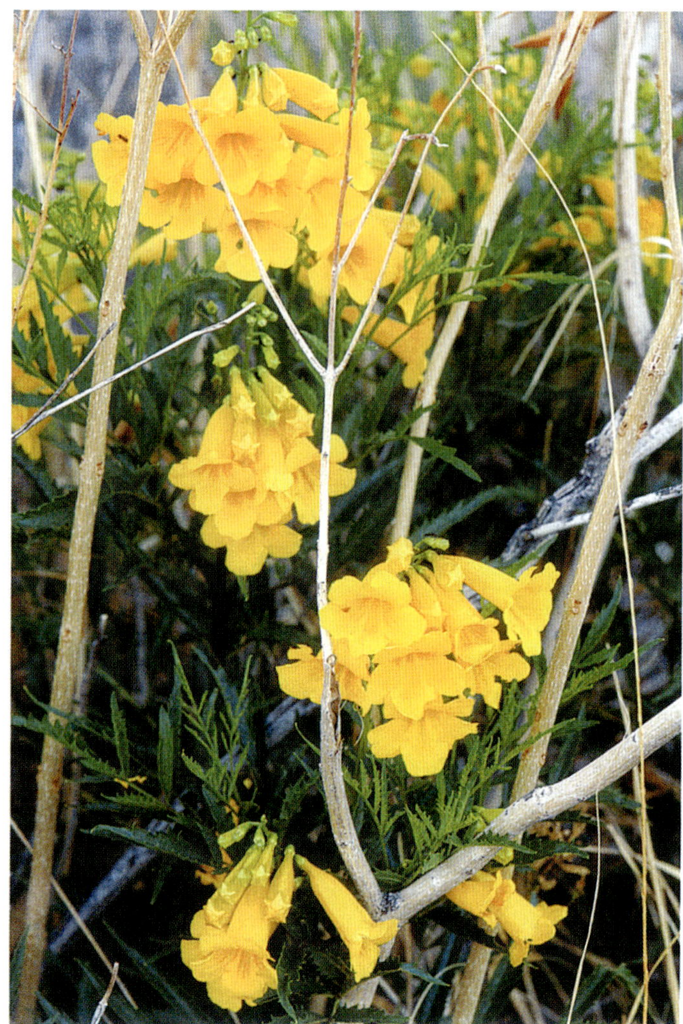

ALSO KNOWN AS tronadora, esperanza, and yellow bells, trumpetflower is an erect shrub three to six feet high with many slender stems. Up to nine inches long, the thin leaves are odd-pinnate: each leaf has one to five pairs of leaflets and a longer leaflet at the tip. Narrowly lance-shaped, the leaflets are hairless and smooth but sharply serrated on the edges.

Trumpet-shaped yellow flowers two inches long appear in broad clusters at the stem tips. The "trumpets" open into a two-lipped border, with a two-lobed upper lip and a larger three-lobed lower lip. Two distinct ridges extend from the lower lip into the flower's throat. As much as eight inches long, the slender fruit pods are tan and smooth.

Trumpetflower thrives in desert lowlands on sandy and gravelly hillsides of desert mountains and brushy arroyos and rocky slopes of higher mountains. In Texas the shrub is frequent throughout the Trans-Pecos (especially in Jeff Davis, Presidio, and Brewster counties) but infrequent in the southern Rio Grande Plains and elsewhere. Far-reaching outside of Texas as far west as Arizona and as far south as Argentina, it is also seen in Florida, Hawaii, and the Caribbean.

In the park this shrub blooms profusely from summer to fall near Sotol Vista and on Burro Mesa. Trumpetflower's colorful displays can be viewed in the Chisos foothills (between Alamo and Burro springs, near Paint Gap, and along Oak Creek) and across the park from the Maverick Badlands to Boquillas Canyon to the Dead Horse Mountains (at Telephone Canyon).

Note: The genus name *Tecoma* is from the Aztec word *Tecomaxochitl,* for this and other plants with vase-like flowers. The species name *stans* means "erect." This species was described by Swedish botanist Carolus Linnaeus (as *Bignonia stans*) in 1763. Charles Plumier, a French monk who became royal botanist to Louis XIV, probably first discovered this plant on early botanical expeditions to the French Antilles and Central America from 1689 to 1695. Plumier's greatest work, on American plants, was published posthumously in 1756. This species was reclassified as *Tecoma stans* by Carl Kunth in 1818. The intergrading variety in the park is *Tecoma stans* var. *angustata.*

Cory Cryptantha

Cryptantha palmeri

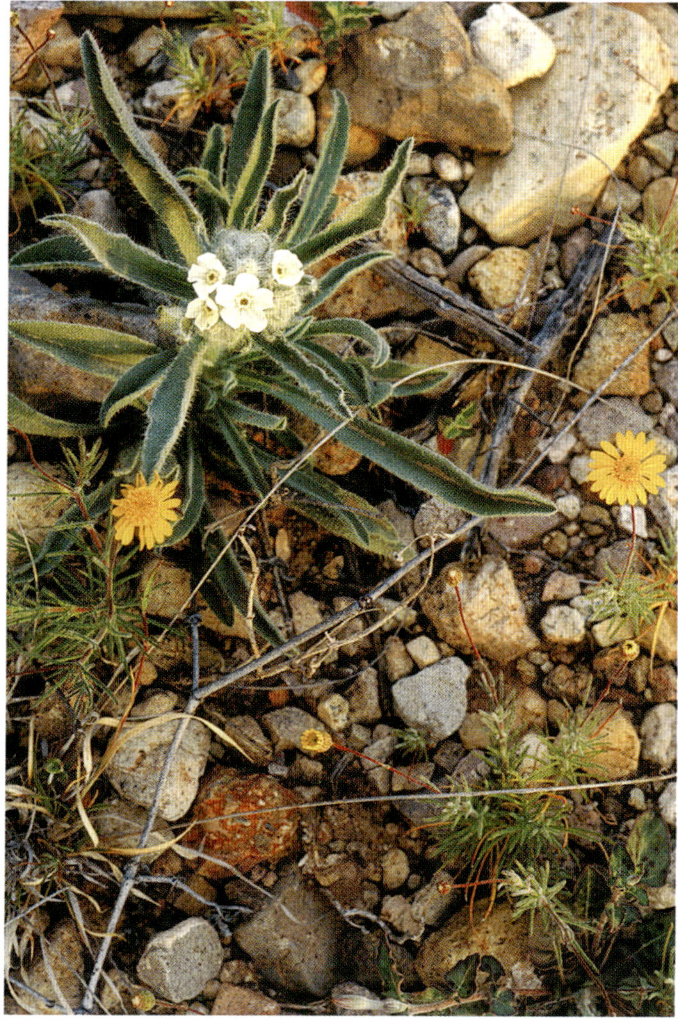

CORY CRYPTANTHA, or Palmer's catseye, is an erect biennial or short-lived perennial up to one foot high, usually with a single rigid stem but sometimes several branches. Covered with thin, appressed hairs and long, bristly, white hairs, this herb has a gray, shaggy appearance.

The long, narrow leaves at the base of this plant may reach six inches, but the oblong leaves along the stem are much shorter. Like the stems, the leaves are hairy, covered with white matted wool.

Small white flowers only one-fourth inch across appear in spring. They form on top of the curved stem, in coiled clusters that gradually unfurl as the fruits develop. Small clusters may form at the base of leaves as well. These pleasant five-lobed flowers may have a yellow ring of hairy outgrowths in the throat.

This herb occupies gravelly soil on limestone hills and rocky slopes, from the Rio Grande Plains to the Panhandle and throughout much of the Trans-Pecos to southern Hudspeth County. Populations reach west to the lower Pecos River basin of New Mexico and south into northern Mexico (Coahuila and Nuevo León).

Cory cryptantha has a wide but sparse distribution in the park. Blooming from February to July, it has been reported near Persimmon Gap at the north end of the park, on Mesa de Anguila in the extreme southwest, between Castolon and Smoky Creek in the south, and in the Dead Horse Mountains to the east. The easiest place to find this herb is near Government Spring and from there to the Grapevine Hills. I have also stumbled across this forget-me-not on the abandoned service road leading to Rough Spring.

Note: This genus was named *Cryptantha* from Greek words meaning "hidden flower" because many species are cleistogamous (self-fertilizing), with tiny flowers that rarely open. In 1939 Ivan Johnston named this species *Cryptantha coryi* for Southern Methodist University botanist Victor Cory, but it is now regarded as synonymous with an older species *Cryptantha palmeri.* Cory coauthored with Harris Parks the first *Catalogue of the Flora of Texas* in 1936. He and Parks, an apiculturist with the Texas Agricultural Experiment Station, collected this species in Brewster County in 1935.

Leafy Heliotrope

Heliotropium confertifolium

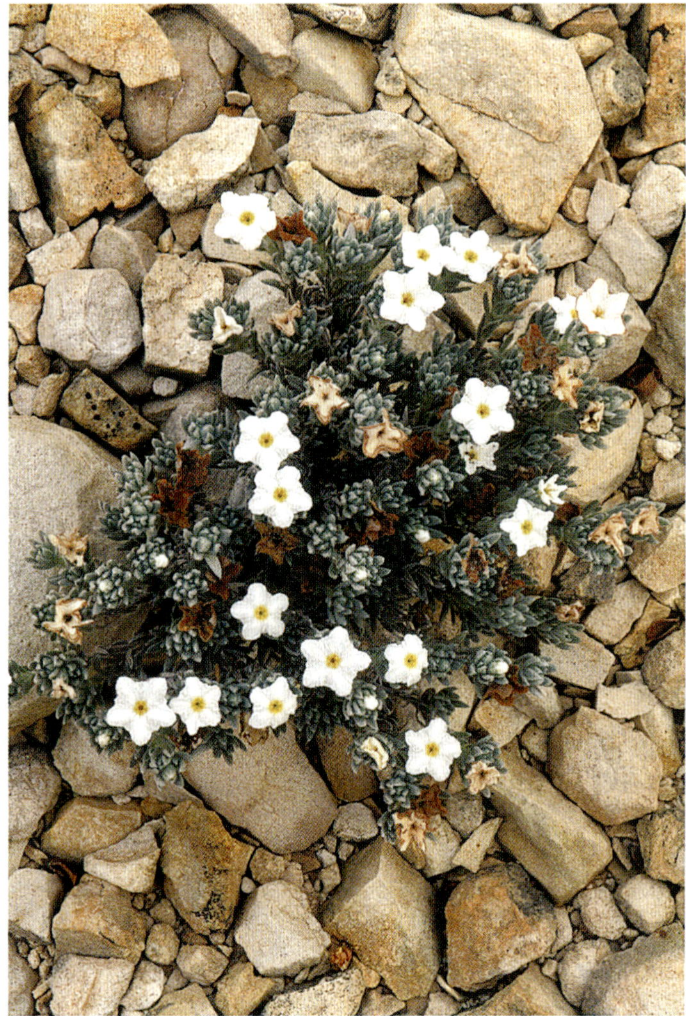

LEAFY HELIOTROPE, or crowded heliotrope, is a compressed low perennial one to six inches high, with many leafy, crowded stems. Intensely gray or silvery, this heliotrope may form dense clumps or rounded, cushionlike mounds, with long, prostrate stems and many shorter, erect branchlets. Only one-fourth inch long, the alternate, narrowly lance-shaped leaves grow in tight, overlapping bundles along and especially near the tips of the stems. Noticeably curled along the edges, the blades are coated with an assortment of thin and rough, closely pressed, and spreading, silky hairs.

Small white flowers form in small, sometimes congested clusters at the branch tips. Each bloom, no more than one-third inch across, is funnel-shaped with an expanded, five-lobed border and distinctive yellow throat. They bloom from spring to fall after good rains.

Leafy heliotrope spreads over rocky and gravelly desert scrub habitats and low desert mountains, especially on gypseous limestone soils. In Texas this heliotrope is dispersed mostly along the Rio Grande, from Presidio County to Del Rio and southeast to the Lower Rio Grande Valley. It also occurs in Coahuila and south to San Luis Potosí in northern Mexico.

In the park leafy heliotrope is most often witnessed in gyp habitats in and near the Dead Horse Mountains. This species has been identified at Boquillas Flats, Dog Flats, and Tornillo Flat; near Muskhog Spring; and at Sue Peaks. But it might be spotted almost anywhere: from San Vicente on the Rio Grande to Glenn Spring, west of the Chisos at Ward and Onion springs, and even in the Basin of the Chisos.

Note: The genus name *Heliotropium* is from the Greek *helios* (sun) and *trope* (turning) because of the ancient belief that the flowers turn toward the sun, or alternatively, bloom at summer solstice. The species name *confertifolium,* from Latin, connotes "with dense leaves." This species was collected by Charles Wright in 1851 and described by John Torrey in 1859 (as a variety of another species *Heliotropium limbatum*).

Greeneye Heliotrope

Heliotropium glabriusculum

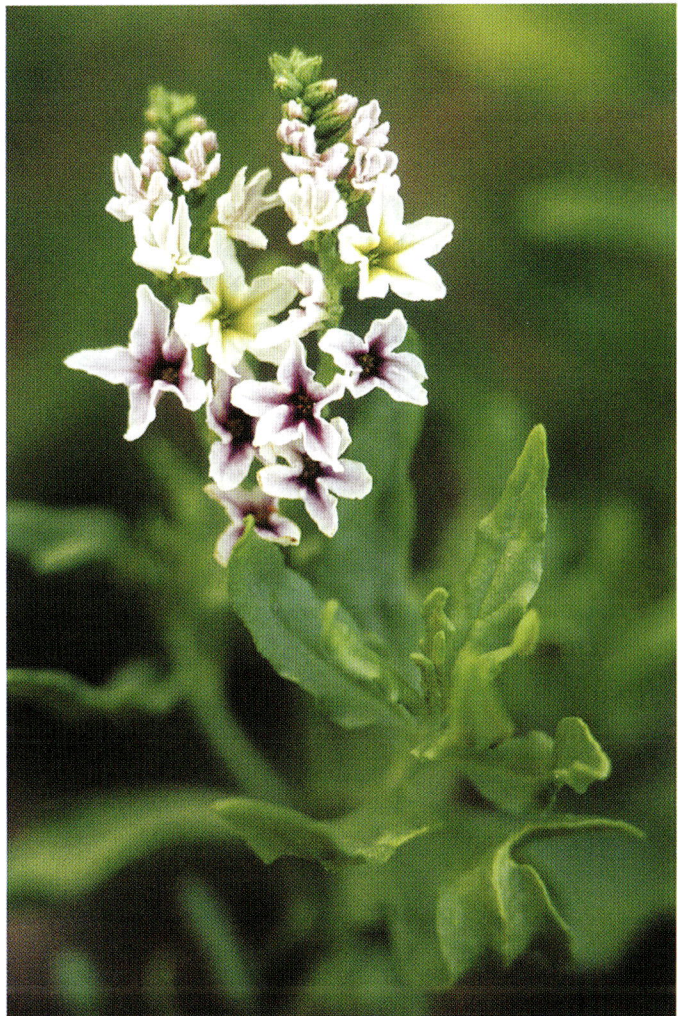

Also known as cola de alacrán, greeneye heliotrope is a low perennial herb four to at most twelve inches high. It spreads widely via underground rhizomes to form leafy colonies several feet across. Often sprawling or reclining, the weak gray stems are covered with unusual, forking hairs, which attach at their middle to the stems.

The alternate, slightly fleshy, leaves, up to two inches long, are narrowly lance-shaped with conspicuous curly or wavy edges. They may be slightly swollen or wrinkled and covered with mineralized dots.

Fragrant white flowers with distinctive greenish-yellow or purple "eyes" form primarily in summer. Appearing in single or paired clusters with up to twenty flowers per cluster, each blossom is funnel-shaped with five protruding and frequently wrinkled lobes. Distinctively one-sided and initially curved or coiled, the clusters gradually unwind as the flowers open.

Greeneye heliotrope quickly pops up on clay flats and alluvial depressions, on sandy and silty soils, and often at disturbed sites or along roadsides. This forget-me-not is primarily a resident of the southeastern Trans-Pecos (from Brewster through Pecos and Terrell counties to Val Verde County) as well as northern Mexico (Chihuahua and Coahuila south to Zacatecas).

In the park greeneye heliotrope can be pervasive along the Terlingua Ranch Road by the Rosillos Mountains. It also colonizes Dog Flats after soaking summer rains.

Note: The species name *glabriusculum* is from Latin words meaning "rather smooth," referring to the relatively smooth leaves. Like leafy heliotrope, this species was described by John Torrey in 1859 (as *Heliophytum glabriusculum*) from specimens collected by Charles Wright and Charles Parry on the U.S.-Mexico boundary survey in 1851–52. By 1859 Torrey already had four decades of experience naming new plants from the American West. In 1822 he had described plants collected by David Douglas on an expedition to the headwaters of the Mississippi, and in 1823 he had identified Edwin James's new and rare plant collections from the Rocky Mountains.

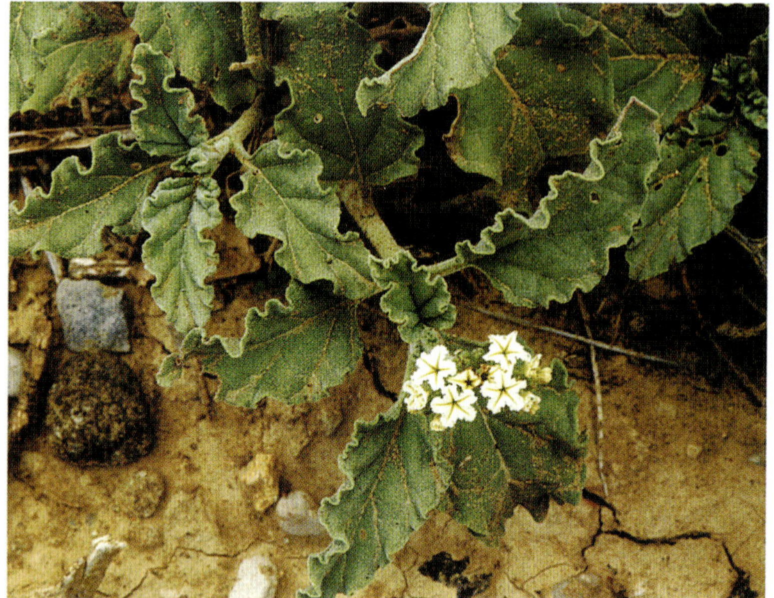

Soft Heliotrope

Heliotropium molle

Soft heliotrope is often a low perennial with ascending or sprawling stems up to one foot long. But with an extensive underground root system, this herb may spread to form colonies many feet across, and the densely hairy stems may extend as much as five feet across the ground. The large alternate leaves, up to three and one-half inches long, are triangular or broadly egg-shaped, and covered with rough hairs. The leaf surface is lined with branching veins, and the leaf margins are scalloped and conspicuously wavy.

Blooming from spring to fall, depending on moisture availability, creamy white, funnel-shaped flowers with yellow throats are organized into short, one-sided, coiled clusters that uncoil as the flowers open. The clusters form in pairs with as many as thirty flowers in each cluster. Each funnel-shaped blossom opens into five broadly rounded and distinctly fringed, petal-like lobes.

Soft heliotrope forms large stands on clay soils, alluvial flats, roadside depressions, and other sites where water collects after rains. In the United States confined to Brewster County and an area along the Rio Grande in Presidio County, this herb is more extensive throughout the adjacent Mexican states of Chihuahua and Coahuila and south to Durango.

In the park look for soft heliotrope south of Persimmon Gap at Dog Flats, along the Terlingua Ranch Road, and at Harte Ranch north of the Rosillos Mountains.

Note: The species name *molle* means "soft." John Torrey described this species (as *Heliophytum molle*) in 1859 in the botany appendix to the *Report on the United States and Mexican Boundary.* Ivan Johnston, leading Boraginaceae specialist during a long career at Harvard, placed it in the *Heliotropium* genus in 1939. These plants were collected by the botany team (Charles Parry, John Bigelow, Charles Wright, and Arthur Schott) on the U.S.-Mexico boundary survey in 1851–52, one by Bigelow at Presidio.

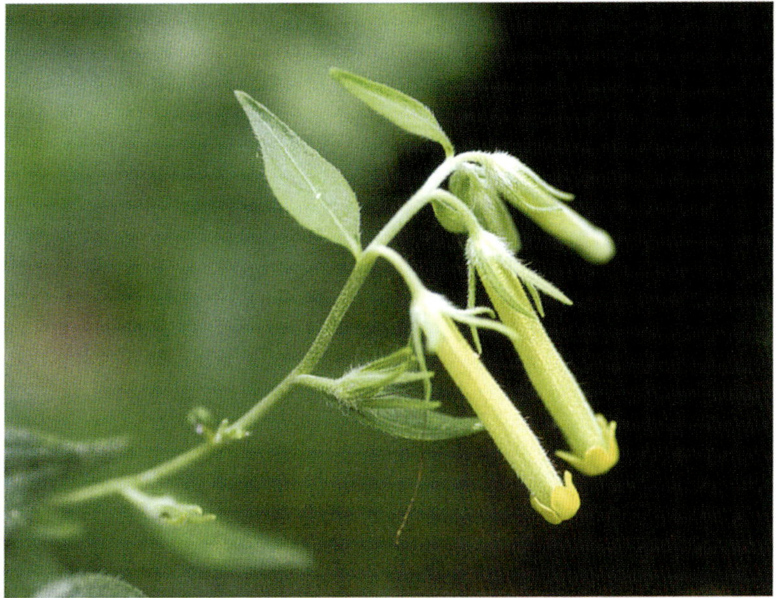

Green Gromwell

Lithospermum viride

GREEN GROMWELL, or green puccoon, is a captivating perennial herb as much as three feet high with few or many sprawling stems. One to three inches long, egg- to lance-shaped leaves alternate along the stem, with the largest near the mid-stem. Coated with rough bristly hairs, the leaf surfaces have an obvious midrib and notable veins.

Blooming from late spring to fall, the greenish-yellow, trumpet-shaped blossoms consist of a slender tube one inch or more in length with a short, brighter yellow border that curves backward on the tube like a cap. Dangling in loose, elongated clusters at the tips of branches, the attractive flowers are often sterile, but they are followed by tiny, unnoticed blooms that fertilize themselves in bud and produce many seeds.

Green gromwell is often hidden in rocky, wooded, and shady canyons in mountains, in limestone and igneous soils. In Texas this herb is limited to the Glass, Del Norte, and Chisos mountains of Brewster County and the Guadalupe Mountains in Culberson County. It also grows in southern New Mexico, southern Arizona, and the Mexican states of Coahuila and Nuevo León.

This member of the borage (Boraginaceae) family had not been confirmed in the park since 1937, when Barton Warnock found a plant on Pulliam Bluff. In 2004 it was rediscovered in two of the most secluded canyons in the Chisos. Green gromwell is classified as imperiled in Texas and critically imperiled in Arizona (NatureServe 2005).

Note: The genus name *Lithospermum* is from the Greek *lithos* (stone) and *sperma* (seed), referring to the hard nutlets. The species name *viride* means "green." This species was collected by Edward Greene near Silver City, New Mexico, in 1880 and described by him in 1881. An Episcopal priest in Silver City, Greene became the first botany professor at the University of California, Berkeley.

Mexican Navelseed

Omphalodes aliena

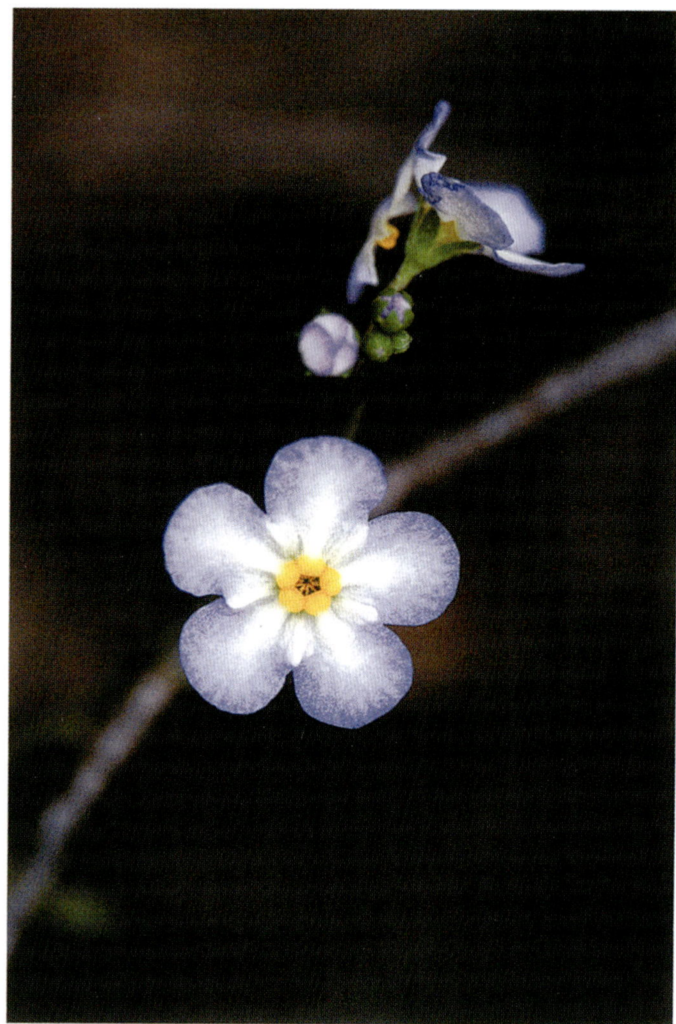

Mexican navelseed, or Mexican navelwort, is a spindly, rather sparsely branched annual up to two feet high, but usually much smaller, with ascending or reclining stems. Most of the egg- or heart-shaped leaves grow low on the branches. From one-half to one and one-half inches long, the leaves form on stalks about as long as the blades.

You will be delighted to find this charming herb in bloom. In early spring, the dainty flowers appear on slender stalks, widely spaced in long, narrow clusters. About three-eighths inch across, the delicate, wheel-shaped flowers are their own special color of blue, and their yellow throat is a perfect accent. Like many other flowers in the forget-me-not (Boraginaceae) family, this blossom has tiny yellow bulges (appendages) that form a ring in the throat, occasionally constricting the flower tube.

Mexican navelseed prospers on gravelly and rocky limestone slopes and in fissures of dry limestone rocks in desert mountains. It is native to the southern Trans-Pecos (Presidio east to Del Rio) and adjacent Mexico (Coahuila, Nuevo León, and probably Chihuahua).

In the park this annual is most prevalent in the Dead Horse Mountains, at Boquillas Canyon and near the tunnel, and at Dog Canyon far to the north where the Dead Horse and Santiago mountains meet. It has also been documented outside the Dead Horse Mountains along the River Road. This photograph was taken on Mesa de Anguila in the park's southwest corner in March.

Note: The genus name *Omphalodes* is derived from a Greek word meaning "navel," a reference to the shape of the nutlets. Members of this Eurasian genus, a favorite of Marie Antoinette, are referred to as navelseed because of the navel-like depression in the nutlets. This species was collected at Monterrey, Mexico, in 1880 by the English-born, self-taught botanist and archaeologist Edward Palmer. At the time, Palmer, who made numerous trips to Mexico from 1878 to 1910, was engaged in a primarily archeological exploration for the Peabody Museum.

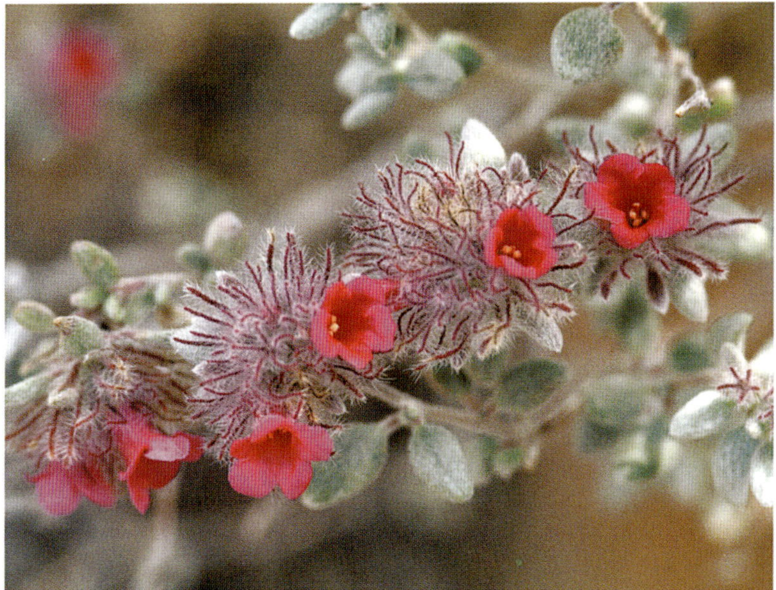

Plume Tiquilia

Tiquilia greggii

Also known as Gregg tiquilia and yerba del cenizo, plume tiquilia is a compact, often rounded little shrub about one foot high with many twiggy, forking stems. The branchlets are stiff and leafy, and both stems and leaves appear gray, covered with fine white down. Three-eighths inch long, the short-stemmed leaves are oval to elliptic.

Flowers form at the ends of leafy twigs in woolly, rounded spheres. Bell-shaped with five rounded lobes, the tiny magenta or pink blooms appear almost embedded in the woolly globes. The clusters get their appearance from the calyx (external green base) of the flower, which develops threadlike lobes with long smoke-colored, feathery hairs. This shrub blooms especially from May to September after good rains.

Plume tiquilia does well on limestone exposures in desert scrub and on rocky slopes of lower mountains. Wide ranging in the Trans-Pecos, it extends west into southern New Mexico and south into Mexico as far as the northern parts of Zacatecas and San Luis Potosí.

This shrub is broadly distributed across the park, throughout the Dead Horse Mountains, in the Grapevine Hills, on foothills surrounding the Chisos, and at the north end of the park from Persimmon Gap to Dog Flats.

Note: In 1805 Christian Persoon renamed the genus *Tiquilia* after the native South American word for its flower (he reclassified a plant from the Hipólito Ruiz–José Pavón expedition to Peru and Chile). Orphaned at an early age and sent to Germany, South African–born Persoon became a physician and fungus expert. An enigmatic, mysterious individual, he died in poverty in Paris. The species was described by John Torrey (as *Ptilocalyx greggii*) in 1855 and named for Josiah Gregg, the jack-of-all-trades merchant and historian of the Santa Fe and Chihuahua trails. For many years, the plant was known as *Coldenia greggii.*

Texas Largeseed Bittercress

Cardamine macrocarpa var. *texana*

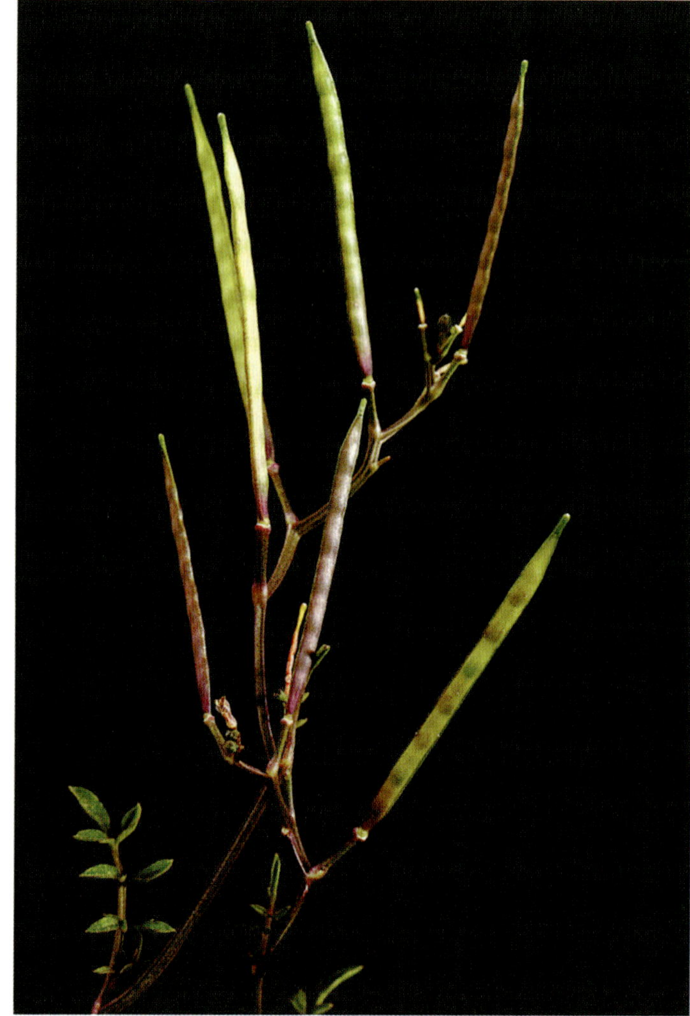

Texas largeseed bittercress is an annual or biennial herb eight to sixteen inches high, with a few flimsy but much-branched stems. Partly erect to partly reclining, the smooth stems are angled with a short thin wing at each angle. The compound leaves are divided into toothed or shallowly lobed leaflets arranged on each side of a common stalk.

Blooming in spring and summer, this herb produces very minute flowers with white or greenish, strap-shaped petals less than three-eighths inch long and about one-sixteenth inch wide. As you might imagine, they are seldom noticed. The fruits (called siliques) produce much more dramatic displays. Sitting atop distinct stalks (pedicels), the narrow, many-seeded capsules are straight and flattened, about one and one-half inches long, with two valves and a row of seeds bulging on each side.

This member of the mustard (Brassicaceae) family seems to seek out shady and moist habitats in mountain woodlands and protected canyons at moderate to high elevations. It is known only from the Chisos and Davis mountains in the Trans-Pecos, Kinney and Uvalde counties in the western Edwards Plateau, and Coahuila, Mexico.

In the park Texas largeseed bittercress is often overlooked. It is isolated in the Chisos: mostly in the Basin, Upper Green Gulch, and Pine Canyon at middle elevations and on the south slopes of Lost Mine Peak, the Boot Spring Trail, and in Upper Cattail Canyon at higher elevations. I have also photographed this photogenic plant in secluded Maple Canyon. The rare mustard is classified as imperiled and very vulnerable to extirpation both globally and in Texas.

Note: The genus name *Cardamine* is derived from the Greek word *kardamon,* a term used by Dioscorides for a species of cress. The species name *macrocarpa* means "large fruit." Reed Rollins, director of the Gray Herbarium at Harvard University from 1948 to 1978 and a specialist in the mustard family, described variety *texana* in 1940. This plant was first collected in 1933 by Victor Cory in the Chisos Mountains.

Mesa Greggia

Nerisyrenia camporum

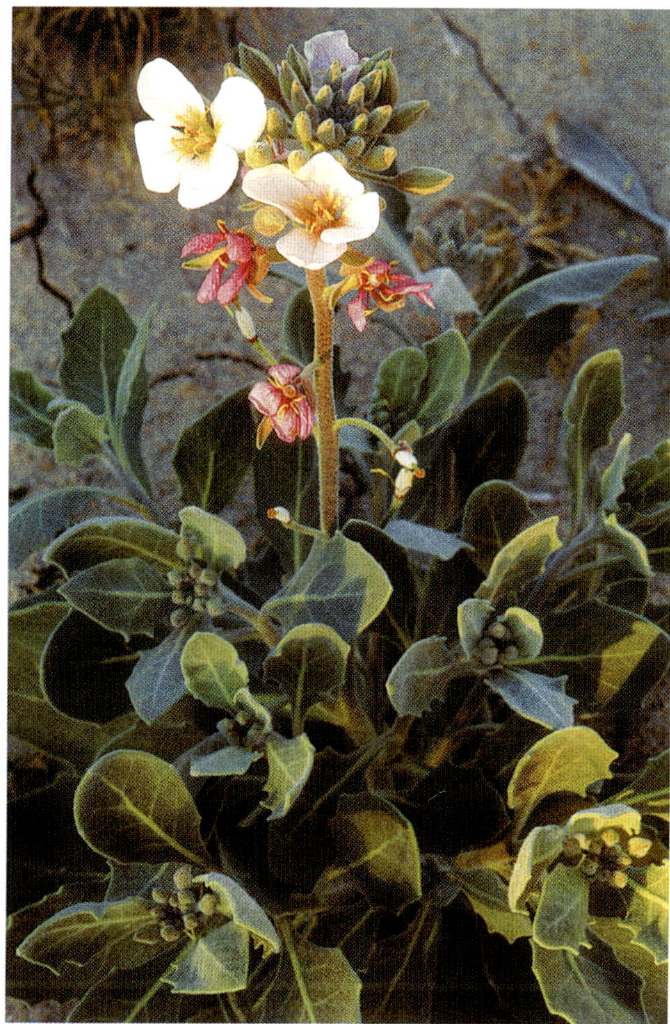

MESA GREGGIA, or bicolor mustard, is a perennial herb eight to twenty-four inches high, with stems branching from taproots at the base. The gray-white stems are densely cloaked with spreading, treelike hairs. Sometimes exceeding two inches in length, the grayish leaves are broadly spoon-shaped or spatulate; coarsely scalloped, lobed, or wavy along the edges; and coated with soft, velvety hairs. They grow on short, thick stems.

White flowers bloom profusely as early as January and continue into fall after good rains. Intensely fragrant, the blossoms appear in elongated clusters at the tips of branches, generally well above the leaves. Perhaps three-fourths inch across, each flower has four broad petals that soon fade to delightful shades of lavender, rose, and purple. The often curved, flattened fruit is a silique (a slender, many-seeded capsule with two valves splitting from the bottom), characteristic of the mustard (Brassicaceae) family.

Mesa greggia usually covers dry sandy and gravelly flats of the Chihuahuan Desert and climbs into arid grasslands, limestone hills, and outwashes. In Texas this perennial can be nearly ubiquitous across the Trans-Pecos (El Paso to Del Rio) and Rio Grande Plains to Starr County. It is also present in southern New Mexico and northern Mexico (Chihuahua to Tamaulipas and south to Durango and San Luis Potosí).

One of the most omnipresent herbs in the park, mesa greggia may stretch from Persimmon Gap and Dog Flats in the north to the Chisos foothills (Dugout Wells, Lone Mountain), and to the Rio Grande, from Santa Elena to Mariscal and Boquillas canyons.

Note: The genus name *Nerisyrenia* may derive from the Greek *neros* (wet, fresh) and *syreon* (a plant described by Pliny) (Quattrocchi 2000). The species name *camporum* means "camphor-scented," referring to the flowers. This species was collected in 1849 by Charles Wright in West Texas and by Josiah Gregg in Mexico and described (as *Greggia camporum*) by Asa Gray in 1852. Gray named the genus for Gregg, but the name was later changed to *Nerisyrenia* because the term *Greggia* had already been used in 1788 to honor another Gregg.

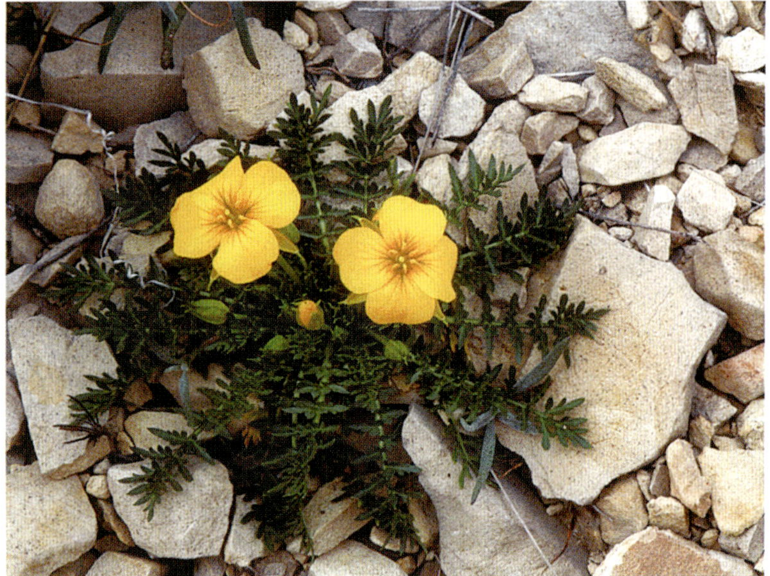

Texas Selenia

Selenia dissecta

TEXAS SELENIA is a low winter annual, usually stemless or with a short flowering stem just above the leaves. Mostly prostrate, the stalked leaves are finely dissected, divided into slender opposite, pointed lobes. Up to four inches long, the smooth blades radiate in a concentric circle from the central root.

Bright yellow, aromatic flowers that age to orange form on short stalks or more rarely on a flowering stem. The cheerful blossoms can be so numerous that they form a low canopy over the basal leaves. Up to one inch across, each flower has four wedge-shaped, rounded petals with dull green streaks at the base. Sometimes almost two inches long, the thick inflated pods (siliques) stand strictly erect, forming a curious circle around the central root.

Texas selenia flourishes on sandy or loamy soils, alluvial flats, and roadside depressions, constructing expansive yellow carpets when in bloom. In Texas this annual can be locally common throughout the Trans-Pecos south and east from Culberson County. It is also encountered in the Permian Basin, the southeastern corner of New Mexico, and northern Mexico (Coahuila and Nuevo León).

In the park Texas selenia blooms profusely in early spring at Dog Flats, near Dog Canyon, and farther south at Tornillo Flat. It has also been noted off the Grapevine Hills Road and along the River Road between Castolon and Johnson Ranch.

Note: In 1825 British-born naturalist Thomas Nuttall, while curator of the Harvard Botanical Garden, named the genus *Selenia* from the Greek *selene* (moon), perhaps because of the resemblance of the fruits to those of the genus *Lunaria.* The species name *dissecta* refers to the dissected leaves. New York botanist John Torrey described this species in 1855 in the *Pacific Railroad Report (Part 2),* which covered plants collected on Capt. John Pope's eastern route from Doña Ana, New Mexico, to the Red River.

Earlobe Mustard

Sisymbrium auriculatum

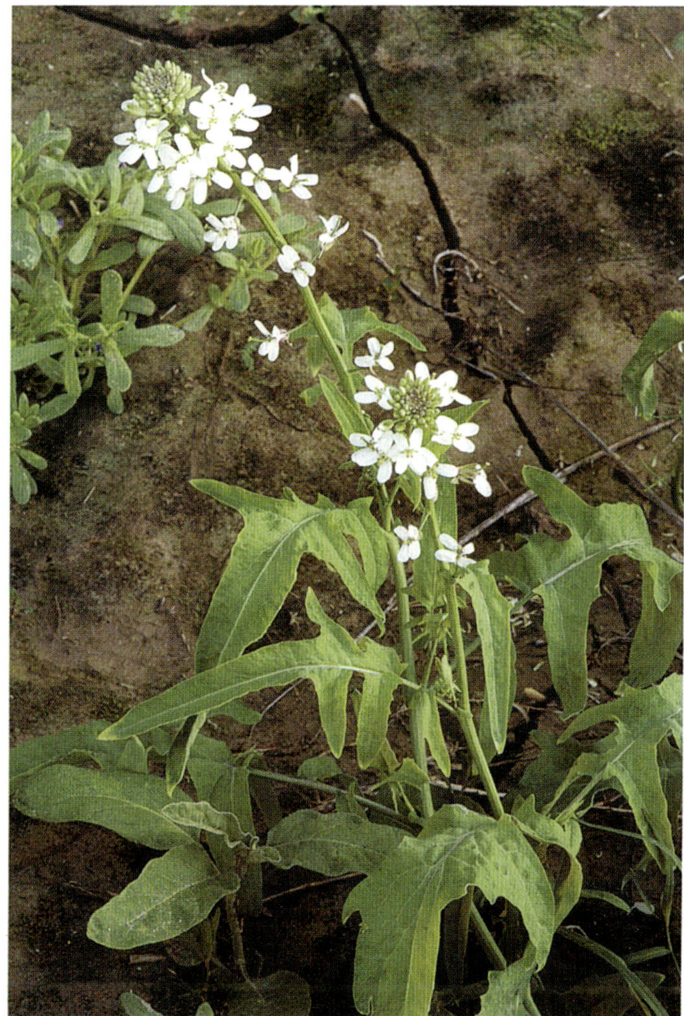

Earlobe mustard is a robust annual or biennial herb three feet high or more with tall, mostly erect but sometimes arching and wand-like stems. Large, weak leaves, occasionally as much as six inches long and half as wide, hang or droop from the lower part of the smooth stems almost like branches. The blades are deeply and irregularly lobed but frequently have a long lance-shaped terminal segment. Smaller, less deeply lobed leaves extend farther up the stem. Earlike bulges at the base of the leaf stalk often clasp and wrap around the stem.

No more than one-fourth inch wide, the flowers form dense clusters near the tips of the stems. Blooming from spring to fall, the blossoms have four white petals shaped a little like butterfly wings, each with a distinct claw at the base. It is usually difficult to find ant-free blooms.

Earlobe mustard springs up quickly on floodplains and sites subject to frequent flooding, beside sandy washes or creeks, on open clay or gravelly flats, and at the edges of dense brush. This mustard is scattered throughout Trans-Pecos Texas (in Hudspeth, Culberson, Presidio, Brewster, and Pecos counties) and northern Mexico (from Chihuahua to Nuevo León and south to San Luis Potosí).

In the park earlobe mustard frequents the Rio Grande floodplain, especially the boat access area below Santa Elena Canyon, the mouth of Blue Creek near Cottonwood Campground, and the Hot Springs area at the mouth of Tornillo Creek.

Note: The term *Sisymbrium,* from the ancient Greek name *sisumbrion* for mustardlike plants, was used by Swedish botanist Carolus Linnaeus to name this genus in 1753 in his masterwork *Species Plantarum.* Known as the father of taxonomy because of his comprehensive system of classifying organisms, Linnaeus wrote *Systema Naturae,* his first description of plants based on their sexual parts, in 1735 at the age of twenty-eight. The species name *auriculatum* means "eared." Asa Gray described this species in 1852 from a specimen collected by Charles Wright in 1849.

Lyreleaf Twistflower

Streptanthus carinatus ssp. *carinatus*

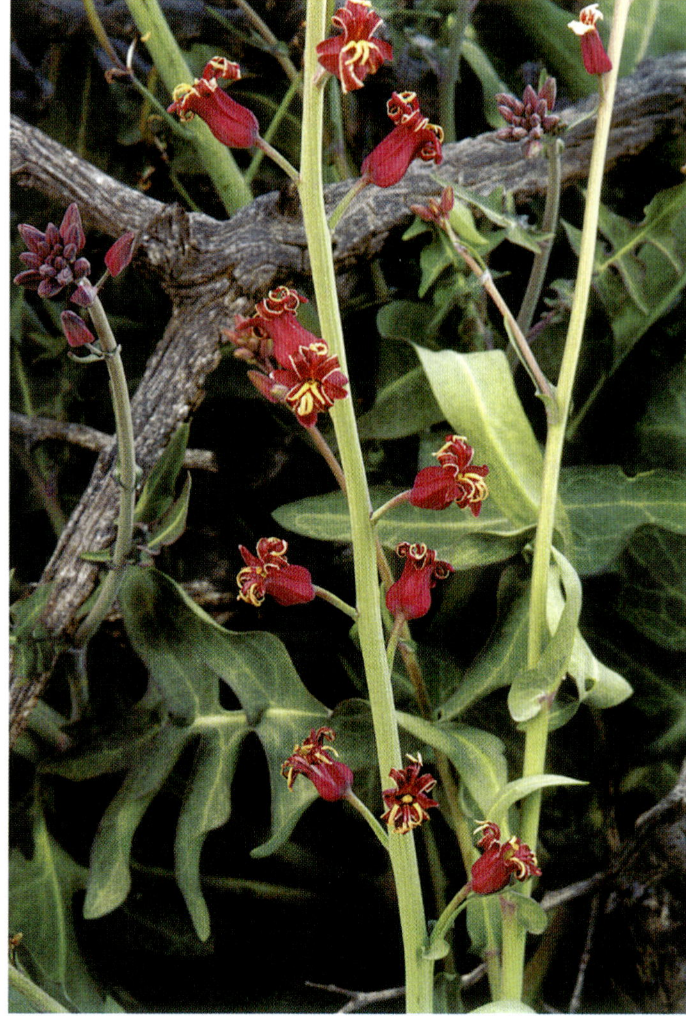

LYRELEAF TWISTFLOWER is one of three twistflowers found in the park. Growing to two feet high or more, this mustard is an erect annual with one or several pale green stems. The upright stems are smooth and rather fleshy, and both stems and leaves are coated with a white powdery substance.

Up to four inches long, the lower leaves are deeply cut into fingerlike lobes. The narrower, oblong upper leaves are stemless with distinctive small lobes at the base that clasp and wrap the stem.

Small urn-shaped flowers, maroon or dull red, appear on stout stalks off the upper part of the stems. Blooming in early spring, each flower consists of a closed, urn-shaped calyx (the external part of the flower, usually green but in this case maroon) and four reduced or almost non-existent petals—slender, twisted, wavy, and curly—at the constricted tip of the "urn."

Lyreleaf twistflower appears in sandy and gravelly soils on desert flats, in dry arroyos, and at the base of gravelly and rocky hillsides in mountain foothills. Species *carinatus* is widespread but spotty in distribution from Arizona to Trans-Pecos Texas in the United States and from Baja California to Coahuila in Mexico. Subspecies *carinatus* is more circumscribed, to southern New Mexico, Trans-Pecos Texas, and Chihuahua and Coahuila, Mexico. It is considered vulnerable in Texas.

This mustard is a widely dispersed but infrequent inhabitant of the park. It has been observed at Santa Elena Canyon to the southwest; Persimmon Gap, the Rosillos Mountains, and Tornillo Flat to the north; and at sites surrounding the Chisos, such as Avery Canyon, Chilicotal Spring, and Talley Mountain.

Note: The genus name *Streptanthus* is from Greek words *streptos* (twisted) and *anthos* (flower). The species name *carinatus* means "keeled," an allusion to the winged seeds or the flowers. Asa Gray described this species in 1853 from a specimen collected by Charles Wright the previous year "on road to Texas, about 60 miles below El Paso."

Cutler Twistflower

Streptanthus cutleri

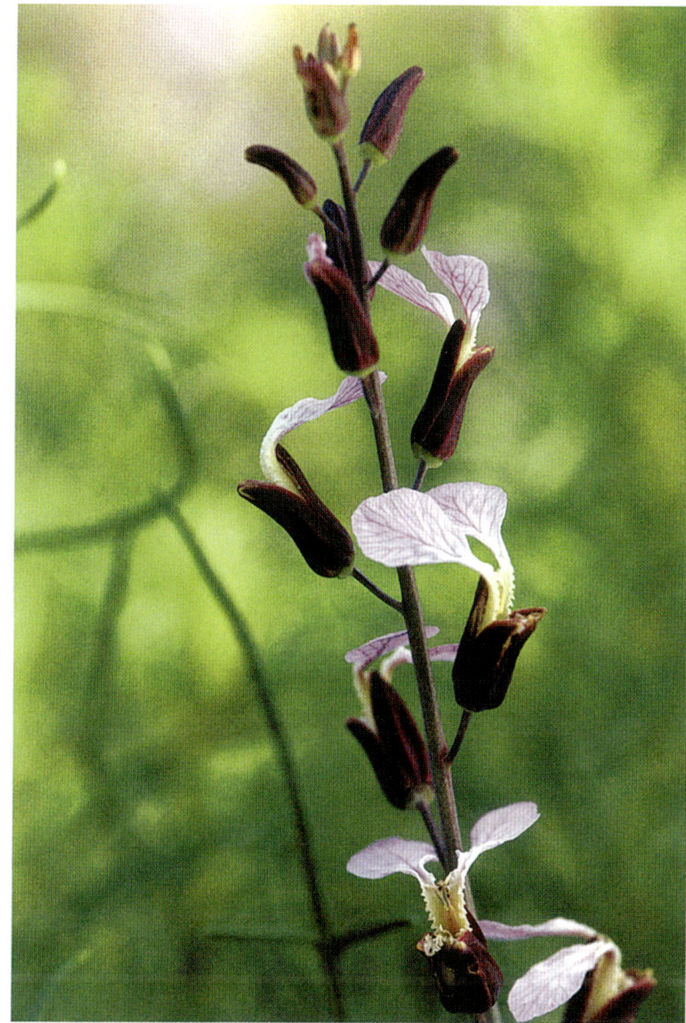

Cutler twistflower is a delightful winter annual one to three feet high with unusual purple flowers and large odd leaves. The lower leaves, up to eight inches long, are often purple-tinged and sharply incised almost to the midrib with a long terminal lobe. Similar but smaller leaves with long stalks run up the stems.

The irregular flowers have two large and erect upper petals, like broad wings, that are pale purple to lavender or almost white. They emerge from a dark purple, vaselike base (the calyx). The lower petals are absent. These unusual flowers may bloom as early as late January.

This mustard has a very narrow, constricted range in and near the Dead Horse Mountains of Brewster County and adjacent Coahuila. It occurs at Black Gap Wildlife Management Area and at Reagan Canyon outside the park and apparently was plentiful in past years: Warnock (1970) described the species as abundant from Dog Canyon to La Linda.

Due to sparse populations over a small geographic range, Cutler twistflower is classified as imperiled both globally and in Texas. It seems to prefer limestone habitats in desert mountains and canyons, usually at the base of rocky slopes or in gravelly washes.

In the park Cutler twistflower blooms from February to April, often at Dog Canyon, Boquillas Canyon, and near the tunnel. I have noticed this annual near Ernst Tinaja, in Passionflower Canyon, and even along the road to Dagger Flat. A third twistflower, broadpod twistflower (*Streptanthus platycarpus*), was seen in the park in 1956 at Dog Flats but has not been relocated in recent years.

Note: This species was described by Victor Cory in 1943 and named for economic botanist Hugh Cutler, with the Missouri Botanical Garden, who collected it in Maravillas Canyon at Black Gap in 1937. In 1936 Cory collaborated with Harris Parks, who was later curator of the Tracy Herbarium at Texas A&M University, on the *Flora and Fauna of the Big Thicket Area.*

Gregg Keelpod

Synthlipsis greggii

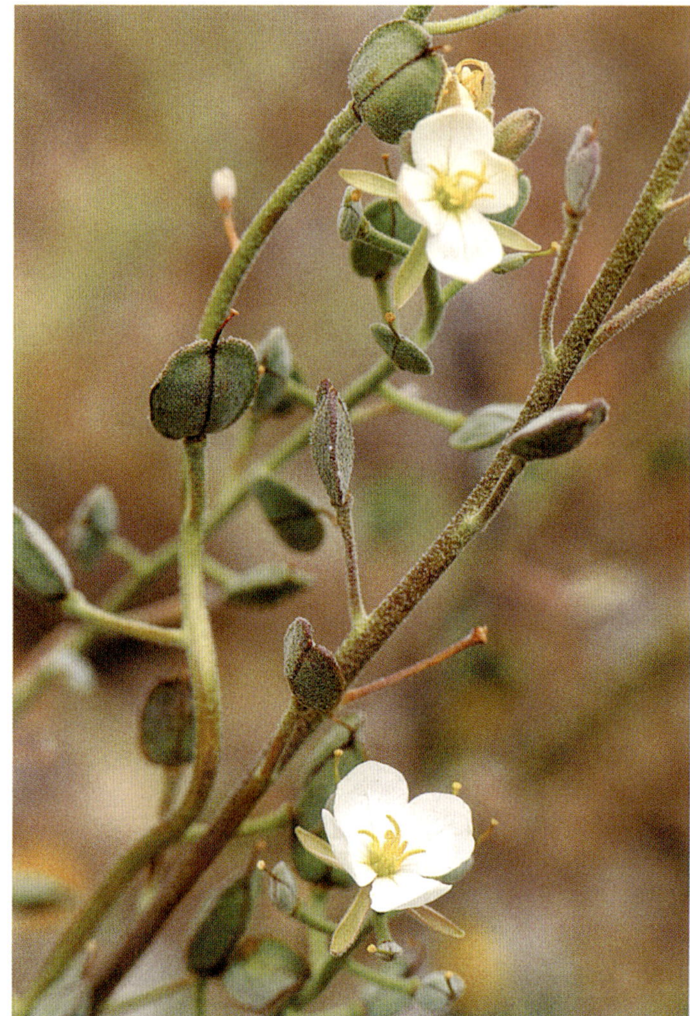

Closely resembling mesa greggia (*Nerisyrenia camporum*), Gregg keelpod is an annual or biennial mustard with spreading and often sprawling stems. Ranging from about eight inches to more than two feet high, this herb has many weak stems covered with white branching hairs. Although quite common in places, it is not nearly as extensive as mesa greggia.

The mostly egg-shaped leaves are highly variable, from under one inch to over five inches in length, and very irregularly lobed or toothed. Like the stems, the leaves are cloaked with branching hairs.

In small, elongated clusters, the flowers are white with four oval petals less than one-half inch long. Like the flowers of mesa greggia, these blossoms will gradually fade (but not as quickly) to varying shades of pink, rose, or lavender. The peak bloom occurs in early spring. This species is named for its half-inch oval pods, which are flat and densely hairy with winged margins shaped somewhat like the keel of a boat.

Gregg keelpod sprawls over sandy flats near alluvial depressions and gravelly soils in open sun and also within the protection of desert shrubs. This Mexican species crosses into the United States in the Trans-Pecos (in primarily Brewster but also Presidio and Culberson counties) and Rio Grande Plains regions of Texas but is most common in the Lower Rio Grande Valley. It is more far-reaching in Mexico (from Chihuahua to Tamaulipas and south to Hidalgo and San Luis Potosí).

In the park Gregg keelpod colonizes Dog Flats south of Persimmon Gap and off the Terlingua Ranch road, in association with spectaclepod and mesa greggia.

Note: The genus name *Synthlipsis,* from the Greek *syn* (with) and *thlipsis* (pressure), refers to the compressed pods. This species was described by Asa Gray in 1849 and named in honor of Josiah Gregg, who first collected it at Saltillo, Mexico, in 1847. A pioneer of the Santa Fe and Chihuahua trails and author of *Commerce of the Prairies* (1845) describing his travels, Gregg practiced medicine for a short time in Saltillo.

Texas Thelypody

Thelypodium texanum

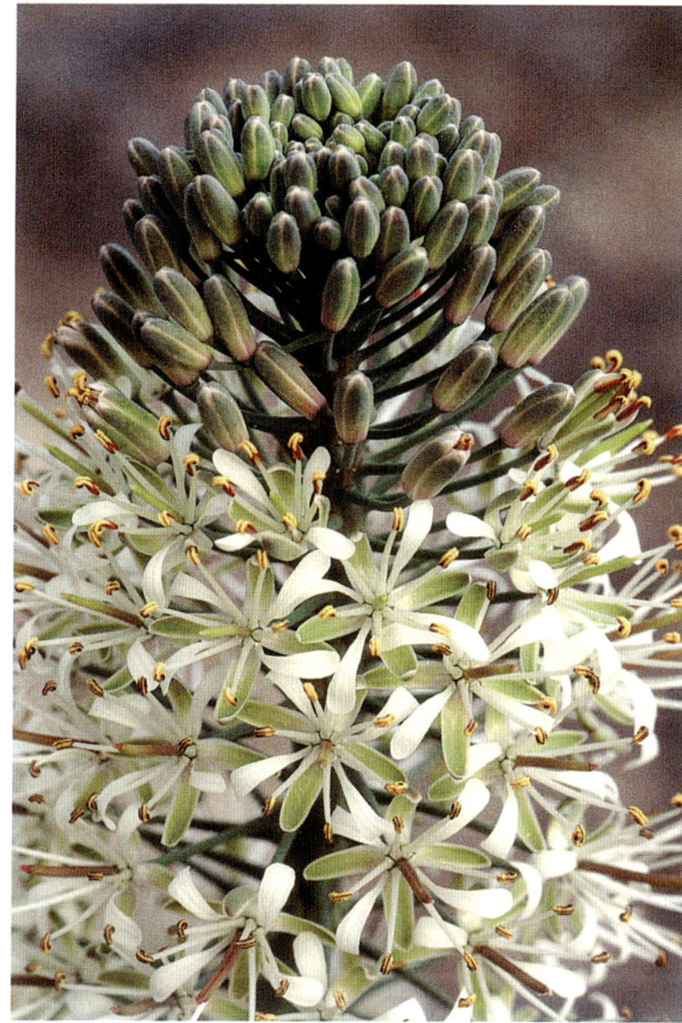

TEXAS THELYPODY is an erect annual one to two feet high, usually with a single sturdy stem closely branched from near the base. Typical of the mustard (Brassicaceae) family, the long, broadly oblong leaves are deeply dissected and toothed.

A delightful harbinger of spring, the white flowers appear in a compact, terminal spire as much as one foot long. On stalks about one-half inch long, they are distinctive for their slender, almost ribbonlike white petals and protruding purplish anthers.

Sometimes more than two inches long, the nearly round or slightly flattened fruit is a silique (the slender, dry, many-seeded pod peculiar to the mustard family, which splits open at maturity). In this case, it is noticeably swollen or constricted.

Texas thelypody occupies low gravelly slopes, rocky hillsides, and gypseous clay flats, especially near creeks and intermittent streams. This mustard is endemic to the Big Bend in Brewster and southeastern Presidio counties near the Rio Grande. Globally and in Texas, it is classified as vulnerable throughout its range.

In the park this annual is easiest to find at low elevations near the Rio Grande. It has been recorded near Boquillas Canyon and Hot Springs to the east and at Santa Elena Canyon to the west. Texas thelypody also inhabits gravelly arroyos from Hot Springs to Glenn Spring, and Dog Flats south of Persimmon Gap. Wright thelypody (*Thelypodium wrightii*), a related species, is a much taller plant found sparingly on rocky, igneous slopes in the Chisos. It has long, narrow leaves and produces round clusters of white flowers at the tips of freely branching stems.

Note: The genus name *Thelypodium* is derived from the Greek *thely* (female) and *podion* (foot), a reference to the distinct stalk of the fruits, or siliques. This species was collected by Southern Methodist University botanist Victor Cory in 1937 along Terlingua Creek in Brewster County and described by him that same year (as *Sisymbrium texanum*). A prodigious botanical collector, Cory also wrote extensively on Texas flora.

Texas False Agave

Hechtia texensis

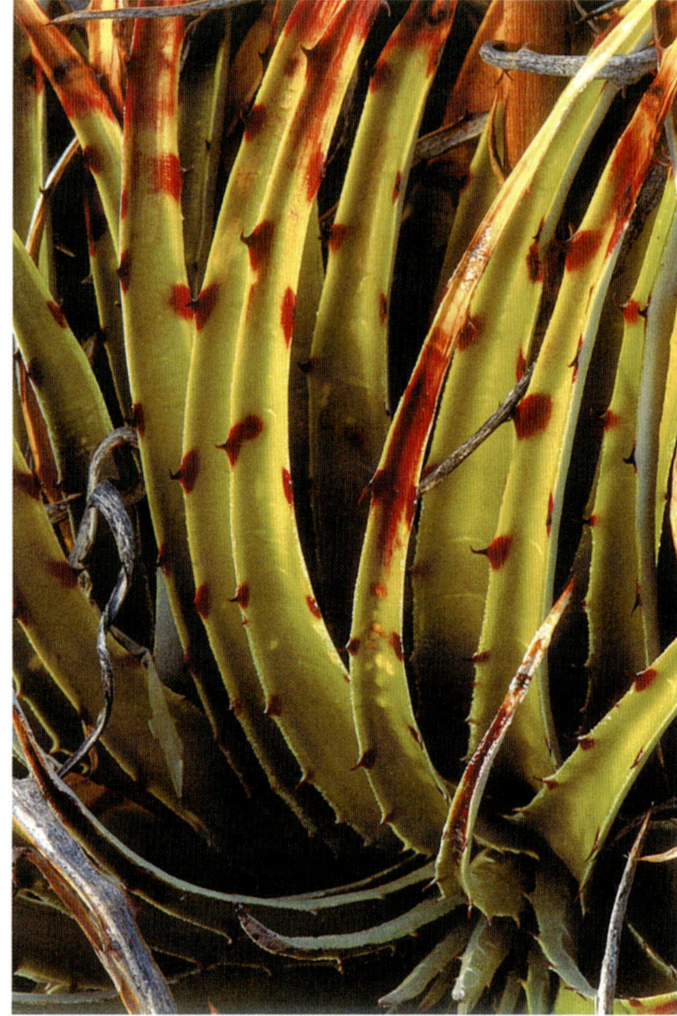

TEXAS FALSE AGAVE is a leafy succulent shrub (vegetatively similar to and often interspersed with lechuguilla) that forms circular clusters up to three feet wide. Occasionally, large colonies can take over entire hillsides.

This stemless shrub produces a flowering stalk up to four feet high, which towers above a crowded basal rosette of spine-tipped, waxy leaves. Often a foot long, these stilettos are stiff but curved, with sharp, curving teeth along the edges. The fleshy, pale green leaves may develop a deep red blush or red blotches in late fall and winter.

Male and female flowers appear in sparsely branched, elongated clusters on separate plants. The dainty white flowers are fragrant and quite remarkable with three distinctive, triangular petals. They bloom from February to May.

A member of the pineapple (Bromeliaceae) family, *Hechtia texensis* spreads over dry limestone bluffs and south-facing slopes. It is restricted to southern Brewster and Presidio counties in the United States, and Chihuahua, Coahuila, and Nuevo León in Mexico.

In the park Texas false agave may dominate limestone hills above the Rio Grande, especially from Boquillas Canyon to Mariscal Canyon. It has also been reported at locations far from the river: McKinney Spring in the Dead Horse Mountains and Glenn Spring, Dominguez Spring, and the Sierra Quemada near the Chisos. The photographs were taken on Mesa de Anguila in February.

Note: The genus *Hechtia* was described in 1835 by Johann Klotzsch, curator of the Royal Herbarium in Berlin, and named in honor of Julius Gottfried Hecht (d. 1837), a privy government counselor to the king of Prussia. This species was collected by Valery Havard in 1883 on limestone bluffs in the "Great Bend of the Rio Grande," and described in 1885 by Sereno Watson, curator of the Gray Herbarium at Harvard.

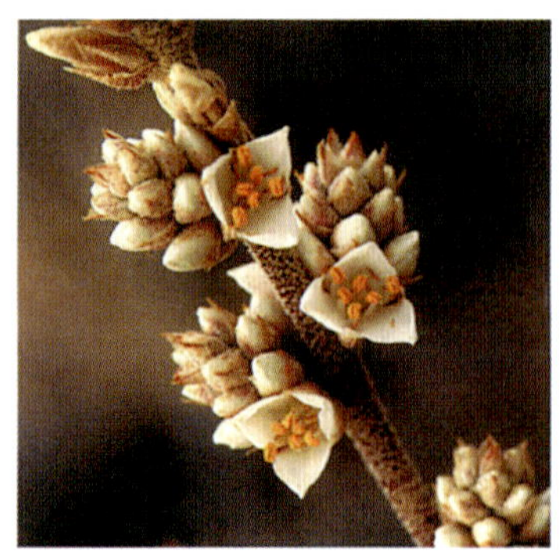

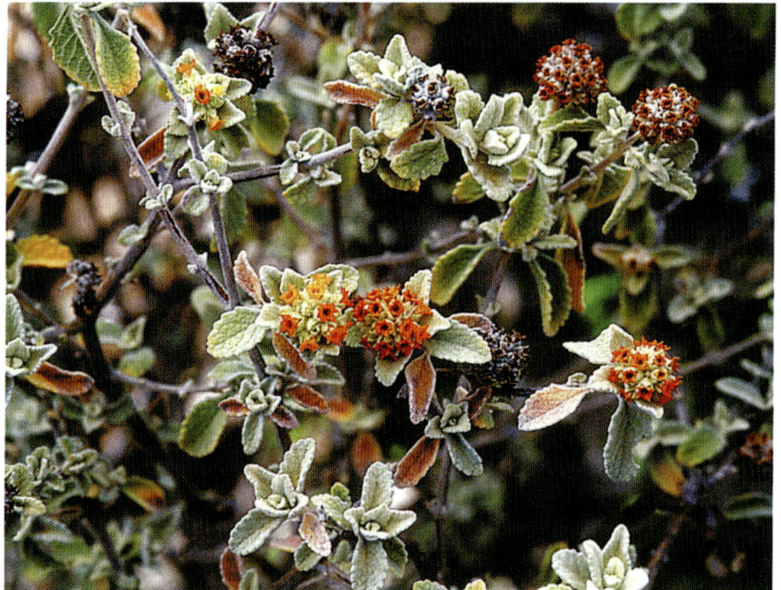

Woolly Butterflybush

Buddleja marrubiifolia

WOOLLY BUTTERFLYBUSH, saffron, or azafrán is a large and conspicuously grayish-white shrub as much as five feet high, either widely spreading or more compact and rounded, with many branches. The slender woody stems sometimes shred conspicuously, and young stems are covered with velvety hairs.

From one-half inch to more than two inches long, the thick, velvety leaves are widely variable, egg-shaped to oblong to elliptic, with coarsely scalloped edges. The paired, short-stemmed blades are cloaked with branching, tree-shaped, often brownish hairs.

In spring and summer, small yellow flowers fading to orange are clustered in round, compact heads one-half inch across. Each blossom has a tiny tube with a four-toothed border. The flower heads have been used in Mexico to color butter and cheese.

Woolly butterflybush fares well in full sun on desert mountains and foothills in limestone soil and in brushy habitat along drainages. Confined to Brewster and Presidio counties in the United States, especially near the Rio Grande, it ranges into northern Mexico (from eastern Chihuahua to western Nuevo León and south to parts of San Luis Potosí and Zacatecas).

In the park woolly butterflybush is most prominent in the Dead Horse Mountains—at McKinney Spring, in Telephone Canyon, and elsewhere. It has also been located at Dominguez Spring south of the Chisos and at Santa Elena Canyon.

Note: In 1753 Carolus Linnaeus named the genus *Buddleja* for the British amateur botanist and vicar in Essex, Adam Buddle, who had published the influential work *English Flora* in 1708. The species name *marrubiifolia* means "with leaves like *Marrubium*," or horehound. George Bentham described this species in 1846 from a plant collected by Jean Louis Berlandier near Monterrey, Mexico, in 1828.

Escobilla Butterflybush

Buddleja scordioides

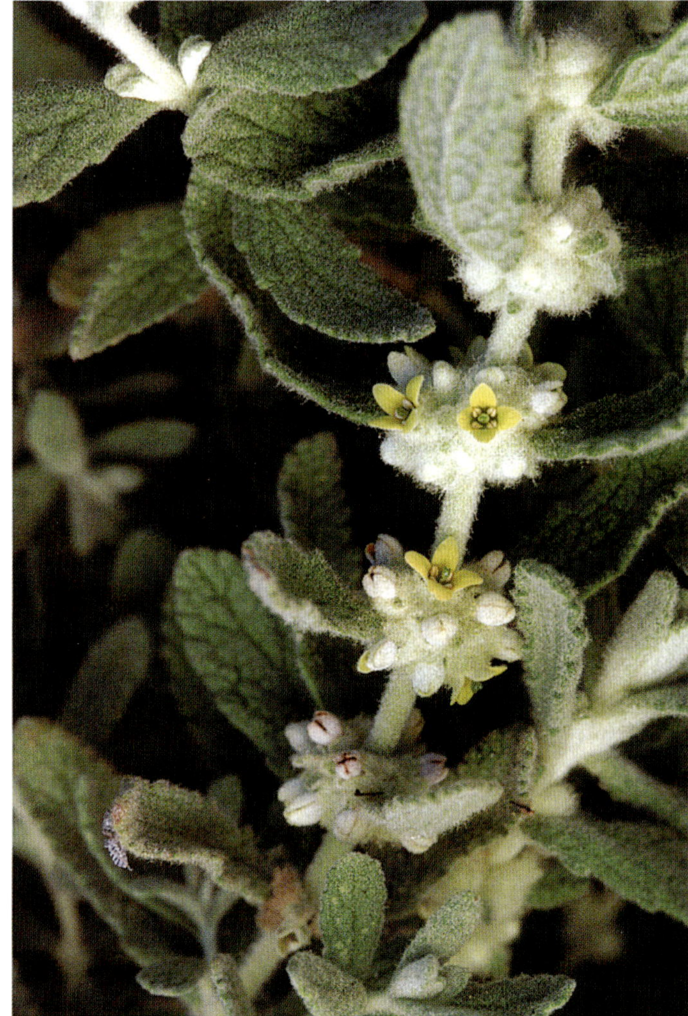

ESCOBILLA BUTTERFLYBUSH, also known as hierba de las escobas and golondrilla, is a compact, rounded or mound-shaped shrub usually two to four feet high but occasionally much taller. Aromatic and very enticing to butterflies, the entire plant is cloaked with densely matted, often rust-colored velvety hairs.

The coarsely serrated, opposite leaves are mostly oblong or narrowly so, usually less than one inch long, but rarely almost two inches in length. They are stemless, distinctly wrinkled and veined, and covered with yellow-gray hairs. Native Americans made a tea from the leaves as an indigestion remedy.

Blooming from summer to fall, the flowers are clustered into densely woolly, globular heads, no more than one-third inch across, distributed along the upper stems. Each individual bell-shaped flower is minuscule, with four pale yellow, rounded lobes sticking out like little teeth. Less than one-eighth inch long, the fruit is an oblong or broadly oval, two-valved capsule with many tiny seeds.

Escobilla butterflybush grows on clay and gyp flats, near alluvial depressions and in gullies beside roads, in open sun and dense scrub brush. It is distributed throughout the Trans-Pecos from Del Rio to El Paso, west into southern New Mexico, and south deep into Mexico to Jalisco, Guanajuato, and Hidalgo.

Uncommon to rare in the park, this dense shrub sometimes lines the highway between Panther Junction and Persimmon Gap, and near the Terlingua Ranch turnoff. It also fills depressions along the Terlingua Ranch road north of the Rosillos Mountains.

Note: Escobilla is Spanish for "little broom." The species name *scordioides* is a reference to the plant's strong scent. This species was collected by German explorer Baron von Humboldt and French botanist Aimé Bonpland near Mexico City and described by Carl Kunth in 1818. In the seven-volume Latin work, *Nova Genera et Species Plantarum,* Kunth described more than forty-five hundred species collected by Humboldt and Bonpland during their five-year expedition.

Berlandier Lobelia

Lobelia berlandieri var. *brachypoda*

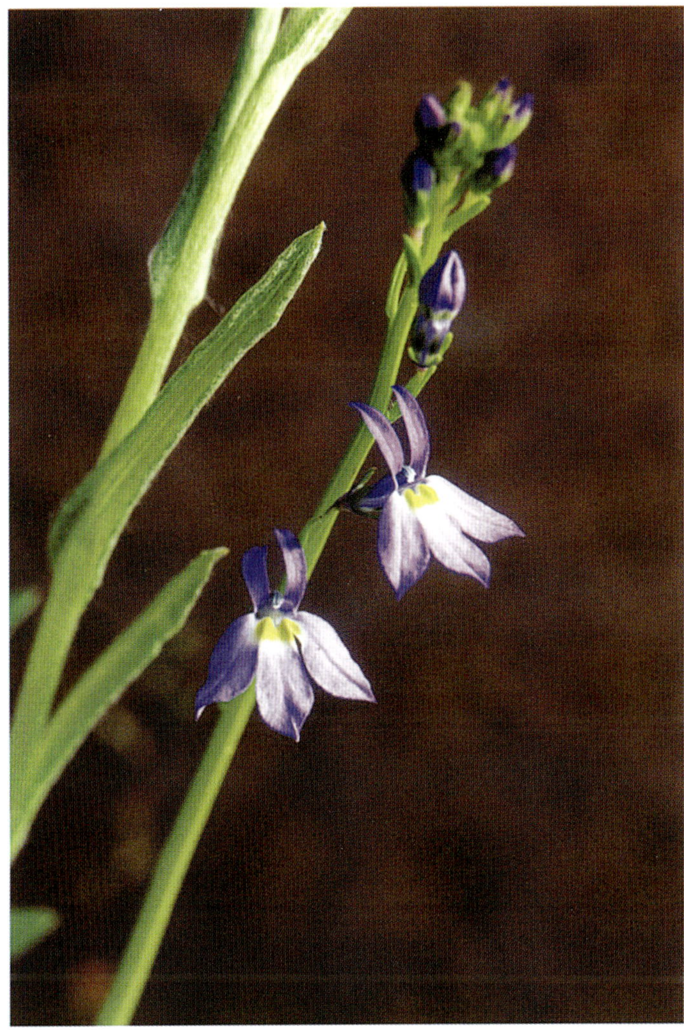

BERLANDIER LOBELIA is an erect annual up to two feet high, usually with several smooth green stems. This variety has highly variable leaves, growing mostly along the stem, that may be broadly egg-shaped in the lower reaches and narrowly lance-shaped upward. Up to two inches long, the leaves are relatively smooth and very finely scalloped along the edges.

This member of the bluebell (Campanulaceae) family develops unusual pale purple flowers near the tips of the stems, in narrow clusters up to ten inches long. Appearing on short, relatively straight stalks, the blossoms are distinctive: two-lipped, with the upper lip split deeply to the base and curling backward, and the broad lower lip divided into three petal-like, pointed lobes. The base of the lower lip has a conspicuous white eye with two yellow bulges or humps in the center.

This annual thrives in both sandy and clay soils at moist sites, including grassy and rocky areas near springs and seeps, in alluvial depressions, and in locations subject to flooding. In Texas it is known primarily in southern Brewster County, the Rio Grande Plains, and portions of the Texas Hill Country. The herb occurs at scattered sites in Coahuila and Chihuahua, Mexico.

Berlandier lobelia has been identified at only a few locations in the park: at Fresno Spring, along Fresno Creek, and in Fisk Canyon to the south of the Chisos, and at Glenn Spring to the east. I have photographed this delicate herb blooming in spring and summer along Terlingua Ranch Road north of the Rosillos Mountains and at Rough Spring just north of the Chisos.

Note: This genus is named for Belgian born Matthias de l'Obel (1538–1616), whose Latin name was Lobelius. A physician to William of Orange, he later became botanist to King James I of England. This species was described by Swiss botanist Alphonse de Candolle (son of Augustin) in 1839 and named for Jean Louis Berlandier, who collected it near Tampico, Mexico. The varietal name is from the Greek *brachys* (short) and *podion* (foot), probably referring to the short flower stalks. Variety *brachypoda* was collected by Charles Wright in 1849.

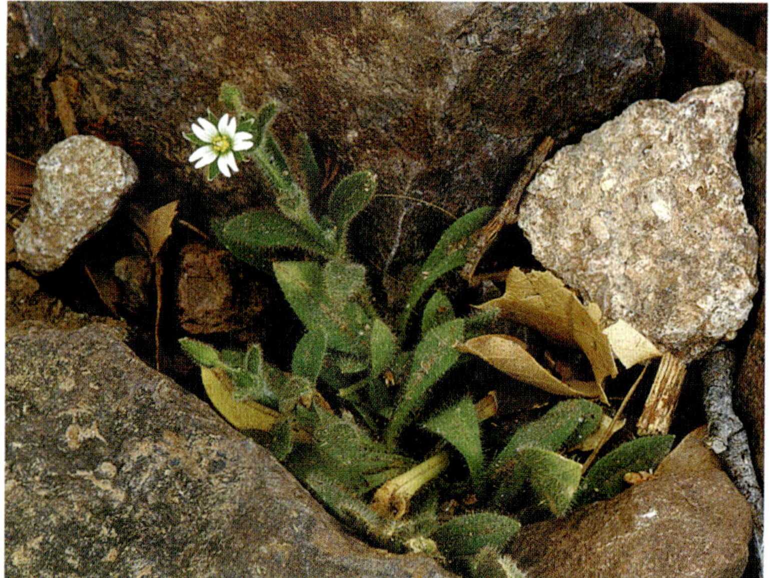

Trans-Pecos Chickweed

Cerastium axillare

TRANS-PECOS CHICKWEED is an annual up to one foot high, usually much lower and erect but sometimes nearly prostrate. Often with only a few weak, sparingly branched stems, this herb is noticeably sticky-hairy throughout. The paired leaves, up to one inch or more in length, are highly variable on the same plant but almost always narrow, most commonly broadest above the middle and tapering at the base.

Delicate white flowers appear, singly, in the leaf axils (upper angles between the leaves and the stem). Each flower has five conspicuously notched petals, frequently deeply cut to near the middle, and five green sepals at the base that are noticeably longer than the petals. The plant blooms from spring to early fall.

Trans-Pecos chickweed is best suited to igneous soils, in shaded woodlands and canyons but also on open rocky or grassy slopes in mountains. Its range includes portions of Trans-Pecos Texas (in Brewster, Presidio, Jeff Davis, and El Paso counties), New Mexico, and Chihuahua, Mexico.

In the park this member of the pink (Caryophyllaceae) family is relatively scarce in the Chisos, confirmed only on the east slopes of Pulliam Bluff above Moss Well, on the east slopes of Lost Mine Peak in Upper Pine Canyon, and along Boot Creek above Boot Spring. The photograph was taken on the Pinnacles Trail in April.

Note: The genus name *Cerastium* is from the Greek *keras,* meaning "horn," referring to the fruit shape. The species name *axillare* means "axillary," alluding to the axillary flowers. This species was described in 1967 by Donovan Correll, a prolific botanical author, collector, and orchid specialist, from a specimen he collected in Little Aguja Canyon in the Davis Mountains.

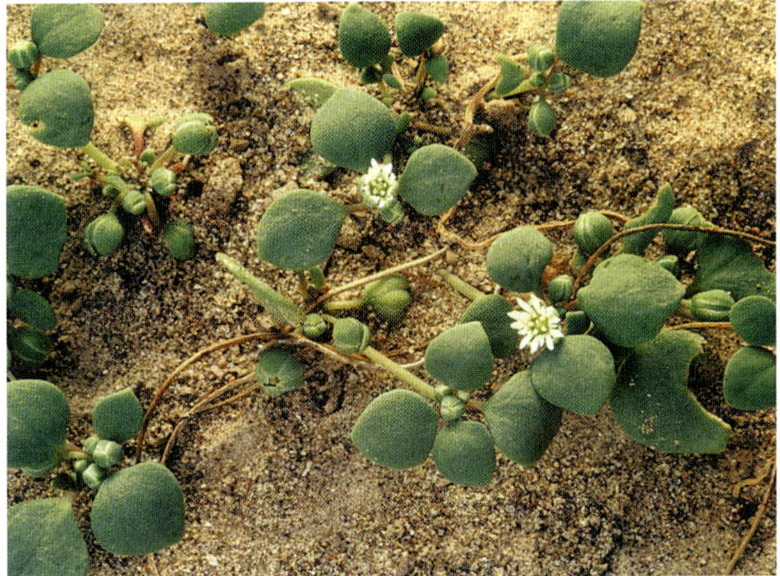

Thickleaf Drymary

Drymaria pachyphylla

Also known as inkweed, this succulent annual may suddenly appear after rains or floods on sand-gravel bars and barren sandy banks along streams in the Trans-Pecos. The prostrate stems often form amusing radial or circular patterns in the sand, resembling inkblots. Mostly oval leaves, about as wide as long, are arranged in small but crowded, circular groups at nodes along the stems.

One-fourth inch wide, each small white flower has five four-lobed petals, each deeply notched, with two small lobes in the middle. The flowers open widely only in bright sunlight. In spite of its charm, this delightful but short-lived annual is considered a nuisance by some because it is toxic to sheep and cattle.

Thickleaf drymary has a long blooming period from January through October. It forms mats across the Trans-Pecos from Brewster County northwest to Hudspeth County. The herb is also present in southern New Mexico, southeastern Arizona, and northern Mexico (in portions of the states of Coahuila, Durango, and Nuevo León).

In the park thickleaf drymary has been documented at Dog Flats and Harte Ranch near the Rosillos Mountains, but the best place to view this annual is probably near the Rio Grande west of Cottonwood Campground and along Alamo Creek. This ephemeral herb lasts only a brief time, until the soil dries out or the next summer rain washes it away.

Note: The genus name *Drymaria* may be derived from the Greek *drymos* (forest), alluding to the habitat of the first species. The species name *pachyphylla* means "thick-leaved." This species was collected by Elmer Wooton south of White Sands, New Mexico, in 1897 and described by him and Paul Standley in 1913. Wooton, pioneering botanist at New Mexico A&M (later New Mexico State), was instrumental in the creation of the Jornada Experimental Range.

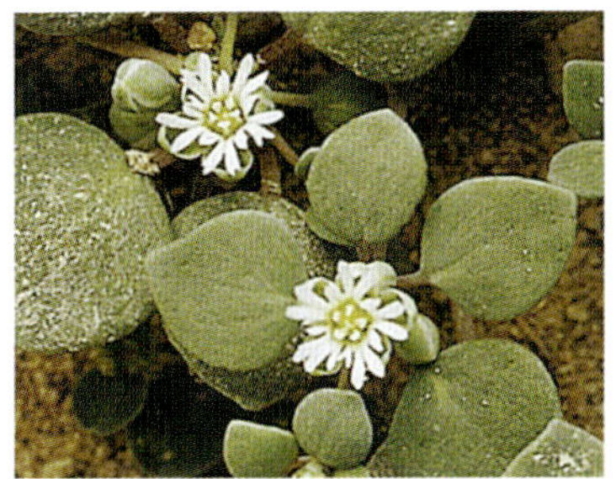

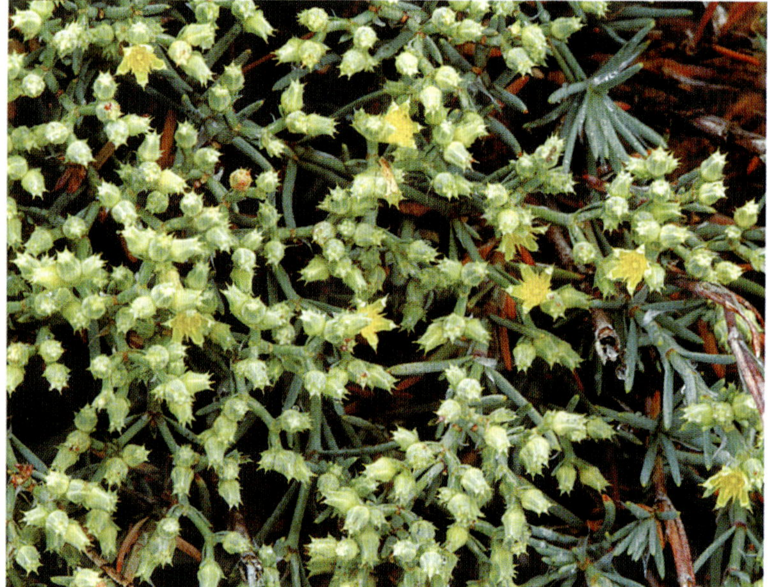

James Nailwort

Paronychia jamesii

JAMES NAILWORT is a low perennial herb two to eight inches high that grows in tufts or dense clumps like turf. A few or many hairy stems rise from thick, woody taproots and branch extensively. Narrowly linear, stalkless leaves, up to one inch long, may form tight bundles along the stems. The blades are distinctly leathery and often abruptly pointed.

Blooming in summer and fall, very atypical, bell-shaped flowers appear in crowded, much-branched clusters of up to seventy blooms. The petals are absent or reduced to bristles, and the flowers consist mostly of greenish-yellow oblong sepals. The sepals have whitish or clear margins that are thin and dry and a cone-shaped, bristly tip.

James nailwort does well in a variety of habitats from desert scrub to high woodlands, from dry hilltops to clay flats and sand dunes, and especially on rocky ledges and outcrops in limestone soil. Its broad distribution reaches from Wyoming and Nebraska to Arizona, New Mexico, and Texas. The perennial is also wide-ranging throughout the western half of Texas, from the Panhandle to the Trans-Pecos, and in Mexico from northeastern Chihuahua to western Nuevo León.

In the park James nailwort blankets exposed rocky outcrops along the Lost Mine Trail and at the top of the Chinese Wall Trail. This low herb has also been observed near Panther Spring and at Onion Spring.

Note: The genus name *Paronychia* is from the Greek *para* (beside) and *onyx* (fingernail) because of the use of these plants in the treatment of whitlow, a nail disease. John Torrey and Asa Gray described this species in their pioneering early work *A Flora of North America* (1838). They named the species for botanist Edwin James, a student of Torrey who collected it in 1819–20 on the Long Expedition to the Rocky Mountains. Later, the abolitionist James settled in Iowa and ran a station on the Underground Railroad.

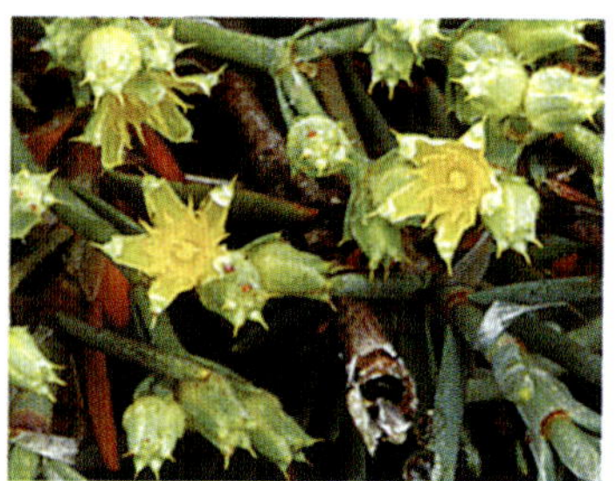

Mexican Starwort

Stellaria cuspidata

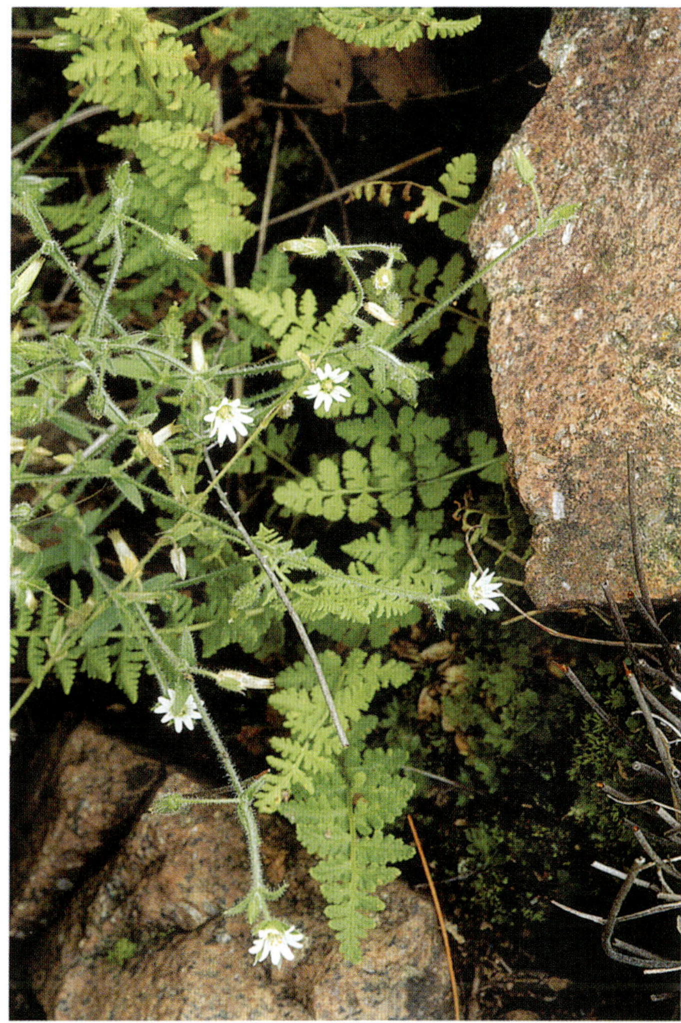

Rarely more than two feet tall, this herb is usually much shorter with weak, repeatedly forking stems. Somewhat gluey to the touch from glandular hairs, the stems may spread widely, covering the ground. The leaves are variable in shape, from triangular to egg-shaped or oblong, and often heart-shaped at the base. They can be an inch or more long on thin stems that are sometimes even longer.

Mostly in summer and early fall, small white flowers about one-half inch wide appear in sparse, leafy clusters. Held on a threadlike stalk longer than the bloom itself, the flower has five petals, each distinctly notched at the tip. Occasionally the petals are so deeply cleft that the flowers seem to have ten petals.

Also known as mountain starwort, Mexican starwort spills over mountain slopes, usually at high elevations in wetter, more protected areas. This spreading herb is frequently embedded in moist, shady pockets among rocks and boulders and near waterfalls and springs. It is a South American and Central American species that is widespread across Mexico from Chiapas to Chihuahua but barely enters Trans-Pecos Texas and part of New Mexico. This starwort is seen in most Trans-Pecos mountains, from Brewster to Hudspeth counties.

In the park Mexican starwort is encountered only sporadically in the Chisos: on the east slopes of Lost Mine Peak, in Green Gulch, on the east slopes of Pulliam Bluff above Moss Well, and especially in Pulliam Canyon and on Emory Peak.

Note: The genus name *Stellaria* is from the Latin *stellaris* (starry), a reference to the starburst shape of the flowers. The species name *cuspidata* means "sharp-pointed." Carl Ludwig Willdenow, director of the Berlin Botanical Garden, described Mexican starwort in 1816 from a specimen collected in Ecuador by Baron von Humboldt and Aimé Bonpland. After mapping seventeen hundred miles of the Orinoco River, in 1802 Humboldt and Bonpland arrived in Ecuador, where they amassed a large collection of plants and took time out to climb eighteen thousand feet up Mount Chimborazo.

Rough Mortonia

Mortonia sempervirens ssp. *scabrella*

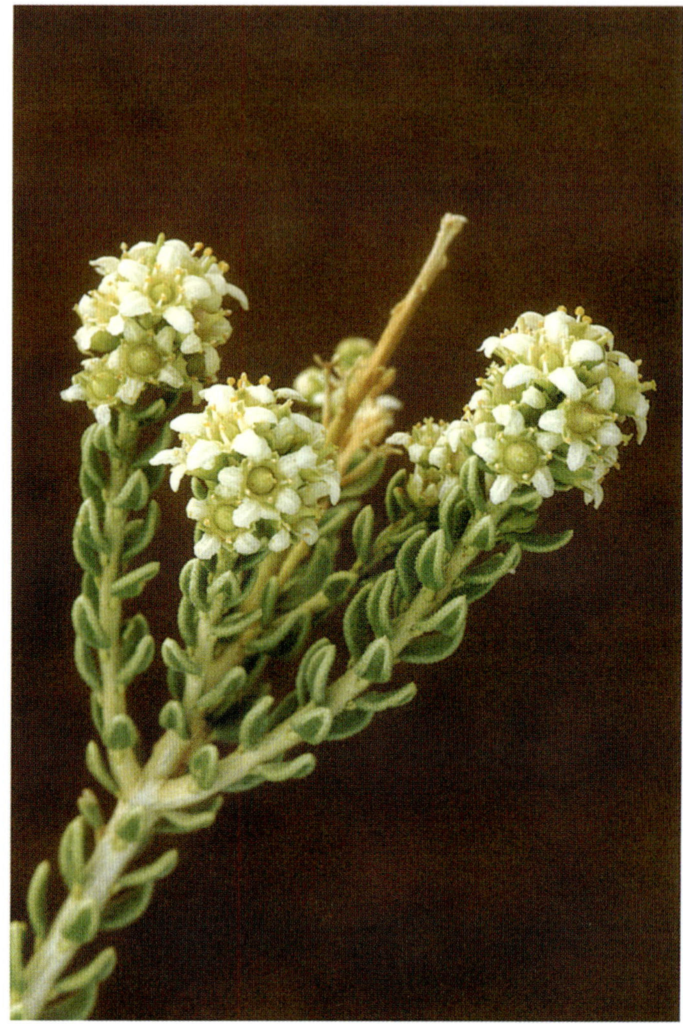

ALSO KNOWN AS tickbush and sandpaperbush, rough mortonia is a compact, stiffly upright evergreen shrub about three to four feet high, with slender and very brittle white stems. The thick oval or egg-shaped leaves, about three-eighths inch long, are coated with stiff hairs and rough as sandpaper. Pale yellow-green in color, the leaves tightly overlap and press closely against the stem.

Creamy white or pale yellow flowers, less than one-fourth inch wide, form in small clusters at the branch tips. They have five rounded, waxy-looking petals. Blooming from spring to fall, these creamy flowers add to the overall pallid appearance of this strange shrub. The fruit is a small oblong achene (a hard, dry, one-seeded fruit).

Rough mortonia occurs primarily on rocky limestone hills, bluffs, and ledges, and in gypseous soils, usually at low to mid elevations. This shrub is widely scattered, but the distribution is spotty. It stretches across Sonora and Chihuahua, Mexico, and throughout much of the Trans-Pecos, especially from Brewster County to El Paso near the Rio Grande, to the southwestern corner of New Mexico and the southeastern corner of Arizona.

In the park rough mortonia is most conspicuous along the Telephone Canyon Trail, although it was spotted on a volcanic dike west of Ward Spring some years ago. At least some Telephone Canyon plants are a rare subspecies (ssp. *sempervirens*) with smaller leaves. I photographed this peculiar shrub in August on Cottonwood Creek near Chisos Pens.

Note: The genus *Mortonia* is named for Samuel Morton, anatomy professor at Pennsylvania Medical College and in 1850, president of the Academy of Natural Sciences. A prominent American paleontologist, Morton died in 1851 after assembling a famous collection of human skulls and attempting to prove that some races were superior based on their braincase capacity. The species name *sempervirens,* from Latin words meaning "always alive," refers to the evergreen leaves. The subspecies name *scabrella* means "rough." Rough mortonia was collected by Charles Wright in New Mexico in 1852 and described (as *Mortonia scabrella*) in 1853 by Asa Gray.

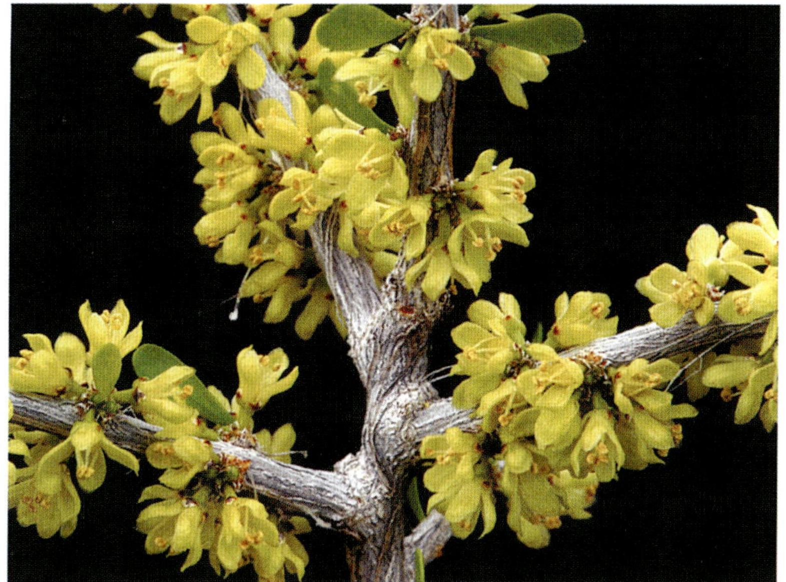

Desert Yaupon

Schaefferia cuneifolia

DESERT YAUPON, capul, or pañalero is an intricately branched, evergreen shrub with stiff, almost spinescent twigs. It is about three to six feet high with light gray bark and often zigzag spur branches. Pale green leaves, one-half inch long, grow in bunches on the short lateral branchlets. Smooth, firm, and slightly leathery, the stalkless leaves are shaped like teardrops.

In spring or summer, four-petaled, greenish-yellow flowers grow singly or in small clusters at the base of leaves. The inconspicuous male and female blossoms appear on separate plants. But there is beauty in numbers: when the fragrant blossoms burst into bloom simultaneously, stimulated by a soaking rain, they are an arresting sight. After the flowers fade, the branches are brightened with shiny, scarlet, pea-sized fruits (drupes).

Desert yaupon is most prevalent in the Rio Grande valley and extends north to San Antonio and west to Brewster County. It also covers large portions of northern Mexico (Chihuahua to Tamaulipas, south to San Luis Potosí, and also Baja California). This shrub prospers in various soils, especially on rocky hillsides in brush.

In the park look for desert yaupon on the road from Nugent Mountain to Glenn Spring. Other sites include the Basin and Green Gulch, Upper Juniper Canyon and Dominguez Spring, K-Bar and numerous additional locations in the Dead Horse Mountains.

Note: Yaupon, from the Catawba (Native American) word meaning "tree leaf," refers to an evergreen holly shrub in the southeastern United States. The genus *Schaefferia* is named for the German botanist, zoologist, and theologian Jacob Schaeffer, famous for his 1768 treatise proving that the heads of snails grow again after decapitation. The species name *cuneifolia* denotes "wedge-shaped leaf." This species

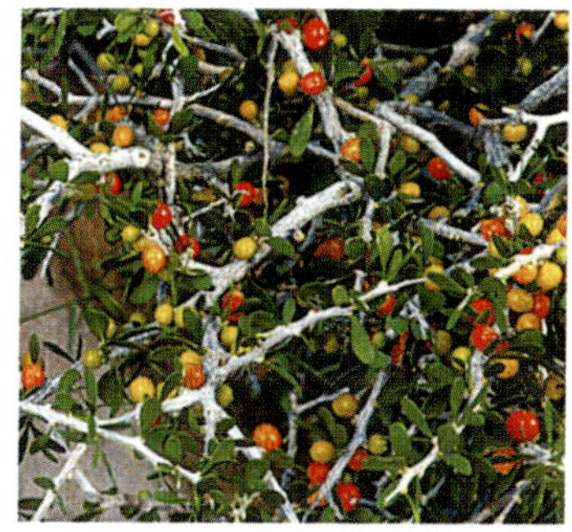

was collected by Charles Wright during the Graham boundary survey and described by Asa Gray in 1852.

Mexican Clammyweed

Polanisia uniglandulosa

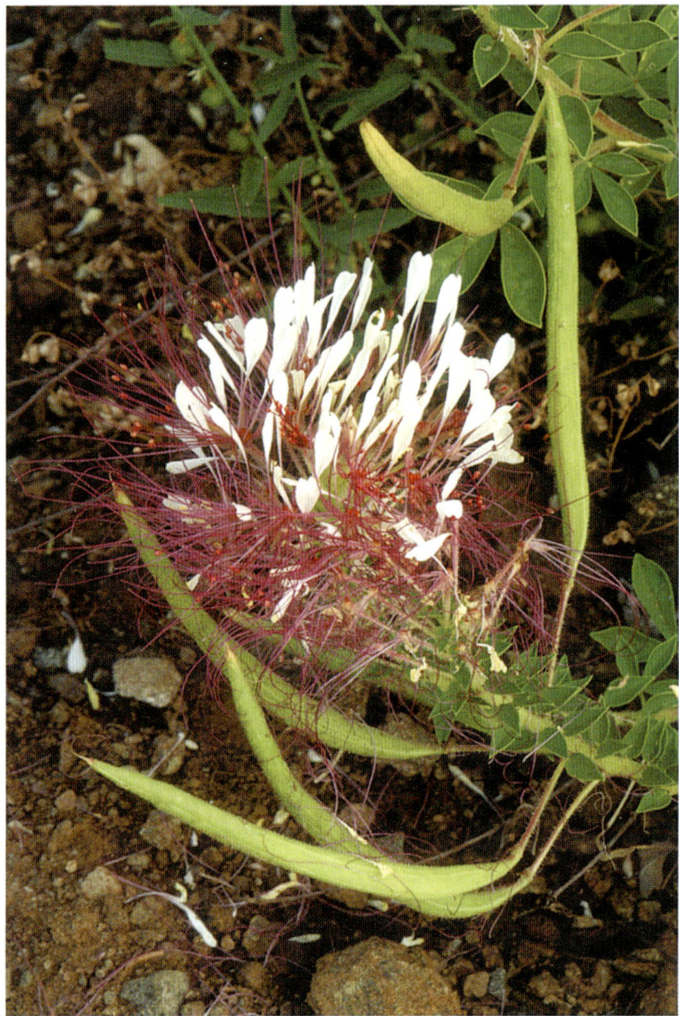

Also known as one-gland clammyweed and hierba del coyote, Mexican clammyweed is a robust annual up to two and one-half feet high with mostly unbranched, upright stems. The long-stalked leaves are divided into three elliptic to spatula-shaped leaflets, each about one and one-half inches long. This foul-smelling herb is coated with hairy glands that secrete a sticky liquid so that the entire plant feels damp or clammy.

In summer and early fall this annual produces spectacular multicolored flowers. In crowded clusters, the white flowers have four slender, upright petals that fade to sulphur-yellow as they age. Twenty or more long, whiskerlike stamens, in shades of purple and pink, extend well beyond the petals. The fruit is an erect, narrow pod up to four inches long, somewhat bladdery or inflated and dotted with hairy glands.

Mexican clammyweed inhabits sandy washes and loose, rocky hillsides above arroyos. This herb also thrives in disturbed areas as well as open flats and woodlands. It is widely dispersed throughout most of the Trans-Pecos from Big Bend National Park to El Paso and also in New Mexico and Mexico as far south as Oaxaca.

In the park Mexican clammyweed has been reported in the Chisos, in Upper Green Gulch, on Casa Grande, at the base of Bailey Peak, and in Oak Creek and Pine canyons. It has also been found at Harte Ranch to the north and at Fresno Spring south of the Chisos. I came across this bizarre annual on the Dodson Trail in August.

Note: The genus name *Polanisia,* from the Greek *polys* (many) and *anisos* (unequal), refers to the ways in which the stamens differ from those of the genus *Cleome.* The species name *uniglandulosa* means "one gland." This species was described in 1797 by Antonio Cavanilles, from a drawing by the Spaniard Martín de Sessé y Lacasta and the Mexican naturalist José Mociño. Sessé and Mociño led a natural history survey (1788–1803) of Spanish colonies in the Caribbean, Central America, and Mexico. Cavanilles, the preeminent Spanish taxonomic botanist of the eighteenth century, wrote and illustrated the six-volume *Icones et Descriptiones Plantarum* (1791–1801).

Trans-Pecos Spiderwort

Tradescantia brevifolia

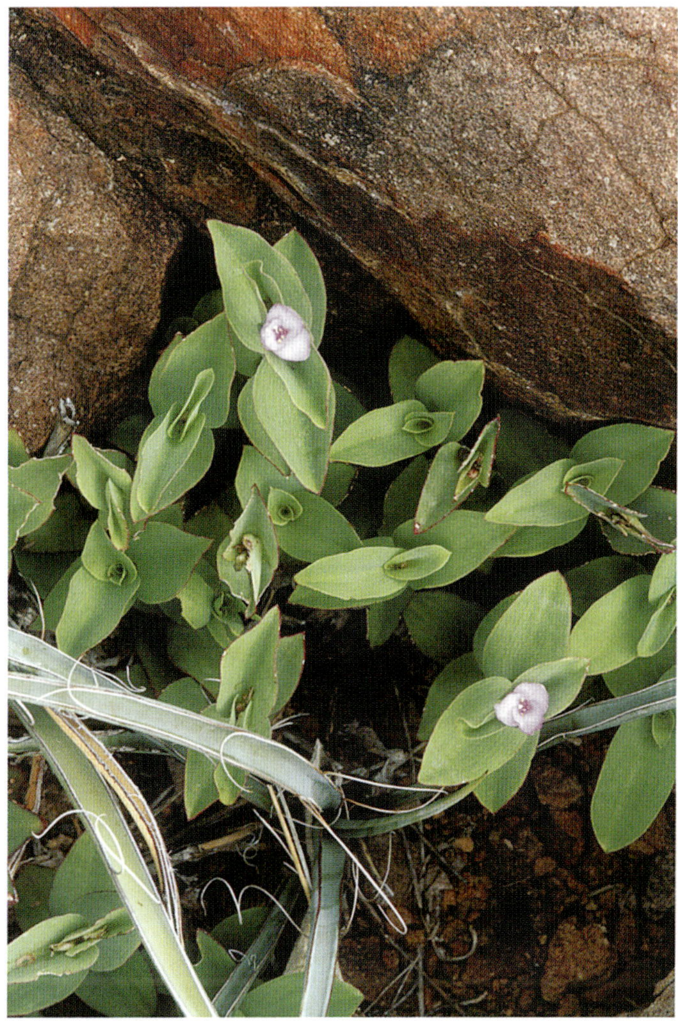

TRANS-PECOS SPIDERWORT is a succulent perennial herb with smooth, leafy shoots extending from underground stems. As much as one foot long, the green or occasionally deep purple shoots bear broadly egg-shaped or oblong leaves in spiraling clusters. Very rounded and often clasping the stem at the base, the stalkless leaves are less than three inches long and notably stout and thick.

Characteristic of the spiderwort (Commelinaceae) family, the pale rose or pink flowers have three broad, rounded petals that are united at the base and bracketed by two floral leaves. On densely hairy stalks, they bloom in summer and fall.

This perennial frequents wetter sites in igneous and also limestone soils. It snakes through crevices of rocky ledges and steep cliffs, among boulders or at the base of rockslides, especially near springs and seeps, creeks and arroyos. The herb is well adapted to mountainous terrain in the southern Trans-Pecos (in Jeff Davis, Presidio, and Brewster counties), to the east in Val Verde County, and in northeastern Mexico (in parts of Coahuila and Nuevo León and eastern Durango).

Trans-Pecos spiderwort is plentiful throughout the Chisos at mid to high elevations from the Basin to Pine Canyon and Casa Grande and on trails to the South Rim and Emory Peak. It has also been noted in the Rosillos Mountains. A similar species, canyon spiderwort (*Tradescantia leiandra*), with longer leaves and purplish flowers, also occurs in the park.

Note: The genus *Tradescantia* is named for John Tradescant the younger (1608–62) and the elder (c. 1570–1638). Both were travelers, botanical collectors, and gardeners to King Charles I of England. John the younger made three trips to Virginia and introduced the tulip tree, pitcher plant, and many other species to England. The two Tradescants developed a garden and collection of curiosities from all over the world, which became the Museum of Garden History. The species name *brevifolia* is a reference to the relatively short leaves. This species was described by John Torrey in 1859 (as a variety of *Tradescantia leiandra*).

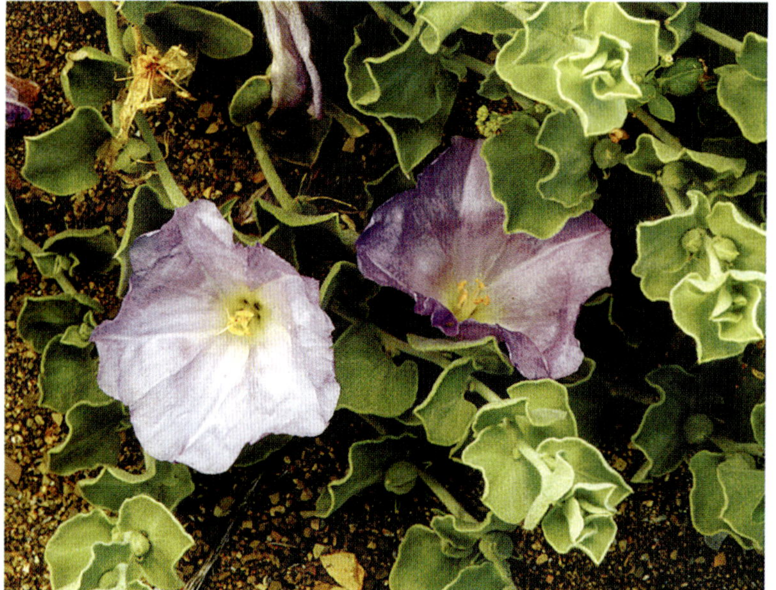

Bigpod Bonamia

Bonamia ovalifolia

BIGPOD BONAMIA is a rare morning glory limited exclusively to a few sites in Big Bend National Park. Although not legally classified as an endangered species, in Texas and worldwide bigpod bonamia is considered critically imperiled and in danger of extinction.

Rather than twining or sprawling over bushes like most morning glories, bigpod bonamia usually spreads over sand, on dunes, in sandy washes, or even on seldom-traveled backcountry roads. It grows from a woody crown and is easily recognized by the silvery or bluish-green, densely hairy oval leaves and funnel-shaped, wide-spreading flowers.

The glorious blooms are light blue to light purple or lavender on the outer edges, and even lighter to white in the deep throat. One and one-half inches wide, the solitary flowers often rest just above ground level, hidden beneath or between leaves that form a mat over the sand. In the park these plants typically bloom in early summer. Clumps or colonies of bigpod bonamia may bloom for days, but any single blossom opens in midmorning and scarcely lasts the day.

For years bigpod bonamia was known only from two locations in a canyon along the Rio Grande, but a couple of new sites have been discovered. One of these is outside the Rio Grande floodplain, so the hope is that this reclusive plant has gained a better foothold and a wider range than originally thought.

Note: The genus is named for François Bonamy (1710–86), French professor of medicine and botany at the University of Nantes. The species name *ovalifolia* refers to the oval leaves. This species was collected by Charles Parry in 1852 on the U.S.-Mexico boundary survey and described (as *Evolvulus ovalifolius*) by John Torrey in 1859. Parry found this morning glory on the "Rio Grande, below San Carlos," in Mexico. The site has never been relocated.

Creeping Rockvine

Bonamia repens

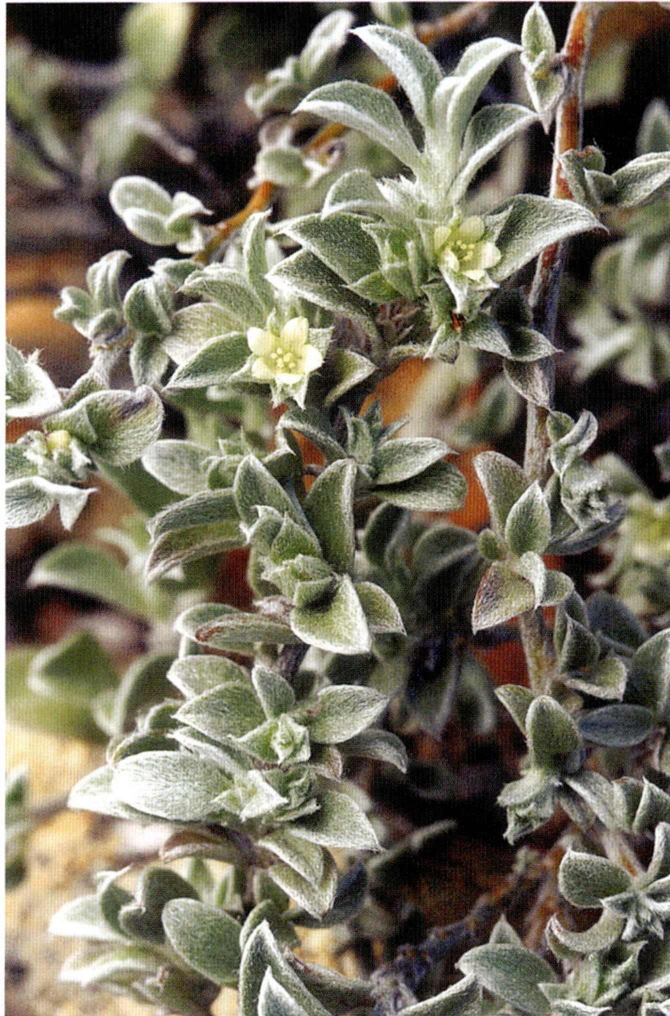

Creeping rockvine is a usually trailing perennial herb with stems up to eighteen inches long that may radiate outward in many directions and occasionally root at the nodes. The elliptic or egg-shaped leaves, on very short stalks (petioles), are normally less than one-half inch long and tightly clustered on short shoots off the main stems. Both leaves and stems are blanketed with silky hairs.

Sometimes tinged with green or yellow, the tiny, creamy white flowers are delicate and attractive. The five-lobed, bell-shaped, solitary blooms are nestled, sometimes almost hidden, among the leaves. They appear slightly succulent or waxy.

This rockvine inches its way along rocky crevices and between boulders on limestone hills. Concentrated primarily in the Chinati, Glass, and Dead Horse mountains (in Presidio, Brewster, and Terrell counties), it is distributed eastward to Del Rio and south through portions of northern Mexico to San Luis Potosí. Worldwide this herb is considered vulnerable throughout its range and imperiled in Texas.

In the park good examples of creeping rockvine have been located on the Telephone Canyon Trail in years past and at a location "11 miles down the Old Ore Road." I have hiked by this morning glory in the Chisos foothills in early September, on the west end of the Dodson Trail near Smoky Creek, and also in the Dead Horse Mountains in the hills above Ernst Tinaja.

Note: French botanist Louis Du Petit-Thouars described the *Bonamia* genus in 1804 from a plant he found in Madagascar. Born into a wealthy family, Du Petit-Thouars was imprisoned by French revolutionaries and in 1792 banished to southeastern Africa, where he spent the next decade collecting plants, especially orchids. The species name *repens* means "creeping." This species was collected for the first time in the United States by Barton Warnock in the Glass Mountains in 1941 and was described by the eminent American botanist Ivan Johnston that same year (as *Petrogenia repens*). Some botanists think this morning glory should remain in the genus *Petrogenia.*

Silver Ponyfoot

Dichondra argentea

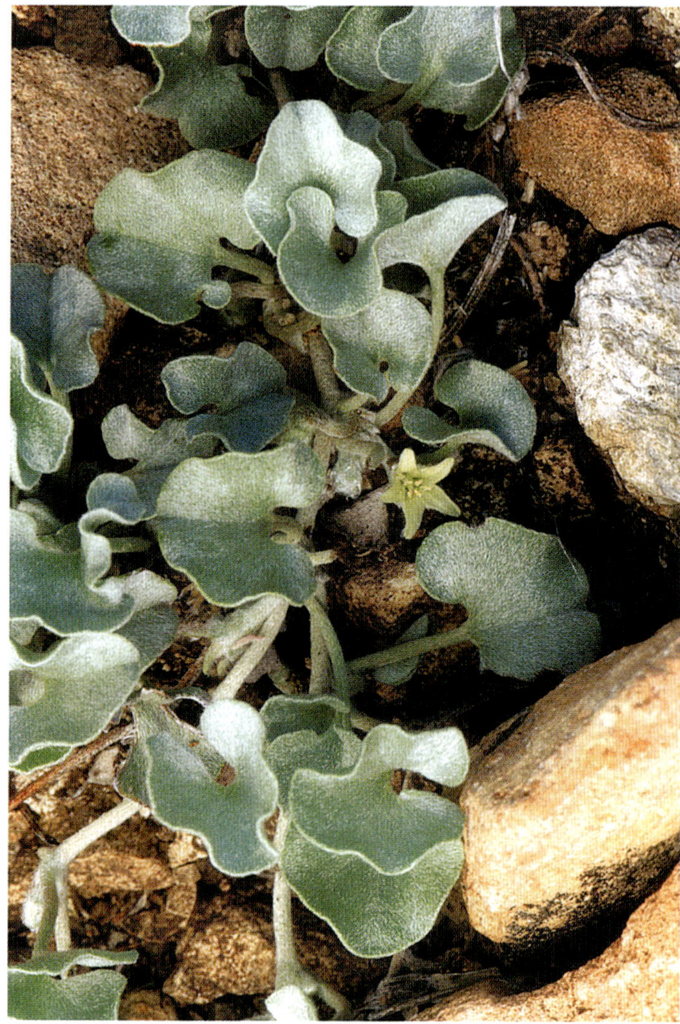

SILVER PONYFOOT, or oreja de ratón (mouse ear), is a creeping and trailing perennial herb no more than two inches high with many branching stems that root at the joints and form dense mats. Usually eight to sixteen inches long, the stems are coated with white, flattened, and sometimes shaggy hairs. The long-stemmed, silvery leaves are almost circular or kidney-shaped, perhaps one inch long and even wider, with a rounded tip and deeply notched, heart-shaped base. Fine, straight, silky hairs are tightly pressed against the surface of these blades.

Not much more than one-eighth inch long, pale yellow-white or greenish-white flowers are borne at the base of leaves along the stem. Nestled among the leaves, the solitary flowers usually hang downward on curving stalks. Blooming from spring to fall, the bell-shaped blossoms have five pointed lobes with long silvery hairs on the edges and outer surface.

Silver ponyfoot constructs mats over dry rocky and gravelly slopes from low hills to higher mountains, on clay and sandy flats and in brushy canyons, on limestone and igneous soils. It ranges widely from Arizona and New Mexico into Trans-Pecos Texas (El Paso to the Big Bend) and is even more far-reaching in Mexico (Chihuahua east to Tamaulipas and south to Chiapas). This delightful but often unnoticed morning glory reaches its geographic limit in Argentina and Bolivia high in the Andes.

In the park silver ponyfoot is uncommon in the Chisos Mountains. It has been recorded at Bois d'Arc Spring, on the lower east slopes of Crown Mountain, and higher in the Chisos between the Blue Creek and Colima trails. This photograph was taken in Lower Juniper Canyon.

Note: The species name *argentea* means "silvery." This species was collected by Baron von Humboldt and Aimé Bonpland at Tolima, Columbia, in about 1800 and described by German taxonomist Carl Willdenow, director of the Berlin Botanical Garden, in 1809. Willdenow, who edited a new multivolume edition of Carolus Linnaeus's *Species Plantarum* after Linnaeus's death, was a friend of Humboldt and a decisive influence on his interest in the New World.

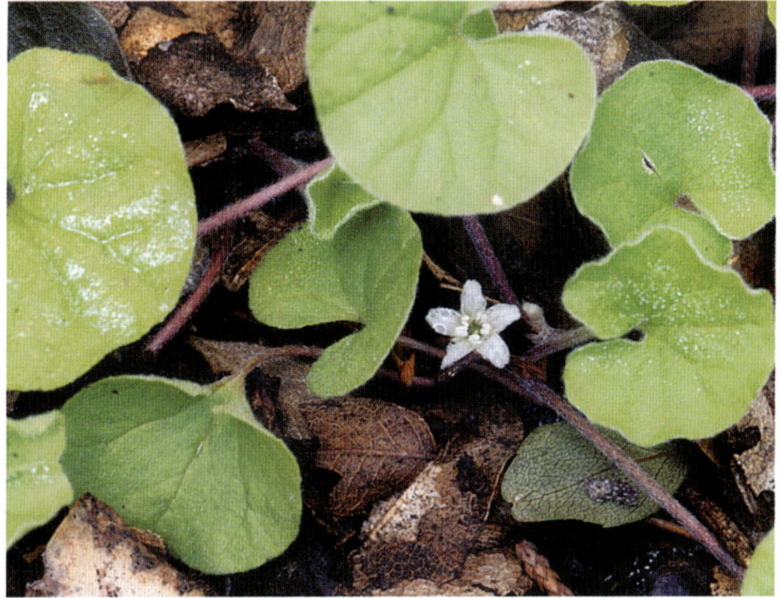

New Mexico Ponyfoot

Dichondra brachypoda

NEW MEXICO PONYFOOT is a trailing perennial herb that spreads across damp soil, rooting at the joints of the stems. The ear-shaped leaves grow from long stalks, forming mats of little "earlobes" several feet across, blanketing the ground. Covered with thin hairs, the leaves are gray-green and often slightly yellowish.

This morning glory, no more than three inches high, has small flowers that are seldom seen. The five-petaled flowers hang beneath the leaves on vinelike stalks and usually point toward the ground. Mostly solitary, they are white or yellowish-white and bell-shaped.

Blooming in summer, ponyfoot colonizes moist rocky soils, especially in shade. It clings to the wet banks of creeks and deep canyons. This creeping herb is very common in Trans-Pecos mountains (in Pecos and Terrell counties and across the Big Bend to Culberson County) and also in New Mexico, Arizona, and Mexico (from Chihuahua and Coahuila to San Luis Potosí and even Oaxaca).

In the park New Mexico ponyfoot is pervasive in the Chisos: in the Basin, lower Oak Creek Canyon, on the lower east slopes of Crown Mountain, at Laguna Meadow, and in Upper Cattail Canyon. I photographed this plant many times before finally finding it in bloom on the Laguna Meadow Trail in August. It is quite a trick to spot the tiny flowers, so this attractive ground cover seldom gets its picture taken.

Note: The genus name *Dichondra* originates from the Greek *di* (two) and *chondra* (grain), a reference to the distinctly lobed fruit. The species name *brachypoda* is derived from the Greek *brachys* (short) and *podion* (foot), probably referring to the relatively short flower stalks. This species was described by botanists Elmer Wooton and Paul Standley in 1913 from a specimen they collected in the Organ Mountains of New Mexico.

Crestrib Morning-Glory

Ipomoea costellata

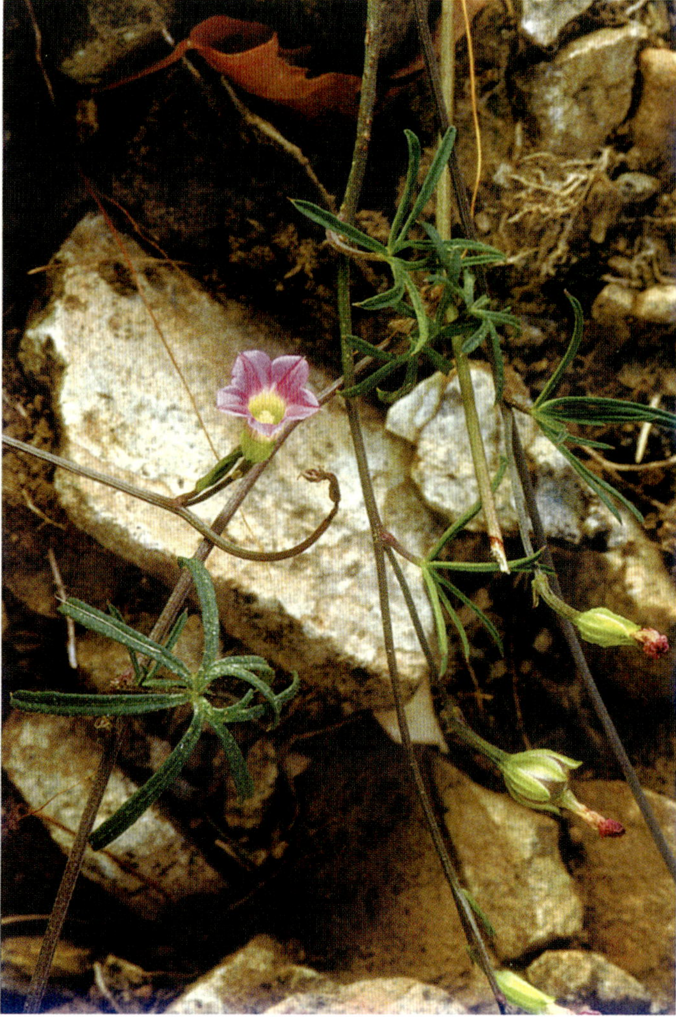

MOST MORNING GLORIES in Big Bend sport very large and showy flowers, and their displays are a pleasure to discover on early morning hikes. Stumbling across this miniature morning glory, so easily missed, is extra special.

Crestrib morning-glory is a low annual, with stems occasionally as much as one and one-half feet long. At first erect, it trails or twines over rocks or other plants as it ages. The leaves are deeply divided, usually into seven to eleven narrowly linear leaf segments, most of them radiating from a central point. The lateral segments are shorter and deeply cut.

Long, slender stalks off the trailing stems support one to three pink or rose, funnel-shaped flowers with white or yellow-white throats. Each of five petal-like lobes has a distinct midstripe or crest, and the sepals forming the green, external base of the flower may also be noticeably ribbed.

Blooming from July to October, crestrib morning-glory occupies rocky igneous slopes in Trans-Pecos mountains (from Brewster County to El Paso). This delicate herb is also present in the Texas Hill Country and near Laredo in the Rio Grande Plains. Although often overlooked, it is notably wide ranging, west through southern New Mexico to Arizona, and south throughout northern Mexico (from Tamaulipas to Baja California and south to Durango and San Luis Potosí).

You might spy this morning bloomer in Green Gulch or the Basin, below Bailey Peak, or higher in the Chisos above Boot Spring or near the top of Mount Emory. Occasionally, it twines through the rockslides below the Window (but above Oak Creek) and near the Oak Spring Trail.

Note: The genus name *Ipomoea* originates from the Greek *ips* (bindweed, or worm) and *homoios* (like) because of the plants' wormlike, twining habit or their resemblance to bindweed (the genus *Convolvulus*). The designated type specimen was collected by Charles Wright in New Mexico in the Rio Grande Valley near El Paso in 1849. This species was described by John Torrey in 1859 in the botany appendix of the *Report on the United States and Mexican Boundary.*

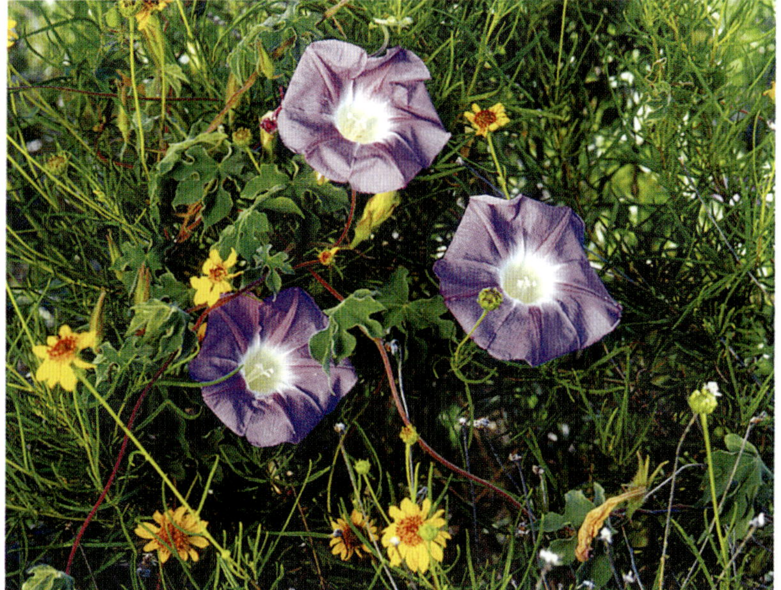

Blue Morning-Glory

Ipomoea lindheimeri

Blue morning-glory, or Lindheimer morning-glory, is a perennial twining vine that trails across rocks and drapes over shrubs and brush often at dry and open sunny sites. This vine is readily distinguished by its large leaves—up to three and one-half inches long and wide—which are deeply cut into three to seven irregular lobes. Covered with short hairs, each lobe is often sharply and abruptly constricted at the base.

Each glorious flower opens for a single day, and then only fleetingly in early morning before the sun rises high. Appearing on long stalks, the blooms are notably variable in color, from pale blue to bluish-purple, with five petal-like lobes united to form a funnel-shaped tube. Blooming mostly in summer and fall, the pleated blossoms have delicate white throats.

Blue morning-glory flourishes on rocky soils and often in rocky and wooded canyons, along arroyos and creek beds, and near springs, from foothills to the highest mountains. This species is scattered throughout much of the Trans-Pecos (from Hudspeth County to Del Rio) and Edwards Plateau to the Hill Country near Austin. It also occurs in parts of New Mexico and Arizona, and in northern Mexico (from Chihuahua to Tamaulipas).

In the park you should find this vine in the Chisos Mountains and foothills from Green Gulch and the Basin, Pine Canyon and Ward Spring, to higher elevations at Laguna Meadow, Boot Canyon, and Emory Peak.

Note: The type specimen for this plant was collected by Charles Wright in 1849, and Asa Gray described the species in 1878. Gray named the plant for Ferdinand Lindheimer, who had collected the vine even earlier near New Braunfels. Nearly fifty Texas plants are named for Lindheimer, who immigrated to Texas because of political unrest in his native Germany and who later became known as the "Father of Texas Botany."

Longpetal Echeveria

Echeveria strictiflora

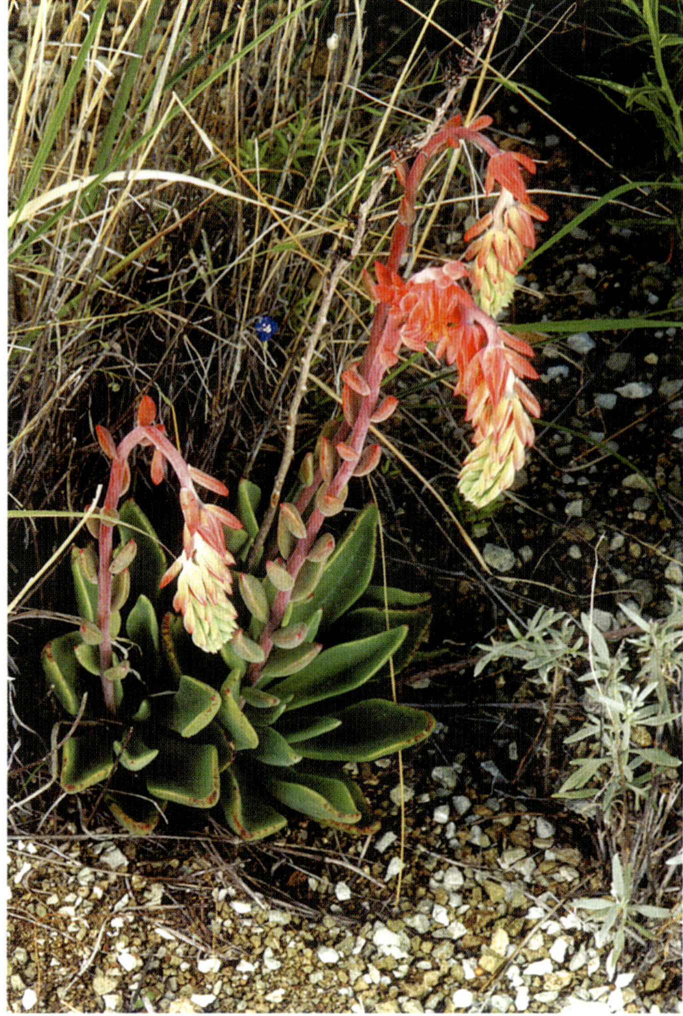

ALSO KNOWN AS live forever and desert savior, longpetal echeveria is a perennial herb usually one to one and one-half feet high when in bloom with a prominent flowering stalk. Thick fleshy leaves up to four inches long form a rosette at the base. The pale green or gray-green basal leaves are spatula-shaped. Leaves along the stem are much reduced and swollen.

Nodding flower clusters sit atop reddish flowering stalks as much as sixteen inches high. Blooming after summer rains brighten the landscape, the flowers are orange-red but yellow inside, with petal-like lobes united in the shape of a five-sided, nodding pyramid. Impressed with their symmetrical leaves and distinctive long-stalked flowers, enthusiasts have bred many ornamental hybrids of this large genus of succulents.

This member of the orpine or stonecrop (Crassulaceae) family is usually seated on rocky ledges or lodged between rock crevices in the foothills and mountains of Trans-Pecos Texas (in Brewster, Presidio, and Jeff Davis counties) and northern Mexico (from eastern Chihuahua to western Nuevo León).

Rather sparse in the park, longpetal echeveria often blooms from June to August at Ward Spring and in the hills above Lower Juniper Spring in the Chisos. I have witnessed striking displays at several other delightful and secluded little springs in the Chisos foothills. This fleshy herb has also been seen in the Rosillos Mountains and on top of the Dead Horse Mountains.

Note: Described by Augustin de Candolle in 1828, the genus *Echeveria* was named for botanical artist Atanasio Echeverría y Godoy, who accompanied the 1788–1803 Sessé and Mociño expedition to Mexico and painted watercolors of many of the plants. Sessé and Mociño returned to Spain in 1803 with an extensive herbarium and about eighteen hundred botanical sketches, but Sessé died in 1808, and the expedition results were not published until 1887, as *Flora Mexicana* and *Plantae Novae Hispaniae.* The species name *strictiflora* refers to the often "tightly closed flower." This species was collected by Charles Wright in 1849 in the Davis Mountains west of Limpia Canyon and described by Asa Gray in 1852.

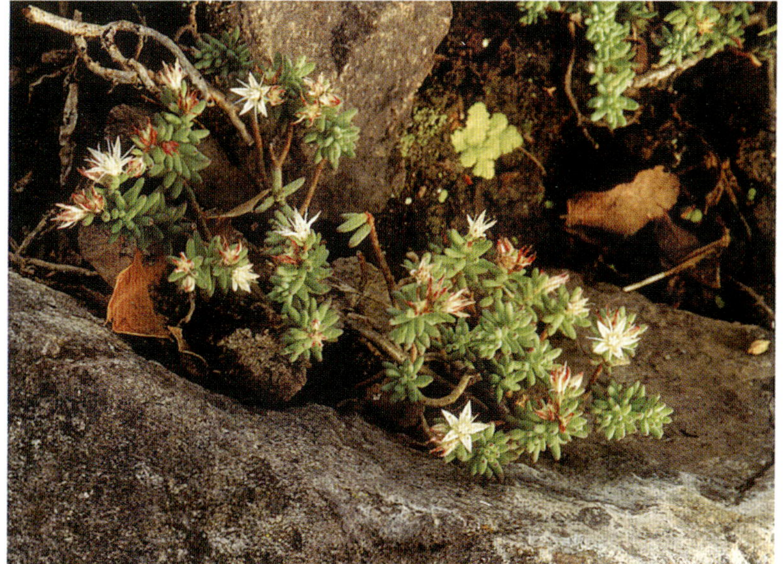

Havard Stonecrop

Sedum havardii

HAVARD STONECROP is endemic to higher elevations in the Davis and Chisos mountains of Trans-Pecos Texas and the adjacent Mexican state of Coahuila. Globally and in Texas, it is classified as imperiled and very vulnerable to extirpation.

This rare herb is a trailing perennial notable for its sprawling reddish stems, which may extend as much as five inches or more. Branching profusely from the base, the stems are covered with protruding tubercles and smooth, fleshy, tightly crowded and typically overlapping leaves. The mostly stemless linear leaves, about one-fourth inch long, may appear slightly swollen or inflated.

A few flowers, with five lancelike and widely spreading white petals and often reddish tips, cluster at the ends of the low-growing branches.

These succulents form clusters at the base of cliffs or trail over talus slopes and through narrow rock crevices in mountain woodlands. Search for Havard stonecrop ensconced among rocks or boulders at wetter, shadier sites in the Chisos Mountains. It blooms from June to September at springs and in moist Chisos canyons and drainages.

Note: The genus name *Sedum* may be derived from the Latin *sedeo* (to sit), referring to the way these succulents sit on rock ledges. In 1905 Joseph Rose, botanist with the U.S. National Herbarium, named this species for Valery Havard, who collected it in the Chisos Mountains. Havard, a French-born medical doctor and self-trained U.S. Army botanist, studied and collected plants in Montana, North Dakota, and Colorado as well as Texas. His *Report on the Flora of Western and Southern Texas* was published in 1885, and he later became assistant surgeon general of the United States. Numerous Texas plants bear his name, including the Big Bend bluebonnet (*Lupinus havardii*) and Havard agave (*Agave havardii*).

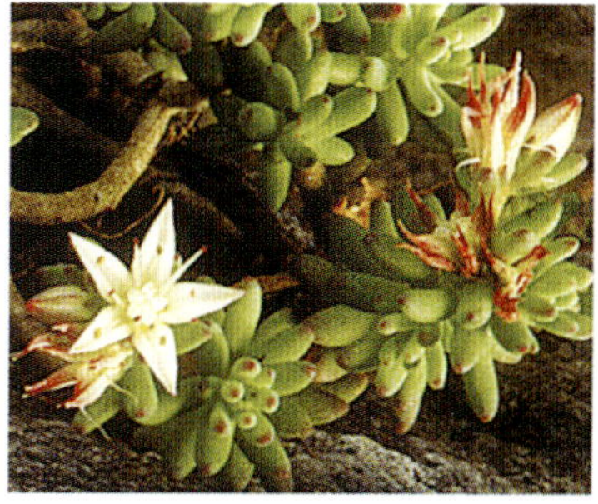

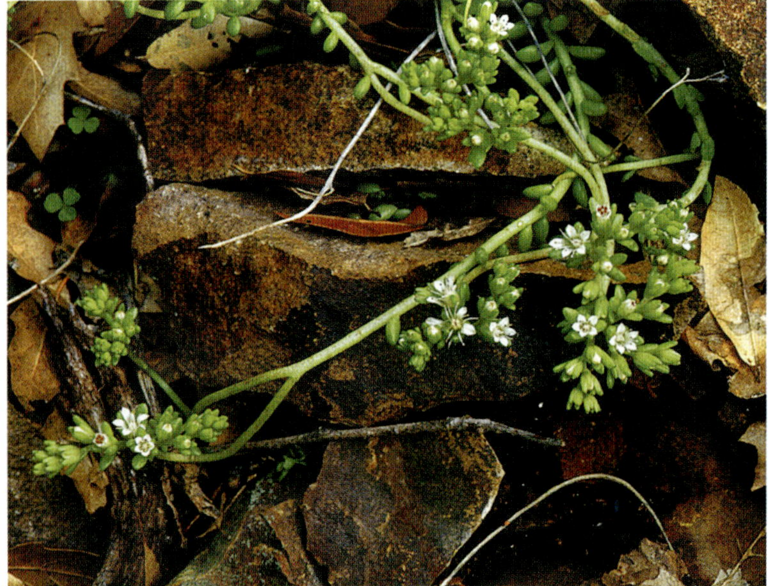

Wright Stonecrop

Sedum wrightii

WRIGHT STONECROP is a fleshy perennial herb that may approach eighteen inches in length. Widely branching, the weak stems may lie close to the ground or straggle far from the basal leaves. The tiny leaves, one-third inch long, are smooth and succulent, often bunched and overlapping.

Occasionally tinged with rose, the dainty white flowers have five short, rounded petals. No more than one-fourth inch wide, the blossoms grow on very short stalks in small, tight, leafy clusters.

Blooming in summer and fall, this succulent prefers rocky soils from foothills to high mountains, especially rocky outcrops in moist, shady canyons. Although classified as vulnerable in Texas (NatureServe 2005), Wright stonecrop is encountered, sparingly, throughout much of the Trans-Pecos, from the Guadalupe Mountains south to the Chisos Mountains. Its range extends farther east into Val Verde and Uvalde counties, west into New Mexico, and south into northern Mexico, to Zacatecas and San Luis Potosí.

Wright and Havard stonecrop plants are often confused, but the stems of Havard stonecrop bear conspicuous tubercles. Also, the flower petals of Wright stonecrop are broader and less pointed at the tips, and the blooms may appear at the ends of long, arching stems. In the park this infrequent herb is most prominent in Boot Canyon and similar habitats, and near springs high in the Chisos.

Note: In 1852 Asa Gray named this species for Charles Wright, who had collected it in 1849. As botanical collector on military survey expeditions in 1849 and 1851, Wright made a major contribution to the botany of Texas, New Mexico, and Arizona. Less known is his subsequent work as botanist on the Ringgold expedition to Japan and northeastern Asia, and his eleven years, beginning in 1856, of productive botanical explorations in Cuba.

Stonecrop

Villadia squamulosa

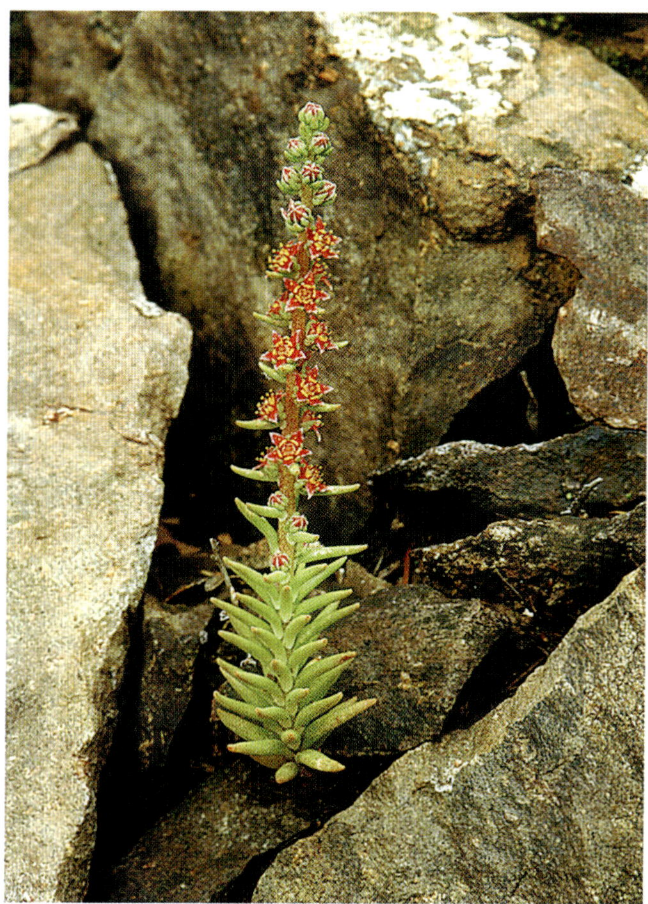

Also known as rat's-tail succulent, stonecrop is an erect, succulent perennial up to one foot high with one or sometimes several smooth, slender stems that are perhaps a little like a rat's tail. Fleshy, cylindrical leaves up to one inch long radiate from the stem and taper to a slender point. Crowded along the stem, they appear distinctly swollen or inflated.

In an elongated flower cluster up to six inches long, star-shaped flowers are nestled into compact groups of one to three. Each small, fleshy flower has five spreading, pointed lobes that join into a short tube. The lobes are mostly ruddy red, with pale green to whitish tips and margins, and very prominent yellow scales or glands bearing the nectar. The flowers may open from late spring to fall.

Stonecrop sits on rocky cliffs and talus slopes and in crevices of boulders at high elevations, usually in cooler, shadier, and wetter niches. About twenty-five mostly Mexican species of this genus are recognized, but this is the only one entering the United States. Its distribution spans the montane regions of northwestern Mexico—from Zacatecas through Durango to Chihuahua—and just barely Big Bend National Park.

In the park stonecrop is restricted to the slopes and canyons of the Chisos. It blooms especially in September along the upper Pinnacles Trail and on the trail to the South Rim above Boot Spring.

Note: This genus is named for Mexican scientist Manuel Villada (1841–1924), editor of the journal *La Naturaleza.* The species name *squamulosa* is "scaly" in Latin. This species was described by Sereno Watson, curator of the Gray Herbarium at Harvard, in 1887 (as *Cotyledon parviflora* var. *squamulosa*). The plant had been collected by Cyrus Pringle in 1886 on Portrero Peak in Chihuahua. Horticulturist and botanical collector, Pringle got his start breeding potatoes and grapes on his family farm in Vermont and searching for rare plants in the hills nearby.

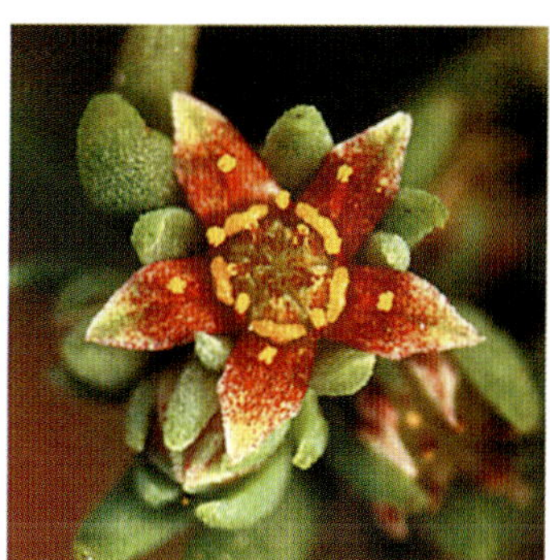

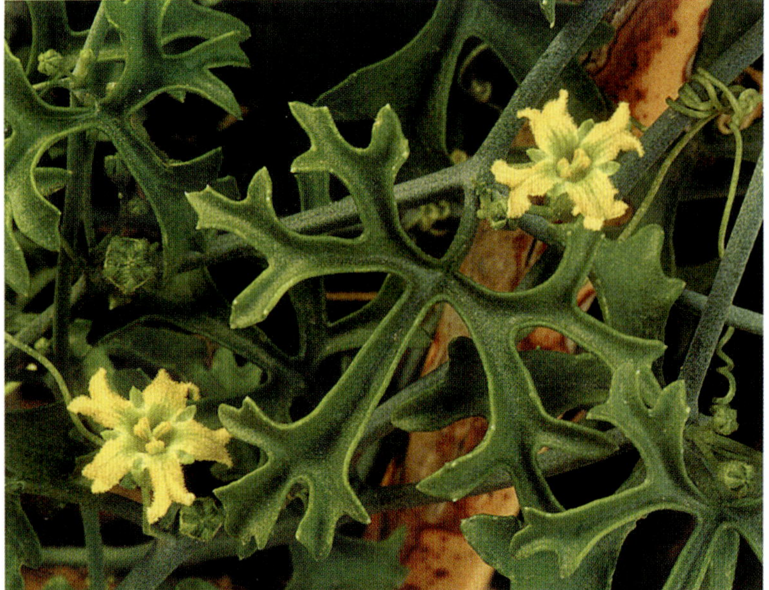

Slimlobe Globeberry

Ibervillea tenuisecta

Slimlobe globeberry, or balsam apple, is a fleshy perennial vine with narrow, smooth stems up to ten feet long. Branching extensively, this vine climbs in other plants, especially allthorn (*Koeberlinia spinosa*), via coiling tendrils. The thick leaves are two to three inches wide and nearly as long. They are deeply cleft into three to five slender lobes that are also lobed or coarsely toothed.

Blooming mostly in summer, the greenish-yellow male and female flowers appear on separate plants. Usually in small clusters, the funnel-shaped male blooms have five hairy petals with distinctly ruffled edges. The female blossoms are solitary. The fruit is a fleshy, melonlike, thin-skinned globe, green with white stripes until ripe, and when ripe it becomes a bright orange-red ornament decorating the vine.

This vine drapes over desert scrub and brush on rocky hills and clay flats, along desert arroyos, near creeks and springs, and in floodplains. Slimlobe globeberry is native to the Trans-Pecos and the western Edwards Plateau in Texas as well as southeastern Arizona and parts of northern Mexico (from Sonora to Coahuila and south to Zacatecas and San Luis Potosí).

Slimlobe globeberry is widely dispersed throughout the park. It has been reported in the Chisos at Gano and Dominguez springs and K-Bar Ranch, in the Dead Horse Mountains at Dagger Flat and elsewhere, in the southwest from Castolon to Terlingua Creek, and in the north at Harte Ranch.

Note: The genus *Ibervillea* may be named for Iberville Parish in Louisiana or for Pierre d'Iberville, a Canadian-born French naval officer who in 1698 led an expedition to Louisiana and the lower Mississippi. The species name *tenuisecta* (finely cut) refers to the leaves. This species was collected by Charles Wright in New Mexico in 1851 and described by Asa Gray (as *Sicydium lindheimeri* var. *tenuisectum*) in 1852.

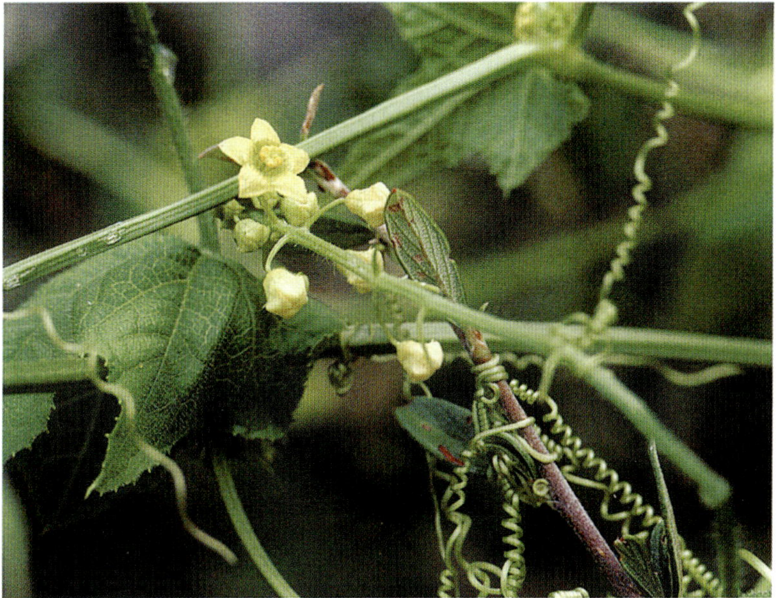

Smooth Bur Cucumber

Sicyos glaber

Smooth bur cucumber is an annual vine with slender, fleshy stems. It climbs over shrubs and trees, occasionally as high as fifteen feet or more, on forking, coiling tendrils. The large, rough leaves are round to egg-shaped in outline and somewhat heart-shaped at the base. About three inches long and wide on two-inch-long stems, they are divided, sometimes deeply, into three or five triangular lobes with wavy or finely toothed edges.

Separate male and female blossoms appear on the same plant, sometimes from the same leaf axil. Only one-fourth inch across, the pale yellow or white, bell-shaped flowers seem almost an afterthought. They have five oval lobes that tend to spread widely and curve backward. Unlike the fruit of most bur cucumbers, the dry, green fruit is smooth, spineless, and only one-fourth inch long.

This vine clings to shrubs on shady slopes and in moist canyons in igneous soil. A member of the gourd (Cucurbitaceae) family, it is confined to higher elevations in the mountains of southeastern New Mexico and Trans-Pecos Texas. In the Trans-Pecos, mesic canyons in the Chisos and Guadalupe mountains are the only known habitats. Globally and in Texas, smooth bur cucumber is vulnerable throughout its range.

Although rare in the park, this annual has been observed above Boot Spring, along Boot Creek toward the South Rim. I photographed it there at the end of August, but September may be the ideal time to hunt for the climbing vine.

Note: The genus name *Sicyos* is from the Greek *sikyos* (cucumber). The species name *glaber* means "smooth." This species was collected by New Mexico botanist Elmer Wooton in the Organ Mountains of southern New Mexico in 1897 and described by him the following year. In 1915 Wooton coauthored with Paul Standley the first account of the state's vascular plants, *Flora of New Mexico.*

Arizona Cypress

Cupressus arizonica

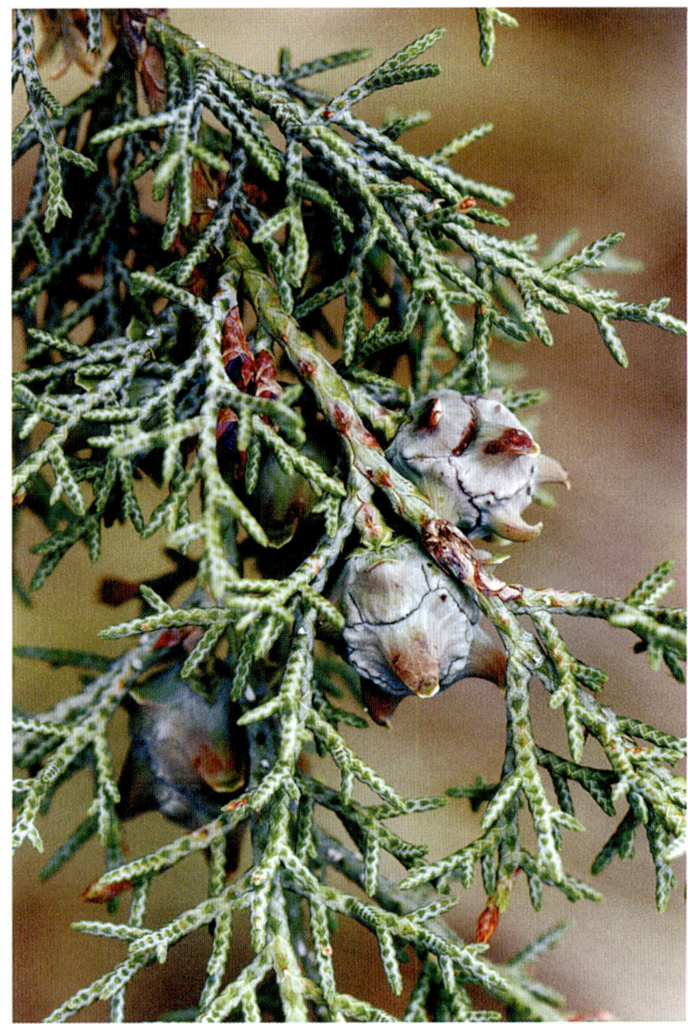

ARIZONA CYPRESS, cedro blanco, or pinabete is an imposing evergreen tree with a single straight trunk, rarely seventy feet high or more and usually much smaller. Distinctly resinous and aromatic, this cypress has many short, rigid branches that form a dense, conical crown. The gray bark of younger branches is mostly smooth, but the dark gray or nearly black bark on older trunks and branches becomes rough, fibrous, and shredding.

Rounded and overlapping, minuscule scalelike leaves only one-sixteenth inch long are tightly pressed against the branches. The sharp-pointed blades are gray-green or blue-green, often with a white waxy coating. Male and female cones appear on the same tree. The male cones are very small and oblong to cylindrical. The rounded female seed cones, as much as one inch wide, form on short stalks at the tips of branchlets. They are blue-green and covered with six to ten shieldlike, interlocking scales, each with an obtrusive hornlike projection. At first leathery, the cones become hard, brown, and dry as they mature in the second year.

In Texas, Arizona cypress is isolated at higher elevations of the Chisos Mountains in Big Bend National Park. Considered critically imperiled in Texas (NatureServe 2005), it is more extensive in southeastern Arizona, southwestern New Mexico, and northern Mexico (from Sonora to Coahuila and also northern Durango).

In the park this towering tree grows only in or near Boot Canyon in the Chisos. The state champion tree in Boot Canyon is 112 feet high with a circumference of 134 inches.

Note: The genus name *Cupressus* is the Latin name for the Italian cypress (*Cupressus sempervirens*). The species *arizonica* was described in 1882 by Edward Greene, an Episcopal priest in Silver City, New Mexico. Greene collected the specimen in 1880 in the San Francisco Mountains near Clifton, Arizona. Later giving up the priesthood, Greene became a botanist at the University of California, Berkeley, and subsequently at the Smithsonian Institution. In 1909 the Smithsonian published his *Landmarks of Botanical History (Part I)* on the botany of ancient Greece and Rome and the German herbalists.

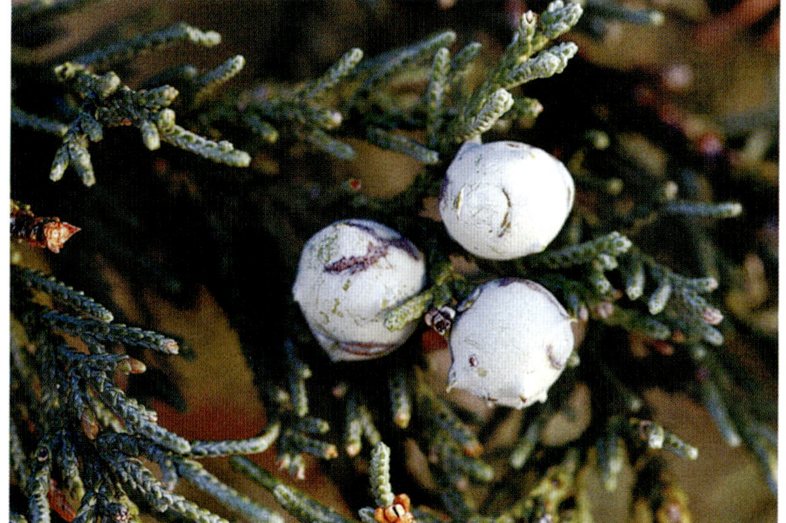

Alligator Juniper

Juniperus deppeana var. *deppeana*

Many botanists discourage the use of common names for plants, but a good common name can be useful in plant identification. This works for alligator juniper, a tree thirty to sixty feet high, because its checkered bark resembles an alligator's back. The thick bark splits into quadrangular plates strikingly similar to alligator skin.

Closely appressed, the tiny blue-green leaves look more like scales than leaves. Male and female fruits are borne on separate plants, and female trees produce miniature, reddish-brown cones with fleshy scales. The cones are often covered with a bluish-white, waxy coating, and the bluish leaves and fruits give the tree a distinctive cold aura.

The cones (known as juniper berries) provide a feast for birds and mammals. Native Americans ate them raw and cooked them with food. In Chihuahua the leaves were used to treat rheumatism and neuralgia.

Also known as checkerbark juniper, enebro, and abori (Tarahumara), this tree is common at mid to high elevations in the Trans-Pecos, especially the Davis and Chisos mountains. With oaks and pinyon, it dominates open rocky slopes and grassy hills. Alligator juniper ranges west to central Arizona and south to Mexico City and Puebla.

In the park this juniper is well established in Green Gulch, the Basin, on Casa Grande, the Lost Mine Trail, at Laguna Meadow, and in Boot Canyon.

Note: The genus name *Juniperus* is the old Latin name for juniper. In 1840 Ernst Steudel, Swiss botanical bibliographer, named this species for Ferdinand Deppe, a German naturalist, explorer, and painter who collected in Mexico and Central America from 1824 to 1829. Deppe traveled with German physician Christian Schiede, who had emigrated to Mexico in 1828, and the two collected this species in Veracruz, Mexico, that same year.

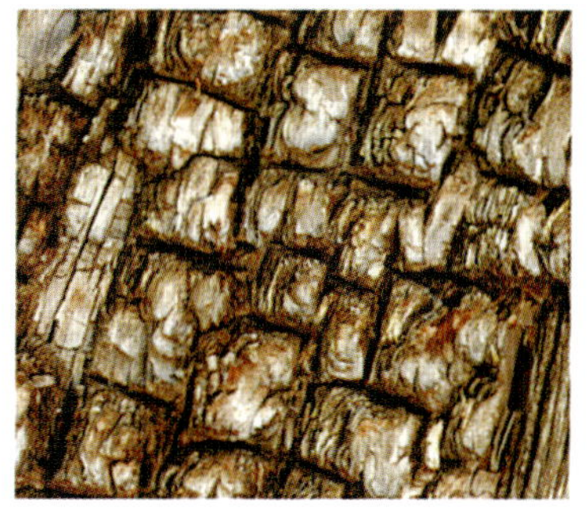

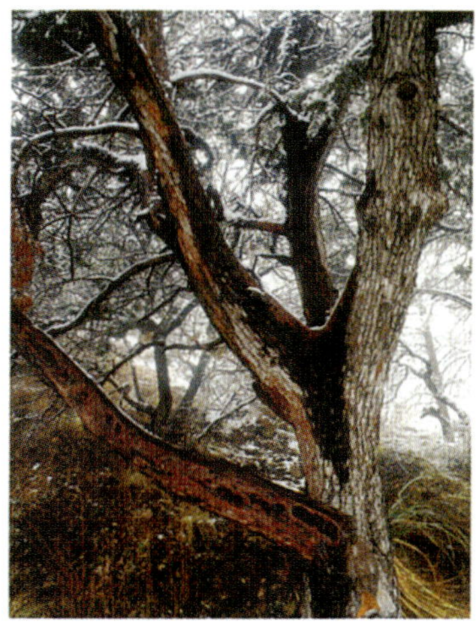

Drooping Juniper

Juniperus flaccida

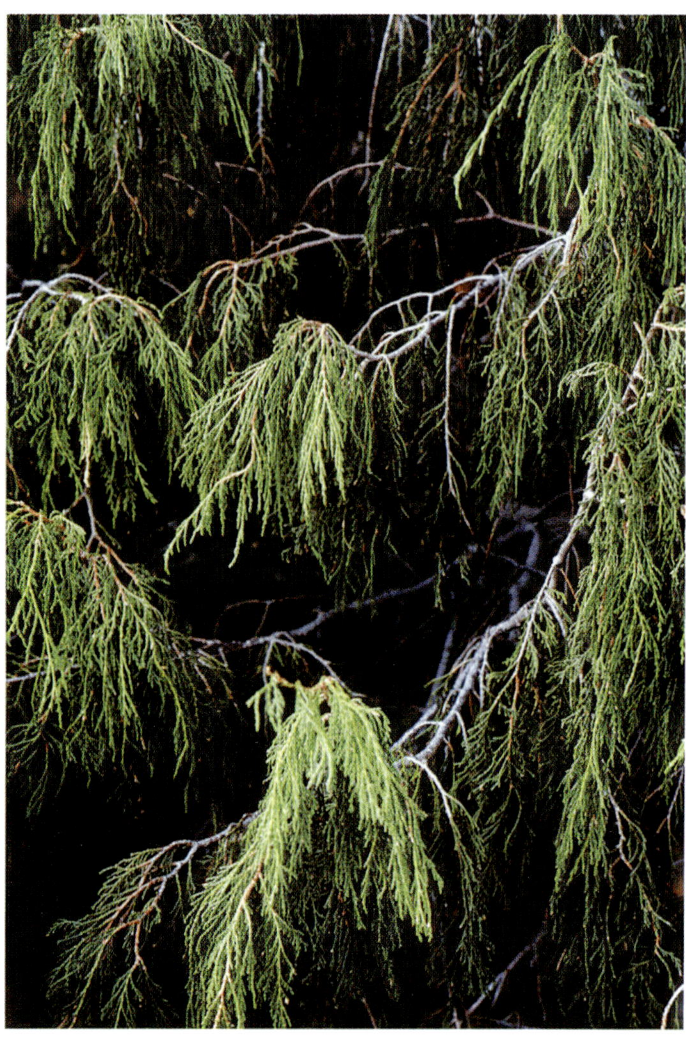

Also known as weeping juniper, tascate, and sabino, drooping juniper is a small or medium-sized evergreen tree up to thirty-five feet tall or more with a globose crown. It bears distinctively drooping branches and cinnamon-colored, furrowed bark that shreds in long, wide strips. Even the green leaves are drooping and graceful. The scale leaves, tiny leaves no more than one-eighth inch long, partially overlap and often press tightly to the branchlets. The whip leaves (leaves on new growth), about three times longer, are slender and taper to a small, sharp point.

In juniper, the male and female reproductive organs usually appear on separate plants. The juniper fruits are conspicuous little cones so fleshy that they resemble berries. Maturing in the fall, the one-half-inch-wide globelike seed cones are reddish-brown or purplish-brown but so waxy that they appear bluish-white. The thick, resinous coat often covering the fruits is distinctly sticky to the touch.

Drooping juniper reaches the United States only in the Chisos Mountains, where it is a fixture above five thousand feet on rocky and wooded slopes. It is widespread in Mexico, from northeastern Sonora to Tamaulipas and south to Oaxaca, and into Guatemala.

In the park these trees are unmistakable in Upper Green Gulch and the Basin, and along the Lost Mine Trail. They also droop over Pulliam Bluff and most of the mountain trails (Laguna Meadow, Colima, Emory Peak, Boot Canyon). The magnificent national champion tree, fifty-five feet high, is found in Upper Juniper Canyon.

Note: The species name *flaccida* means "flaccid," alluding to the drooping habit. This species was described by German botanist Diederich von Schlectendal in 1838 and collected by Carl Ehrenberg in the Mexican state of Hidalgo. Brother of the famous microscopist Christian Ehrenberg, Carl was an enterprising botanist specializing in cactus, who collected in the Virgin Islands, Haiti, and Mexico in the 1820s and 1830s. Schlectendal described many of Ehrenberg's discoveries and in 1831 coauthored a book on the trees of Mexico with the German naturalist and poet Adelbert von Chamisso.

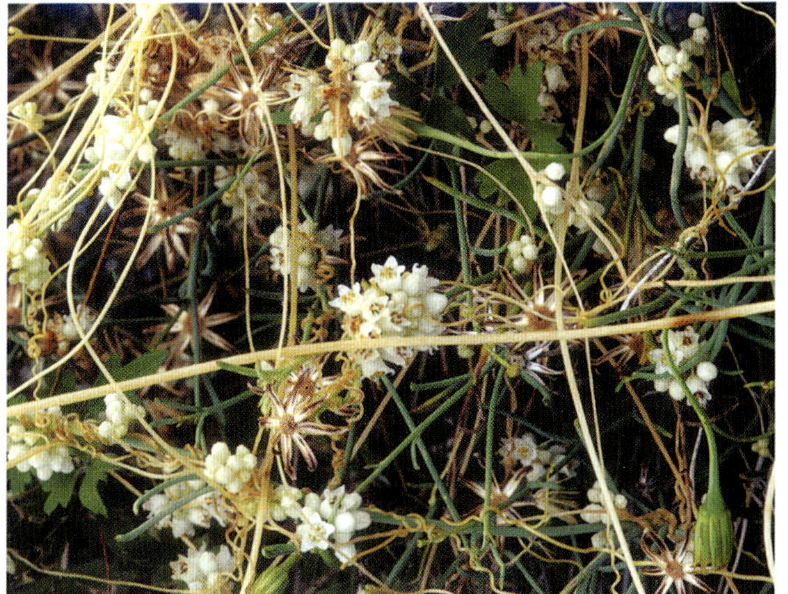

Pretty Dodder

Cuscuta indecora

PRETTY DODDER has inspired many graphic names: complimentary ones like love vine and angel hair, and pejorative ones like witches' shoelaces and strangle vine. This herb is a twining parasitic vine with fleshy orange, stringlike stems.

Although considered both rootless and leafless, pretty dodder does produce much reduced, scalelike leaves. Lacking in chlorophyll, dodder is unable to manufacture nutrients from sunlight and instead extends highly specialized suckers or rootlike connections that attach to the stems of a wide variety of other plants.

Mostly in summer and fall, small white flowers only one-eighth inch long bloom in scattered but sometimes dense clusters along the stems. Each bell-shaped blossom opens into five grainy lobes with tips that are pointed but notably turned inward. Typically, fringed scales or small nipplelike projections are located at the base of the flower tube.

This parasite envelops other plants along roadsides, in disturbed areas, and also on open flats and in dense thickets. It is wide ranging throughout most of the United States and most of Texas as well as Coahuila and Nuevo León in northern Mexico.

In the park pretty dodder is sometimes abundant in the Chisos foothills, along the road from Panther Junction to the Basin turnoff, and in Green Gulch, but this twining vine may show up almost anywhere, including in the Rosillos Mountains or at Santa Elena Canyon.

Note: The genus name *Cuscuta* is the ancient Latin name for dodder, presumably from the Arabic *kuskut.* The species name *indecora* implies "unattractive." The common name dodder refers to the weak or "doddering" stems. Jacques Choisy, Swiss botanist, clergyman, and philosopher at Geneva, described this species in 1841 from a specimen collected by Jean Louis Berlandier near Matamoros, Mexico, in 1830.

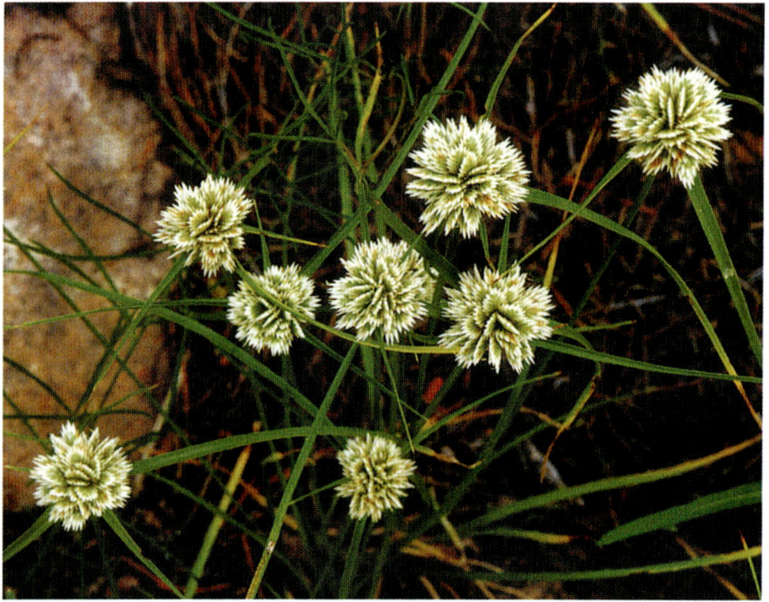

Texas Flatsedge

Cyperus seslerioides

Texas flatsedge is a low perennial that establishes turflike clumps or mats across the ground. The short bare stems of this small sedge, about four to nine inches high, are straight and stiff. This sedge has only a few flat, basal leaves—one to five per stem—that are very narrow and grasslike.

The stems are topped with dense, round flower heads, white with touches of green or brown, that resemble little whirligigs or pinwheels. These heads are a delightful example of nature's handiwork: they are multiplane hemispheres of twenty to sixty tiny, dense spikelets, each usually containing ten to twelve flowers.

You might expect to find this grassy plant in wet, marshy environments, but in Texas this sedge apparently is limited to the Chisos Mountains, where it is scarce in moist, shady canyons, on high ridges, and along open grassy slopes. These little stalks may pop up suddenly after soaking summer rains. Blooming from June to September, Texas flatsedge is sometimes seen in Arizona, but it is more plentiful in Mexican mountains and parts of Central and South America.

In the park check for this sedge near the top of the Emory Peak Trail, in Upper Cattail Canyon, and along the trail to the South Rim.

Note: The genus name *Cyperus* is from the Greek *kupeiros,* the name of the Eurasian species (*Cyperus longus*) first placed in the genus. Although known as Texas flatsedge, this sedge was discovered by Humboldt and Bonpland along the Orinoco River in Venezuela in 1800. On this trip Humboldt also discovered the electric eel and the Humboldt Current off the west coast of South America. Carl Kunth, who lived in Paris from 1815 to 1828 while describing Humboldt and Bonpland plants and building a personal collection of more than seventy thousand specimens, described this species in 1815.

Torrey Ephedra

Ephedra torreyana var. *powelliorum*

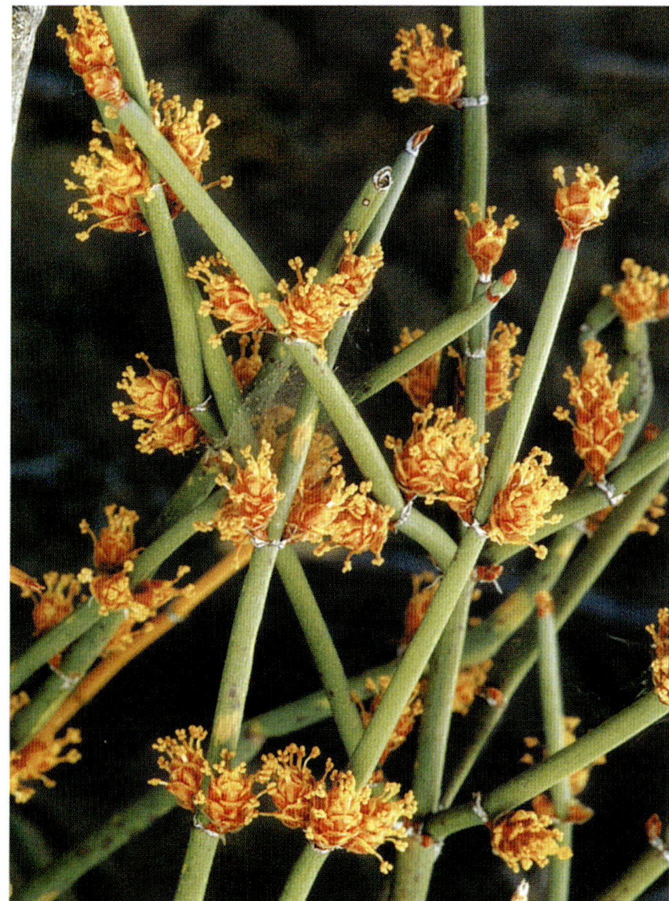

Also known as Torrey jointfir and Mormon tea, Torrey ephedra is an intricately branched shrub up to three feet high with many gray-green, jointed stems. Straight and rigid, the cylindrical branches are smooth, with fine longitudinal grooves. With age, they become ashy-gray and fissured. Reduced to sheathing scales no more than one-fifth inch long, three brownish leaves are clustered at each stem node.

Male and female cones appear on separate plants. Spherical male cones, up to one-third inch long, grow at each node. Each cone consists of pale yellow, papery scales (bracts) arranged in threes and prominent, protruding anthers. Egg-shaped female cones, up to one-half inch long, form at each node on female plants. Each beaked cone consists of thin, papery scales, arranged in threes, which are orange in the center and transparent on the margins.

Torrey ephedra is distributed from Nevada, Utah, and Colorado through Arizona and New Mexico into the Texas Panhandle and the Trans-Pecos, and northeastern Chihuahua. Variety *powelliorum* is localized at Tornillo Flat near upper Tornillo Creek in Big Bend National Park, and adjacent Chihuahua. This variety, with fewer more robust branches, thrives on gypseous clay and gravelly flats in desert scrub and sandy badlands.

In the park look for this shrub seven miles north of Panther Junction on the Persimmon Gap Road.

Note: The genus name *Ephedra* is the ancient Roman name used by Pliny for plants in the horsetail (Equisetaceae) family with similarly jointed stems. This species was described in 1879 by Sereno Watson and named for pioneering American botanist John Torrey. The type specimen had been collected by Charles Wright in 1852. Variety *powelliorum* was described by Tom Wendt, curator of the University of Texas Plant Resources Center, in 1993 and named for A. M. Powell, authority on the flora of Trans-Pecos Texas.

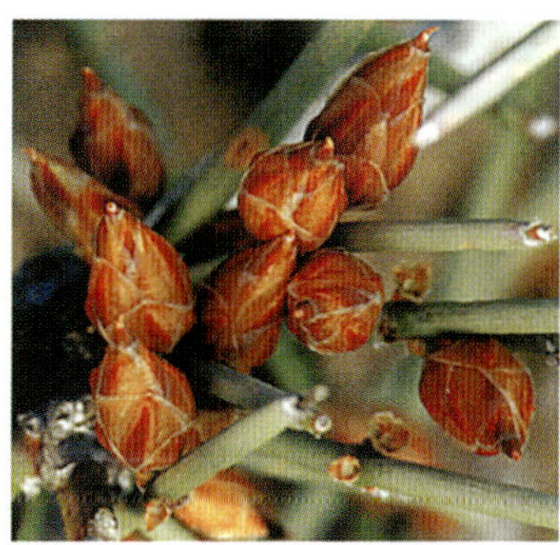

Texas Madrone

Arbutus xalapensis

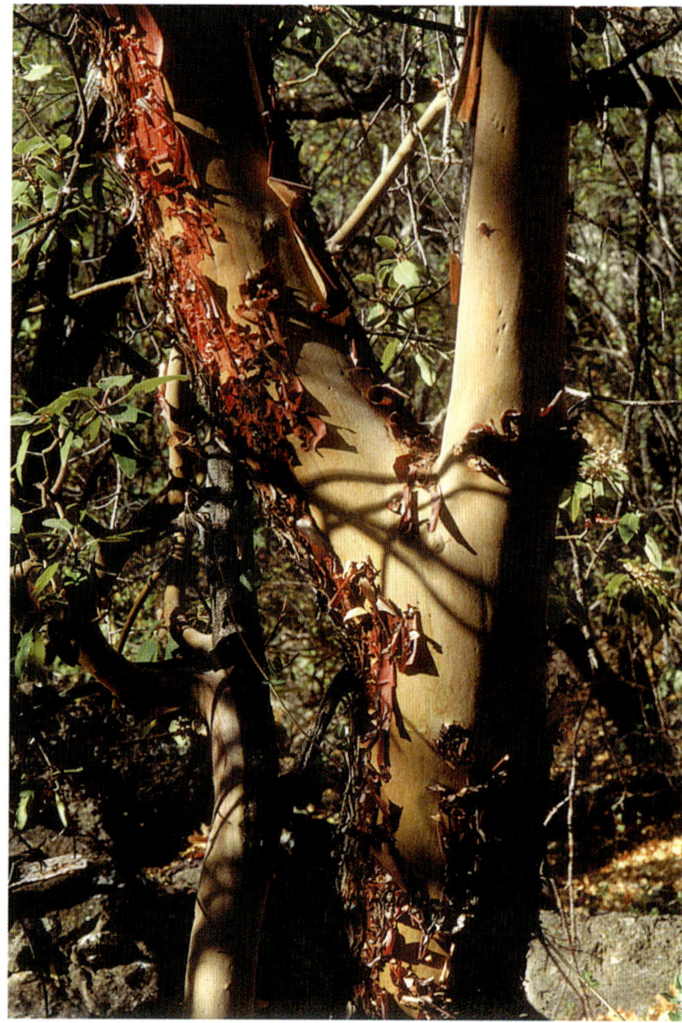

ALSO KNOWN AS naked Indian, lady's leg, and madron or madroño, Texas madrone is an exquisite evergreen tree usually no more than twenty feet high with twisting and curving branches and pink, peeling bark. The older layers of bark unravel in paper-thin, curling sheets, leaving a smooth new layer in shades of white, pink, and red.

The shiny, dark green leaves up to four inches long are oblong to egg-shaped with rather long, stout stems. Often bunched near the tips of stems, the blades are thick and leathery.

White flowers, sometimes blushed with pink, form in clusters up to four inches wide. Each flower has five lobes fused into the shape of a tiny, narrow-rimmed urn. The peak bloom usually takes place in March and April. Bright red, berry-like fruits, notably coarse-textured or grainy, follow the blooms in about six weeks.

Texas madrone does well on limestone and igneous soils, especially in wooded canyons and on rocky, forested slopes. In Texas the tree occurs in the Edwards Plateau and the Trans-Pecos, where it is centered in Brewster and Jeff Davis counties. Populations extend into southern New Mexico and widely across Mexico and south to Nicaragua.

In the park beautiful madrones are easily accessible in Green Gulch and the Basin. These striking trees are resident in Pine, Oak Creek, and Juniper canyons, and in higher mountains between Boot Spring and the South Rim, and on Emory Peak.

Note: The common name madrone, from the Spanish *madroño,* means "strawberry tree," referring to the red fruit. The genus name *Arbutus* is the classical name of the European wild strawberry (*A. unedo*) described by Linnaeus in 1753. Species *xalapensis* is named for the city of Xalapa in the Mexican state of Veracruz, where the first specimen was collected by Baron von Humboldt and Aimé Bonpland in 1804.

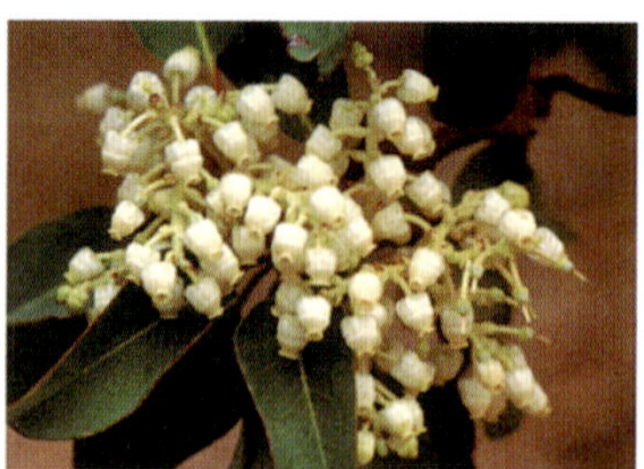

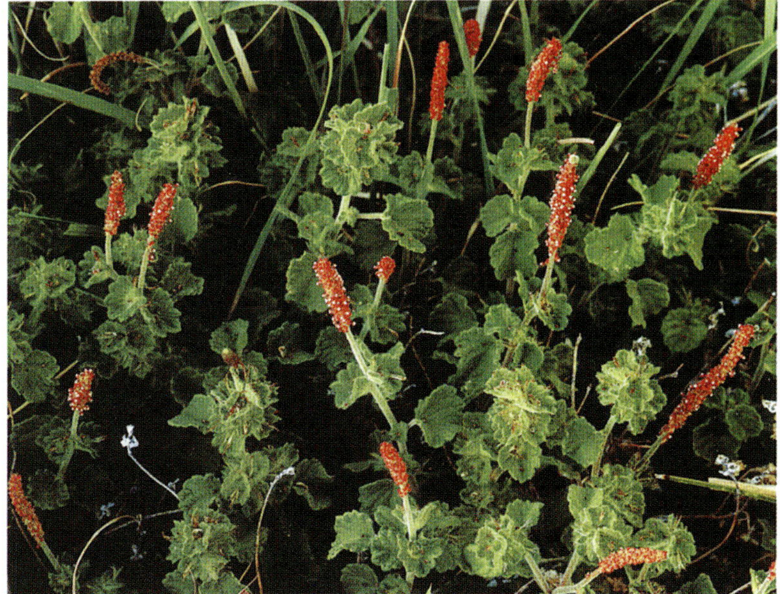

Round Copperleaf

Acalypha monostachya

Round copperleaf is a perennial herb up to sixteen inches high, with many sprawling, densely hairy stems. It grows in leafy, rounded clumps or spreading low mounds. About one-half inch long and wide, the leaves are round or kidney-shaped with scalloped or wavy edges. The stalks of these thin, hairy leaves may be longer than the blades.

From spring to fall, male and female petal-less flowers most often bloom on separate plants. Minute male flowers form on slender red spikes as much as three inches long at the tips of stems. Female flowers may form short, compact clusters or appear singly at the base of leaves. Red, hairlike extensions of the styles, sometimes over one-half inch long, give the female blossoms a distinctly feathery appearance.

Round copperleaf sprawls over dry rocky and gravelly flats, and in desert scrub, grasslands, and brush, mostly in limestone soil at low to moderate elevations. In Texas this herb is far reaching, from Brewster County into portions of the Edwards Plateau, and south along the Rio Grande and across the southern Texas brush country to Brownsville. Its broad range also encompasses much of Mexico as far south as Puebla and Oaxaca.

In the park round copperleaf is widely scattered from desert to foothills. It has been identified on desert flats and in desert washes south of Persimmon Gap, and in the Dead Horse Mountains, at Muskhog Spring, and along Heath Creek in Telephone Canyon.

Note: Linneaus named this genus *Acalypha* from the Greek *akalephe* (nettle) because of its nettlelike leaves. The species name *monostachya,* from the Greek *stachus* (ear of grain or spike), relates to the flower spikes. This species was described in 1800 by Antonio Cavanilles, Spanish clergyman and director of the Royal Botanic Gardens in Madrid. Another species, *Acalypha hederacea,* described by John Torrey in 1858 is now considered synonymous.

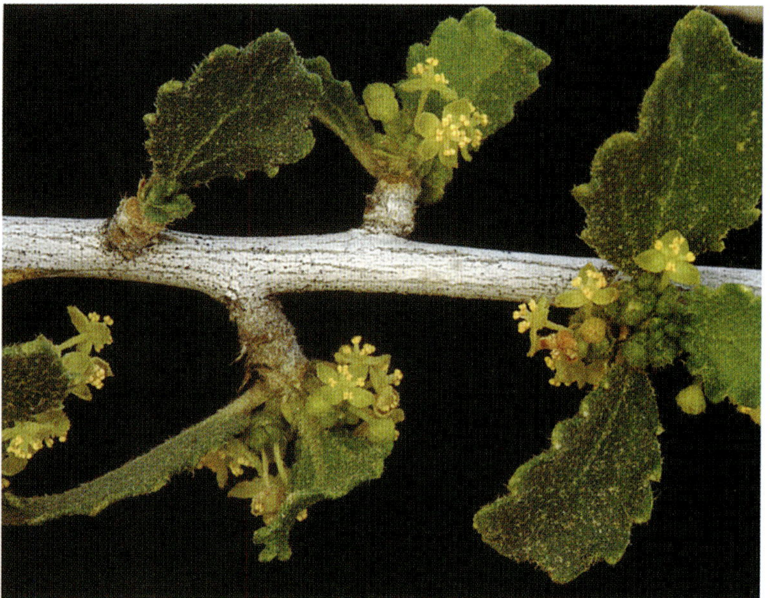

Desert Myrtlecroton

Bernardia obovata

Desert myrtlecroton is a small shrub two to four feet high with many branches and short, stiff branchlets. The short-stemmed leaves are obovate (egg-shaped but wider at the upper end), scalloped on the edges, and rounded at the tips. Grayish-green from a sparse coating of hairs, the leaves are alternate or grow in tight bundles along the branchlets.

The flowers of this member of the spurge (Euphorbiaceae) family are, to use Barton Warnock's words, "too small to be beautiful" (Warnock 1974). Male and female flowers, on separate plants, lack petals but have pale, greenish-yellow sepals resembling petals. Blooming in summer and fall, the male flowers form small groups but the female flowers are solitary.

This shrub is most prevalent on limestone soils in desert scrub and chaparral and near the base of dry, rocky slopes in desert mountains. It occupies much of the Trans-Pecos (from Hudspeth County to Del Rio) and parts of northern Mexico (Chihuahua and Coahuila).

Desert myrtlecroton is present throughout the park, especially from the Chisos eastward to the Dead Horse Mountains and the Rio Grande. It has been confirmed near Sam Nail Ranch and at Ward Spring west of the Chisos, at Grapevine Hills on the north, and in Boquillas Canyon and the Dead Horse Mountains at McKinney Spring in the east.

Note: In 1754 Philip Miller, superintendent of the Chelsea Physic Garden in London and editor of *The Gardeners Dictionary*, named the genus *Bernardia* for Bernard de Jussieu. Jussieu was superintendent of the Royal Garden at Versailles near Paris and creator of an early system of plant classification. The species name *obovata* refers to the obovate leaves. This species was collected by John Moore and Julian Steyermark in 1931 in the Chisos Mountains and described by Ivan Johnston in 1940.

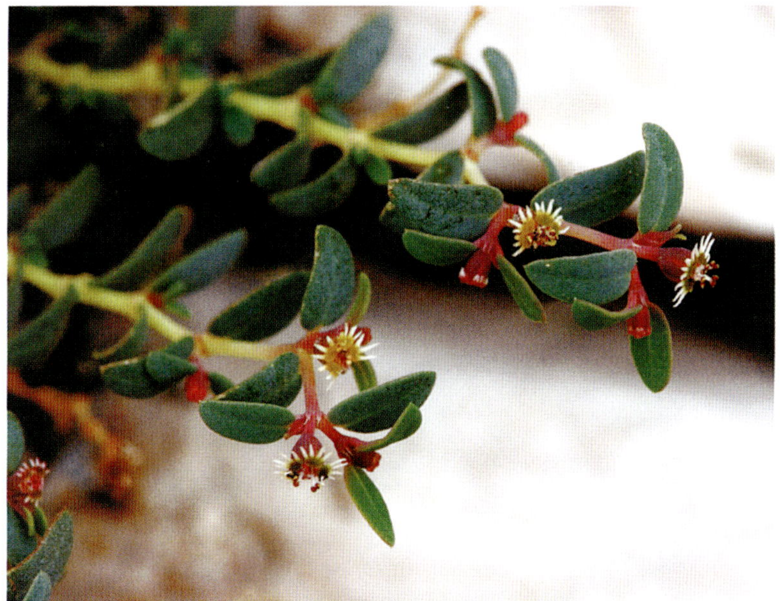

Three-Tongue Spurge

Chamaesyce chaetocalyx var. *triligulata*

THREE-TONGUE SPURGE is a low perennial, no more than eight inches high, with weak, ascending, and spreading stems from a small woody crown. The branches may become quite woody as they age, developing gray furrowed bark. Less than one-fourth inch long, the paired leaves are lance-shaped or oblong, about three or four times as long as wide.

Each urn-shaped flower group (cyathium) consists of a solitary, protruding female flower surrounded by twenty-two to thirty-five male flowers. Four yellowish-green and fleshy oval glands form the rim of the floral cup, each with three to five bristlelike white appendages extending from the sides.

Three-tongue spurge is rooted in rock crevices or amidst rock debris on steep limestone cliffs in desert mountains. In the United States this rare perennial is restricted to extreme southern Brewster County near the Rio Grande. It has also been documented in adjacent Coahuila, Mexico, and at five remote sites in western Coahuila.

In the park three-tongue spurge is located exclusively on the steep pediment slopes of Boquillas Canyon. Although regarded as secure worldwide, it is classified as critically imperiled in Texas. Another more common variety of this species (*Chamaesyce chaetocalyx* var. *chaetocalyx*), with gland appendages not deeply cut and fewer male flowers, is also found in the park.

Note: The genus name *Chamaesyce,* from the Greek *chamai* (low) and *sykon* (fig), describes a prostrate plant with figlike fruits. The species name *chaetocalyx,* from the Greek *chaete* (bristle), refers to the bristly calyx hairs. The variety name *triligulata* means "three-tongue." This variety was described (as *Euphorbia fendleri* var. *triligulata*) by California botanist and spurge specialist Louis Wheeler in 1936, and it had been collected in 1931 by John Moore and Julian Steyermark at Boquillas Canyon. Steyermark, an expert on Central and South American flora, also authored *Flora of Missouri* (1963).

Zigzag Croton

Croton pottsii var. *thermophilus*

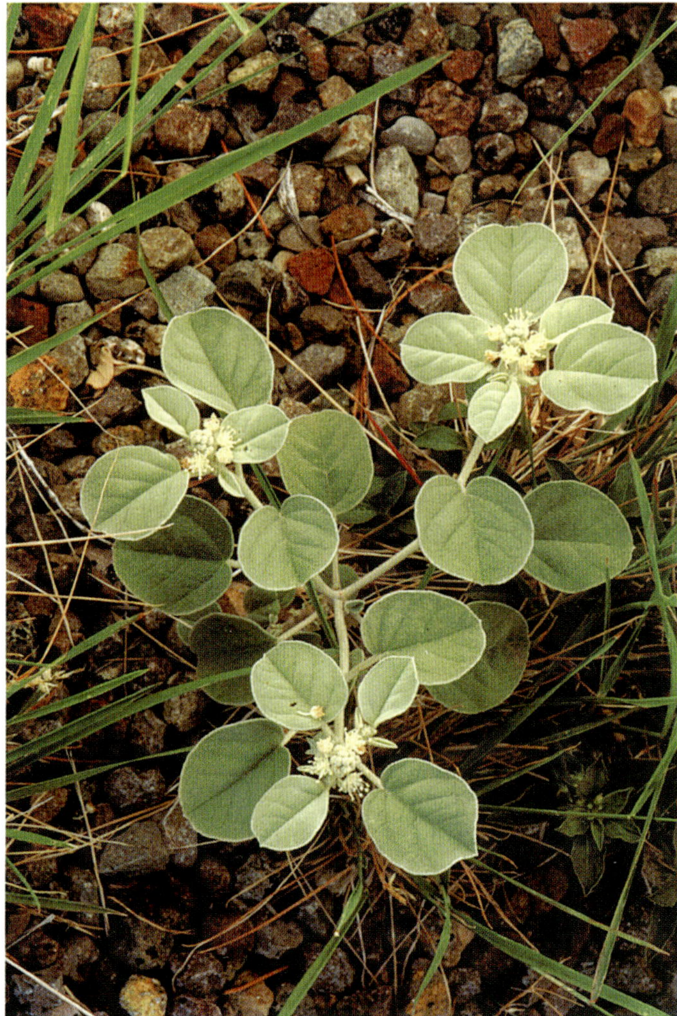

NAMED FOR the tendency of its stems to branch in a zigzag fashion, zigzag croton is a perennial herb up to one foot or more in height with only a few whitish stems. Erect or sometimes sprawling and nearly prostrate, the leafy stems are blanketed with scurfy or scaly hairs. Compared to the more common variety of this species (var. *pottsii*), the leaves are relatively short—no more than three-fourths inch long—and usually very rounded. The lower leaves, especially, appear on long stalks.

Male and female flowers, borne on cuplike disks, form on the same plant, often in the same small clusters no more than one-half inch across. The male flowers may have five distinct sepals and petals and five bulging orange glands. Female flowers may have five sepals but lack petals. Usually blooming from late spring to fall, the small, inconspicuous flowers are rather typical of the spurge family.

In the United States zigzag croton is confined to southern Brewster County in desert scrub habitats, on hot and dry open flats, and rocky and brushy low hills, in limestone and especially sandy soils. It crosses into adjacent Coahuila, Mexico.

This rare croton variety is known almost exclusively from the Dead Horse Mountains to the east and northeast of the park. It is probably easiest to spot in the hills above the tunnel, at Boquillas Canyon and on the low gravelly hills above Boquillas Flats, on the Marufo Vega Trail and along the Old Ore Road, or near Tornillo Creek.

Because of its extremely circumscribed geographic distribution, zigzag croton is classified as imperiled worldwide and critically imperiled in Texas.

Note: The genus name *Croton* is from the Greek *kroton* (tick), an allusion to the seed shape in some plants. Species *pottsii* was named by Johann Klotzsch, curator of the Royal Herbarium in Berlin, in 1853 (as *Lasiogyne pottsii*) in his description of plants collected on the round-the-world voyage of HMS *Herald*. The variety name *thermophilus* translates from Greek as "heat-loving." This variety was described in 1959 by Marshall Johnston, coauthor of the *Manual of the Vascular Plants of Texas*.

Candelilla

Euphorbia antisyphilitica

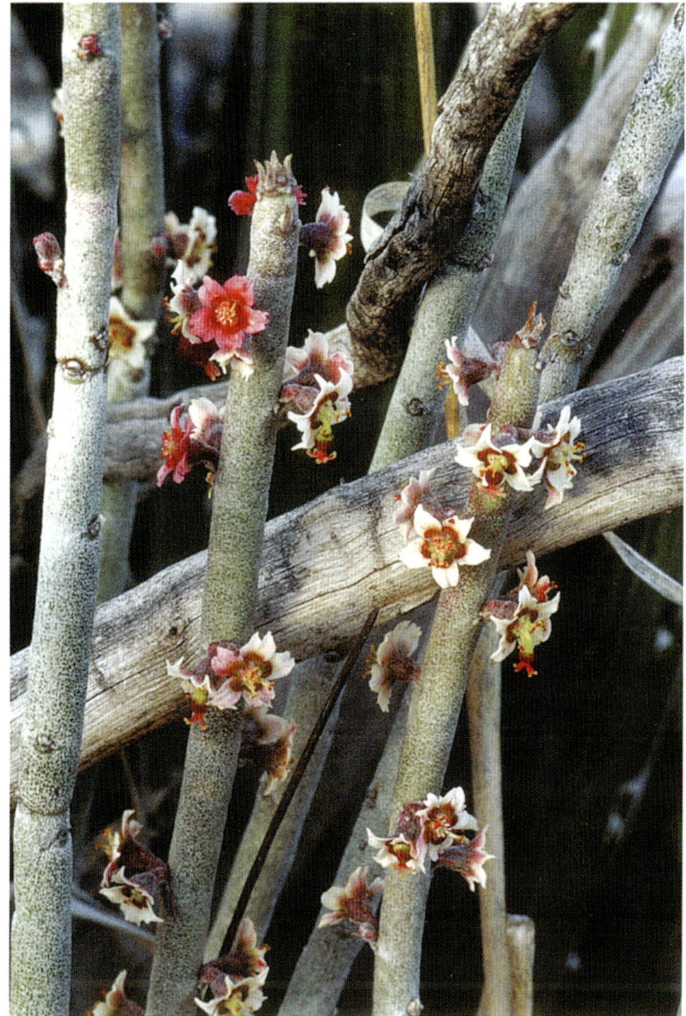

ALSO KNOWN AS wax plant, candelilla is a low succulent subshrub one to two feet high, with round, fleshy stems growing upright in dense clumps. It produces a wax that reflects sunlight and preserves moisture, so the slender stems may appear gray or even white.

Candelilla can form clusters greater than six feet wide with more than one hundred stems. It may seem to grow in a ring because stems in the center die first, leaving only the outer circle. This succulent produces leaves, but they are only a few millimeters long and fall early.

Candelilla blooms from spring to fall, often after rains. The waxy stems seem to sprout petite flowers with white petals and reddish centers. But each "flower" is really a bell-shaped floral cup with five purplish-brown glands around the rim, each with a white petal-like appendage. The floral cup contains many male flowers and a single central female flower. Lacking petals, each male flower is a single stamen. The encircled female flower is a single protruding pistil.

Although overharvested in much of the Trans-Pecos, candelilla remains pervasive at disjunct locations, especially in Presidio and Brewster counties. It can be observed rarely in the lower Rio Grande Plains and commonly in northern Mexico (as far south as Zacatecas and San Luis Potosí and at scattered locations elsewhere in Mexico).

In the park candelilla frequents low limestone hills in the Chihuahuan Desert, mostly to the east and south of the Chisos. This spurge has been reported near Glenn, McKinney, Muskhog, and Hot Springs and on Chilicotal, Mariscal, and the Dead Horse mountains.

Note: Candelilla is Spanish for "little candle." The wax has been used to make not only candles but also phonograph records, chewing gum, insulation, carbon paper, and shoe, floor, and car wax. Reportedly, the genus name *Euphorbia* originated in ancient Mauretania when King Juba II named a plant from Mount Atlas for his physician Euphorbus. The species name *antisyphilitica* refers to its use in Mexico as a venereal disease remedy. This species was collected in Mexico by Baron Karwinski and described by Munich botanist Joseph Zuccarini in 1832.

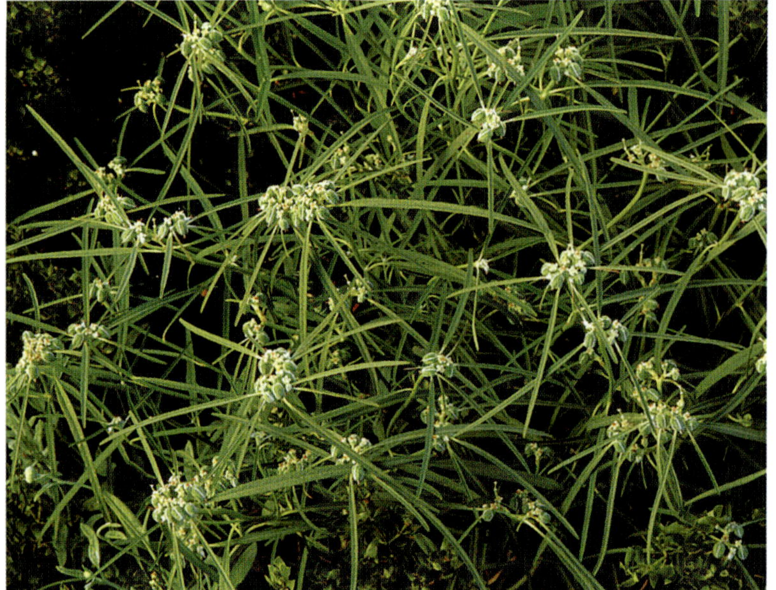

Woollyflower Spurge

Euphorbia eriantha

WOOLLYFLOWER SPURGE, or beetle spurge, is a stout annual up to two feet high with a thick main stem and numerous ascending branches. The leafy stems may be noticeably swollen and hairy at the nodes. Up to three inches long, the slim leaves usually appear in bundles below the flower clusters. Narrowly linear, the leaves are tapered at the tip with a pronounced midvein. The smooth margins of the blades may be conspicuously tinged with purple.

Blooming in spring and summer, the odd flower groups (cyathia) appear singly or in compact clusters of up to twenty near the stem tips. Each whitish group resembles a single flower but actually consists of a solitary female flower surrounded by twenty-five to thirty-five male flowers, all in a densely hairy floral cup. Perched on the rim of the cup are three or four round, fleshy glands, each with strap-shaped appendages that curve over and hide the glands.

Woollyflower spurge spills across gravelly hills in desert habitats, and alluvial flats and floodplains. In Texas this annual is nearly isolated in southern Presidio and Brewster counties. It also grows at discontinuous locations on the Mexican border as far west as California, and in northern Mexico, from Coahuila west to Sonora and Baja California.

In the park woollyflower spurge may cover large portions of the floodplain along the Rio Grande from Boquillas Canyon to Johnson Ranch and west to Castolon and Santa Elena Canyon. It also spreads into desert mountains, especially Talley Mountain, Tuff Canyon, and the Rattlesnake Mountains off the Old Maverick Road.

Note: The common name "spurge" is from the Old French word *espurge,* because some plants were used as a purgative in the Middle Ages to treat black bile or melancholy. The species name *eriantha* means "woolly-flowered." This species was described by George Bentham in 1844, in his account of specimens collected on the round-the-world voyage of HMS *Sulphur.* Bentham is famous for his floras of Hong Kong and Australia, and his extensive classification of all seed plants.

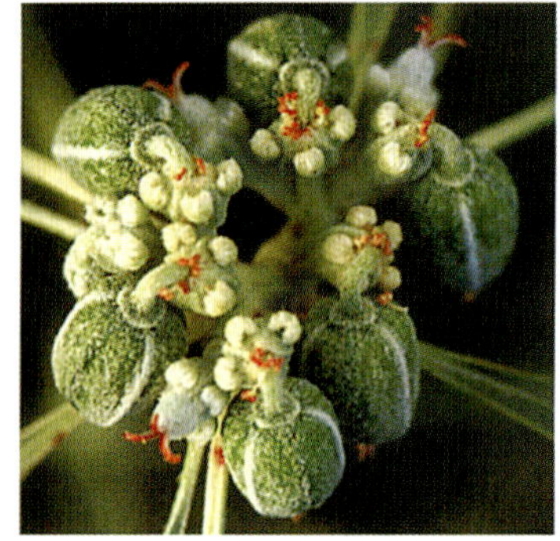

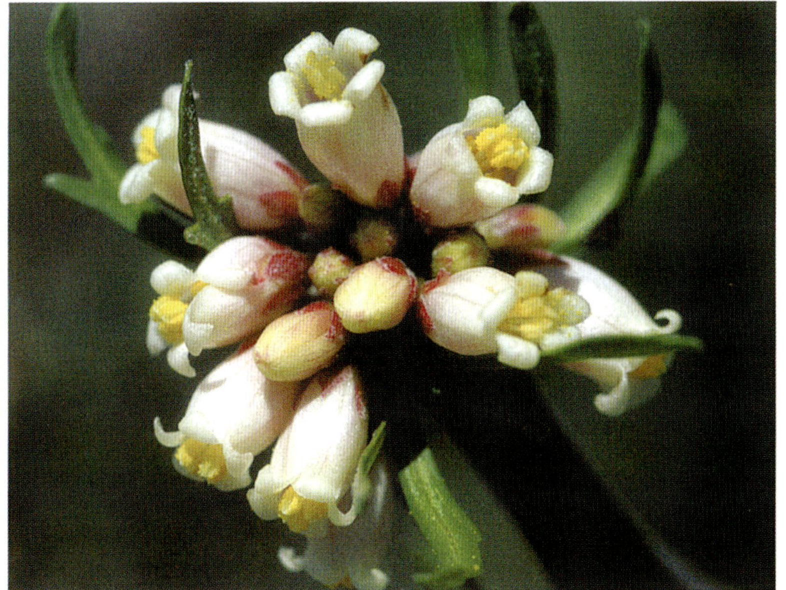

Leatherstem

Jatropha dioica var. *graminea*

Also known as sangre de drago (dragon's blood) and rubber plant, leatherstem is a low succulent usually less than two feet high with rubbery red-brown stems. Sometimes forming dense colonies from underground runners, the stems are fleshy and flexible yet thick and tough. The narrow leaves are spatula-shaped and usually less than two inches long.

In spring and early summer, white flowers cluster near the stem tips. Supported by distinctive red sepals, the urn-shaped flowers have five lobes that flare and curve backward at the tips. Male and female flowers are dioecious (on separate plants). The fruit is a leathery, beaked capsule with oil-rich seeds prized by white-winged doves.

Variety *graminea* inhabits gravelly limestone soils in hotter desert reaches near the Rio Grande, especially in Brewster and Presidio counties. It also colonizes northern Mexico (parts of Chihuahua and Coahuila and northern Zacatecas). Variety *dioica,* with shorter, wider leaves, is widespread from Del Rio southeast to Brownsville.

Leatherstem is predominant at lower park elevations from Persimmon Gap in the north to Hot Springs on the east, and it advances into many sites encircling the Chisos.

The roots were used to make red dye, and reportedly the juice was used by Native Americans to treat sore gums. The juice is now marketed in Blood of the Dragon Styling Gel, which supposedly encourages healthy hair growth.

Note: The genus name *Jatropha,* from the Greek *iatros* (physician) and *trophe* (nourishment), refers to the plant's medicinal use. The species name *dioica* means "dioecious," and the variety name, *graminea,* signifies "grasslike." This variety was described in 1945 by botanist Rogers McVaugh, an expert on Mexican flora, from a specimen collected at Jimulco, Coahuila, in 1939.

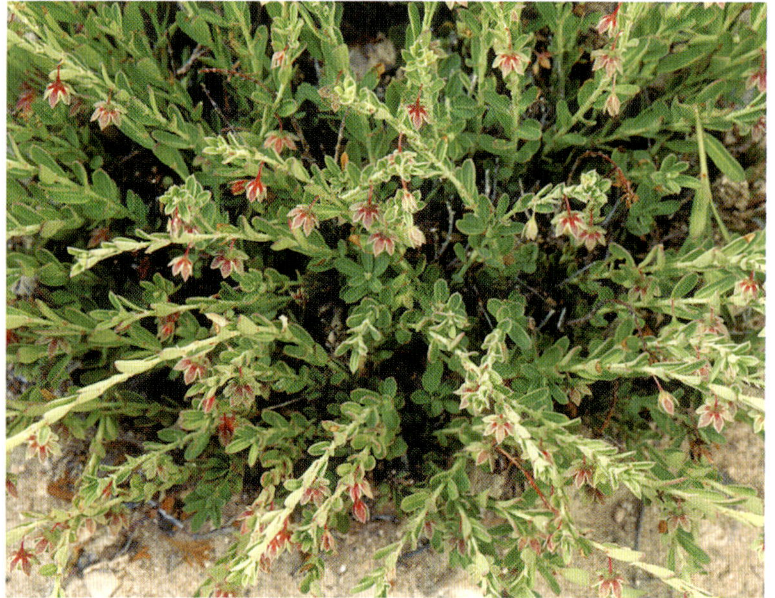

Knotweed Leafflower

Phyllanthus polygonoides

KNOTWEED LEAFFLOWER is a low perennial herb, seldom more than six inches high, with many crowded stems. Growing mostly from the base, the slender branches are erect but limber and arching. They bear narrow oblong leaves, less than one-half inch long, some with short, abrupt tips. The leaves are arranged in spirals and often pressed against the stems.

From spring through fall, separate male and female flowers usually appear on the same plant, often in the same small clusters. In groups of one to three at the base of leaves, the tiny blooms are no more than one-eighth inch across. Lacking petals, the greenish-yellow blossoms consist of six sepals and a segmented or cuplike disk. Frequently tinged with red externally, they hang from reddish stalks like minuscule colored leaves.

This perennial prospers in rocky limestone soil and also in sandy and igneous abodes in desert mountains and foothills. It is widely dispersed in the Trans-Pecos and throughout Texas and parts of northeastern Mexico (from Coahuila to Tamaulipas and south to Zacatecas). The herb also ranges west to Arizona, east to Louisiana, and north to Oklahoma and Missouri.

Knotweed leafflower is encountered throughout the Chisos foothills and the Dead Horse Mountains. You might notice it along Oak Creek, in Pine Canyon, or far north at Dagger Flat.

Note: The genus name *Phyllanthus* is from the Greek *phyllon* (leaf) and *anthos* (flower) because some species bear flowers on leaflike, flattened stems. The species name *polygonoides* implies "resembling the *Polygonum* (knotweed) genus." This species was collected by Thomas Nuttall on the Red River in Arkansas in 1819 and described in 1826 by German botanist Kurt Sprengel. Nuttall, an English botanist who spent thirty-three years in the United States, participated in extensive expeditions west of the Mississippi River. Sprengel authored a monumental history of botany, *Historia Rei Herbariae,* in 1807–8.

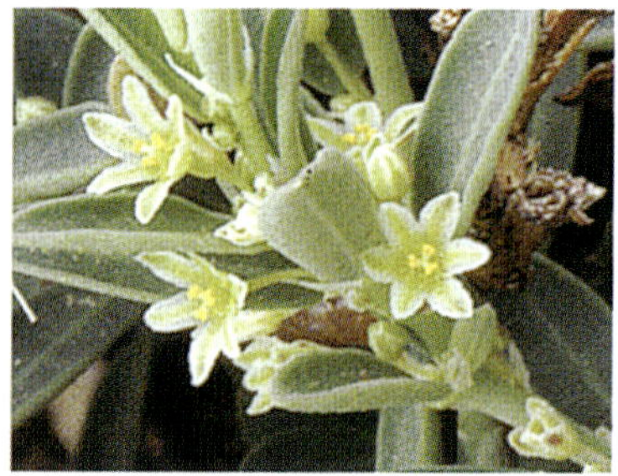

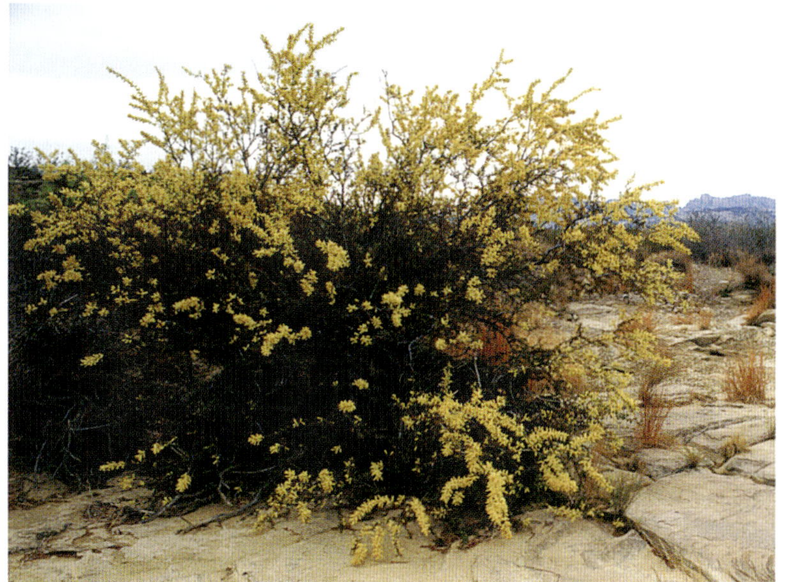

Blackbrush Acacia

Acacia rigidula

In arroyos on the east end of the River Road, these brushy, dark acacias loom. With limbs in shades of white, gray, and black, they sit on ledges above the washes, the tallest living things for miles. In very early spring, their limbs are brightened with fragrant yellow, oblong flower clusters so thick they hide the branches below. Then for a few short days, the washes come alive, a palette of soft pastel colors, with the Chisos Mountains a distant backdrop. Later, this legume produces narrow, swollen beans, constricted around the seeds, reddish brown or black, and up to four inches long.

Blackbrush acacia is a thorny shrub, usually less than ten feet tall, often branching profusely at the base. The stout branches bear pairs of straight, sturdy spines and dark green, shiny leaves up to one inch long. The twice-compound leaves usually have one pair of leaflets (pinnae) divided into two to four pairs of smaller leaflets.

Also known as chaparro prieto and gavia, blackbrush acacia flourishes in the southeastern Trans-Pecos, Rio Grande Plains, and southern Coastal Prairies from Big Bend to Brownsville and the Gulf Coast. The shrub is also extensive in Mexico, from Sonora to Tamaulipas and south to Michoacán and Veracruz. In the United States its western limit is the park, on rocky limestone soils near the Rio Grande. Blackbrush acacia is most prominent near Mariscal Mountain, Solis, and San Vicente.

Note: The name *Acacia,* from the Greek *akis* (sharp point), was used in the Bible to describe a thorny Sinai tree. The species name *rigidula* refers to the rigid branches. This species was named in 1842 by George Bentham, who described plants from botanical expeditions for the Kew Botanic Gardens near London. Blackbrush acacia was collected by Scottish naturalist Thomas Drummond, who spent the years 1833–34 botanizing south-central Texas. Earlier, from 1825 to 1827, Drummond had explored western Canada.

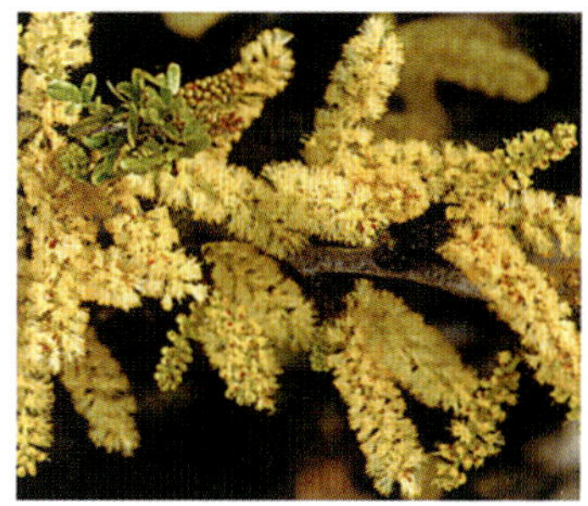

Roemer Acacia

Acacia roemeriana

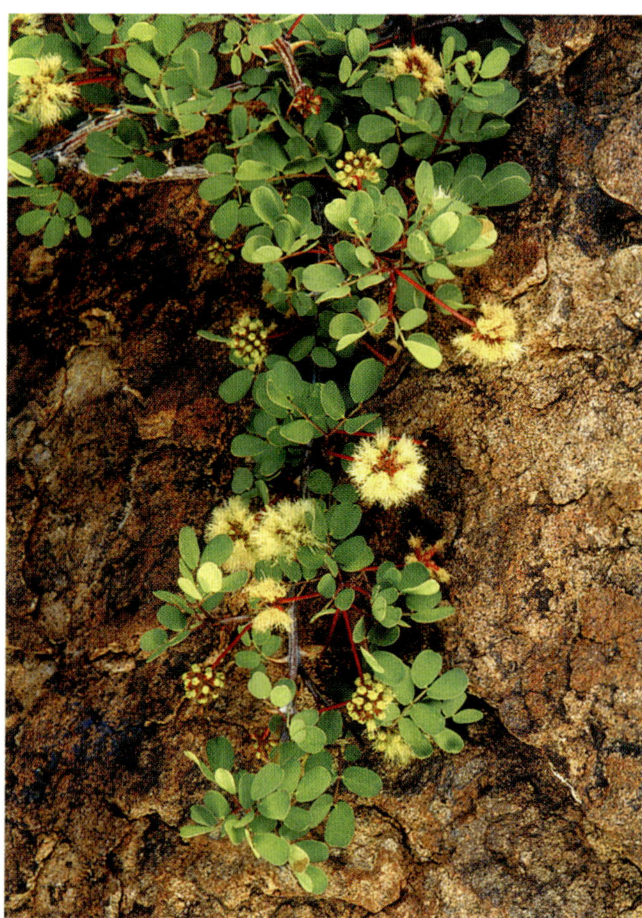

ROEMER ACACIA, also called catclaw and uña de gato, is a shrub about six feet high but occasionally fifteen feet in moist habitats. This diffuse, often gangly shrub has slender branches armed with flat, recurved prickles. The compound leaves, two inches long, are bipinnate: divided into one to four pairs of larger leaflets (pinnae), which are themselves divided into three to twelve pairs of smaller leaflets. Each oblong leaflet is hairless, veined, and rounded at the tip.

In spring, fragrant cream-colored blossoms appear in globes one-half inch across. On a slender red stalk, each spherical head contains many tiny flowers, and each flower has as many as one hundred protruding stamens. The reddish-brown fruits are flat, oblong pods, up to four inches long, with leathery valves. Thin, almost transparent, and dotted with seeds, they have a whimsical, comical appearance.

Roemer acacia is found in desert scrub, chaparral, and mountain foothills, on dry limestone cliffs and in rocky canyons at low to mid elevations. In Texas this shrub is concentrated in the southern Trans-Pecos (Jeff Davis and Presidio counties to Del Rio) and the Edwards Plateau. It occurs in southeastern New Mexico and northern Mexico (Chihuahua to Nuevo León).

In the park Roemer acacia is most conspicuous in the Chisos, below the Window and along Oak Creek, and in Blue Creek and Pine canyons, but it has been recorded from the Rosillos foothills to Hot Springs to Santa Elena Canyon.

Note: This species is named for Ferdinand Roemer, German naturalist and geologist who from 1845 to 1847 collected scientific specimens in Texas. He published the first geological map of Texas and two German-language books on the state's natural history and geology. Adolf Scheele, author of seven articles on Ferdinand Lindheimer's Texas plants in the European journal *Linnaea,* named this species for Roemer in 1848.

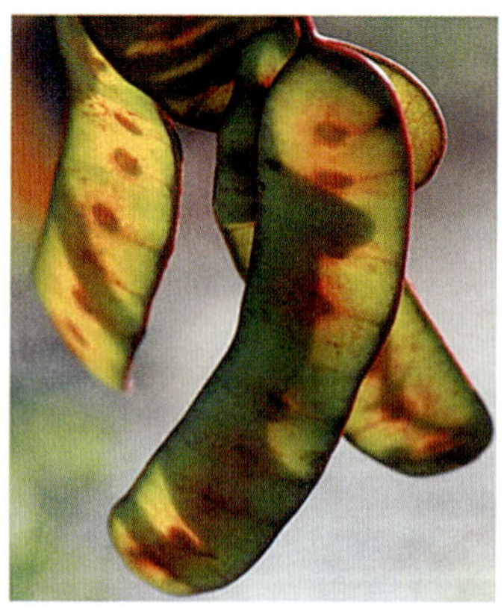

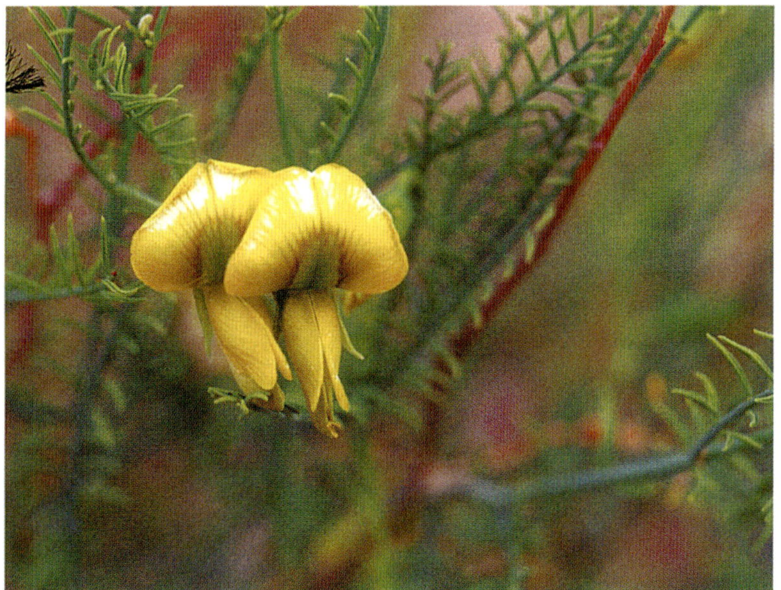

Littleleaf Brongniart

Brongniartia minutifolia

LITTLELEAF BRONGNIART is a delicate, flimsy low shrub one to four feet high with thin wandlike stems. The many thornless branches, mostly smooth and sometimes zigzag, turn dark red as they mature and lose their bark.

This shrub is named for its smooth feathery leaves, one to three inches long, which are divided into very tiny and narrow leaf segments. Each alternate leaf is divided into twenty-one to forty-five short leaflets, which are linear in shape and usually slightly curling.

The yellow flowers are distinctively marked with a green, or less often, ruddy red "eye" on the rounded upper petal (banner). Usually, solitary blossoms one-half inch wide form on thin stems about as long as the blooms. Blooming mostly from May to September, the rather sparse flowers are butterfly-shaped, characteristic of the legume family.

Globally, littleleaf brongniart is imperiled, and in Texas, critically imperiled. In the United States it maintains a foothold in a small area of Big Bend National Park near the Rio Grande, and its distribution stretches into parts of Chihuahua and Sonora, Mexico. This feathery-foliaged shrub appears in sandy washes and desert arroyos, and on igneous desert mountains in association with lechuguilla and creosote.

In the park littleleaf brongniart is limited to a small swath from San Vicente to Mariscal Canyon along the Rio Grande and north to Mariscal Mine and Glenn Spring.

Note: The genus is named for Adolphe Brongniart, botanist for the Museum of Natural History in Paris from 1833 to 1876 and pioneer in the study of plant fossils. The species name *minutifolia* refers to the minute leaflets. This species was described in 1885 by Sereno Watson—who named many new species discovered by explorers of the American West—from a specimen collected by Valery Havard in 1883 south of the Chisos.

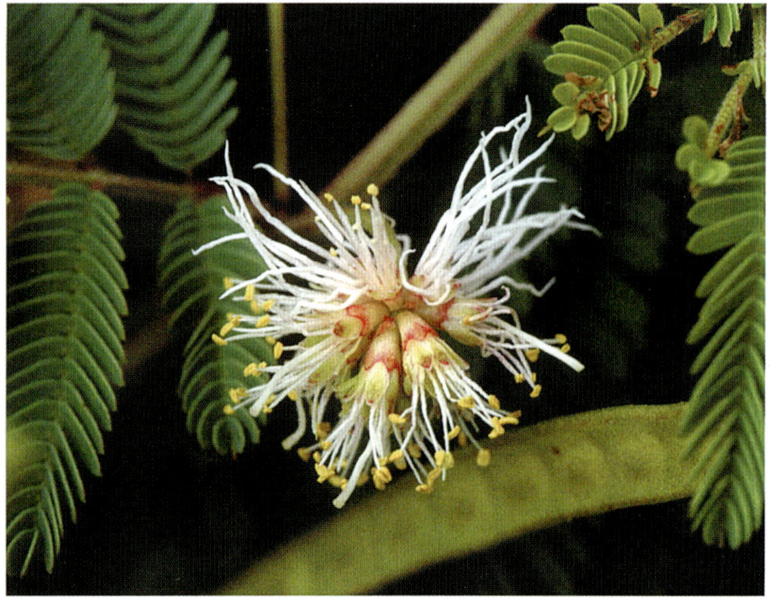

Dwarf Fairy Duster

Calliandra humilis var. *humilis*

DWARF FAIRY DUSTER is a low perennial herb with sprawling, weak stems four to eight inches long. The compound leaves are two to four inches long, twice divided into three to eleven pairs of larger leaflets (pinnae), with each pinna again divided into six to twenty pairs of smaller oblong leaflets.

At the end of a stalk as much as three inches long, the flower head is actually a collection of four to ten cream-colored flowers, each with many longer, protruding stamens. These intricate flowers bloom in late spring and summer. The fruit is a narrow, flattened pod about two inches long and conspicuously erect that turns reddish-brown.

Also known as false mesquite, this herb sprawls across open rocky hillsides, grasslands, woodlands, and roadsides, mostly above four thousand feet. Variety *humilis* is distributed throughout Trans-Pecos mountains (in Brewster, Presidio, and Jeff Davis counties), southern and central parts of New Mexico and Arizona, and Mexico (from Coahuila to Zacatecas and Jalisco).

In the park dwarf fairy duster has been noted in Green Gulch and on the Laguna Meadow Trail, but it is most evident on the west of the Chisos, near Burro Mesa and along the Ross Maxwell Road from the Sam Nail Ranch to the Wilson Ranch Overlook.

Note: The genus name *Calliandra,* from the Greek *kallos* (beautiful) and *andra* (stamen), refers to the protruding stamens. The species name *humilis* (humble) alludes to the low-growing habit. This species was described in 1846 by George Bentham and collected in the Mexican state of Zacatecas by Irish naturalist Thomas Coulter. After collecting in Mexico from 1824 to 1834, Coulter returned to Ireland in 1840 with fifty thousand specimens for Trinity College in Dublin. Variety *humilis* was long regarded as a distinct species, *Calliandra herbacea,* which was described by Asa Gray in 1849.

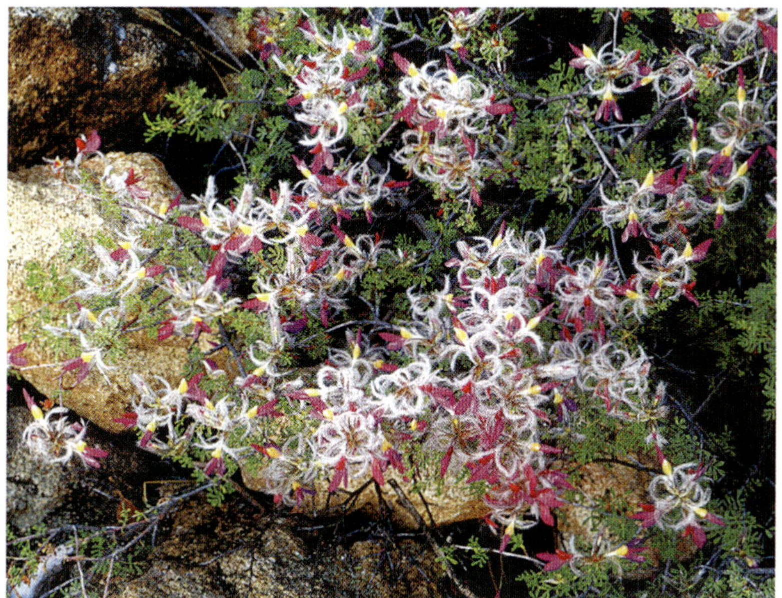

Feather Dalea

Dalea formosa

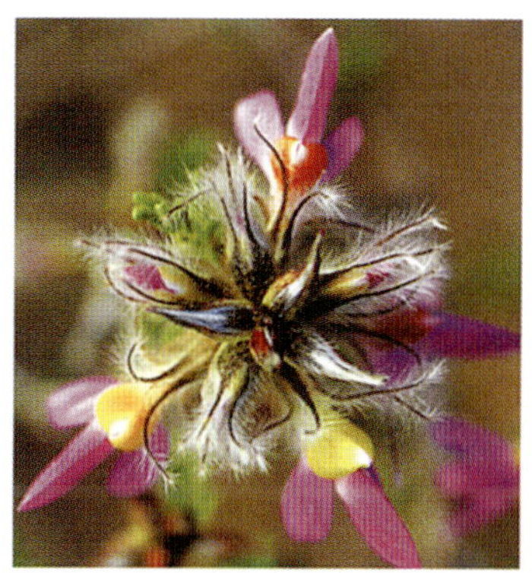

ALSO KNOWN AS feather plume and yerba de Alonso Garcia, feather dalea is a low shrub up to three feet high with many stiff, crooked stems. Because the smooth, woody stems branch haphazardly, the shrub often appears bedraggled or disheveled. The leaves, up to one-half inch long, are usually divided into three to six pairs of oblong, often folded, leaflets and a terminal leaflet. Each blade is dotted below with oil glands.

Blooming in spring and later after good rains, fragrant flowers form in clusters of two to nine blooms. Each blossom has five irregular petals: the upper petal (banner) is yellow but fades to rust red; the two lateral petals (wings) and two partly united, lower petals (keel) are magenta. The lobes of the hairy tube (calyx) supporting the petals are feathery and grayish-white with dark glands, giving the flowers a smoky, plumelike appearance.

Feather dalea is situated on gravelly flats in desert scrub and dry rocky, sunny slopes in limestone and igneous soils. This shrub is wide ranging, from southeastern Colorado and the Oklahoma Panhandle through the Texas Panhandle, western Edwards Plateau, Trans-Pecos, and Rio Grande Plains. It reaches into New Mexico, Arizona, and northern Mexico (from Sonora to Coahuila).

This member of the legume (Fabaceae) family produces remarkable flower displays at many park locations, from Persimmon Gap to desert flats north of the Chisos (Tornillo Flat, Dagger Flat), to the Chisos (Green Gulch, Oak Creek), and to Hot Springs.

Note: The species name *formosa* signifies "finely formed." This species was described by John Torrey in 1827. It had been collected in 1820 by his student Edwin James during the Long Expedition to the Rocky Mountains.

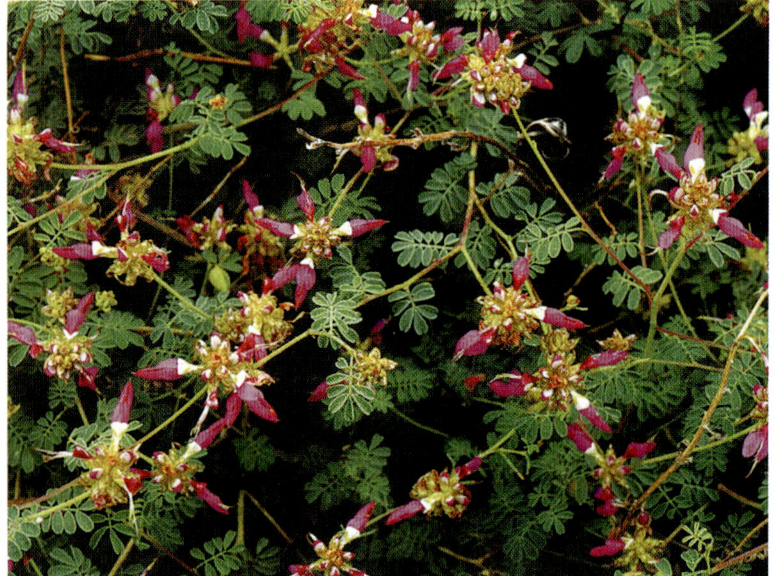

Black Dalea

Dalea frutescens

Black dalea is a low shrub up to three feet high, rounded and dome-like or spreading to five feet across, with smooth, widely branching, slender stems. The densely leafy stems bear short-stemmed, compound leaves up to one inch long, which are divided into nine to nineteen small leaflets. Each leaflet is spoon-shaped, rounded, or shallowly notched at the tip and dotted with oily glands underneath.

Blooming in summer and fall, flowers similar to those of feather dalea but without the feathery plumes appear in short headlike or oblong spikes of up to thirty blossoms. Each multicolored blossom has a white upper petal (banner) that ages to red, plus two lateral petals (wings) and two partly united lower petals (keel) that are pinkish-purple or magenta.

This often weak shrub prefers dry, open, sunny sites on rocky slopes and flats, in grassland and brush, often on limestone soils at moderate to high elevations. Its wide range extends from Oklahoma to Austin, west through the Edwards Plateau and Trans-Pecos into southeastern New Mexico, and south across northern Mexico (from Chihuahua to Nuevo León and south to northern Zacatecas).

In the park black dalea is well established in the Chisos foothills and on higher mountain trails. It is most prevalent on the more open flanks of the Chisos in the Paint Gap Hills, at Sam Nail Ranch, and near Dugout Wells; in the Chisos in Green Gulch and the Basin, and in Juniper and Oak Creek canyons at moderate elevations; and on the Laguna Meadow and Pinnacles trails and in Boot Canyon at higher elevations.

Note: The species name *frutescens* connotes "somewhat shrubby." This species was described by Asa Gray in 1850, from a specimen collected by Ferdinand Lindheimer in 1846 on the Colorado River in central Texas.

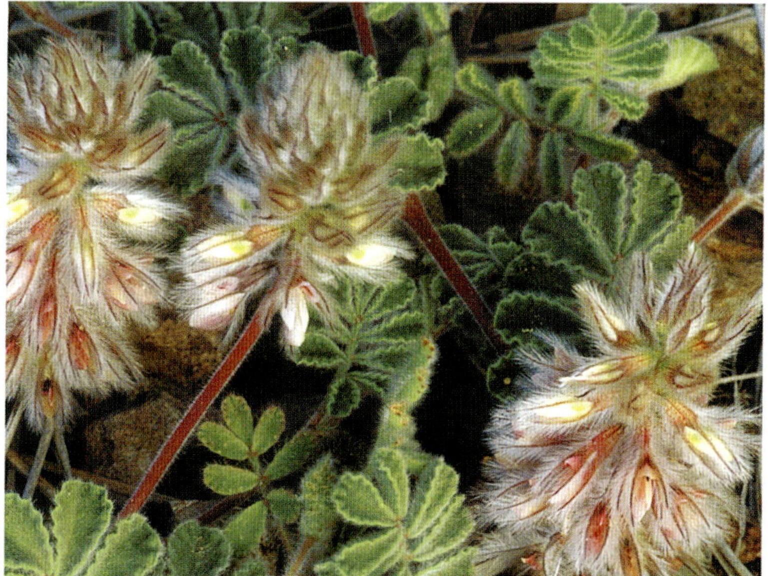

New Mexico Dalea

Dalea neomexicana

New Mexico dalea is a low perennial herb, spreading and straggling, with slender, densely hairy stems as much as eight inches long. The compound leaves, up to nearly two inches long, have seven to fifteen small shaggy leaflets with very wavy, scalloped edges. Each leaflet is wedge-shaped with a blunt, often notched tip. The backside of the leaflet is dotted with reddish-black glands.

Blooming profusely from spring to fall, the small flowers are borne in silky, densely hairy terminal spikes. The blossoms are multicolored: the two lateral petals (wings) are usually creamy white, while the upper petal (banner) and the two lower, united petals (keel) are more commonly pale rose to rust red. The calyx (external base) of each flower bears soft, spiraling hairs that become conspicuously feathery with age.

New Mexico dalea forms dense mats on roadsides, in gravelly and clay soils, and on desert flats and grasslands. Populations span southern Arizona and New Mexico, mostly the central and southern Trans-Pecos, and northern Mexico (from Sonora to Tamaulipas and also Durango).

In the park this perennial occupies clay flats, desert mountains, and the Chisos foothills. It has been verified at Dog Flats and the Rosillos Mountains in the north, at Boquillas Canyon and Hot Springs to the east, near Mariscal Mine and Solis in the south, and in the Chisos foothills (near Lone Mountain, Burro Mesa, and Gano Spring, and in Lower Juniper Canyon).

Note: In 1758 Carolus Linnaeus named the genus *Dalea* for Samuel Dale (1659–1739), an English apothecary, physician, and amateur botanist who helped finance American botanical expeditions in the early 1700s and authored a treatise on medicinal plants. Asa Gray described this species (as *Dalea mollis* var. *neomexicana*) in 1852 from a specimen collected by Charles Wright on his 1849 western Texas expedition.

Bearded Dalea

Dalea pogonathera var. *pogonathera*

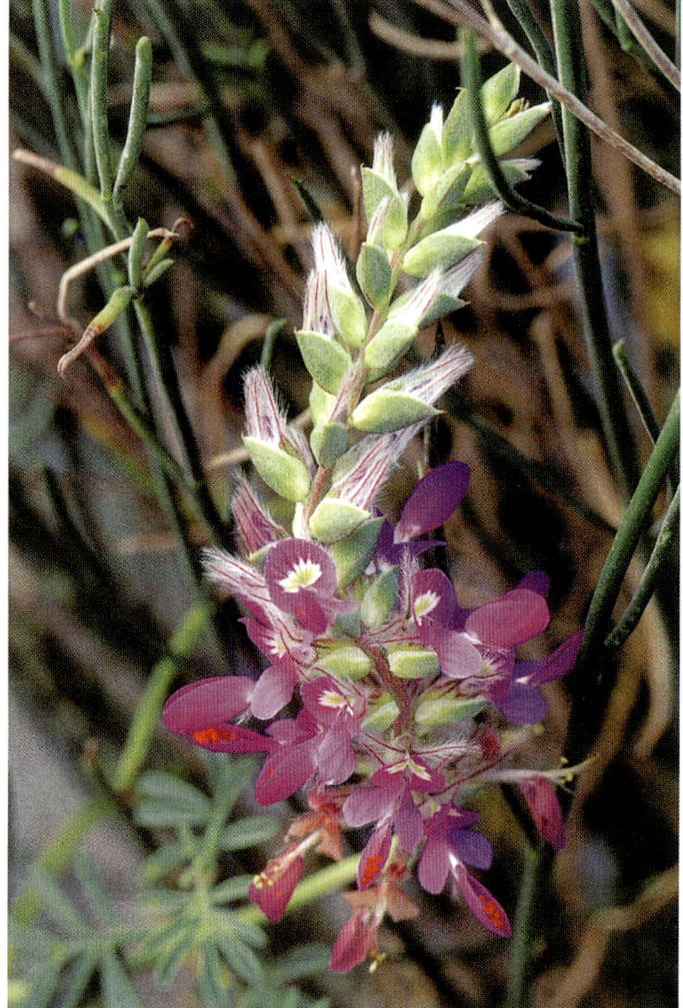

Bearded dalea, or hierba del corazon, is a low perennial herb with numerous weak stems, usually four to twelve inches long. Sometimes erect but often diffuse and sprawling, the stems are hairless and relatively smooth. The stalked leaves, up to one inch long, are odd-pinnately compound: divided into usually two or three pairs of leaflets with a single leaflet at the tip. Each yellowish-green leaflet is linear or narrowly oblong in shape and dotted with red-black glands underneath.

From spring to fall as many as forty bearded flowers appear in dense spikes up to three inches long. Each butterfly-shaped flower has five irregular petals: a broad, rounded upper petal (banner) that is purple with a striking white and greenish-yellow eye; two purple lateral petals (wings); and two united, protruding petals (keel). The threadlike, silky-haired lobes of the calyx (external green base of each blossom) give these flower spikes their distinctive feathery appearance.

Bearded dalea grows on grassy flats and rocky slopes, and also in brushy arroyos, depressions on roadsides, and alluvial flats and floodplains. Two varieties are recognized in Texas. One (var. *pogonathera*) is largely confined to the Trans-Pecos, the other (var. *walkerae*) to southern Texas. Variety *pogonathera* is far-reaching: west to Arizona and south across northern Mexico (Sonora to Nuevo León and south to Zacatecas and San Luis Potosí).

When in bloom, bearded dalea is not easily overlooked in the park, in low desert, or on mountain foothills. It seems to do best in floodplains, from Terlingua Creek and Santa Elena Canyon to Smoky Creek east of Castolon and farther east to Hot Springs. This legume can also be plentiful in the Chisos foothills (at Oak Spring and Chisos Pens on the west and in Avery Canyon to the north).

Note: The species name *pogonathera* is derived from Greek words *pogon* (beard) and *athera* (awns or bristles), referring to the plumelike flowers. This species was collected by Dr. L. A. Edwards, U.S. Army surgeon, near Monterrey, Mexico, during the U.S.-Mexico War and described by Asa Gray in 1849.

Wright Dalea

Dalea wrightii var. *wrightii*

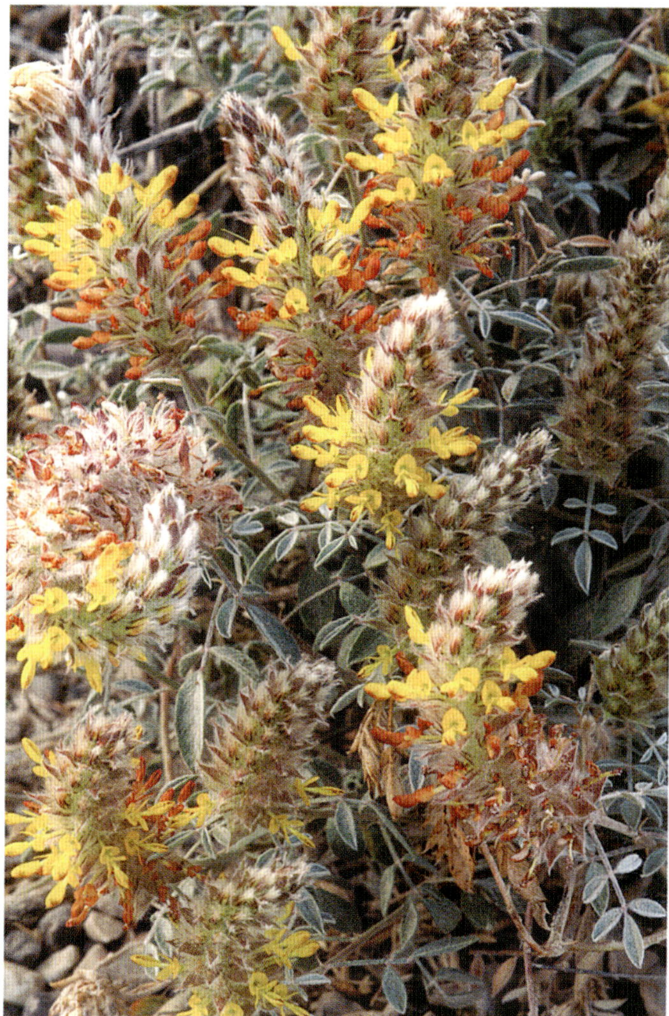

WRIGHT DALEA is a low perennial herb usually less than eight inches high, with erect, silky-hairy stems growing in small clumps and each bearing a single flower spike. The compound leaves, up to one and one-half inches long, are divided into five leaflets, each narrowly egg-shaped or elliptic and covered with grayish-white down.

Blooming from spring to fall, the flowers appear in dense spikes sometimes almost one inch wide and nearly two and one-half inches long. Each small blossom is yellow but fades to shades of orange or brownish-pink so that the spike seems delightfully multicolored. The conspicuously fuzzy-furry appearance of the spikes results from the threadlike, feathery lobes of the calyx (the external, usually green outer part of each flower), which are much longer than the flower tube. Like other *Dalea* species, each flower has a rounded, in this case fanlike, upper petal (banner); two oblong, lateral petals (wings); and two united lower petals (keel).

Wright dalea forms silvery clumps on rocky and gravelly limestone hillsides, and in grasslands, desert flats, and recesses that collect water after intermittent rains. Variety *wrightii* is scattered throughout the Trans-Pecos (except Hudspeth and Reeves counties), southwestern New Mexico, southeastern Arizona, and northern Mexico (northeastern Sonora and parts of Chihuahua and Coahuila).

In the park Wright dalea is most frequently seen in the Dead Horse Mountains and the foothills of the Chisos. Seek out this colorful, feathery herb at McKinney Spring, Ernst Tinaja, Telephone Canyon, near Hot Springs and Boquillas in the Dead Horse Mountains, or at Oak or Ward springs in the Chisos Mountains. It is also present in desert mountains to the east (Chilicotal) and south (Mariscal) of the Chisos.

Note: This species was described by Asa Gray in 1852 and named for Charles Wright, who had collected it on an 1849 western Texas expedition. Wright found the herb on dry hills near Barilla Spring in Jeff Davis County and on mountains near El Paso.

Texas Kidneywood

Eysenhardtia texana

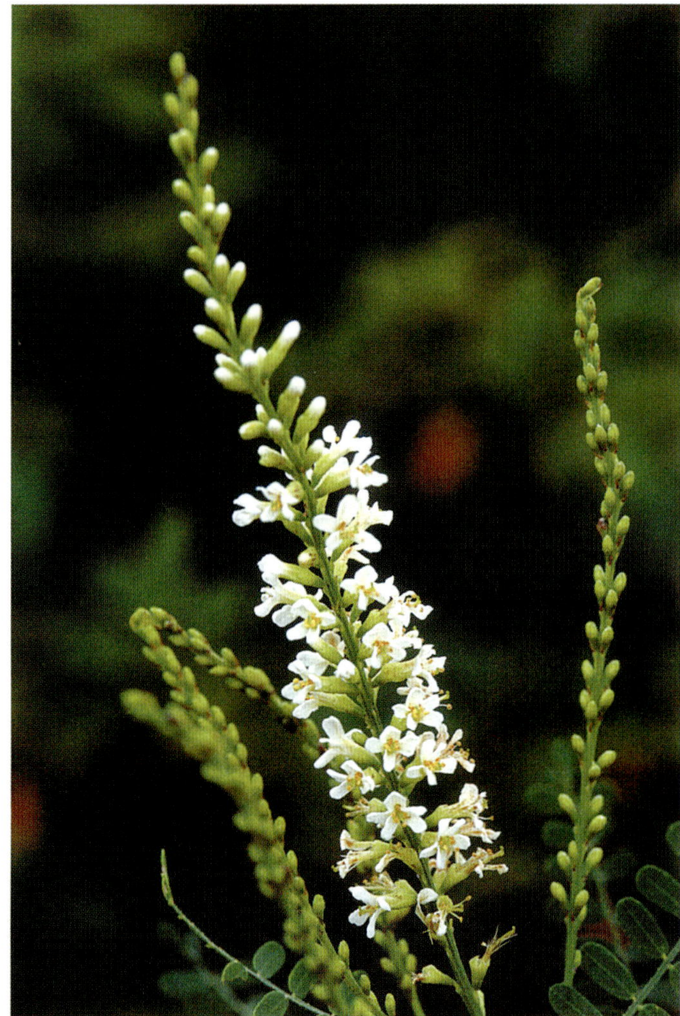

ALSO KNOWN AS vara dulce and rock brush, Texas kidneywood is a spineless aromatic shrub six to ten feet high with multiple trunks and slender gray branches. It has an open, irregular shape. The lacy compound leaves, up to three and one-half inches long, are divided into fifteen to fifty-three small, oblong leaflets. If crushed, the oil glands on the leaves emit a pungent odor.

Also covered with glandular hairs, the tiny white flowers form in narrow, spikelike clusters near the branch tips. Sometimes exceeding five inches in length, these clusters are intensely fragrant, one of Big Bend's most tempting bee and butterfly attractions. They can bloom repeatedly from late spring through summer into fall, especially after drenching summer rains. The fruit is a small, erect brown pod, less than one-third inch long, narrowly oblong in shape and dotted with glands.

Texas kidneywood lines arroyos and brushy canyons primarily on rocky limestone soils, at low to mid elevations, in open sun or partial shade. This legume covers large portions of southern, central, and southwestern Texas as well as northeastern Mexico as far south as Puebla and Veracruz. In the Trans-Pecos it can be quite common from Jeff Davis and Presidio counties east to Del Rio.

In the park Texas kidneywood is easiest to find in Green Gulch and the Basin of the Chisos. It is also encountered in the lower canyons of the Chisos; in Pine, Oak Creek, and Juniper canyons; and far to the north at Dog Canyon and Devil's Den. I have witnessed surprisingly good examples in sparser habitats near McKinney Spring in the Dead Horse Mountains and along the creek below Glenn Spring.

Note: Some related Mexican species were used to treat kidney and bladder problems, hence the common name "kidneywood." In 1824 Carl Kunth named the genus *Eysenhardtia* for Carl Wilhelm Eysenhardt, botanist at the University of Königsberg who took part in Otto von Kotzebue's 1815–18 voyage to the South Sea and Bering Straits seeking a Northeast Passage. This species was collected by Ferdinand Lindheimer near New Braunfels in 1846 and described by German botanist Adolf Scheele in 1848.

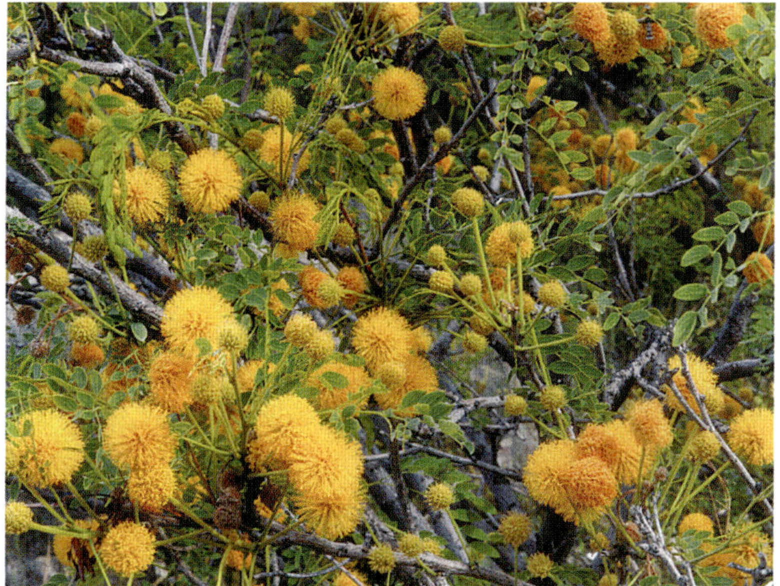

Littleleaf Leadtree

Leucaena retusa

ALSO KNOWN AS goldenball or notched leadtree, littleleaf leadtree is a shrub or small tree sometimes twenty-five feet but usually less than fifteen feet tall with one or several trunks and brittle wood. It has gray or cinnamon-colored bark that is smooth on young branches but broken and flaky on older trunks. The leaves, up to seven inches long, are twice compound: divided into about three to four pairs of larger leaflets (pinnae), which are again divided into about five or six pairs of much smaller, green leaflets.

This tree produces spectacular flower displays in spring and sometimes in summer and fall after good rains. The golden flowers appear in dense, globelike heads one inch wide. Each ball contains many blossoms, each with five narrowly oblong petals and ten protruding stamens. The fruits are narrow, flattened brown pods up to ten inches long. Formed at the end of long, sturdy stalks, the straight pods are thin and papery.

Littleleaf leadtree ranges throughout the Trans-Pecos from Fort Davis east to Del Rio, into the western Edwards Plateau, and more sparsely to Austin. It also enters northern Coahuila and a small part of Chihuahua in northern Mexico.

In the park this tree is probably most accessible in Green Gulch and along the road to Dagger Flat. It has been confirmed in other disparate locations: fifteen miles south of Persimmon Gap, near Sue Peaks in the Dead Horse Mountains, and at the mouth of Mariscal Canyon on the Rio Grande.

Note: The genus name *Leucaena* is from the Greek *leukos* (white), the flower color of some species. The species name *retusa* (notched) refers to the tip of the leaflets. This species was collected by Charles Wright in 1849 in the Nueces River bottom in Uvalde County in the Texas Hill Country and described by George Bentham in 1852.

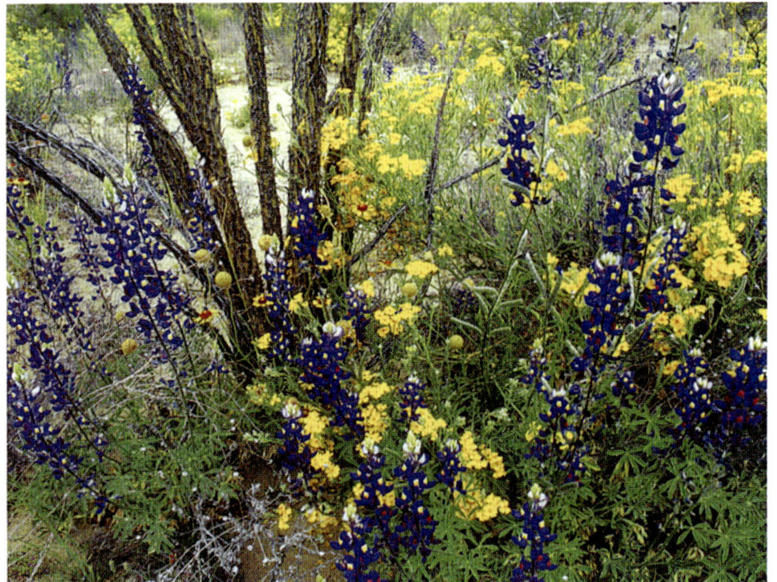

Big Bend Bluebonnet

Lupinus havardii

Big Bend bluebonnet is taller and sturdier than Texas's other bluebonnet species, with flowers a deeper blue. Also known as Chisos bluebonnet, wolf flower, and conejitos (the flowers resemble "little rabbits"), this regal annual can attain astonishing heights in the Big Bend country: a bluebonnet four feet high is not unusual in good years.

The bluebonnet is a key ingredient of the Big Bend experience, as characteristic of this singular landscape as giant dagger and lechuguilla. Along with mesa greggia, it is one of the first heralds of spring. Big Bend bluebonnet may color the hills blue for miles, especially near the south Tornillo Creek bridge; above Tuff Canyon, with Santa Elena Canyon looming in the distance; near the Desert-Mountain Overlook with desert marigold; and on the Persimmon Gap road to Panther Junction.

The bluebonnet is named for the resemblance of its flower to a lady's bonnet. Plants of the genus *Lupinus*, from the Latin *lupus* (wolf), were also called "wolf flowers": since they grew where nothing else did, according to folklore, they robbed the soil of nutrients, devastating it as would wolves. Bluebonnets actually add nitrogen to the soil, enriching it.

This winter annual forms rosettes of compound leaves in late fall. Each long-stemmed leaf usually has seven leaflets and is covered with silky hairs underneath. The flowers appear in showy spires up to eighteen inches long. Each dark blue, five-petaled flower, one-half inch wide, has one distinctive petal (banner) with a cream-colored spot that turns yellow and then red with age. After blooming, the plants produce narrow, hairy beans.

Big Bend bluebonnet is restricted mostly to the southern parts of Brewster, Presidio, and Hudspeth counties in the Trans-Pecos, and to Chihuahua, Coahuila, and Nuevo León in Mexico.

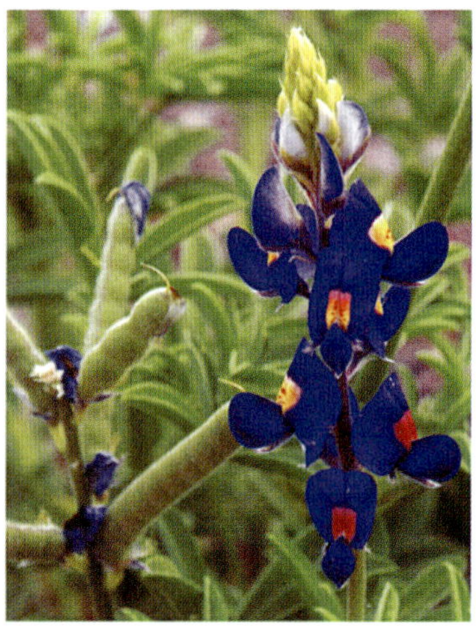

Note: This species was described by Sereno Watson, curator of the Gray Herbarium at Harvard, in 1882. Watson was also botanist on Clarence King's Geological Exploration of the 40th Parallel. He named this species for Valery Havard, the U.S. Army surgeon who, while stationed at a nearby military outpost, collected the plant on hills near Presidio in 1881.

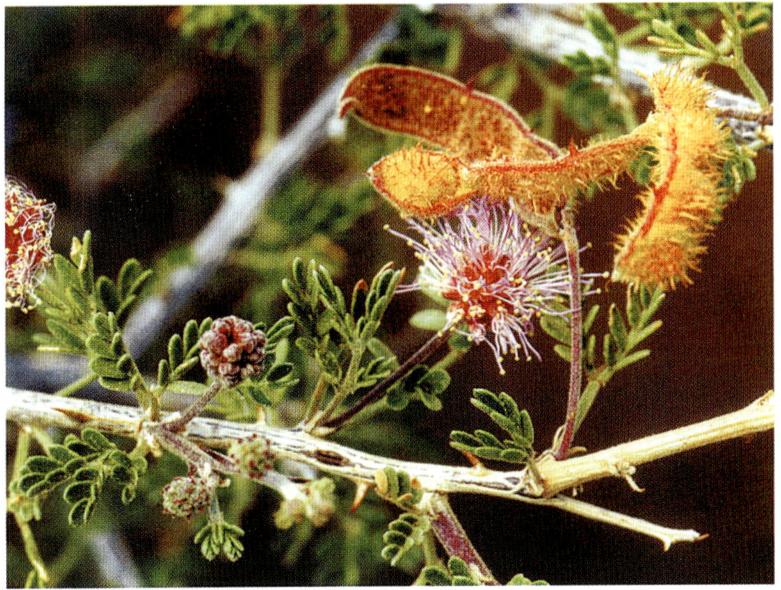

Emory Mimosa

Mimosa emoryana var. *emoryana*

Emory mimosa is a low shrub usually two or three feet high, often scrawny or bedraggled in appearance. The slim branches are armed with short, recurved prickles and may be noticeably lined or grooved. Each grayish-green, compound leaf is divided into one or two, or occasionally three, pairs of pinnae (larger leaflets), which are divided again into three to six pairs of smaller oblong leaflets. These leaflets are covered with fine, silky hairs.

Clusters of pink flowers appear in compact globes one-half inch across. Each fragrant flower has five tiny petal-like lobes obscured by twice as many long protruding, pinkish stamens. They bloom in late spring and summer. Emory mimosa is most easily distinguished from other mimosas by its yellowish-brown, fuzzy pods, which are covered with minute yellow bristles and may also bear hooked prickles along the sides. Sometimes well over two inches long, these pods are often partially flattened and more or less constricted around the seeds.

This legume straggles across brushy desert flats and low rocky slopes in desert mountains, in limestone and igneous soils, at elevations below forty-five hundred feet. Relatively isolated in the Big Bend, primarily in Brewster and Presidio counties in the United States, it is also found in northeastern portions of Chihuahua and Durango, Mexico.

In the park Emory mimosa may be most apparent north of the Chisos, near Dugout Wells and from Government Spring to the Grapevine Hills. It also occurs west of the Chisos in the Paint Gap Hills, at Dripping and Gano springs, and near Sam Nail Ranch; to the south it grows at Dominguez Spring and along Smoky Creek.

Note: The genus name *Mimosa* may be from the Greek *mimos* (to mimic), referring to the movement of the leaves in response to touch in some species. This species was described by George Bentham in 1875 from a specimen collected by John Bigelow in 1852. Bentham named the plant for Maj. William Emory, a cartographer with the Corps of Topographical Engineers, who directed the boundary survey until 1853 and then surveyed the Gadsden Purchase until 1857.

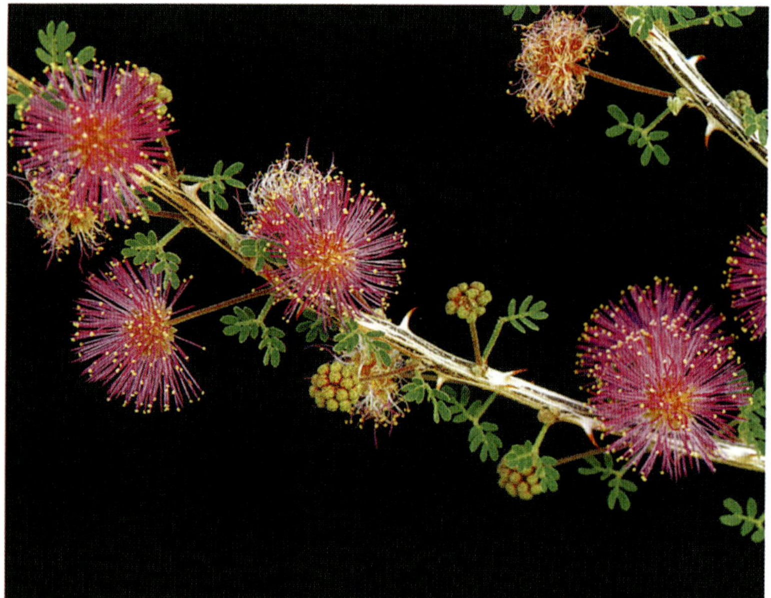

Turner Mimosa

Mimosa turneri

LIKE EMORY MIMOSA, Turner mimosa is a low shrub, often spindly and straggly, with rather sparse and meandering branches. Bearing stout recurved prickles, the stems are streaked with broad, dark lines. The leaves have relatively few leaflets: each is usually divided into only one pair of pinnae (larger leaflets), which are divided again into two or three pairs of smaller leaflets. The leaflets are mostly smooth and free of hairs.

Tiny pink flowers are tightly clustered in dense, round heads about one-half inch wide. Blooming in late spring and early summer, the flowers usually have five partly united petals and many conspicuous, pale pink stamens. The pods have a relatively smooth surface and only a few scattered prickles along the margins. They may at times be tightly constricted or twisted around the seeds.

This shrub spreads across open, dry desert habitats and skirts desert washes and gullies, mostly on limestone and sandy soils, at elevations below four thousand feet. In the United States, Turner mimosa is limited to the Trans-Pecos, primarily from southern Hudspeth County to southern Brewster County, especially near the Rio Grande. It is also native to portions of Coahuila and Nuevo León in northeastern Mexico.

In the park Turner mimosa has been located near Castolon, Tuff Canyon, and Smoky Creek. The photograph was taken near Woodson's fishing camp on the Rio Grande, off the River Road, in July.

Note: This species was described by Rupert Barneby in 1986, from a specimen he had collected the previous year in Presidio County on the Rio Grande above Lajitas. Barneby, a specialist on legumes at the New York Botanical Garden, named this mimosa in honor of B. L. Turner, a University of Texas botanist who in 1959 published *The Legumes of Texas.*

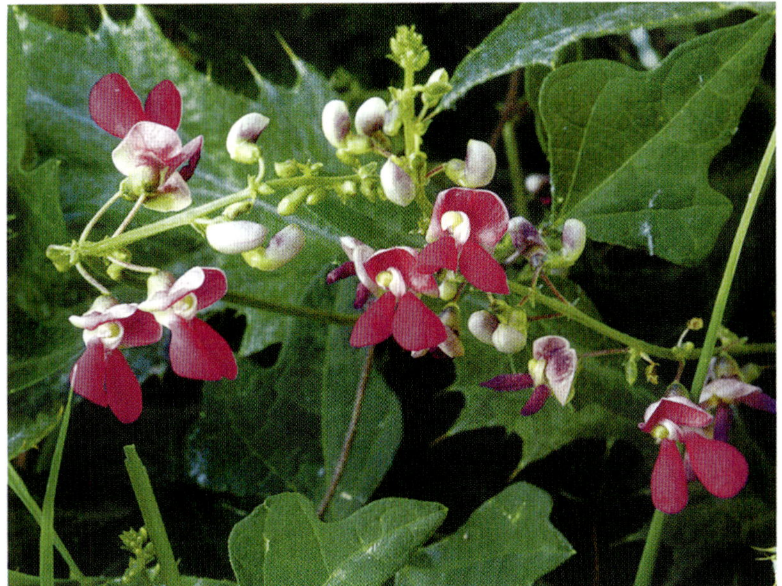

Gray Bean

Phaseolus grayanus

Gray bean or Sonoran bean is a trailing perennial with a few vinelike stems often more than one foot long. Formed on long stalks, the compound leaves are divided into three leaflets. About two inches long and wide, each leaflet is triangular or egg-shaped, typically with a protruding middle lobe and conspicuous bulging lobes at the base. The lobes sometimes have distinctive whitish blotches along the central vein.

In elongated clusters as much as fourteen inches long, the pinkish-purple blooms have the peculiar butterfly shape of most flowers of the legume (*Fabaceae*) family. Each flower has a banner (upper petal), two wings (lateral petals), and a keel (two partly united lower, or central, petals). In this species, the keel is very slender, distinctively twisted or coiled, and the dangling wings are the longest, and lowest, of the five petals.

Gray bean thrives in oak-pinyon-juniper woodlands on rocky slopes, and in chaparral and scrub habitats, on rocky soil, at elevations above fifty-two-hundred feet. In the U.S., this herb is widely dispersed from southern Arizona east through New Mexico to the Trans-Pecos (the Davis Mountains of Jeff Davis County and the Chisos Mountains of Brewster County). In Mexico, it stretches across the Sonoran and Chihuahuan deserts to Zacatecas and central Mexico.

In the park, Gray bean blooms in summer and fall in the Chisos, especially on the higher mountain trails. Look for it in Pine and Upper Pine Canyon, along the Pinnacles Trail, and in Boot Canyon.

Note: The genus name *Phaseolus* (Latin for "bean") is from the Greek *phaselos* (small boat), alluding to the shape of the bean. In 1913, Elmer Wooton and Paul Standley named the species *grayanus* in honor of the eminent American taxonomist Asa Gray. This species was collected by U.S. Army surgeon E. A. Mearns in 1893 during the second United States-Mexico boundary survey. Mearns found the specimen in the San Luis Mountains of Mexico and New Mexico.

Parry Caesalpinia

Pomaria melanosticta

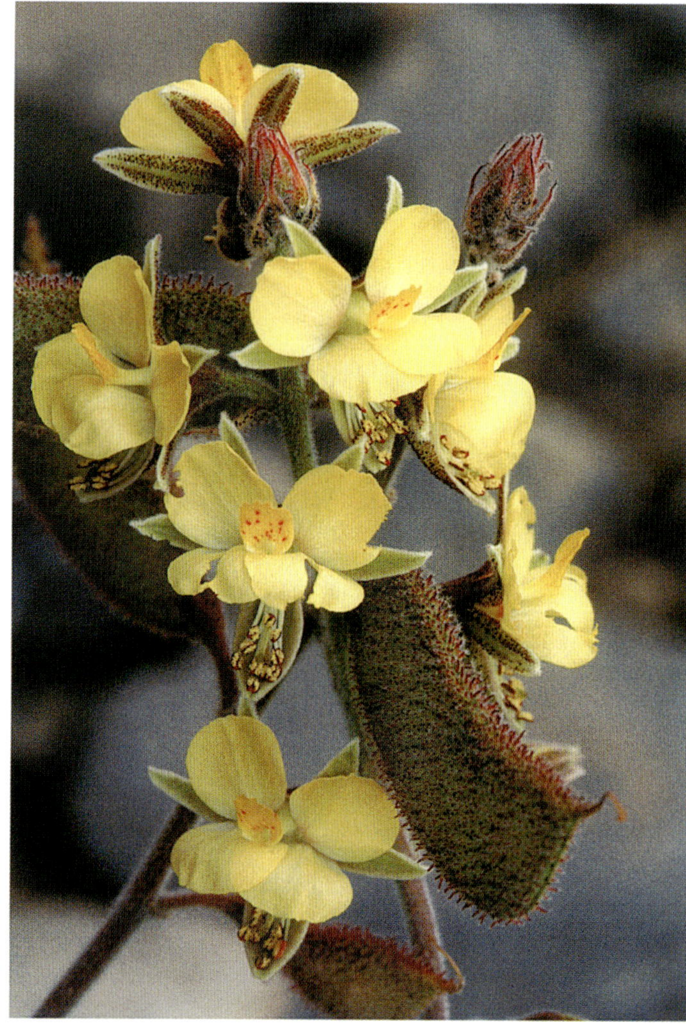

Parry caesalpinia is a low, erect subshrub up to twenty inches high, noticeably cloaked with shaggy hairs. The compound leaves are twice divided, with three to five pairs of primary leaflets (pinnae), which are divided again into two or three pairs of smaller leaflets. Each leaflet, one-fourth inch long, is shaggy-haired and dotted with orange glands underneath.

Primarily in late spring and early summer, yellow flowers one-half inch wide appear in narrow clusters of as many as thirty blooms. Each flower has four broadly rounded petals and a fifth smaller, partially folded central petal (banner) with distinctive red dots. The stamens rest conspicuously on the lowermost sepal below the petals. Flat and broad, the fruit is a crescent-shaped pod, one inch long, with fuzzy reddish-orange hairs protruding from the sides and edges. The flowers and fruits together make quite a show.

Parry caesalpinia frequents rocky limestone slopes and flats in Chihuahuan Desert scrub, desert mountains, and the foothills of higher mountains. In the United States this shrub is centered in southern Brewster County and a few locations in Presidio, Pecos, and Terrell counties. More widespread in northern Mexico, Parry caesalpinia reaches south to Zacatecas, San Luis Potosí, and Querétaro.

In the park this legume is most often observed in the Dead Horse Mountains at such locations as Boquillas Canyon, Ernst Tinaja, and Telephone Canyon. It has also been identified near Glenn Spring, at Santa Elena Canyon, and in the Rosillos Mountains. A good population has developed in the Chisos foothills on the west end of the Dodson Trail.

Note: In 1799 Antonio Cavanilles named *Pomaria* for Jaime Honorato Pomar, professor of medicinal plants at the University of Valencia and presumed author of an illustrated atlas of natural history, *El Codice Pomar* (c. 1590). Described by German botanist Sebastian Schauer in 1847, the species was named *melanosticta* (dark-spotted) for the dark glands dotting the plant. In 1998, in a genus revision, this plant (formerly *Caesalpinia parryi* and *Hoffmannseggia melanosticta* var. *parryi*) was reclassified by Beryl Simpson, director of the University of Texas Plant Resources Center.

Trans-Pecos Senna

Senna pilosior

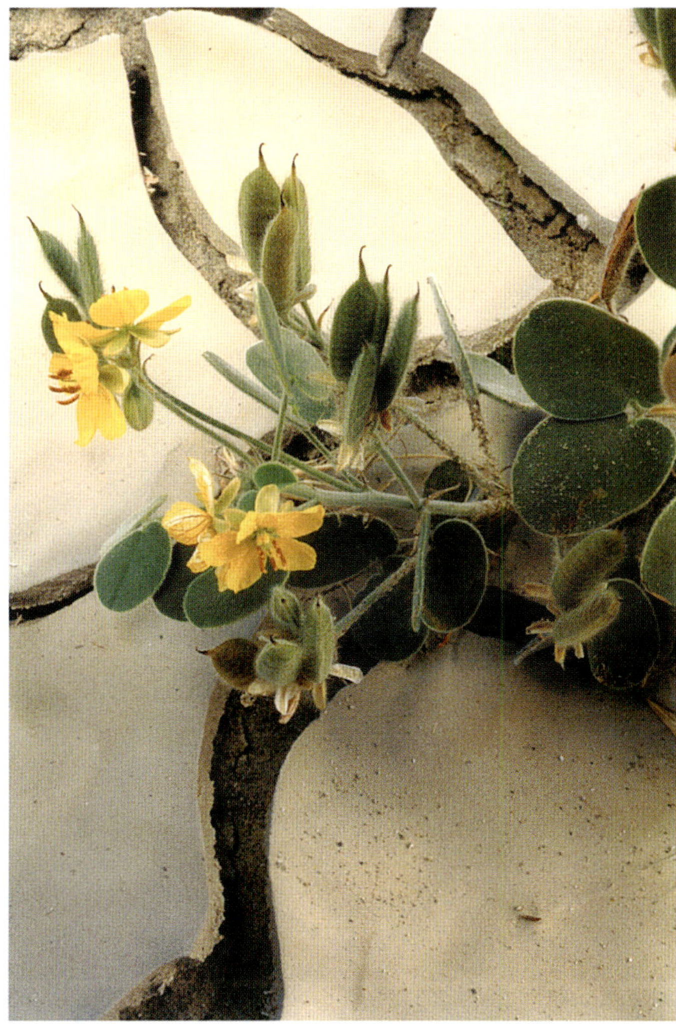

TRANS-PECOS SENNA, sometimes locally known as Durango senna, is a perennial herb up to two feet high, but usually less than half that size, with a few erect or spreading weak stems. The stems are noticeably pale and grayish from a dense covering of hairs.

Like its relative Bauhin senna (*Senna bauhinioides*), which also occurs in the park, this legume has compound leaves with only a single pair of leaflets from a common stalk. Compared to the more elongated leaflets of Bauhin senna, the leaflets of Trans-Pecos senna are very broadly oblong or nearly round, up to one and one-half inches long and half as wide, and notably unequal or lopsided at the base. The stalks of these silvery-haired leaflets may be longer than the blades.

In spring and summer clusters of a few yellow, five-petaled flowers form on long stalks. The thin, early falling petals are often streaked with brown veins. Compared to those of Bauhin senna, the erect fruit pods are relatively straight and also hairy and bristly.

Trans-Pecos senna spills across open desert flats and gravelly limestone soils. It tracks desert washes and intermittent creeks from low elevations to foothills at thirty-five hundred feet. In the United States, this legume is known only in the Big Bend, in Brewster and Presidio counties. It inhabits eastern Chihuahua and Durango and western Coahuila in northern Mexico.

In the park you might notice Trans-Pecos senna near the Rio Grande, at Santa Elena and Boquillas canyons, along Smoky Creek, and near San Vicente and Hot Springs. It has also been reported in the Chisos foothills at Croton Spring.

Note: Philip Miller, superintendent of the Chelsea Physic Garden in London, who introduced about two hundred new plants to Europe, described this genus in 1754. The genus name *Senna* is from an Arabic name (*Sana*) for a bush native to Africa and Arabia. The species name *pilosior* signifies "more hairy," compared to Bauhin senna. This species was collected by Valery Havard in the Bofecillos Mountains of Presidio County in 1883 and described in 1919 by James Macbride, of the Gray Herbarium at Harvard, as a variety of Bauhin senna (as *Cassia bauhinioides* var. *pilosior*).

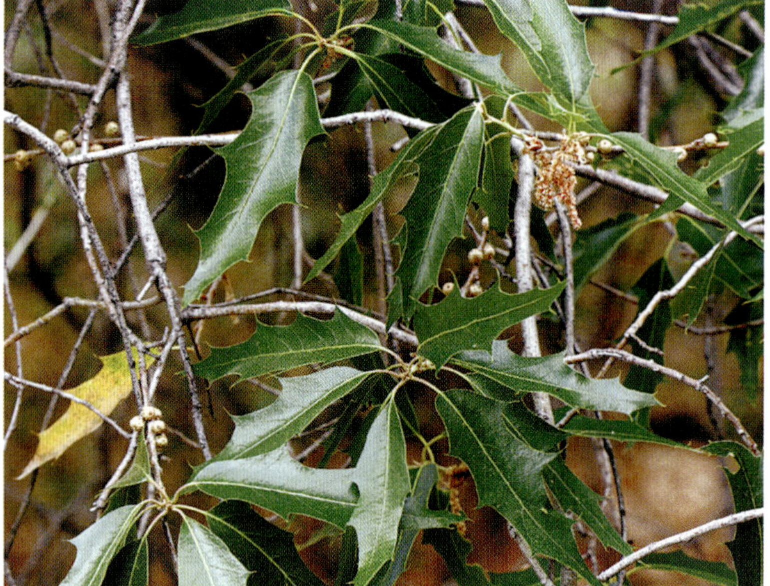

Chisos Oak

Quercus graciliformis

The species name (*gracili-formis*) of this rare, small oak means "graceful," and that is an apt description of its slender, arching branches and dangling leaves. Also known as slender oak, Chisos oak is usually no more than twenty-five feet high. This semi-evergreen tree has gray, often furrowed bark, and twigs that may be a conspicuous reddish-brown when young.

The narrow, lancelike leaves are rather thin, slightly leathery, and shiny. Sometimes as much as four inches long, they have elongated, needlelike tips and bristly toothed lobes along the edges.

Like other oaks, this tree has male and female flowers on the same plant. In spring, male flowers appear in limp, beadlike, cylindrical clusters (catkins) dangling from the branches. The tiny, petal-less flowers are shaped like cups enclosing the stamens. Female flowers appear in shorter, reddish-brown catkins, usually with only two minute tubular florets. Partially enveloped at the base by a shallow cup, the fruit is a narrowly egg-shaped biennial acorn.

Chisos oak grows exclusively in mountain woodlands in rocky, wooded canyons at mid elevations. Classified as critically imperiled worldwide and in Texas, this species is apparently endemic to the Chisos Mountains in Big Bend National Park. Most notable in Upper Blue Creek Canyon, it has also been documented at Oak Creek Canyon, Upper Juniper Spring, and off the Lost Mine Trail. It hybridizes with Emory oak.

Note: The genus name *Quercus* is the classical Latin name for an English oak. This species was described in 1934 by Cornelius Müller, botanist with the University of California, Santa Barbara, and specialist in the oaks of western North America. He collected the specimen in the Chisos Mountains the previous year. A native Texan, Müller published *The Oaks of Texas* in 1951.

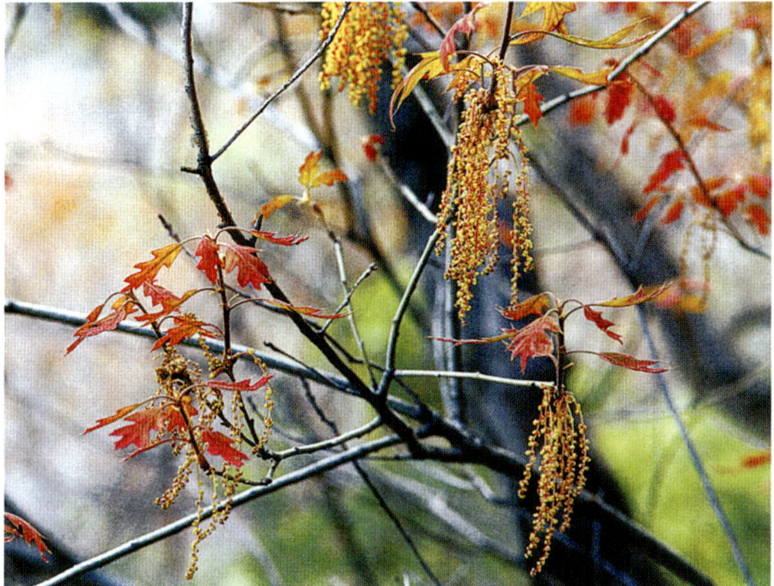

Chisos Red Oak

Quercus gravesii

Chisos red oak, or Graves' oak, is a deciduous tree that can reach a height of forty-two feet. Its old bark is dark and fissured, while the young branches are light gray and smooth. The shiny, dark green leaves are egg-shaped and two to five inches long. Typically, the leaves have several bristle-tipped, broadly spaced lobes on each side and an elongated, toothed terminal lobe.

Male and female flowers bloom in separate spikes (catkins). Male catkins, two inches or more long, are loosely flowered, reddish, and slightly hairy. Female catkins, on red-brown stalks, are one-half inch long, with no more than three flowers. This red oak produces biennial fruits on short stalks. They are light brown, oval acorns seated in a scaly cup.

The yellow, beadlike catkins make attractive spring displays, but the real beauty of this oak is not seen until autumn, when the leaves turn. In late fall the red of this oak's leaves is the dominant color in the landscape, lighting brilliant red fires in every canyon pocket.

Chisos red oak can be abundant in Trans-Pecos mountains, in the Chisos and Glass mountains of Brewster County, and the Davis Mountains of Jeff Davis County. This magnificent oak also forms isolated stands near the Rio Grande in Val Verde County, and in the mountains and arroyos of adjacent Coahuila, Mexico.

This tree probably abuts every trail in the Chisos, from Green Gulch to the Basin, from the Basin to Emory Peak and the South Rim, and down Pine, Juniper, Blue Creek, and Oak Creek canyons. In the Chisos, it hybridizes with Emory oak.

Note: This species was described in 1927 by George Sudworth, chief dendrologist for the U.S. Forest Service. Sudworth named the species for H. S. Graves, dean of the Yale School of Forestry.

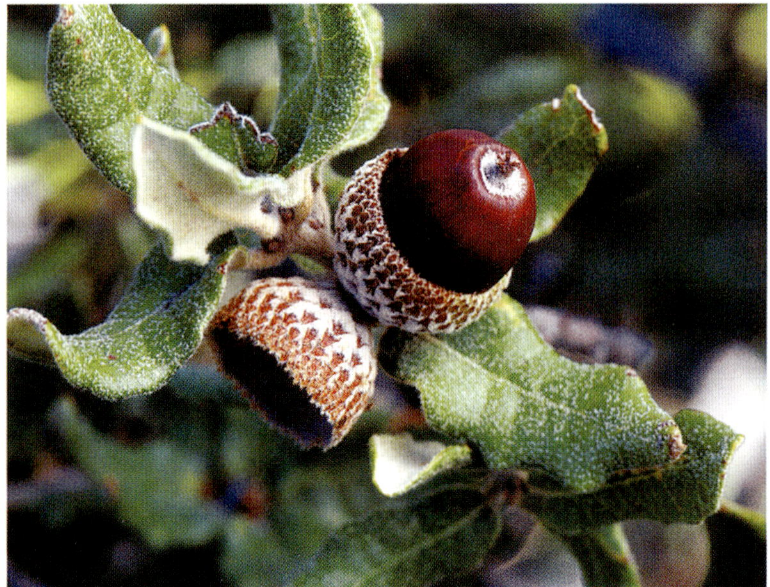

Coahuila Scrub Oak

Quercus intricata

Coahuila scrub oak, or dwarf oak, is a stiff, intricately branched low shrub usually three to four feet high, almost always found in dense groups. The small evergreen leaves, up to one inch long, are oblong to egg-shaped and slightly hairy. Noticeably tough, thick, and leathery, the little leaves are often curled back and wavy on the edges, and covered with white wool below.

The sparsely flowered male catkins are seldom over an inch long. Petal-less, the tiny flowers are shaped like a cup holding pollen-bearing stamens. Female catkins one-third the size of the male appear on short hairy stalks and bear one to five tiny flowers. Resting on hairy stalks, the small brown, oval acorns, sometimes in pairs, are no more than one-half inch long and seated in deep cups.

Coahuila scrub oak is rare in Texas—confined to the Chisos Mountains of Brewster County—but it is more wide-ranging throughout Coahuila, Mexico, as well as portions of Nuevo León, Durango, and Zacatecas.

An impenetrable thicket of this little oak has developed from Laguna Meadow west to Upper Cattail Canyon. The shrub is also well entrenched on rocky and forested slopes from Boot Spring to the South Rim and near the top of Lost Mine Peak.

Note: This species was described (as *Quercus microphylla* var. *crispata*) in 1864 in a monumental work on the development of plant species, *Prodromus Systematis.* Begun by the Swiss botanist Augustin de Candolle, who died in 1841, the last ten volumes were edited by Augustin's son, Alphonse, also a renowned botanist who drafted international rules of botanical nomenclature. This species was collected by Josiah Gregg in about 1848 in Coahuila, Mexico, at the site of a famous Mexican War battle, Buena Vista.

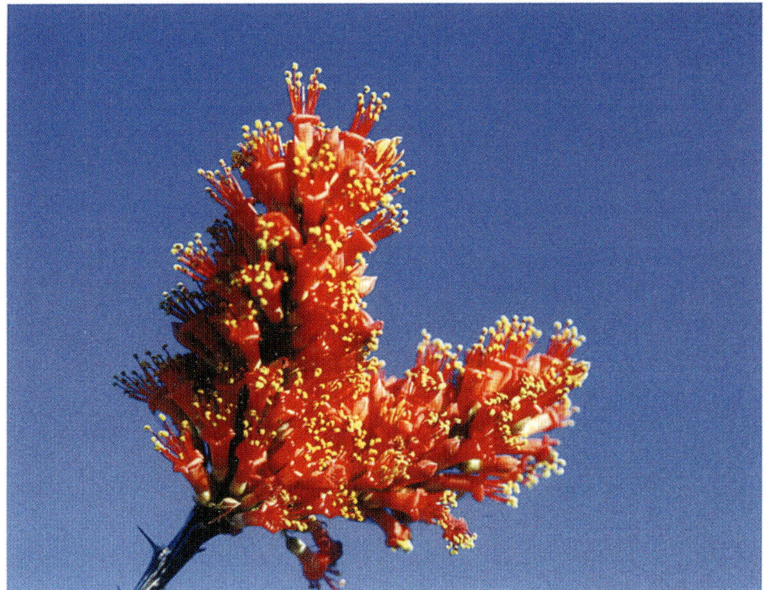

Ocotillo

Fouquieria splendens

Also known as coachwhip, Jacob's staff, albarda, and candlewood, ocotillo is a characteristic species of the Chihuhuan Desert. Often mistaken for a cactus, this trunkless desert shrub has viciously thorny branches that radiate, whiplike, from a central crown. The stout, noticeably grooved stems, up to thirty feet high, are bare and leafless much of the year, with a waxy gray bark.

Dormant during dry periods, these shrubs revive after rains, and the stems are quickly covered with bundles of thick, leathery, spatulate leaves. The leaves soon fall during the next dry spell, and the plant becomes dormant again. With the next rainy spell, new leaves form in the axils of the spines. Four or five crops of leaves may grow each year.

The brilliant orange-red flowers are less susceptible to the cycles. In dense clusters at the stem tips, the waxy flowers bloom in spring, and often in summer with ample moisture. Pollinated by migrating hummingbirds, bees, and foraging birds, each tubular bloom has short, curled lobes at the tip and many protruding stamens.

Ocotillo is distributed on desert scrub and grasslands, arid rocky hillsides, south-facing slopes, and alluvial fans at elevations below five thousand feet. In the United States this whiplike shrub ranges from the Chihuahuan Desert of western Texas to the Mojave and Sonoran deserts in California. In Mexico it stretches from Nuevo León to Baja California and south to San Luis Potosí.

This strange desert denizen appears throughout much of the park. It has been recorded at Glenn Spring, Government Spring, and Grapevine Spring, and on Burro Mesa. Small ocotillo "forests" flourish near the park's west entrance and along the west end of the River Road.

Note: A diminutive of the Aztec-derived *ocote* (a pine used to make torches), the name ocotillo therefore means "little torch," referring to the blazing flowers. This genus is named for Paris professor of medicine Pierre Fouquier (1776–1850). The species name *splendens* (splendid) refers to the flowers. This species was collected by Adolph Wislizenus near Chihuahua City in 1847 and described by George Engelmann in 1848.

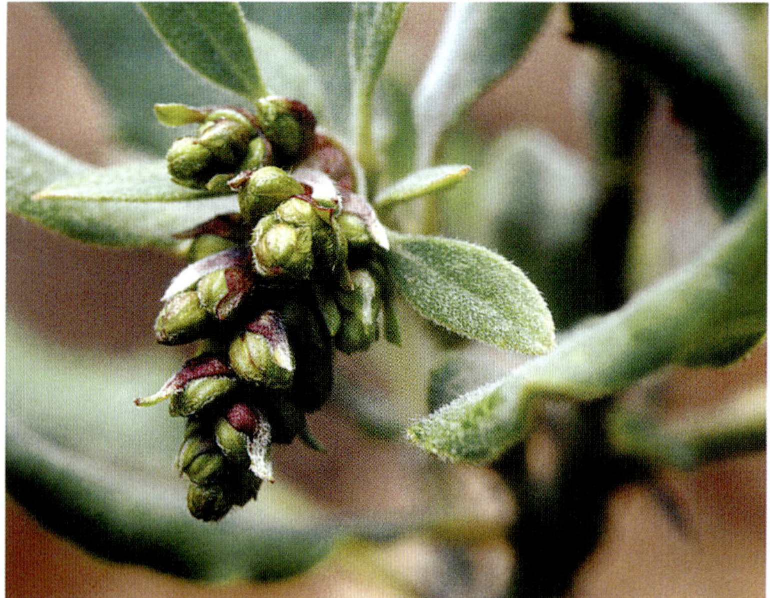

Eggleaf Silktassel

Garrya ovata ssp. *lindheimeri*

ALSO KNOWN AS Lindheimer silktassel and Mexican silktassel, eggleaf silktassel is a dense evergreen shrub that can reach eleven feet in height. Silktassel has thick, somewhat leathery leaves that are broadly elliptic to narrowly egg-shaped and densely hairy and whitish underneath. The Lindheimer variety is distinguished by leaves up to three inches long that are only slightly wavy or entirely flat.

In spring male and female flowers appear in clusters (catkins) on separate shrubs, but they lack petals and are difficult to recognize as flowers. In dangling catkins, the male blooms form on silky stalks that give the shrub the "silktassel" name. Female flowers are nearly stalkless. All flowers are partially enclosed in green or rust-red, leaflike bracts (reduced leaves). In fall berrylike, dark blue fruits cover the female trees.

Once considered endemic to the Texas Hill Country, this silktassel is now recognized in Val Verde County, the mountains of Brewster and Jeff Davis counties in the Trans-Pecos, and the mountains of Coahuila, Mexico. The shrub occupies rocky slopes and shady canyons and ravines in mountain woodlands.

Eggleaf silktassel prospers in the Chisos, from the Basin to the South Rim. It is perhaps most prominent on Pulliam Bluff, in Pulliam Canyon and upper Boot Canyon, and also on the Laguna Meadow Trail, Lost Mine Peak, and Casa Grande.

Note: In 1834 botanical explorer David Douglas named the genus *Garrya* for his friend Nicholas Garry, explorer of the Northwest Territories and deputy governor of the Hudson's Bay Company. This species was described (as *Garrya lindheimeri*) by John Torrey in 1857 and was probably first collected by Ferdinand Lindheimer in 1846. Lindheimer, encouraged by his old Frankfurt friend George Engelmann, collected plants in central Texas from 1843 to 1852.

Narrowleaf Fendlerbush

Fendlera linearis

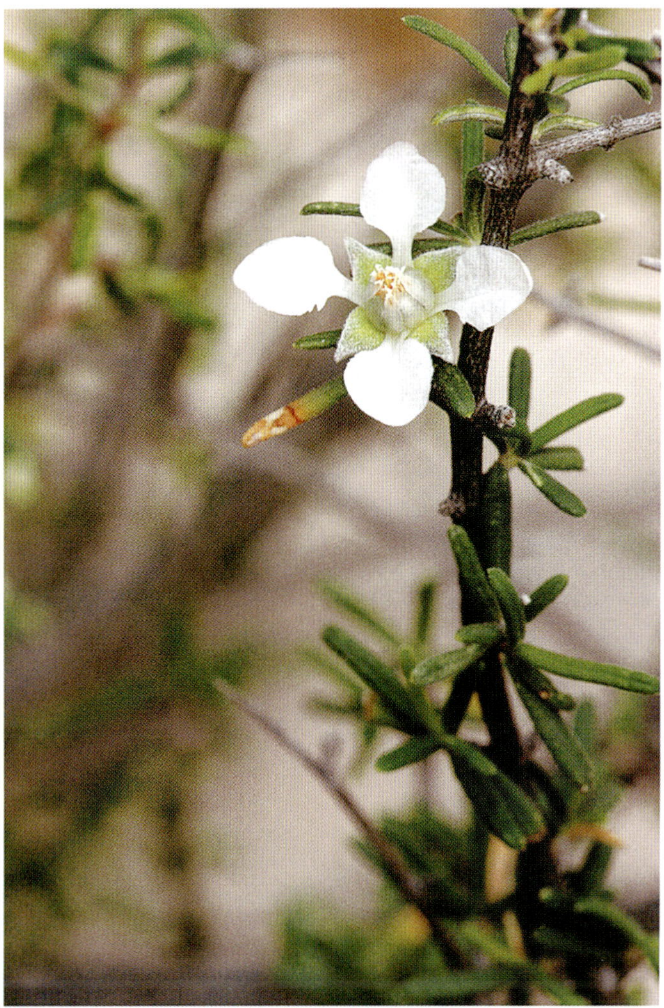

NARROWLEAF FENDLERBUSH, or stiff fendlerbush, is an erect shrub about three feet high with slender but rigid branches. Although stiff, the stems seem to spread rather irregularly and randomly. The very narrow linear leaves, one inch long, often form in small, widely spaced groups on short spur branchlets. Rather thick, the dark green leaves curl under on the edges and often curve backward at the tips.

Perhaps three-fourths inch wide, the unusual, short-stemmed white flowers have four white, spade-shaped petals that are widely spaced and narrowly contracted at the base. The petals spread outward from four conspicuous, pale-green sepals. This shrub blooms in spring and sometimes after good summer rains.

Narrowleaf fendlerbush often sits atop ledges and the edges of cliffs in brushy canyons, or less commonly on open rocky slopes. In the United States, it is restricted to the Big Bend—primarily the Chisos Mountains in Brewster County and the Solitario in Presidio County—but it is also present in northern Mexico, in Coahuila and Nuevo León, and at one site in Chihuahua. Considered vulnerable worldwide, this shrub is critically imperiled in Texas.

In the park narrowleaf fendlerbush has been verified at only a few sites: the lower east slope of Crown Mountain, the southwest slope of Chilicotal Mountain, and the south slope of Pummel Peak. I found this intriguing shrub blooming in late July in a steep, rocky canyon on Burro Mesa.

Note: In 1852 Asa Gray named the genus *Fendlera* for Augustus Fendler, the German immigrant sent by Gray to collect in New Mexico. After collecting in the Santa Fe area in 1846–47, Fendler later botanized in Mexico, Panama, and Venezuela and authored a much-debunked work on world origins, *The Mechanism of the Universe.* This species was collected in 1889 by Cyrus Pringle south of Monterrey, Mexico, and described in 1920 by Alfred Rehder, curator of the Herbarium of the Arnold Arboretum. Rehder, a German-born, self-taught taxonomist, was originally employed to weed the arboretum shrubs. In 1927 he wrote a popular manual on the cultivated trees and shrubs of North America.

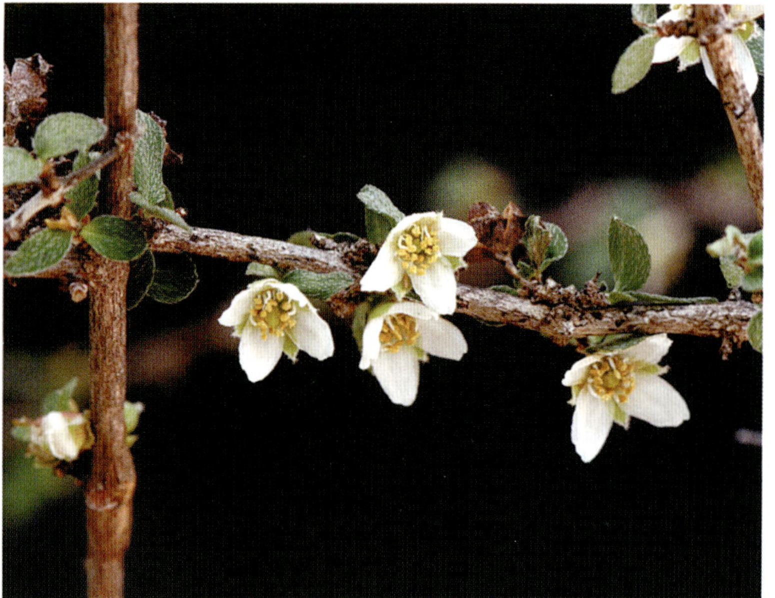

Mearns Mockorange

Philadelphus mearnsii

MEARNS MOCKORANGE is a low, rounded shrub up to three feet high, with many short, dense branches. The rigid stems are reddish-brown to light gray with peeling and flaky bark. Variable in size and shape, the paired leaves are one-fourth to one inch long, oblong to lance- or egg-shaped, with very conspicuous straight, stiff, and coarse hairs pressed against the upper and lower surfaces.

Blooming in late spring and summer, the white, star-shaped flowers are solitary and fragrant, with a distinctly sweet, syrupy odor. Each blossom has four oblong petals, broad at the base and sometimes notched at the tip, and thirteen to twenty-three stamens with bright yellow anthers.

This member of the hydrangea (Hydrangeaceae) family fares well on rocky slopes and limestone canyons at higher elevations, frequently lodged at the base of canyon walls, near boulders, and in dense pinyon-oak woodlands. Imperiled in Texas (NatureServe 2005), the shrub is most evident in the Franklin Mountains near El Paso and the Guadalupe Mountains in Culberson County. Occasionally, populations enter south-central New Mexico and the Sierra del Carmen and elsewhere in Coahuila, Mexico.

Mearns mockorange was unknown in the park until May 2005 when, to my delight, I discovered this small shrub in the upper reaches of Campground Canyon in the Chisos. It was perched at the base of the canyon wall near a large rockslide.

Note: This species was described in 1905 by Walter Evans, an experiment station botanist with the U.S. Department of Agriculture and specialist in Alaskan flora. The species is named in honor of U.S. Army surgeon E. A. Mearns, who collected it in 1892 during the second U.S.-Mexico boundary survey. Mearns, a naturalist and ornithologist, found the plant in Grant County, New Mexico.

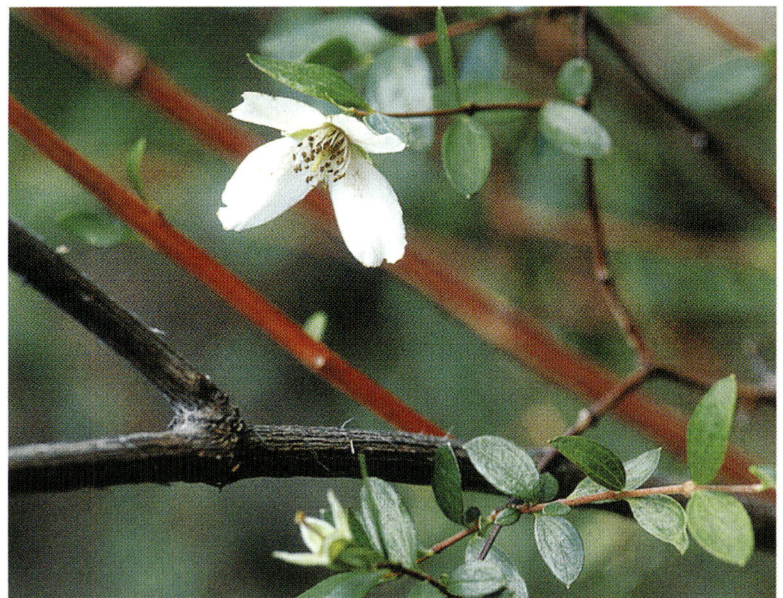

Littleleaf Mockorange

Philadelphus microphyllus

LITTLELEAF MOCKORANGE is a shrub sometimes as much as six feet high but usually smaller, with many branching and bending stems with peeling or flaking bark. The leaves, about one-half inch long, are mostly elliptic or egg-shaped and dark green but with a paler lower surface covered with dense hairs.

Described as having a fruity or orange-blossom fragrance, the flowers are cross-shaped, with a rather delicate or flimsy appearance, and almost always nodding. Mostly solitary, they have many stamens and widely separate, milk-white petals that are oblong in shape and rounded (and frequently notched) at the tips. They bloom from June to September.

This delicate shrub shows up on rocky mountain slopes and in wooded canyons at higher elevations in the southwestern United States from Nevada to Colorado, and California to western Texas. It is concentrated in the Trans-Pecos mountains (in El Paso, Culberson, Jeff Davis, and Brewster counties) in Texas and reaches well into Mexico. Although secure globally, littleleaf mockorange is classified as critically imperiled in Texas (NatureServe 2005).

In the park this uncommon member of the hydrangea (Hydrangeaceae) family is most prevalent on Mount Emory, in Boot Canyon and above Boot Spring toward the South Rim, and along the Chinese Wall Trail. The photograph was taken in Boot Canyon in August.

Note: A popular little shrub, this mockorange was first cultivated in 1883. The leaves contain saponins and can be used to make a gentle, lathery soap. This genus may be named for Ptolemy II Philadelphus, an Egyptian pharaoh and Greek king credited with making Alexandria the cultural center of the Greek world. The species name *microphyllus* connotes "small-leaved." This species was collected by Augustus Fendler in 1847 near Santa Fe, New Mexico, and described by Asa Gray in 1849.

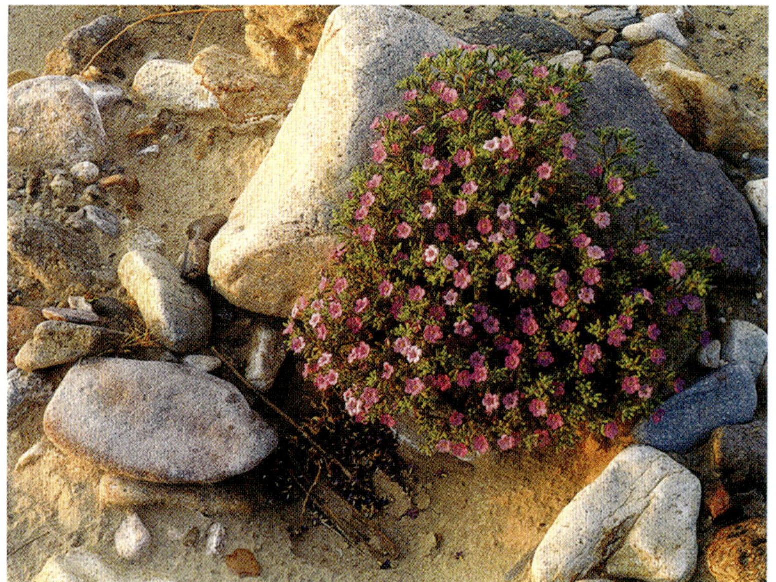

Havard Nama

Nama havardii

HAVARD NAMA is a somewhat fleshy, persistent annual about one foot high. Coated with soft, shaggy hairs, the leafy stems are erect with many short, sturdy branches from the base. The alternate, usually stalked leaves are highly variable: some are narrowly oblong or elliptic, others more spatulate or egg-shaped. They have a tendency to curl under along the edges and curve forward at the tip.

The small funnel-shaped flowers are usually pale pink to deep rose but occasionally lavender or white. Clustered at the branch tips, they may form a neat bouquet covering the plant. They bloom from spring to fall, especially in early spring or after late summer rains.

Havard nama proliferates along desert washes, beside streams, and on gravel bars, clay and alluvial flats, and gravelly slopes in limestone soils. Primarily a resident of the Big Bend country, this herb reaches from Presidio to Del Rio and into parts of Chihuahua and Coahuila, Mexico. Because of its limited geographic distribution, it is classified as vulnerable in Texas (NatureServe 2005).

In the park this member of the waterleaf (Hydrophyllaceae) family is most abundant near the Rio Grande: at Hot Springs, along Tornillo Creek, on the River Road, and at the mouth of Terlingua Creek and Santa Elena Canyon. It is also seen in the Chisos foothills and to the north at Dog Flats and Harte Ranch.

Note: The genus name is from the Greek *nama,* meaning "spring" or "stream," presumably a preferred habitat of these plants. This species was described in 1885 by Asa Gray and named for Valery Havard, who first collected it on the banks of Tornillo Creek in present-day Big Bend National Park.

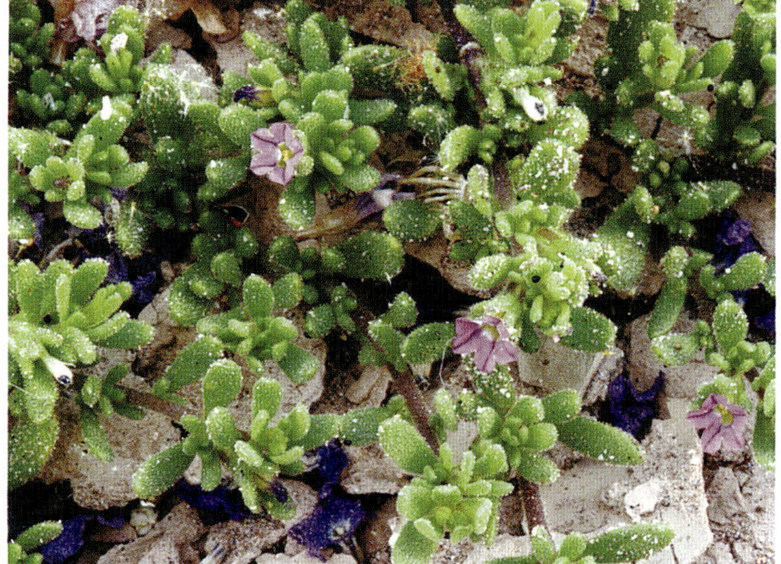

Mat Nama

Nama torynophyllum

MAT NAMA is an unusual member of the waterleaf (Hydrophyllaceae) family, even less common than Havard nama. Its leafy mats blanket the ground after rains, with stems sometimes radiating from a central point like the spokes of a wheel. This densely hairy, prostrate annual has very small leaves in close bundles along the stem, but occasionally a few blades are conspicuously larger, up to one-half inch long. The fleshy, swollen leaves are spatulate or spoonlike, often with curling margins.

The tubular flowers, with five pleated lobes, are very obviously smaller than those of other *Nama* species in the park—just one-sixth inch long. They bloom singly along the branches but may be quite abundant for short periods.

Mat nama covers the mouths of intermittent arroyos flowing into larger streams as well as gravel bars and sandy banks near flowing water and sites of recent flooding. Centered in the Big Bend in southern Brewster and Presidio counties, it also occurs to the east near Del Rio and south in Coahuila and Chihuahua, Mexico.

Although rare, mat nama has been confirmed at the mouth of Tornillo Creek. I have also come across this annual at the mouths of Blue Creek and Alamo Creek by the Rio Grande. It is considered critically imperiled in Texas (NatureServe 2005).

Note: This species was described in 1905 by Jesse Greenman, curator of the herbarium at the Missouri Botanical Garden, from a specimen collected by Carl Purpus in Coahuila, Mexico, in 1903. The German-born, freelance collector Purpus spent decades in Mexico, searching for plants for his brother, Joseph, with the Darmstadt Botanical Garden. Purpus also sold sets of specimens to U.S. and German botanists and herbariums and wrote natural history articles for German magazines.

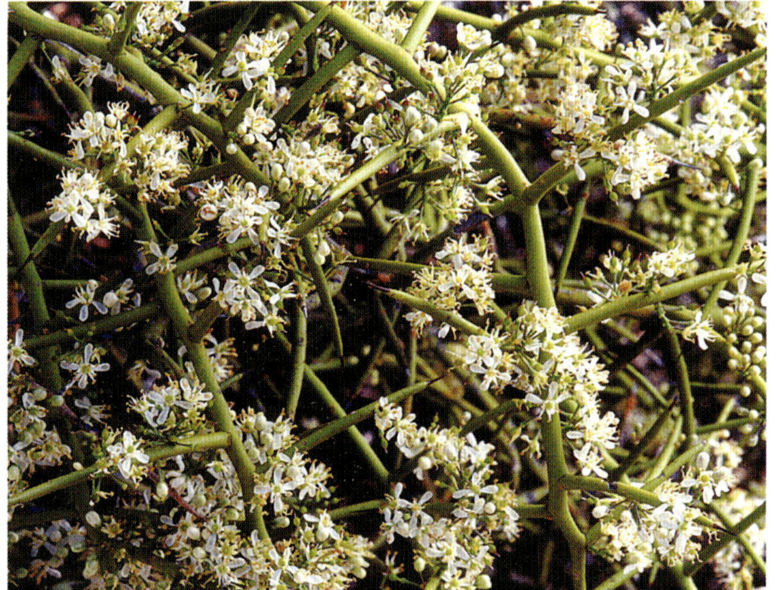

Allthorn

Koeberlinia spinosa

Resembling a dense maze of rigid yellowish-green thorns, this evergreen shrub may attain a height of five feet or more and erect impenetrable thickets. It occasionally grows much higher, becoming treelike with a short trunk. Also known as corona de Cristo (Christ's crown) and junco, allthorn seems a very inhospitable desert plant, yet it is a preferred hiding place for animals and a nesting site for birds. I will forever wonder how these animals wander through the thorns unscathed.

Although apparently all thorns, hence the name, this shrub produces minute leaves that soon fall off. To survive in the harsh desert climate, the plant carries on photosynthesis in its green stems. Blooming from spring to fall, the tiny white or greenish-white flowers are only one-fourth inch wide with four small petals, but they appear in tight clusters along the stems. They are followed by round, clustered berries that turn a shiny black.

Allthorn spans a variety of plant communities—in desert, grasslands, and mountain foothills—from southeastern California throughout southern Arizona and southern New Mexico into Trans-Pecos Texas and the Rio Grande Plains. It is also far reaching in Mexico, from Baja California throughout most of northern Mexico and scattered locations farther south.

In the park allthorn sprawls across desert flats, perhaps most notably off the Dagger Flat road. It is also scattered throughout the Chisos foothills, near Dugout Wells and Gano Spring, south to Castolon on the Rio Grande, and to Lajitas in the extreme southwest.

Note: The genus *Koeberlinia* was named in honor of Christoph Koeberlin (1794–1862), German clergyman and amateur botanist. The species name *spinosa* means "spiny." Joseph Zuccarini described this species in 1832 from collections taken near Mexico City (1826–32) by the Bavarian explorer Karwinski. Zuccarini is best known for describing Philipp Franz von Siebold's extensive collections from Japan in *Flora Japonica* (1835).

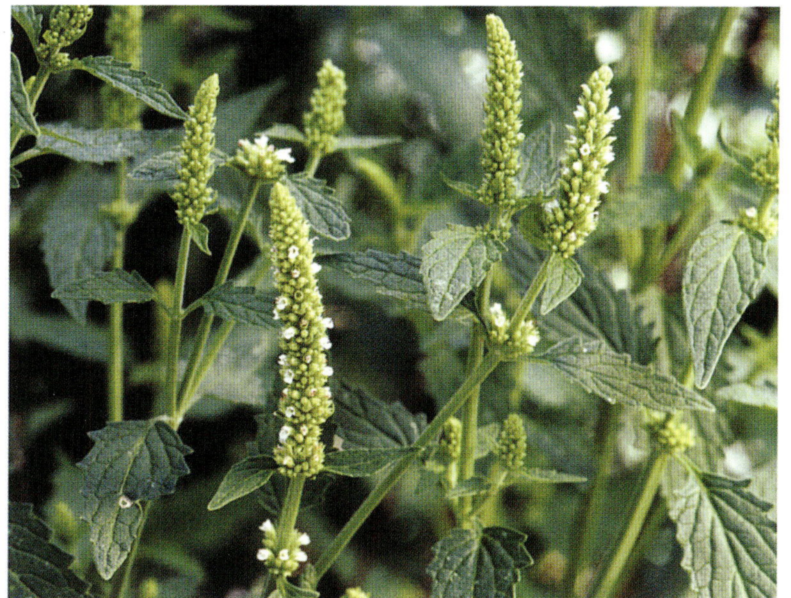

White Giant Hyssop

Agastache micrantha var. *micrantha*

THIS PLANT won't win any awards for its beauty, but the closer you get, the more interesting it is. White giant hyssop is an inconspicuous but aromatic herb two feet high with square, erect stems. Usually one or two inches long, the leaves are triangular or lance-shaped, prominently toothed on the edges, and covered with hairs. Basal leaves, however, may be kidney-shaped with stalks often much longer than the blades. Variety *micrantha* is identified by its relatively thin, weak leaves.

Blooming from July to October, the minuscule white flowers are crowded on narrow, dense spikes up to five inches long. If you look closely, you can see that the tiny flowers, one-eighth inch long, consist of a very short tube opening onto a two-lipped border.

In Texas this member of the mint (Lamiaceae) family is isolated in the Davis and Chisos mountains in the Trans-Pecos. It also grows in north-central New Mexico, north-central Arizona, and northern Mexico (central and eastern Chihuahua and northwestern Coahuila). A closely related variety, *durangensis,* is native to the Mexican states of Durango and Zacatecas.

White giant hyssop prefers locations on rocky, sandy, and moist loamy soils in high woodlands, especially in canyons near water sources, such as springs and intermittent creeks. In the park it blooms profusely on the trail to Boot Spring and above Boot Spring along Boot Creek.

Note: In 1762 Johannes Gronovius, describing plants collected by John Clayton in Virginia, named this genus *Agastache* from the Greek *agan* (much) and *stachys* (ear of grain, spike), referring to the many flower spikes. The species name *micrantha* means "small-flowered." This species was collected by Charles Wright in 1849 in Madera Canyon in the Davis Mountains and described (as *Cedronella micrantha*) by Asa Gray in 1870.

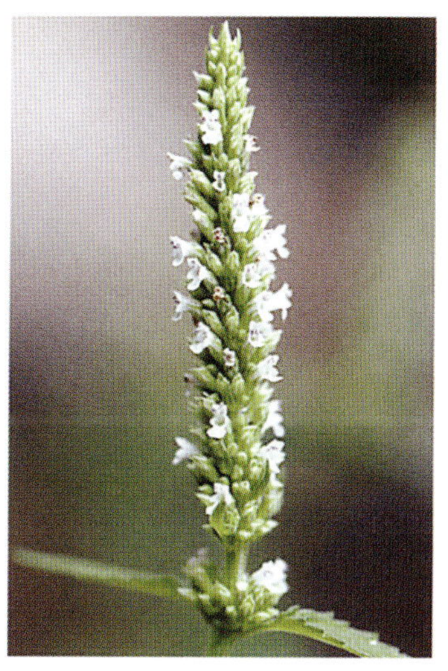

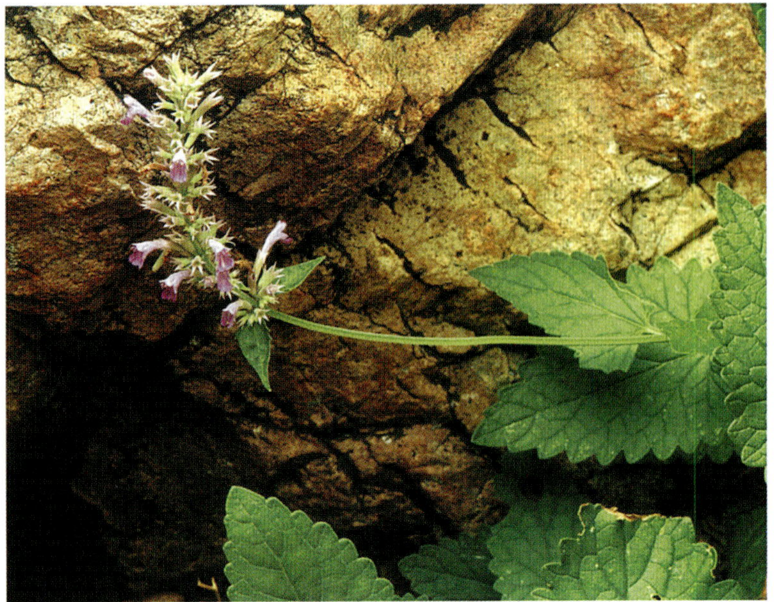

Havard Giant Hyssop

Agastache pallidiflora ssp. *neomexicana* var. *havardii*

HAVARD GIANT HYSSOP is an erect aromatic herb up to three feet high with square stems growing from a woody crown. The stalked leaves are heart-shaped to triangular, occasionally two inches long and almost as wide but usually smaller. They are slightly hairy, noticeably veined on the surface, and bluntly serrated on the edges.

Small pinkish or rose-colored flowers are crowded into compact cylindrical spikes atop the four-angled stems. Blooming from mid-summer to fall, the flowers are typical of the mint (Lamiaceae) family: each consists of a narrow tube which opens into a broad, two-lipped border (limb). The upper lip is erect, two-lobed, and spreading; the lower lip is three-lobed.

Havard giant hyssop is scarce in the Chisos, Glass, Davis, and Guadalupe mountains of Trans-Pecos Texas. Also known as New Mexico giant hyssop, it is encountered farther west in New Mexico and south in the mountains of northern Coahuila, Mexico.

Unseen on official trails in the park, this mint inhabits secluded, moist canyons of the Chisos, in rocky crevices or amidst decaying leaves and trees. Years ago, it was reported in a canyon on the east side of Casa Grande, in bloom in late July. I photographed this sweet-smelling herb near the top of Maple Canyon in September.

Note: The species name *pallidiflora* connotes "pale-flowered." Asa Gray described this species in 1885 (as *Cedronella brevifolia* var. *havardii*). It is one of many Big Bend plants named in honor of Valery Havard, who first collected this giant hyssop in the Chisos Mountains in the summer of 1883. The common name "hyssop" probably is associated with a plant of Eurasian origin (*Hyssopus officinalis*) that has blue flower spikes and sweet-smelling leaves used to make perfume.

Hairy Hedeoma

Hedeoma mollis

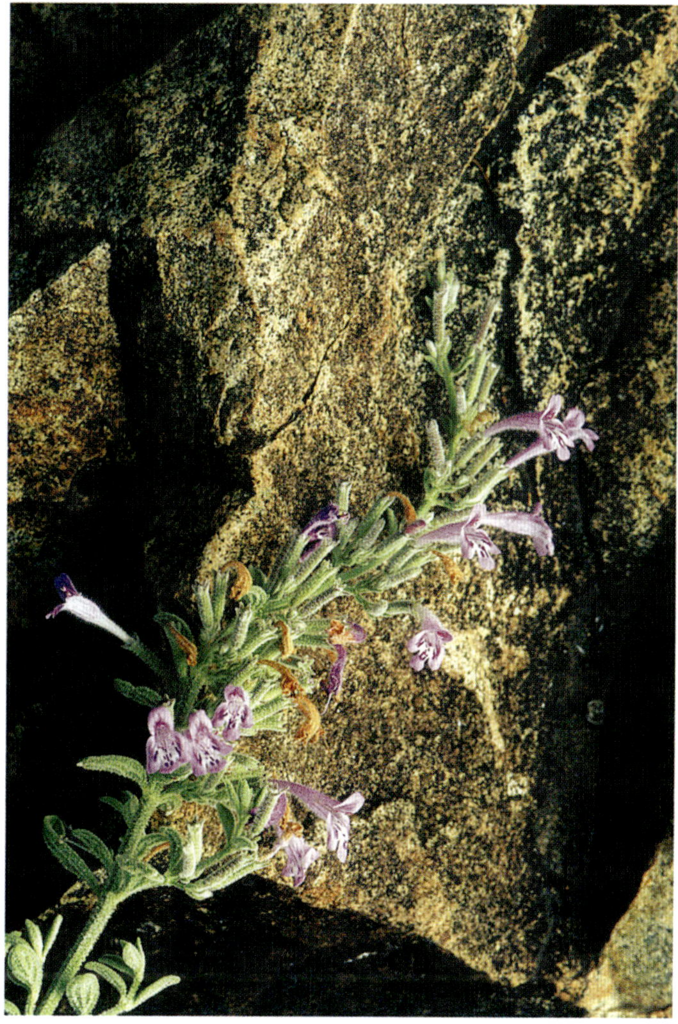

HAIRY HEDEOMA, or hairy false pennyroyal, is a low, weak subshrub less than two feet high that develops from a woody crown. The stems are mostly erect and unbranched, or branched at the base, rather slender, and often sprawling. The entire plant has a distinctive gray coloration because the stems and leaves are covered with grayish-white, soft, woolly, treelike hairs. Usually one-third inch to one inch long, the stalked leaves are elliptic or oval with more or less pointed tips.

The flowers are clustered near the tops of the stems with about ten flowers in each elongated cluster. The pale lavender or magenta flowers consist of a tube opening widely into a two-lipped border. The upper lip is short and straight, the lower broader and three-lobed with distinctive maroon spots or streaks. Even the inside of this flower is hairy.

Blooming from summer to fall, hairy hedeoma is considered endemic to the Trans-Pecos, known only from the Chisos and Glass mountains of Brewster County, the Chinati and Sierra Vieja mountains of Presidio County, and the Davis Mountains of Jeff Davis County. It is classified as vulnerable in Texas and worldwide.

This mint springs up at shady locations in the Chisos, especially in August and September. One of five *Hedeoma* species in the park, it has been documented in the Basin, on the west slope of Casa Grande, in Upper Pulliam Canyon, at the head of Pine Canyon, and along the Pinnacles Trail. I have noticed hairy hedeoma along the Lost Mine Trail, near water in Maple Canyon, and at the base of the Window.

Note: The genus name *Hedeoma,* from the Greek *hedys* (sweet) and *osme* (scent), was an ancient name for a mint. The species name *mollis* means "soft," a reference to the soft hairs that cover the foliage. This species was described (as *Hedeoma molle*) by John Torrey in 1859 from specimens collected by John Bigelow and others during the U.S.-Mexico boundary survey.

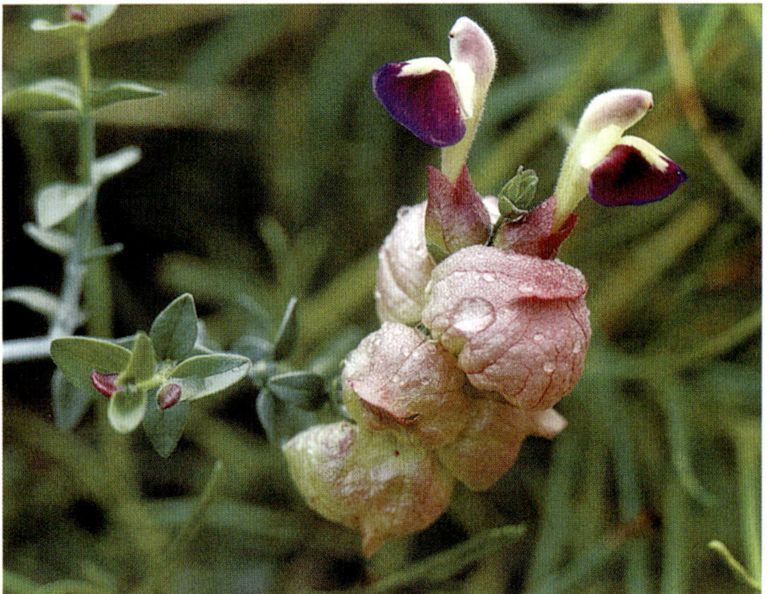

Bladder Sage

Salazaria mexicana

BLADDER SAGE, or paperbagbush, is a rounded shrub up to four feet high with stiff, spine-tipped branches and wiry, flexible short spurs. Pale green or white and often almost leafless, the naked stems may form a dense, jumbled mass. The sparse leaves are egg- or lance-shaped, up to one inch long, and on short stalks. They are slightly thick and leathery.

The two-lipped, usually paired flowers are curious in shape. They have a creamy white tube that opens into an arching, pale lavender upper lip and a broad, dark purple-violet lower lip. The flowers bloom in spring and summer, often after summer monsoons.

Bladder sage is named for its bladdery fruits shaped like inflated paper bags. The calyx (green external base) of the flower expands into a paper-thin globe enclosing four nutlets. By the time the nutlets mature, the globe is inflated, ready to waft in the wind.

Bladder sage fills sandy desert washes, silty flats, and gravelly arroyos in mountain foothills. A Mojave Desert denizen, it also spreads eastward into the Sonoran and Colorado deserts. A disjunct population is based in the Chihuahuan Desert in Brewster and Presidio counties and in Coahuila and northeastern Chihuahua, Mexico.

In the park this shrub is locally common from Mesa de Anguila in the extreme southwest to Santa Elena Canyon, Castolon, and Smoky Creek farther east along the Rio Grande. It spills into the Rattlesnake Mountains and flats surrounding the Chisos.

Note: This genus was named for José Salazar Ylarregui, Mexican scientific commissioner and chief surveyor for the U.S.-Mexico boundary survey and a signatory of the Gadsden Purchase Treaty of 1853. The species was described by John Torrey in 1859 from plants collected in 1844 by explorer John Frémont near Las Vegas, Nevada. From 1842 to 1854, Frémont led five western expeditions and became known as "The Pathfinder."

Mountain Sage

Salvia regla

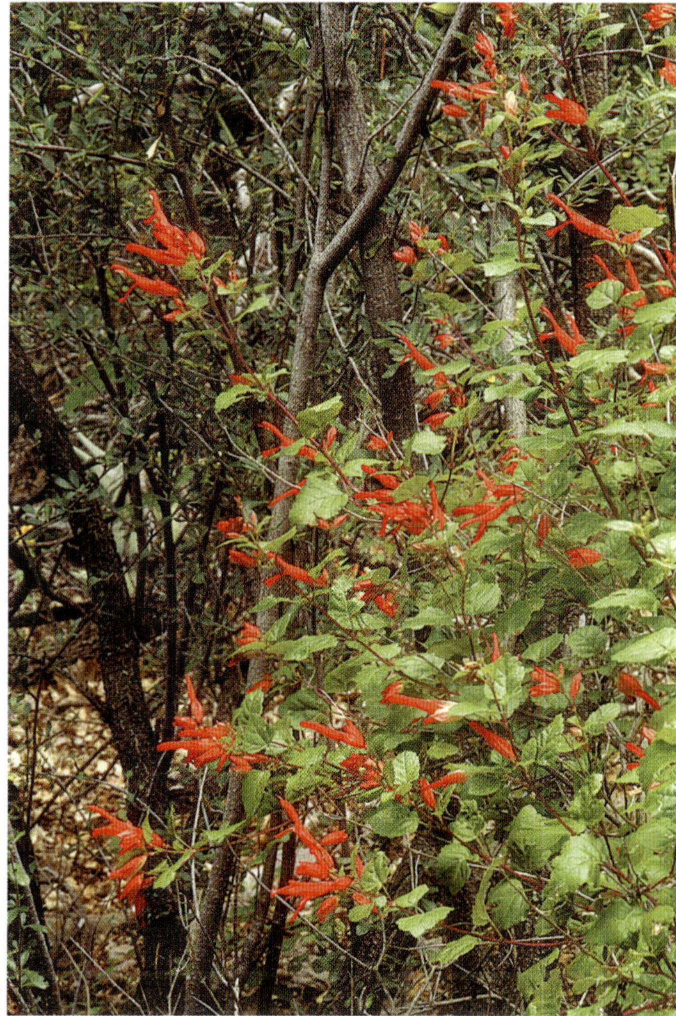

MOUNTAIN SAGE, or royal sage, is a distinctly fragrant, much-branched, leafy shrub two to six feet high. Up to two inches long and wide, the variable leaves have been described as triangular, egg-shaped, or kidney-shaped. Coarsely toothed on the edges, the dark green blades bear prominent orange glands underneath.

In late summer and fall, after many other mountain plants have bloomed, these shrubs take over the forest understory, coloring it crimson. In clusters of one or two near the stem tips, the scarlet flowers split into two long flaring lips. Up to two inches long, they have apparently evolved to attract resident and migrating hummingbirds.

Mountain sage occupies wooded and rocky slopes in shady canyons at mid to high elevations. The shrub is widespread across Mexico from Oaxaca north to all Mexican states bordering the United States. Critically imperiled in Texas (NatureServe 2005), it crosses into the United States only in the Chisos Mountains.

In the park mountain sage produces unforgettable flower displays along the road in Upper Green Gulch and the Basin and on many mountain trails. At times it seems ubiquitous in the Chisos, in Pine and Pulliam canyons, and on Lost Mine Peak and Casa Grande. Look for it also on the Laguna Meadow and Pinnacles trails, from Boot Canyon to the South Rim, and on Emory Peak.

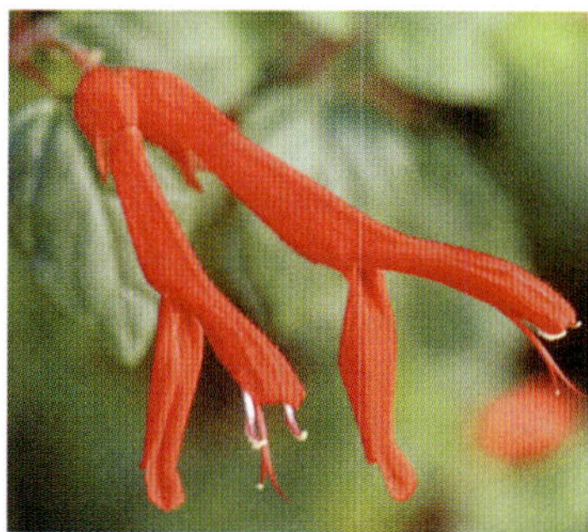

Note: The genus name *Salvia* is the Latin term for the herb sage, from *salvus* (saved), referring to presumed medicinal properties. The species name *regla* is unclear; it may be named for Regla in the Mexican state of Hidalgo, where it was first collected. This species was described in 1799 by Antonio Cavanilles, Spanish clergyman, botanical illustrator, and eminent taxonomist.

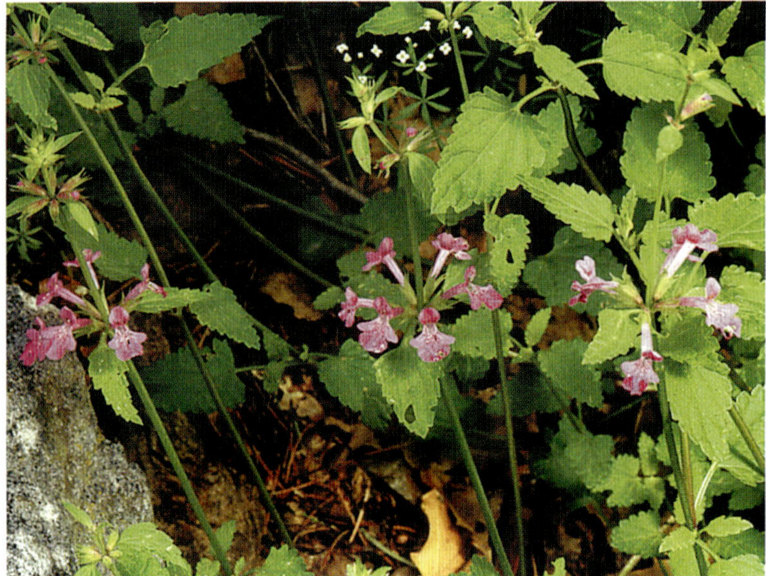

Rock Betony

Stachys bigelovii

ROCK BETONY, or rock hedgenettle, is a leafy perennial herb with upright, clumped stems up to two feet high that sometimes produces colonies many feet across. The paired leaves, up to two and one-half inches long, are mostly triangular or egg-shaped with broad, rounded teeth on the margins. The lower leaves, especially, grow on stems (petioles) almost as long as the blades.

The bright pink, two-lipped flowers form in whorls of two to six, radiating in a circle from the same point on the stem. Each stemless flower has a slender tube that opens into an erect hoodlike upper lip and a longer, broader, three-lobed lower lip. The throat is whitish but with distinct pink dots that guide hummingbirds and other pollinators to the nectar inside. The flowers bloom from late spring to early fall but especially in summer.

Rock betony does best at higher elevations in shady wooded canyons, along canyon walls, under trees, and in moist pockets on rocky slopes. It is dispersed in mountainous terrain of the Trans-Pecos (in Brewster, Presidio, and Jeff Davis counties) and northern Mexico (eastern Chihuahua to Nuevo León and south to San Luis Potosí and Zacatecas).

In the park rock betony colonizes shady, relatively moist locations high in the Chisos. It has been located on Pulliam Bluff, Lost Mine Peak, and Casa Grande; in Pine and Pulliam canyons; and on the Pinnacles Trail and Emory Peak.

Note: The genus name *Stachys,* from Greek, means "ear of grain" or "spike," referring to the spikelike flower clusters. In 1872 Asa Gray named this species for John Bigelow, one of the botanists who collected it on the Emory boundary survey in 1851–52. Later, Bigelow was the first botanist to collect specimens in parts of northern California and the Sierra Nevada.

Berlandier Flax

Linum berlandieri var. *filifolium*

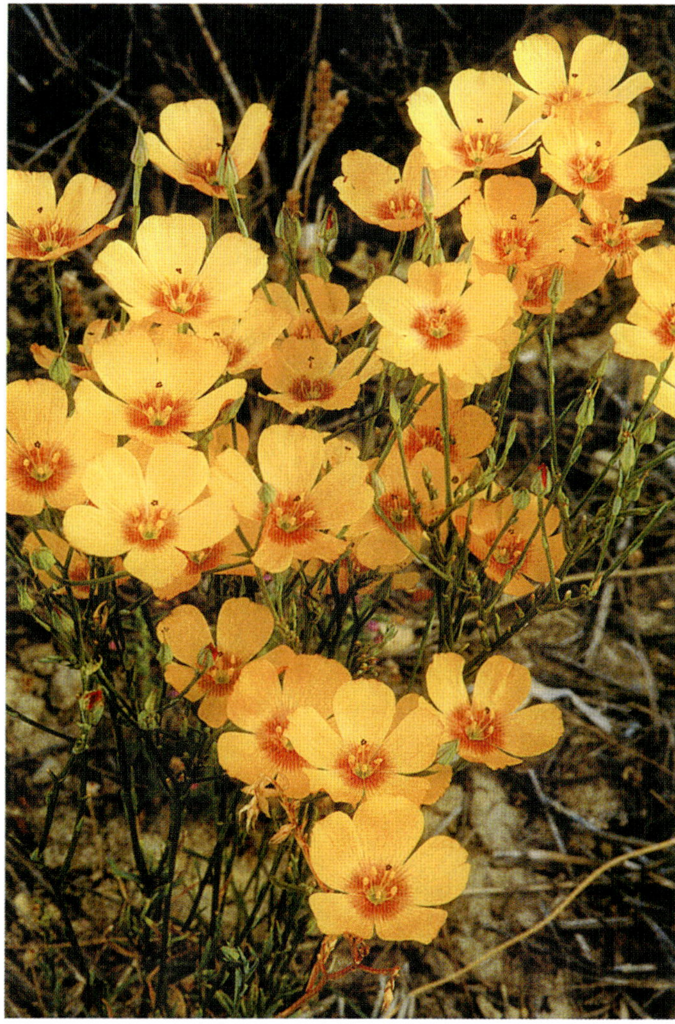

ALSO KNOWN AS bowl flax, Berlandier flax is a wiry annual eight to twenty inches high with smooth green stems. Scattered along the stems, the leaves are very narrowly linear or needlelike, seldom more than one inch long and often much less.

One of half a dozen yellow flaxes in the park, Berlandier flax boasts the largest, most colorful, and attractive flowers. Perhaps most outstanding in March and April, the delicate, bowl-shaped flowers reach one inch across. Each coppery-yellow bloom has five flimsy and early falling petals with distinctive red to maroon bands or veins converging near the base.

This flax is usually situated on dry, open desert flats, alluvial flats, and floodplains on sandy, limy, and gravelly soils; it grows more sparingly on rocky hillsides. Berlandier flax ranges widely throughout the Trans-Pecos (from Culberson County southward and east to Ozona), north to Andrews in the Permian Basin, and as far south as Eagle Pass in the Rio Grande Plains. It also reaches into Coahuila, Mexico.

In the park this herb may be most plentiful from Tornillo Flat to Persimmon Gap and along the Rio Grande from Johnson Ranch on the River Road west to Castolon and Santa Elena Canyon. In the Chisos it has been recorded at Panther Canyon and in the Basin.

Note: The genus name *Linum* is from the Greek word for flax, *Linon.* This species was described in 1836 by Sir William Jackson Hooker, first director of the Royal Botanic Gardens at Kew, and named for Jean Louis Berlandier, who had discovered it near San Antonio, Texas, in 1828. The variety name *filifolium* implies "with threadlike leaves." This variety was collected by Rogers McVaugh in 1947 at Big Canyon on the Rio Grande in Brewster County and described in 1949 by Lloyd Shinners, Canadian-born botanist and herbarium director at Southern Methodist University. McVaugh, an authority on the history of botanical collection, wrote a biography of Edward Palmer and a study of the Sessé and Mociño expedition.

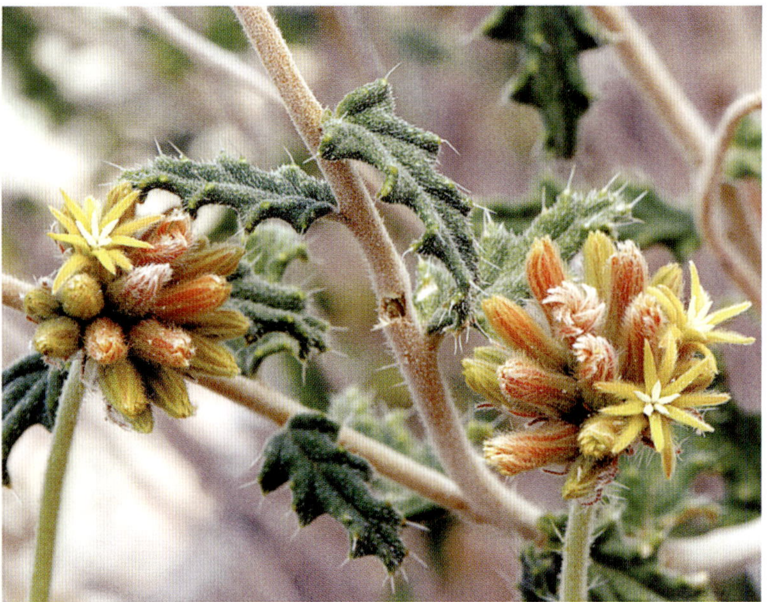

Stinging Cevallia

Cevallia sinuata

Stinging cevallia, stinging serpent, or Shirley's nettle is a low subshrub up to two feet high, densely branched, often rounded, and with sprawling whitish stems. The oblong leaves, one to two inches long, are wavy on the edges and usually deeply toothed or lobed. Both stems and leaves are cloaked with several kinds of hairs: some treelike, some barbed, some glasslike and stinging.

Mostly in summer and fall, the unusual flowers bloom in compact, hemispheric clusters on stalks up to four inches long. Each silky, feathery flower head is different, bearing blooms in varying shades of yellow, pink, orange, and reddish-brown. Each blossom consists of five very slender yellow petals, five yellow sepals similar to the petals, and five prominent stamens. Petals and sepals are fringed with long white hairs.

Stinging cevallia occupies open gravelly sites, clay and alluvial flats and floodplains, and desert washes, but it may be most prolific along roads and in disturbed areas. In Texas, it is distributed throughout the Trans-Pecos, along the Rio Grande to the Gulf, and into the Panhandle and western Edwards Plateau. This subshrub extends west to southeastern Arizona and across northern Mexico (from Chihuahua to Tamaulipas and south to Zacatecas).

At elevations below forty-five hundred feet, stinging cevallia can be nearly omnipresent in the park. It has been observed on Mesa de Anguila to the southwest, at Hot Springs to the east, on Harte Ranch to the north, and around the Chisos at Gano Spring and Dugout Wells.

Note: The name Shirley's nettle honors Shirley Powell, who conducted extensive cellular studies of this stinging plant. The genus and species were described in 1805 by Mariano Lagasca y Segura, disciple of Spanish botanist Antonio Cavanilles and director of the Madrid Botanic Gardens after 1814. With the restoration of absolutism in 1823, Lagasca's herbarium was destroyed by a mob, and he was exiled to England.

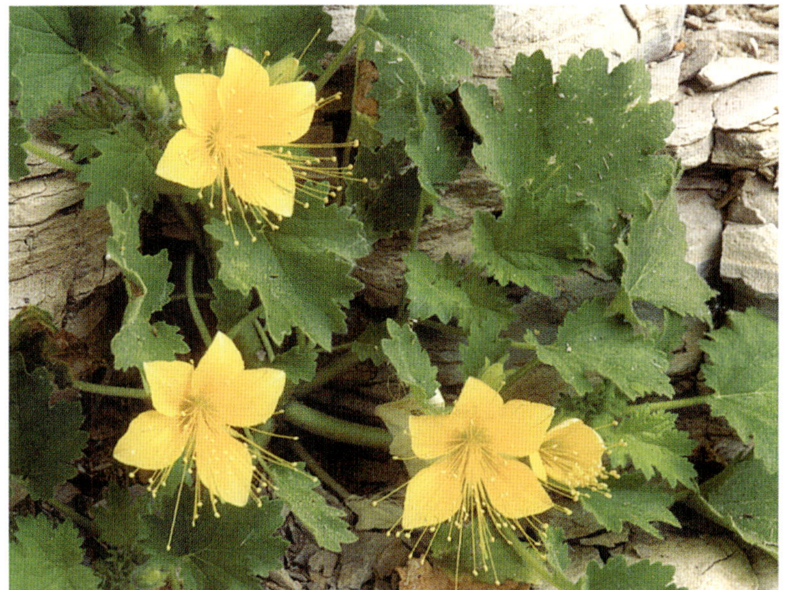

Yellow Rocknettle

Eucnide bartonioides

Yellow rocknettle is a perennial or annual up to one foot high with slightly brittle stems that tend to sprawl across the ground. In broad clumps or rounded mounds, it can take over desert washes and canyon walls. Up to four inches long, the dark green leaves are nearly round in outline, shallowly lobed and scalloped on the edges, and covered with stiff hairs. The leaf stems (petioles) may be longer than the leaves.

Among the most memorable and glorious of Texas wildflowers, the yellow, funnel-shaped blossoms have many thin, yellow stamens that protrude well beyond the petals. Two inches wide, with five spade-shaped petals united at the base, the delicate flowers open only in bright sun. They are not so particular about when they bloom, which is more or less year round, especially in summer after showers.

This herb hangs from rock ledges and limestone cliffs in partial shade and near water or takes root in floodplains and desert arroyos. In the United States yellow rocknettle is found in south-central and southwestern Texas, especially near the Rio Grande. It reaches throughout the southern Trans-Pecos (from Fort Davis and Presidio to Del Rio) to the western Edwards Plateau and throughout northern Mexico (Chihuahua and Coahuila to San Luis Potosí).

In the park this herb is a well-known fixture at Hot Springs, and it can be seen at many sites, especially along the river. Yellow rocknettle is usually present in Boquillas Canyon, along Tornillo Creek, at Ernst Tinaja, and farther south from Johnson Ranch to Castolon. Frequently, small colonies develop near McKinney Spring.

Note: The genus name *Eucnide,* from the Greek *eu* (good) and *knide* (nettle), means "strongly nettlelike." The species name *bartonioides* implies "similar to the genus *Bartonia.*" Both genus and species were collected by the German explorer Baron Karwinski on his 1840–43 trip to Mexico and described by Joseph Zuccarini in 1844.

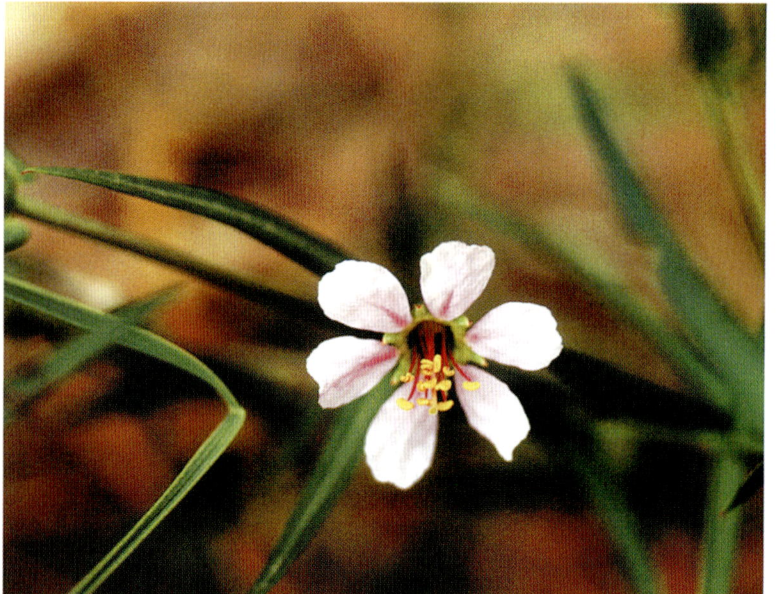

Stalkflower Nesaea

Nesaea longipes

FORMERLY KNOWN AS stalkflower heimia, stalkflower nesaea is a low perennial with woody roots but slight stems sometimes three feet long or more that spread diffusely and drape over other vegetation. Nearly stalkless, the slender linear leaves, up to two inches long, are pointed at the tip and noticeably rolled upward on the edges. The paired leaves may have prominent, earlike bulges at the base.

One-half inch wide, the strikingly attractive, solitary flowers form on threadlike stalks. Blooming primarily from May to July, they usually have six widely separated, rounded petals and conspicuous, protruding red stamens in the throat. The delicate petals are pale pink or suffused with pink and tinged with darker pink or purplish streaks or markings.

Stalkflower nesaea frequents desert springs, streams, and mostly limestone sites where water periodically drains or seeps. In Texas it occurs very infrequently in the southern Edwards Plateau (near Bandera and Kerrville) and at disjunct locations in the southeastern Trans-Pecos (in Brewster, Pecos, Terrell, and Val Verde counties). This perennial also grows in southeastern New Mexico and Coahuila, Mexico.

Globally stalkflower nesaea is vulnerable or imperiled throughout its range and in Texas imperiled and very vulnerable to extirpation. In the park it has been confirmed near Glenn Spring east of the Chisos and to the west at Chisos Pens on Cottonwood Creek.

Note: Carl Kunth named this water-loving genus in 1823, perhaps for *Nesaea,* one of the fifty Nereids, or sea nymphs, in Greek mythology; or the name may originate from the Greek *nesos* (island) because the first species was discovered on the island of Mauritius during the French Bougainville expedition (1766–69).

The species name *longipes* means "long-stalked." This species was collected by Charles Wright in 1849 and described by Asa Gray in 1852.

Propellerbush

Janusia gracilis

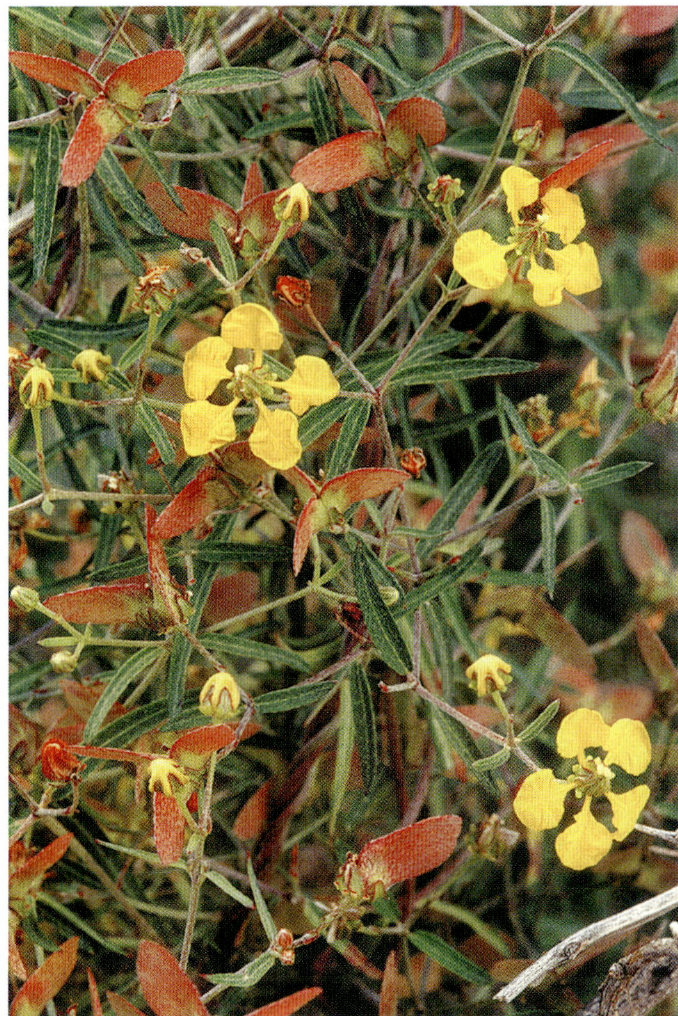

PROPELLERBUSH, or slender janusia, is a cross between a shrub and a vine with woody stems near the base and vinelike, wiry upper stems that twine through and climb over other shrubs. The diffuse stems can reach nine feet in length. Up to one and one-fourth inches long, the short-stemmed leaves are narrowly oblong or lance-shaped, pointed at the tip, and somewhat silky.

This member of the malpighia or Barbados cherry (Malpighiaceae) family produces two types of flowers. The "normal" flower is yellow, aging to dark orange, with five spade-shaped petals that are widely spaced and very narrow at the base. The petals are often distinctly wrinkled and wavy on the edges. The "abnormal" flower is much smaller with only rudimentary petals. Self-fertilizing, it never opens.

An oil-seeking bee pollinates the "normal" blossoms, while gathering nutritious oil from glands near the base of the sepals. The distinctive winged fruit (called a samara) is at first green but matures to orange or brown. Shaped like the blades of a propeller, two or three veined wings protrude from this dry fruit, hence the common name "propellerbush."

Sprawling over open slopes and foothills in dry, rocky limestone and igneous soils, propellerbush stretches from southern Terrell and Brewster counties west to El Paso and throughout southern New Mexico into southern Arizona. It is also wide ranging in northern Mexico (in Coahuila and Durango and west to Sonora and Baja California).

Except in the higher Chisos, this unusual shrub thrives throughout the park: in the Dead Horse and Rosillos mountains, Grapevine Hills, and the Chisos and lower desert mountains. Along the Rio Grande, propellerbush is extensive from Boquillas Canyon west to Terlingua Creek. It blooms from spring to fall, often after good rains.

Note: The genus *Janusia* is named for the Roman god of doorways and archways, Janus, who was depicted in art and on Roman coins as two-faced. Presumably the term refers to the two types of flowers. The species name *gracilis* (slender or graceful) refers to the vinelike stems. Propellerbush was collected by Charles Wright in mountains east of El Paso and described in 1852 by Asa Gray.

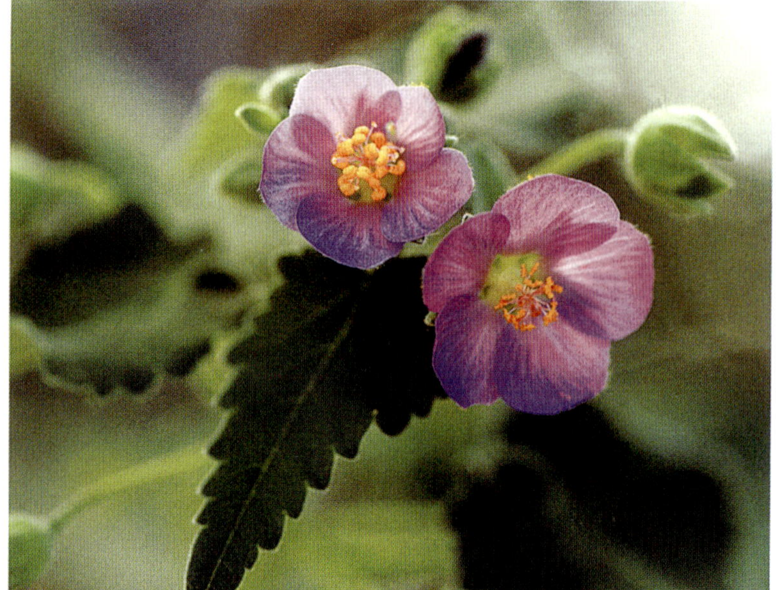

Purple Gaymallow

Batesimalva violacea

PURPLE GAYMALLOW is a slender, often bending shrub seldom over four feet high with distinctly hairy branches. The large leaves on long stalks are usually lance- to egg-shaped and pointed with rounded teeth on the margins. The tops of the leaves are green, but their lower surfaces are noticeably pale and covered with fine star-shaped hairs.

In rich shades from blue to violet, a few lovely flowers form at the base of leaves. This very uncommon shrub of the mallow (Malvaceae) family will sometimes bloom in spring, but it does so much more often in fall in moist and relatively cool shade. The fruit is a distinctive, discoid capsule (schizocarp) that splits into nine one-seeded partitions.

In the United States purple gaymallow is confined to a canyon on the west side of the Chisos in Big Bend National Park, where it flourishes among boulders and in the shade of oaks. It is more common in the states of Coahuila (in the Sierra Maderas del Carmen) and Nuevo León in northern Mexico.

Globally this mallow is classified as imperiled throughout its range and in Texas as critically imperiled and especially vulnerable to extirpation.

Note: In 1975 University of Texas botanist and mallow specialist Paul Fryxell named the genus *Batesimalva* for Cornell botanist David Bates. A year earlier Bates, also a mallow specialist, named the genus *Fryxellia* for Fryxell. Fryxell later wrote an article in the journal *Sida* titled "Very Personal Generic Names . . . : A Contribution to Whimsical Botany." This mallow was described in 1909 (as *Gaya violacea*) by Joseph Rose of the U.S. National Herbarium. (There is no evidence that *Gaya* named anything for Rose.) Cyrus Pringle collected the plant in the Sierra Madre Oriental near Monterrey, Mexico, in 1906.

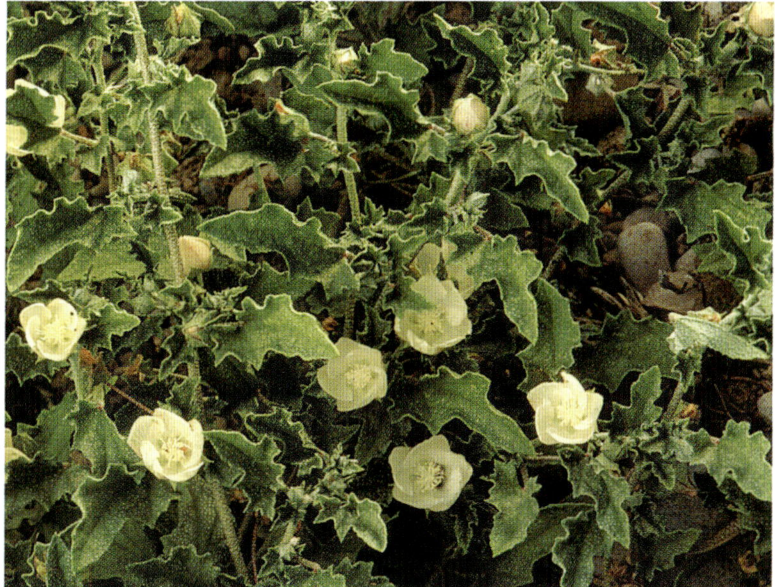

Scurfy Mallow

Malvella lepidota

THIS PERENNIAL prostrate herb, with radiating or trailing branches as much as sixteen inches long, is easy to recognize from the scurfy appearance of its foliage. Scalelike hairs on stems and leaves form a silvery crust much like dandruff. The alternate leaves on slender stalks are arrowlike, wedge-shaped, or triangular, pointed at the tips, wrinkled on the margins, and irregularly lobed toward the base. According to Barton Warnock (1977), Native Americans used the leaves in a cough remedy.

Delicate, almost succulent flowers, creamy white or very pale yellow, are the redeeming quality of this scaly plant. Solitary and cuplike, with five wedge-shaped petals, they may be tinged or blotched with pink or purple. Appearing waxy or chalky, the creamy flowers open around noon and seem to peek out from the base of leaves low to the ground.

Scattered in the upper Rio Grande Plains, parts of north-central Texas, and the Panhandle, scurfy mallow is more frequent throughout much of the Trans-Pecos. Often reclining on clay and alluvial flats and gravelly soils or in roadside depressions, it also spreads into southern New Mexico, southeastern Arizona, and northern Mexico (as far south as Zacatecas, San Luis Potosí, and Querétaro).

Probably most prevalent in the Marathon Basin on the Novaculite Hills to the north of the park, scurfy mallow may invade the Dog Flats and Harte Ranch areas south of Persimmon Gap. Usually blooming in late July and August, it can bloom almost any time from spring to fall after rains.

Note: The genus name *Malvella* is the diminutive of the genus name *Malva.* The species name *lepidota* in Greek means "scaly." This species was collected by Charles Wright in New Mexico in 1851 during Colonel Graham's survey of the U.S.-Mexico boundary and described by Asa Gray (as *Sida lepidota*) in 1852.

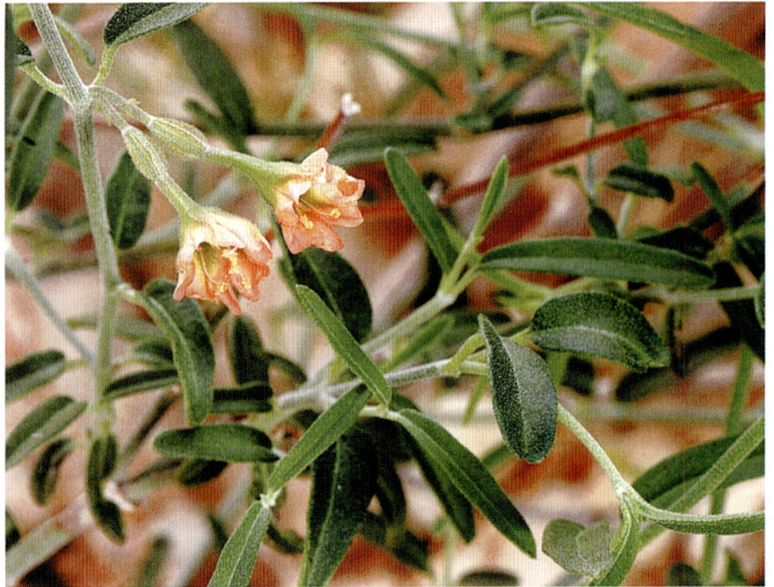

Narrowleaf Moonpod

Acleisanthes angustifolia

A LOW, ERECT SUBSHRUB usually less than sixteen inches high, narrowleaf moonpod has numerous woody lower stems but only thin, sparse upper branches. The short-stemmed, paired leaves, up to three-fourths inch long, are narrowly oblong, thick, and fleshy. Both stems and leaves are covered with white and gland-tipped hairs.

Blooming from spring to fall, tubular flowers open in late afternoon and close the next morning. In shades of orange, brown, green, and yellow, the solitary or paired blooms have short, narrow tubes that flare into a pleated, five-lobed funnel. In this species, nature has a back-up plan: some smaller flowers may be cleistogamous. The whimsical, pendent fruit is as captivating as the flower: it is hard and nutlike; grooved, with thin, finlike, transparent wings protruding from the sides; and decorated at the tip with its own dried calyx.

This moonpod prospers on gypseous clay soils and gravelly and rocky slopes in desert mountains. It is restricted to Presidio and Brewster counties in the Big Bend and to Chihuahua, Coahuila, and Durango in Mexico.

Narrowleaf moonpod appears throughout the park, especially on clay soils. Along the Rio Grande, it is most accessible at Boquillas Canyon, Hot Springs, and Santa Elena Canyon. The subshrub has also been verified at McKinney Spring and Dog Canyon; near Croton, Burro and Wasp springs; and at Talley Mountain.

Note: The genus name *Acleisanthes*—from the Greek *a* (without), *cleis* (close), and *anthos* (flower)—alludes to the lack of a floral cup (involucre) enclosing the flower. The species name *angustifolia* means "narrow-leaved." This species was collected by Charles Parry in 1852 near Presidio during the U.S.-Mexico boundary survey and described (as *Selinocarpus angustifolius*) by John Torrey in 1859. In 2002, as a result of new molecular analyses, Rachel Levin and others included *Selinocarpus* in the genus *Acleisanthes.*

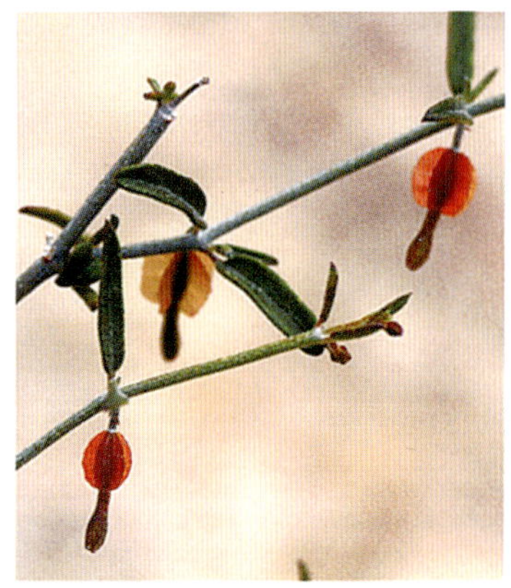

Littleleaf Moonpod

Acleisanthes parvifolia

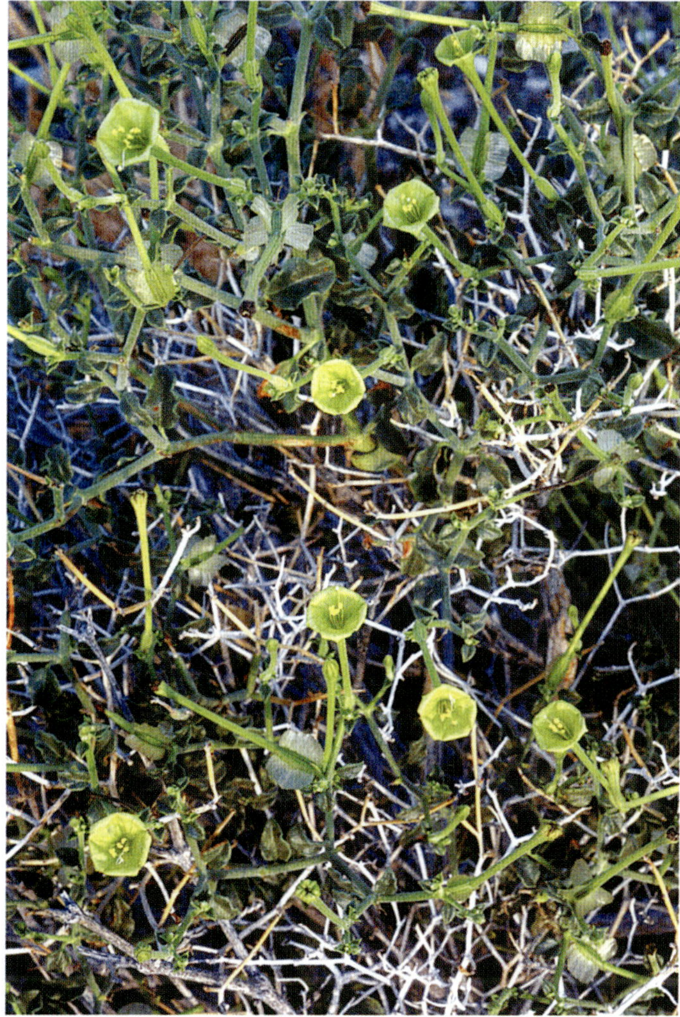

LITTLELEAF MOONPOD is a low subshrub under two feet high with a jumbled maze of slender, brittle, white branches. The crowded stems may be covered with short hairs and a white waxy coating. Less than one inch long at the base, the egg- or lance-shaped leaves are distinctly smaller on the upper branches. Curly or wavy along the edges, the rather sparse little leaves are thick and slightly fleshy.

Pale yellow or greenish-yellow tubular flowers bloom at the tips of branches. Very slender tubes open abruptly onto a pleated border (limb) one-half inch across. Blooming from spring to early fall, they open in late afternoon and close in the bright sun of the following morning.

Moonpods are named for their curious winged fruits (anthocarps). Less than one-half inch long, the compressed fruits have up to five finlike wings, which are very thin and transparent. Backlit by the sun in late afternoon, the fruits resemble glowing miniature ornaments.

This four-o'clock forms jumbled mounds on clay hills and flats, sandy soils, and slopes above washes and streambeds. Vulnerable globally and in Texas, littleleaf moonpod is limited to the southern Trans-Pecos and northeastern Chihuahua, Mexico. In the Trans-Pecos, it is concentrated primarily in southern Brewster County and a few sites in Presidio and southern Hudspeth counties near the Rio Grande.

When in bloom, this dense subshrub is conspicuous at lower elevations in the park: Muskhog Spring off the Old Ore Road, between Burro and Wasp springs near Burro Mesa, at Chisos Pens on Cottonwood Creek, and near Smoky Creek in the Sierra Quemada. I have photographed it blooming in late May and early June above lower Tornillo Creek, in the picturesque sandstone area near upper Tornillo Creek, and in the Maverick Badlands and Rattlesnake Mountains near the park's western boundary.

Note: The species name *parvifolia* means "little leaf." This species was collected by John Bigelow in 1852 near Presidio, Texas, and described by John Torrey in 1859 as a variety of another moonpod species (*Selinocarpus diffusus*).

Red Cyphomeris

Cyphomeris gypsophiloides

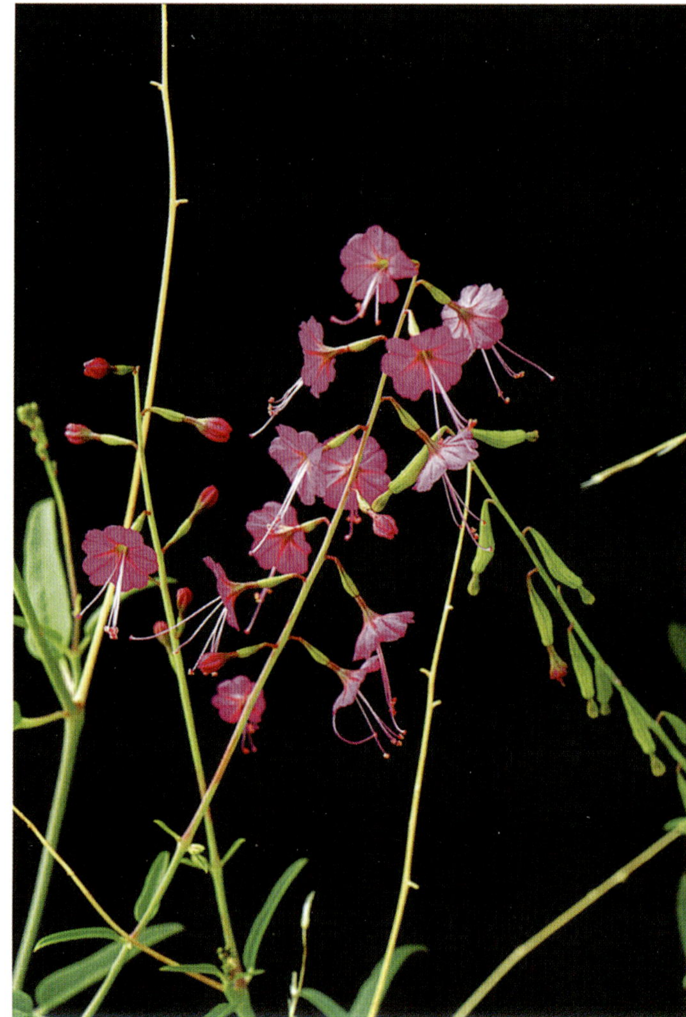

A MEMBER of the four-o'clock (Nyctaginaceae) family, red cyphomeris is a perennial herb occasionally four feet high or more with slender, erect, or sometimes clambering stems growing from a woody base. These branches are conspicuously brittle. On slender stalks, the leaves are opposite and highly variable: upper leaves have a more linear shape, while lower leaves are larger (up to three and one-half inches long) and broadly lance-shaped.

The real beauty of this plant is its glorious red or dark pink flowers appearing in slender, elongated clusters that dangle like a flight of tiny parachutes. The blossoms of red cyphomeris consist of a short, funnel-shaped tube that opens abruptly into a shallowly five-lobed border. Five long stamens protrude beneath this envelope like the cords of a parachute. Seldom more than one-half inch long, the small green fruits are stiff and leathery with an unusual, gibbous shape (similar in outline to a bird's body) that is helpful in identification.

Blooming from summer into fall, red cyphomeris clambers over desert scrub habitats, arid rocky slopes, and mountain woodlands throughout Trans-Pecos Texas. Its geographic range encompasses southern New Mexico and much of Mexico (from Chihuahua east to Tamaulipas and as far south as Puebla and Oaxaca).

This perennial is prominent throughout the Chisos: in the Basin, at Laguna Meadow, near the Boot Spring cabin, and at Ward Spring. Outside the Chisos, it has been reported in the Rosillos Mountains and near Hot Springs. I have witnessed superb examples in Passionflower Canyon in the Dead Horse Mountains in early October.

Note: The genus name *Cyphomeris* is derived from the Greek *kyphos* (bent) and *meris* (part), a reference to the fruit's out-of-kilter shape. The species name *gypsophiloides* means "like the genus *Gypsophila*," a group of gypsum-loving plants. This species was described (as *Lindenia gypsophiloides*) by Belgian botanists Martin Martens and Henri Galeotti in 1843 from a plant collected by Galeotti in the Mexican state of Puebla during the 1830s. Galeotti was a prodigious collector of Mexican plants and later director of the botanical garden in Brussels.

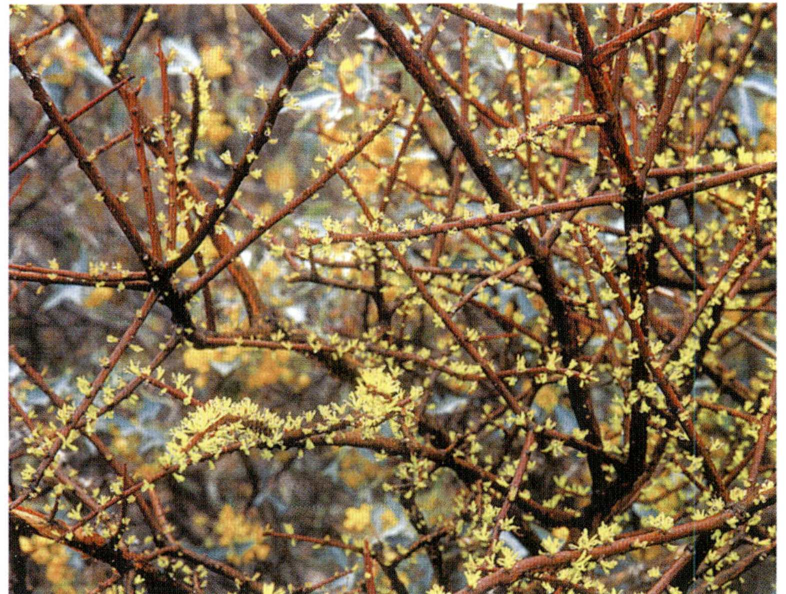

Desert Olive

Forestiera angustifolia

Also known as narrowleaf forestiera and pañalero, desert olive is an evergreen shrub usually three to eight feet high with many stiff, crowded branches. The smooth branches may extend at right angles to the main stems like elbows, but longer branches are curved or arching. They bear linear leaves about three-fourths inch long with smooth, thickened, and curling edges. The leaves often form in small clusters on short spur branchlets.

Inconspicuous pale yellow or greenish-yellow flowers may appear before the leaves in very early spring. They lack petals and have only very tiny sepals, so what you see is mostly the protruding stamens (the male pollen-bearing organs). The unusual flowers form in tight bundles of three to seven blossoms on the knotty branchlets. The fruit is a dark blue to black drupe (a fleshy one-seeded fruit) up to one-half inch long. Edible but not very tasty, it makes a nice meal for birds and mammals.

Desert olive inhabits gravelly arroyos, brushy washes and canyons, rocky hillsides, and desert scrub habitats below thirty-five hundred feet. This shrub is plentiful in southern and southwestern Texas, primarily from the southern and eastern Trans-Pecos to the Rio Grande Plains and the Gulf. It is also widespread in northern Mexico (from Sonora to Tamaulipas and south to Hidalgo).

In the park, desert olive is a customary sight at low to moderate elevations: at Persimmon Gap, Dog Canyon, and Dagger Flat in the north; at many locations in the Dead Horse Mountains; near Oak Creek and Blue Creek in the Chisos Mountains; and from Smoky Creek to Johnson Ranch near the Rio Grande.

Note: The genus *Forestiera* was described in 1810 by French clergyman, botanist, and writer Jean Poiret, author of a seven-volume work on economic plants (1824–29). Poiret named the genus for his botany teacher, Charles Le Forestier, a French physician and naturalist at Saint-Quentin. The species *angustifolia* was collected by Ferdinand Lindheimer at Matagorda Bay, Texas, in 1845 and described by John Torrey in 1858.

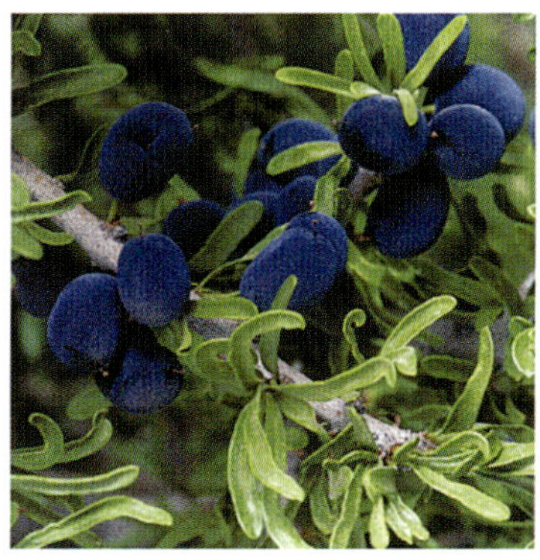

Showy Menodora

Menodora longiflora

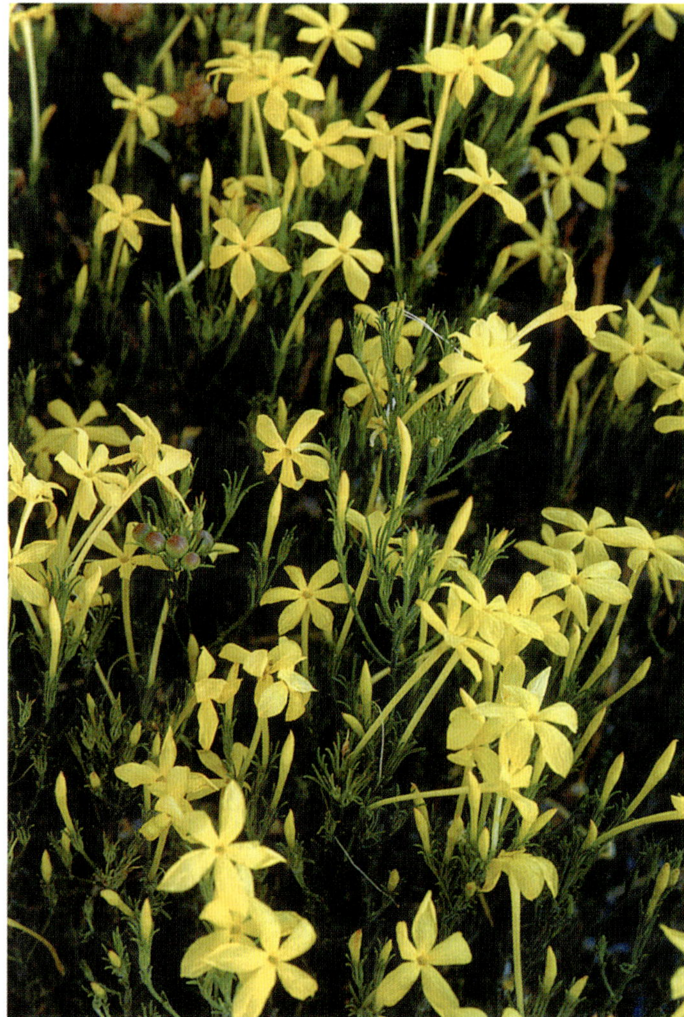

SHOWY MENODORA, or twinpod, is an erect perennial herb or low shrub one to two feet high with many slender, clumped stems growing from a woody base. From one-third inch to one and one-half inches long, the dull green leaves are narrowly oblong in shape with a short, abrupt tip and smooth but curled edges. The slightly leathery leaves taper to a very short, winged stalk.

Blooming from summer to fall, the flowers are the real show: up to two and one-half inches long, bright yellow, exquisite, with a long, slender tube abruptly flaring into five or rarely six flat petal-like lobes. The fruits are twin, spherical capsules one-third inch across. Erect on stalks, they are conspicuously thin-walled, light brown, but semi-transparent and shiny.

This unmistakable herb may cover dry, rocky hillsides or line intermittent streams in rocky canyons on limestone and igneous soils at elevations below sixty-five hundred feet. It is relatively far reaching, from southeastern New Mexico throughout the Trans-Pecos and Edwards Plateau regions of Texas and deep into Mexico (from eastern Chihuahua to Nuevo León and south to Puebla).

In the park showy menodora has been documented in the foothills of the Rosillos Mountains, in Telephone Canyon and Muskhog Spring Canyon in the Dead Horse Mountains, at Oak Spring and the base of Bailey Peak in the Chisos Mountains, and at Glenn Spring.

Note: The *Menodora* genus was first described by Aimé Bonpland in 1812 in Humboldt and Bonpland's *Plantae Aequinoctiales*, an account of species collected on their 1799–1804 expedition to tropical America. Some have suggested that the genus name *Menodora* is from the Greek *mene* (moon) and *doru* (spear) because of the spearlike, half-moon appearance of the fruit on a stiff stalk; others argue that it is from *menos* (strength) and *doron* (gift), possibly because of the strength it gave animals (Vines 1960). Asa Gray described this species in 1852. It had been collected by Ferdinand Lindheimer in 1847 along the upper Guadalupe River in the Texas Hill Country.

Hidalgo Ladies Tresses

Deiregyne confusa

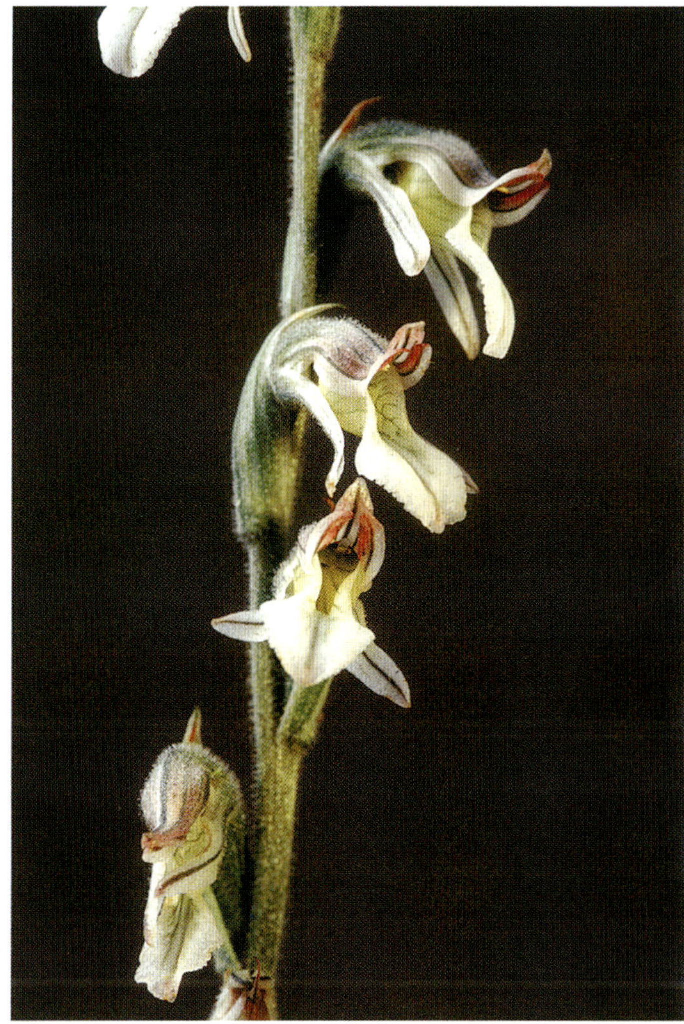

Hidalgo ladies tresses puts up a leafless, sheathed stem up to twenty inches high with a few narrowly lance-shaped leaves at the base. The leaves may reach one foot in length but are ephemeral and sometimes gone well before the orchid blooms.

The hairy flowering spike may be seven inches long with as many as fifteen fragrant, distinctly nodding flowers that hang down one side of the spike in a lax, well-spaced array. They are cream-colored with pink or orange and brown highlights and veined with green. Each flower has three recurved sepals and three petals with the two upper petals forming a hood over the broader lower petal, or lip. Pointed at the tip, the lip is conspicuously constricted or fiddle-shaped near the base.

Until 2004 this rare orchid had been seen only once in the United States—in 1931, high in the Chisos Mountains. I literally stumbled across one in early June 2004 in the Chisos on a rocky forested slope amidst grass. Two days earlier, park personnel had identified another of these delicate orchids even higher in the Chisos. It was a very good year.

From May to July, Hidalgo ladies tresses blooms on or near the base of rocky slopes, especially where water drains or seeps. This orchid was discovered in the Mexican state of Hidalgo, hence the name. It has also been located in the mountains of Coahuila, Nuevo León, and San Luis Potosí and near Mexico City. Hidalgo ladies tresses is also known as confused ladies tresses because the 1931 specimen was originally confused with and misidentified as Durango ladies tresses (*Deiregyne durangengis*), an orchid unknown in the United States.

Note: The common name "ladies tresses" refers to the braidlike, spiral flower clusters. *Deiregyne,* the genus name, is from two Greek words meaning "woman's neck," an allusion to the flower shape. The species name *confusa* means "confused." This orchid was named *Deiregyne confusa* by Leslie Garay in 1980 based on a specimen from the Lagoon of Meztitlán in Hidalgo, Mexico. Garay, botanist at the Oakes Ames Orchid Herbarium at Harvard, codrafted the first *Handbook of Orchid Nomenclature.*

Scarlet Ladies Tresses

Dichromanthus cinnabarinus

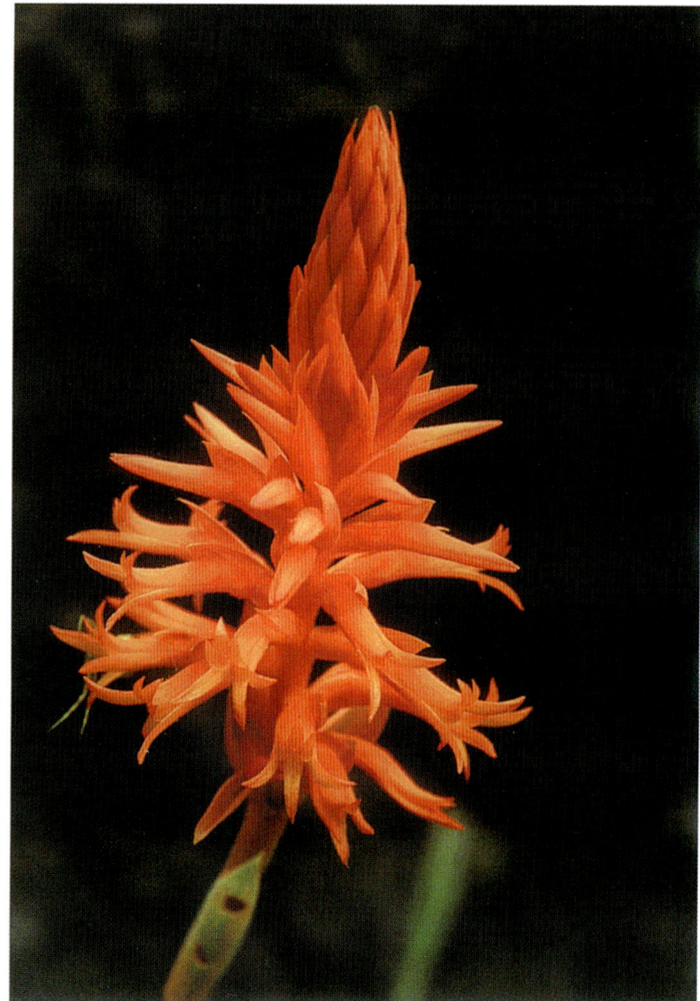

SCARLET LADIES TRESSES, or cinnabar ladies tresses, is a dazzling orchid with colorful spikes easily six inches long with thirty to forty tubular flowers. The color varies, but in the Chisos, flowers are orange to scarlet or cinnabar.

This orchid's bloom stalk sits atop a sheathed stem that extends as much as two feet high, far above three or four large basal leaves. Ladies tresses orchids acquired their apt name because the dense, spiraling flower spikes are so reminiscent, to the fanciful among us, of a woman's braided hair.

In the United States this perennial is isolated in the Chisos, Glass, and Del Norte mountains in Brewster County, where it is scarce. It is encountered more often, however, throughout much of Mexico, from Chihuahua to Oaxaca and into Guatemala. From August to October after the summer rainy season, this orchid puts up its cinnabar spike in volcanic soil high in the mountains, on rocky slopes, in canyons, and on bluffs above creek bottoms.

Scarlet ladies tresses has been observed in Upper Cattail Canyon, along the trail from the Pinnacles to Boot Spring, at Bailey Peak, and recently on talus slopes near the Window. This photograph was taken in Boot Canyon at the base of a steep cliff in late August.

The orchid may be both the most beautiful and most inefficient of plants. It is elaborately adapted for but also highly dependent upon cross pollination by insects. In addition, many orchids cannot germinate unless their seeds are pierced by specific kinds of fungus. Consequently, many are quite rare, highly prized, and a special joy to discover.

Note: The genus name *Dichromanthus* is from Latin words meaning "two-color flower." The flower petals while cinnabar externally are more yellowish internally. The species is named for cinnabar, the mercury ore, because its color is so close to the distinctive flower color of this ladies tresses. This is one of about fifty Michoacán orchids described in 1825 (as *Neottia cinnabarina*) by the Mexican naturalist, priest, and Veracruz state senator Pablo de La Llave, in collaboration with Juan José Martínez de Lexarza, Michoacán politician, scientist, and botanist.

Giant Helleborine

Epipactis gigantea

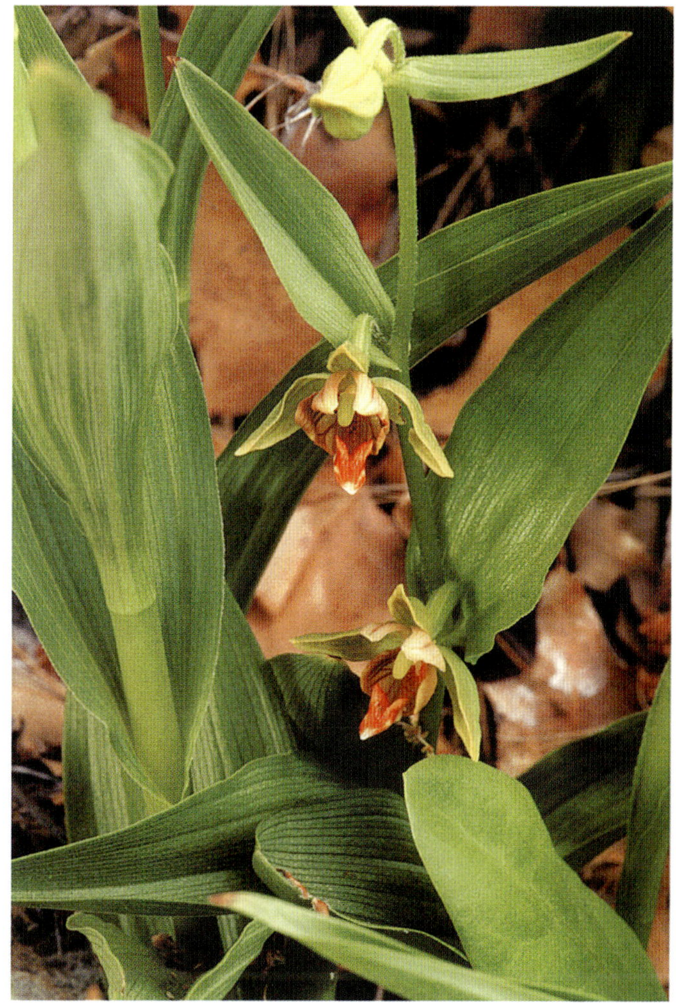

Also known as stream orchid or chatterbox orchid, giant helleborine is a leafy orchid with a single tall and sturdy stem. Up to three feet high, the smooth, hairless stems are erect with as many as fourteen large leaves that clasp the stem. Broadly lance-shaped, the stemless leaves may be eight inches long, noticeably veined, and folded. Spreading from creeping underground roots, this orchid can form large colonies at preferred sites.

From spring into summer, up to fifteen delicate and intricate flowers appear along the stem. Each bloom has three spreading, yellow-green sepals and three smaller, pale pink to orange petals with conspicuous reddish-brown veins. The most distinctive feature of this flower is the three-lobed, hinged, lower lip petal, which projects in a thin orange flap and flutters in the wind like a "chatterbox."

This orchid colonizes open, wet sites near sources of water, usually in partial shade. It is present in most of the western half of the United States from southern British Columbia to Baja California and central Mexico. However, giant helleborine is classified as vulnerable both globally and in Texas and is now rare in a number of western states. In Texas it occurs primarily in the Edwards Plateau and southern Trans-Pecos in Brewster and Presidio counties.

Secluded at a few grottolike habitats and desert "oases" in the park near springs and waterfalls, this helleborine is most often found near a waterfall on the west slopes of the Chisos.

Note: The genus name *Epipactis,* from the Greek *epipaktis,* was applied by Theophrastus to a plant with presumed milk-curdling properties. This species was formerly placed in the *Helleborine* genus, hence the common name. The species name *gigantea* (gigantic) refers to the robust stems. Described in 1839 by Sir William Jackson Hooker, it had been collected by David Douglas, a Scottish botanist sent by Hooker in 1824 on a botanical expedition to California and the Pacific Northwest. Douglas introduced the Douglas fir and other plants to British cultivation but died a suspicious death in 1832 while exploring Hawaii; supposedly he fell into a pit trap and was trampled by a wild bull.

Giant Coral Root

Hexalectris grandiflora

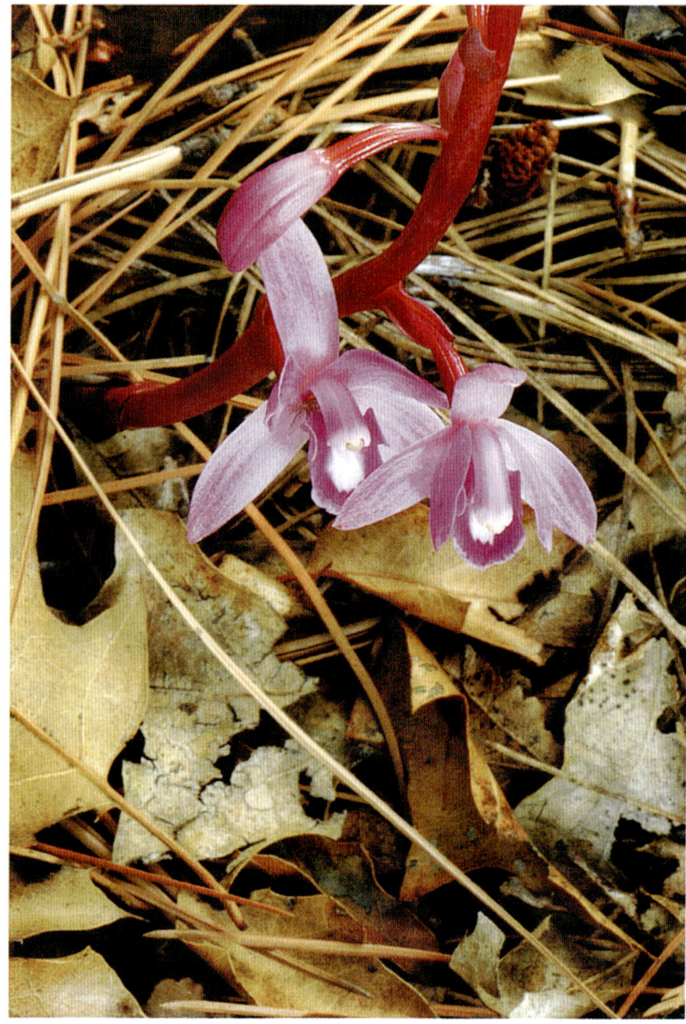

ALSO KNOWN AS Greenman's hexalectris, giant coral root is one of Big Bend's most appealing orchids. Lacking chlorophyll and living off decaying leaf mulch, it produces a leafless flowering stalk from four inches to as much as two feet high. Instead of leaves, three to five sheathing bracts (reduced leaves) enclose the smooth pink or maroon stem.

Blooming primarily in July and August, as many as twenty bright blossoms, a vibrant pink or magenta, form on the flower stalk, each on its own stout stem. Each large flower has three outer petal-like sepals and three inner petals. The lower lip petal is deeply three-lobed with the broad, wedge-shaped middle lobe much larger than the other two. A large white blotch with five white crests marks the center of the showy middle lobe. Unlike some other *Hexalectris* (especially *H. revoluta*), the sepals and petals of giant coral root are only slightly if at all recurved (curved backward).

Giant coral root comes up amidst humus and pine needles in moist canyons and along streams in pinyon-oak-juniper woodlands, often in the shade of oaks and madrones. In the United States this orchid is known only from the mountains of Trans-Pecos Texas in Jeff Davis and Brewster counties. It also grows in the Sierra del Carmen and other mountains of adjacent Coahuila, in Puebla and Oaxaca, and at other locations in Mexico.

In the park giant coral root was recorded in 1950 and again in 2005 in the Chisos in Upper Pine Canyon. This orchid is considered secure globally but is classified as imperiled and very vulnerable to extirpation in Texas (NatureServe 2005).

Note: The species name *grandiflora* connotes "large-flowered." This species was described (as *Corallorhiza grandiflora*) in 1845 by Achille Richard, professor of natural medical history at the School of Medicine in Paris, and Henri Galeotti, director of the botanical gardens in Brussels. In the late 1830s Galeotti brought an extensive collection of Mexican plants back to Belgium and set up a business selling Mexican cacti in Europe. In 1944 orchid specialist Louis Williams, author of *Orchidaceae of Mexico* (1965), moved this species to the genus *Hexalectris.*

Curly Coral Root

Hexalectris revoluta

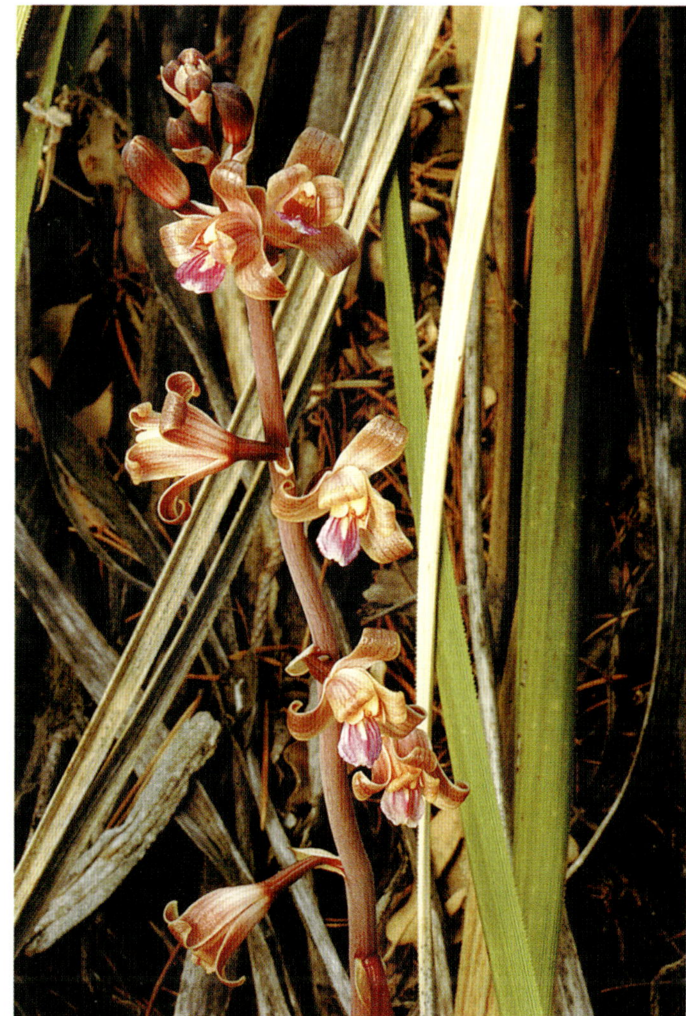

CURLY CORAL ROOT, or Chisos coral root, is probably the rarest of seven coral root species found in the mountains of Trans-Pecos Texas and northern Mexico. It is classified as critically imperiled in Texas and imperiled or critically imperiled globally. Growing alone or in colonies, this coral root is twelve to twenty inches high with a single leafless flower stalk that is pinkish-tan or light maroon. The absent leaves are replaced with three to five sheaths that clasp the stalk.

A single flowering stalk usually bears from five to fifteen colorful flowers that combine shades of tan, pink, purple, and creamy white. Blooming from May to July, each flower has three sepals and three petals that in this species are distinctively and very obviously revolute, or curled back. The delicate blossom has a prominent, three-lobed lower lip that is creamy white with pinkish-purple streaks or ridges.

Presumably without chlorophyll and dependent on decayed organic matter, this orchid lives mostly on leaf mulch and humus in oak-juniper-pinyon woodlands, often very notably in the shade of gray oak trees. In Texas it is apparently confined to the Guadalupe Mountains in Culberson County and the Glass and Chisos Mountains in Brewster County. Curly coral root also survives in the Santa Rita and Baboquivari mountains of southeastern Arizona and at a few locations in Nuevo León in northeastern Mexico.

In the park this orchid has been confirmed on Emory Peak and along the Colima Trail. I have also noticed it in Boot Canyon and above Laguna Meadow.

Note: The genus name *Hexalectris* is probably from the Greek *hex* (six) and *alectryon* (rooster), a reference to the supposed six crests resembling a cock's comb on the lip of the flower. This species was collected by Cornelius and M. T. Müller in 1934 and described by Donovan Correll in 1941. Cornelius Müller, botanist with the University of California at Santa Barbara, found it in San Francisco Canyon in the Sierra Madre Oriental in Nuevo León, Mexico. The first U.S. specimen was collected by Barton Warnock in 1937 in the Chisos Basin.

Crested Coral Root

Hexalectris spicata var. *spicata*

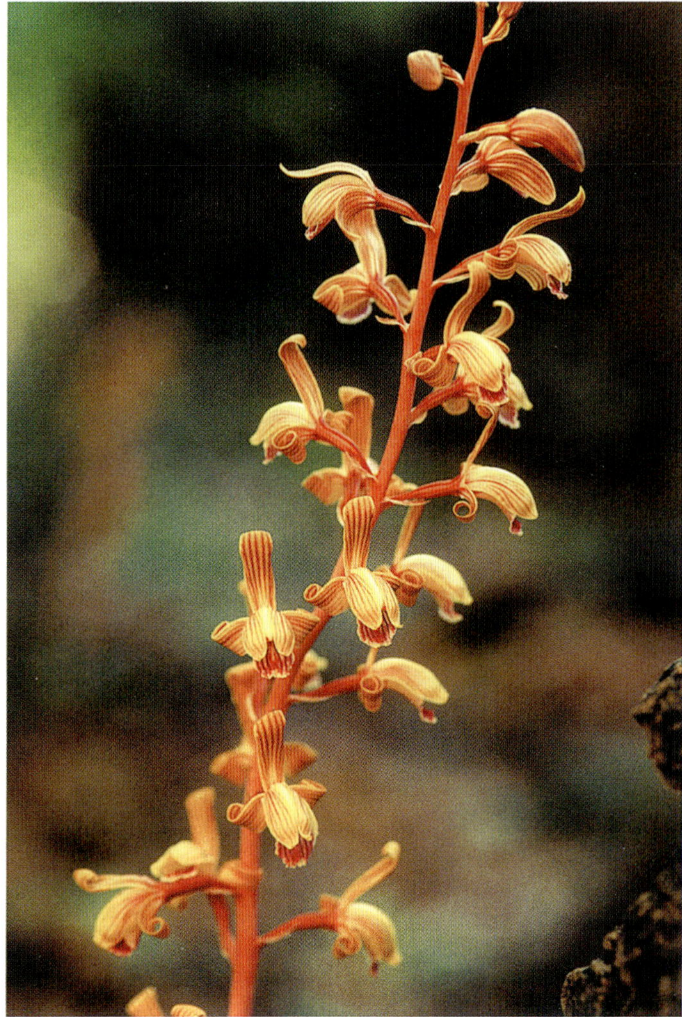

Crested coral root is a stout orchid occasionally as much as two and one-half feet high with a smooth, pale tan or pinkish-tan stem. The flesh-colored stalk is completely leafless, the leaves being replaced by three to five sheathing bracts sparsely distributed along the stem.

Usually in summer in Big Bend up to twenty-five pale yellow-tan flowers bloom along the stem, each on a slightly twisted stalk as much as three-fourths inch long. Often curling backward at the tips, the oblong sepals and petals are lined with dull purple or purplish-brown stripes. Each nodding flower has an extended lower lip petal that is only very slightly three-lobed. The lip is white but suffused with purple with five to seven deep purple crests or raised parallel streaks.

This orchid appears in humus and leaf mulch in the shade of juniper-oak-pinyon forests and on grassy banks by streams and other water sources on primarily limestone soils. Its broad distribution extends from Maryland on the Atlantic seaboard south to Florida, west to central and Trans-Pecos Texas and parts of New Mexico and Arizona, and south into Mexico. Although extensive, it is threatened, endangered, or rare in many states.

In the park crested coral root has been verified in Lower Oak Creek Canyon, at the beginning of the Juniper Canyon Trail, and on moist west slopes of the Chisos, although it might spring up in any of the more protected canyons. Also found in the park is a rarer but less attractive variety of this species, Arizona crested coral root (*H. spicata* var. *arizonica*), which has smaller flowers that usually self-fertilize without opening.

Note: The species and variety name *spicata* means "spiked," referring to the shape of the flower stalk. This variety was described in 1788 (as *Arethusa spicata*) by Thomas Walter in his *Flora Caroliniana*, an accounting of specimens he collected in South Carolina. In 1768 the British-born Walter had immigrated to South Carolina, where he became a plantation owner, politician, and botanical collector.

Texas Purplespike

Hexalectris warnockii

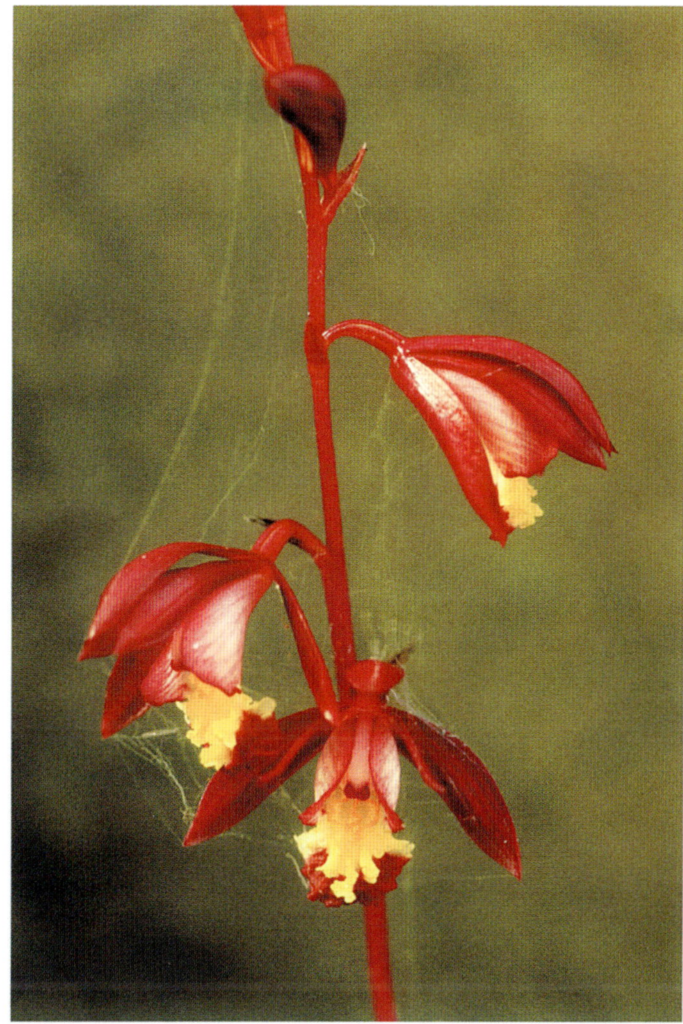

TEXAS PURPLESPIKE is a leafless maroon stalk at most twelve inches high. From June to August after summer rains, the stalk bears as many as ten delicate, nodding flowers in a loose, elongated spike. About one inch wide, the intricate flowers have three freely spreading sepals and three petals that are maroon and cream. The broad, three-lobed, lower lip petal is mostly white with dainty yellow ruffles or ridges on the middle lobe and more maroon at the tip.

This orchid, like other *Hexalectris* species, lives on leaf mulch and decaying organic matter. A symbiotic fungus helps to break down and digest the decomposing materials.

Although Texas purplespike is the most common *Hexalectris* in the park, it is a species of concern and is being closely monitored by the National Park Service. Globally Texas purplespike is classified as vulnerable or imperiled throughout its range and in Texas as imperiled and very vulnerable to extirpation.

In Texas this species is restricted to the Trans-Pecos mountains—from the Guadalupe Mountains south to the Chisos—and to a few disjunct locations in south-central and north-central Texas. It has also been noted in southeastern Arizona at Chiricahua National Park and in Mexico in Coahuila and Baja California.

From June through August, Texas purplespike blooms at moist, shady sites beneath trees in oak-juniper-pinyon woodlands, often near springs or creeks. In the park this rare orchid shows up periodically at Upper Juniper Spring; in Boot, Blue Creek, Pulliam, and Pine canyons; along the Lost Mine Trail; below Bailey Peak; and on moist west slopes of the Chisos.

Note: Texas purplespike was discovered by Cornelius Müller in 1932 in the Chisos Mountains. In 1943 botanists Oakes Ames and Donovan Correll named it for Barton Warnock, the author of three popular works on Trans-Pecos wildflowers, who collected the type specimen in 1937 in Upper Blue Creek Canyon. Ames, a Harvard professor and the youngest son of a Massachusetts governor, is the author of a seven-volume work on orchids illustrated by his wife and collaborator, Blanche.

Wendt's Malaxis

Malaxis wendtii

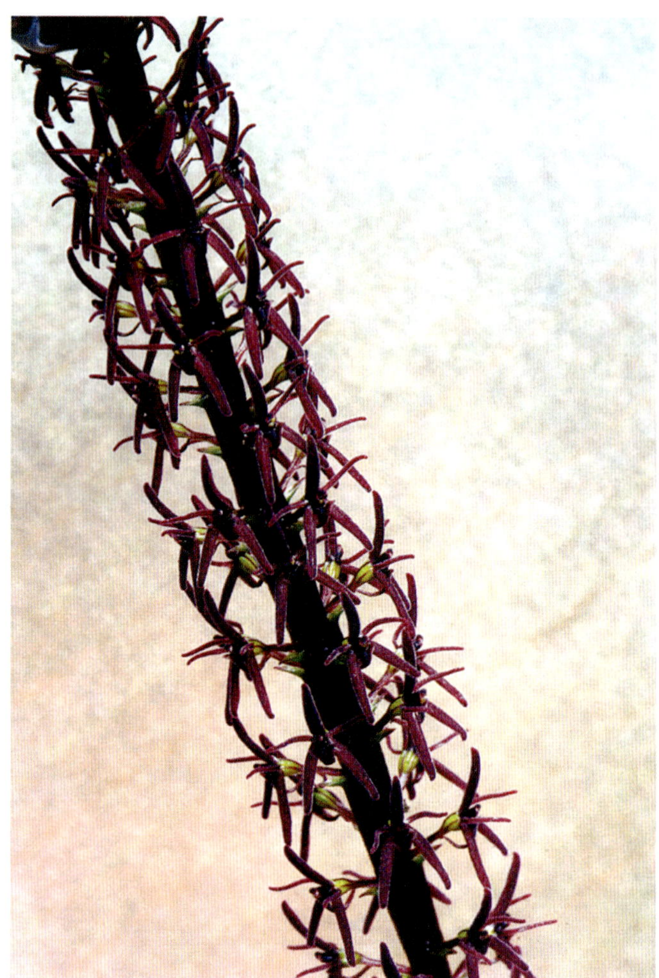

WENDT'S MALAXIS, or Wendt's adder's-mouth, is a small, inconspicuous orchid with miniature flowers that easily blends with the surrounding grassy vegetation. From a bulbous underground base, it puts up a dark maroon flowering scape six to eighteen inches high with a single, low leaf clasping the stem. Fleshy and even rubbery to the touch, the relatively large green leaf is broadly egg-shaped and occasionally more than three inches long.

Slender but dense, the flower clusters can reach ten inches in length, sometimes with more than a hundred minute purple flowers. Quite unusual in appearance, each tiny bloom extends from the scape on its own slender stalk (pedicel). It has three slender, backward-curving sepals; two lateral, threadlike petals; and a projecting, arrow-shaped lower lip petal with earlike bulges at the base. With a magnifying glass, you can see that parts of the flowers are covered with microscopic, nipplelike protuberances (like the bumps on a tongue), which help botanists distinguish this rarity from other close relatives.

In the United States Wendt's malaxis is limited to the Chisos Mountains of Big Bend National Park, where it is rare. In late summer and early fall this orchid blooms below oak-juniper-pinyon canopies in shady, moist canyons. It is also infrequently seen in the mountains of northeastern Mexico in the states of Coahuila and Nuevo León.

In the park this very difficult-to-spot herb occupies rocky slopes beyond Boot Spring. Globally Wendt's malaxis is classified as imperiled throughout its range and in Texas as critically imperiled.

Note: The genus name *Malaxis* is derived from a Greek word meaning "soft," referring to the texture of the leaves. Long misidentified as Ehrenberg adder's-mouth (*Malaxis ehrenbergii*), this orchid was described as a new species in 1993 by the Mexican orchid specialist Gerardo Salazar. Salazar named the species in honor of Tom Wendt, University of Texas botanist and specialist in Mexican rain forest plants, who had collected it in Coahuila, Mexico, in 1975. In 1998 University of Guadalajara botanist Roberto Tamayo and botanist Dariusz Szlachetko at Gdansk University in Poland proposed moving this species into the new genus *Tamayorkis.*

Michoacán Ladies Tresses

Stenorrhynchos michuacanum

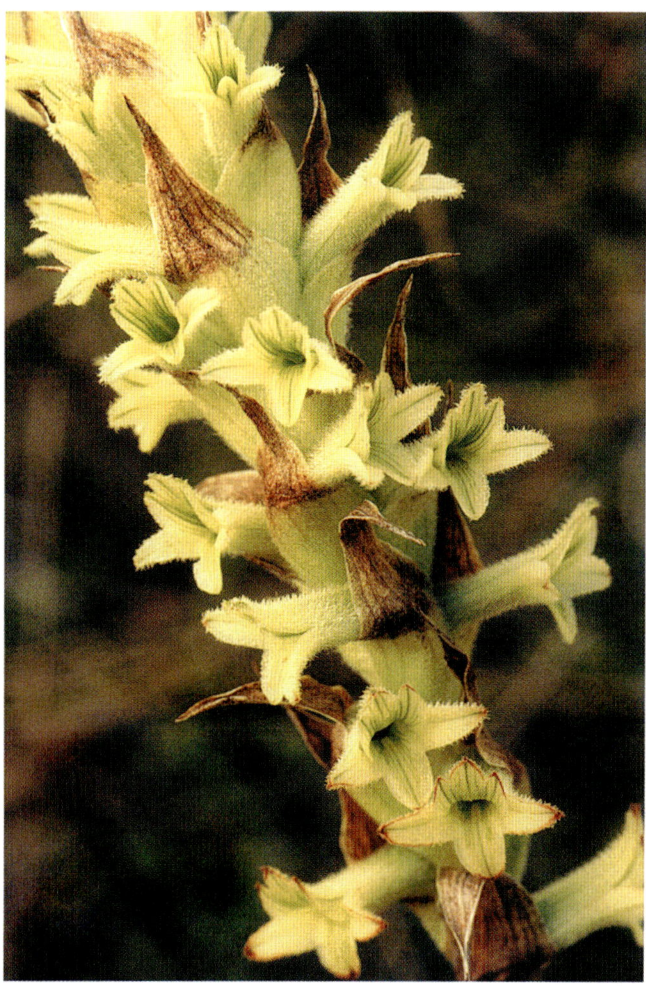

MICHOACÁN LADIES TRESSES is an upright perennial usually one to two feet high with a thick, sturdy stem and fleshy, tightly clustered roots. Thin leaflike sheaths enclose the stems, and up to five narrowly lance-shaped leaves up to six inches long form a rosette at the base of the stem. Stemless and flat, the leaves may die off before the flowers form.

In fall, up to thirty tubular flowers appear in a tight, crowded spike at the top of the stem. The blooms are creamy white or pale yellow with a greenish throat and thin green stripes. The flowers are distinctly hairy: the edges of the petals bristle with long hairs, and the bracts supporting the flowers are covered with twisted hairs. As they age, the long bracts turn brown, so the floral spikes seem to be dying even before all the flowers have bloomed.

This species prefers rocky and grassy open slopes in mountains in oak-juniper-pinyon forests and in canyons where water drains after drenching summer rains. Michoacán ladies tresses is a Mexican species, scattered from Chihuahua to Guerrero and Chiapas, and it enters the United States only in Trans-Pecos Texas and southeastern Arizona. Extremely rare in the United States, it has taken root in the Chisos and Chinati mountains of the Big Bend and in the Huachuca and Santa Catalina Mountains of Arizona.

In the park this ladies tresses has in recent years been relocated in one little-known canyon high in the Chisos, where this photograph was taken in October.

Note: The genus name *Stenorrhynchos,* from the Greek *stenos* (narrow) and *rhynchos* (snout), refers to the beaklike extension of the stigma in these orchids. The species *michuacanum* is named for the Mexican state of Michoacán, where it was discovered. This orchid was described in 1825 (as *Neottia michuacana*) by the Mexican naturalists and politicians Pablo de La Llave and Juan José Martínez de Lexarza. After living in Europe and serving as director of the Madrid Botanical Gardens, La Llave became a senator from Veracruz and later director of the Natural History Museum in newly independent Mexico. Valery Havard found the first U.S. plant in the Chinati Mountains of Texas in 1880.

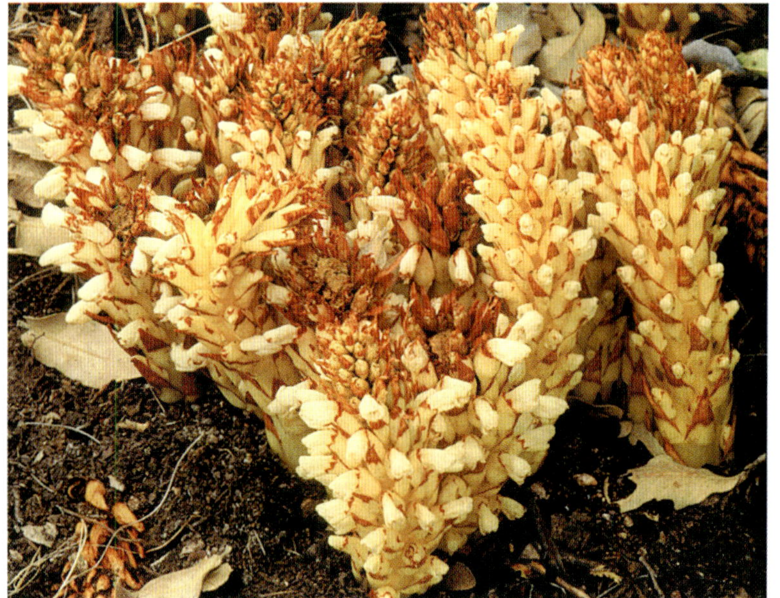

Mexican Squawroot

Conopholis alpina var. *mexicana*

MEXICAN SQUAWROOT, or Mexican cancer-root, is an erect perennial herb up to ten inches high with stubby rigid stems. The fleshy stems, often tightly clumped in small mounds, are pale yellow but turn dark brown with age.

Scaly yellow leaves that lack chlorophyll surround the lower stems. The flowers are crowded into spikes at the top of the stems. Each flower is a curved tube with two lips: the upper lip is notched and protruding, and the lower lip, somewhat shorter, is divided into three lobes.

Mexican squawroot prospers at higher elevations in mountain woodlands, usually beneath oaks, junipers, or pinyons. It is parasitic primarily on the roots of oaks amidst decomposing leaves and twigs in shade. Blooming in spring and early summer, this perennial is uncommon in the Trans-Pecos mountains (in the Guadalupe Mountains south to the Chisos Mountains) and also in New Mexico, southern Arizona, and Mexico south to Oaxaca.

In the park this parasite has been reported high in the Chisos, in Boot Canyon, between Boot Spring and Emory Peak, on the east slopes of Lost Mine Peak, and on the northwest slopes of Pulliam Peak. I photographed it in Upper Cattail Canyon in late April.

Note: A small American genus, *Conopholis* acquired the common name cancer-root from its use as a folk remedy for cancer. From the Greek *konos* (cone) and *pholis* (scale), the genus name refers to the scaly, overlapping leaves. This variety was collected in 1851 by Charles Wright in the Organ Mountains of New Mexico and described as a species (*Conopholis mexicana*) by Sereno Watson in 1883. Watson collaborated with Asa Gray and William Brewer on the first volume of *Botany of California* and then published the second volume himself (1880).

Manyflowered Broomrape

Orobanche ludoviciana ssp. *multiflora*

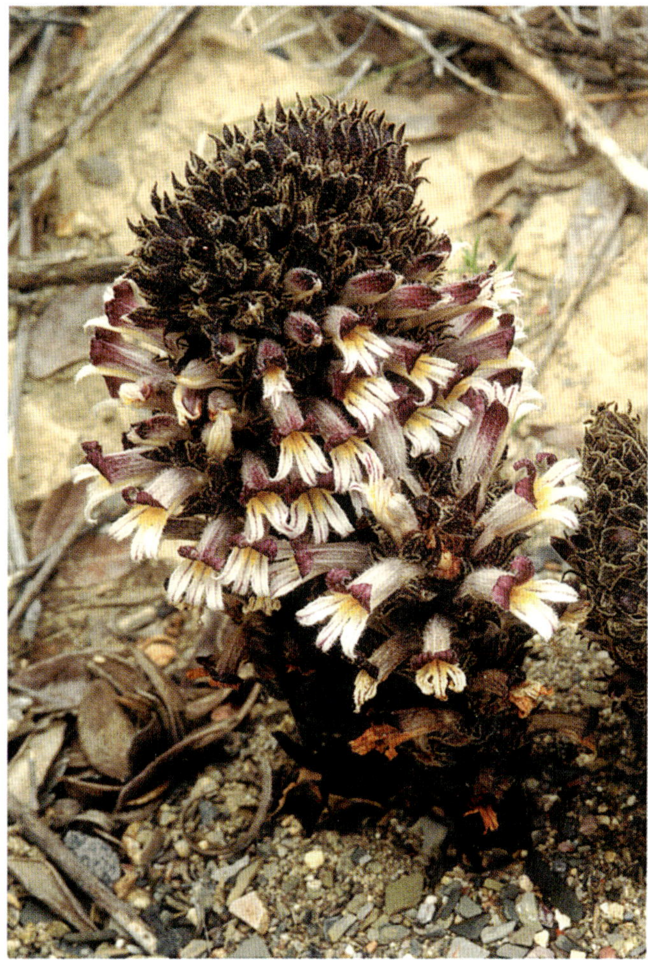

Manyflowered broomrape is a bizarre plant that mostly parasitizes members of the sunflower family. Lacking chlorophyll, it attaches underground to the roots of its host for nourishment. Also lacking true green leaves, this parasite produces inconspicuous, scalelike leaves less than one-half inch long.

The stout spikes of manyflowered broomrape, perhaps six inches high, are usually purplish-brown, but they gradually decay to a rust color in fall and winter. Found in sandy washes and muddy clay flats, this perennial is often caked with sand and dirt like a little mud statue. Dirt sticks to oily secretions from the hairy glands on the stems.

Blooming from spring to fall, the mud-sculptured spires are strangely beautiful—seemingly weighted down with dense, two-lipped flowers in shades of purple, gray, white, and yellow. Each flower has a dark purple, often curled, two-lobed upper lip and a pale, three-lobed lower lip, each rounded lobe with a single, distinctive purple streak.

Manyflowered broomrape is widely distributed in western states from Idaho and Wyoming to Arizona, New Mexico, and Texas. Its range extends through the Trans-Pecos and other parts of Texas, north to Kansas and Missouri, and south into northern Chihuahua.

In the park this herb frequents backcountry dirt roads, the east and west ends of the River Road, and the Glenn Spring Road near Nugent Mountain. You might also spot this bizarre herb along the Smoky Creek drainage into the Rio Grande, under hosts at Dagger Flat, or in desert arroyos near Persimmon Gap.

Note: The genus name *Orobanche* is from the Greek *orobos* (vetch) and *ancho* (strangle), referring to a plant parasitic on vetch. The species name *ludoviciana* implies "from Louisiana." Thomas Nuttall, an English botanist, ornithologist, and prodigious plant collector who lived in Philadelphia for many years, described the species *ludoviciana* in his *The Genera of North American Plants* (1818). Thirty years later, in 1848, he described *multiflora* as a distinct broomrape species (it was reclassified as a subspecies in 2001). Because of his extensive expeditions west of the Mississippi River, Nuttall became known as the father of western American botany.

Chisos Pricklypoppy

Argemone chisosensis

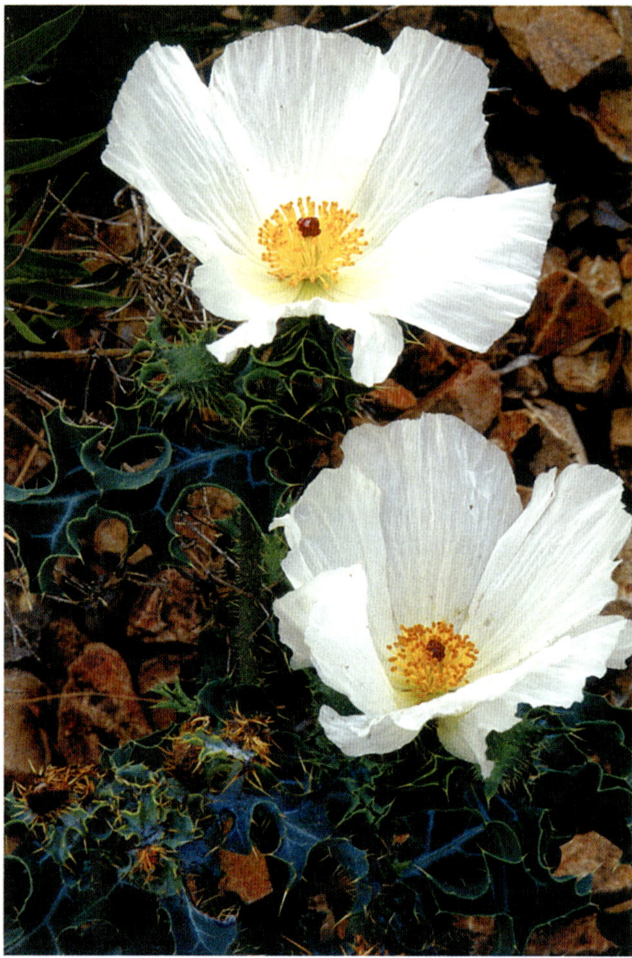

Chisos pricklypoppy is a sturdy biennial or perennial herb up to two and one-half feet high with one or a few sparsely branched, leafy, prickly stems containing a yellow, acrid sap. Stemless and often clasping, the lower leaves are deeply lobed and wavy, with sharp teeth on the edges and dense prickles on the midvein underneath. All leaves are dark bluish-green but whitened with a chalky cast like cabbage leaves.

Up to nearly four inches wide, the flowers have six rounded, cuplike petals with 150 or more yellow or reddish stamens in the center. Drooping and early falling, the flimsy petals are usually white but occasionally a delicate, fragile pink. This spiny, thistlelike herb blooms from spring to early fall.

Chisos pricklypoppy flourishes from low, arid desert flats and disturbed areas to dry, rocky mountain slopes. It is wide ranging throughout Trans-Pecos Texas (from southern Hudspeth and Culberson counties to Del Rio) and northern Mexico (in parts of Chihuahua and Coahuila and northeastern Durango).

This herb pops up at disparate locations in the park, from clay and alluvial flats south of Persimmon Gap to the Chisos (at the Basin and Laguna Meadow) and the Rio Grande (along the River Road, near Johnson Ranch, and at Mariscal Mountain).

Note: *Argemone* is an ancient Greek name of a similar poppylike plant, derived from *argemon* (cataract); the plant was used as a remedy for cataracts. This species was collected by Roxana Ferris and Carl Duncan in 1921 in Juniper Canyon of the Chisos and described by University of Minnesota botanist and herbarium curator Gerald Ownbey in 1958. Ferris was curator of the Dudley Herbarium at Stanford and author of books on wildflowers of Point Reyes and Death Valley.

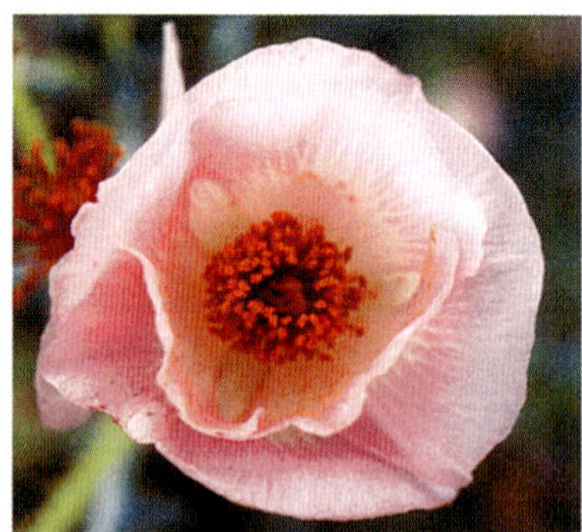

Spreading-Lobe Passionflower

Passiflora tenuiloba

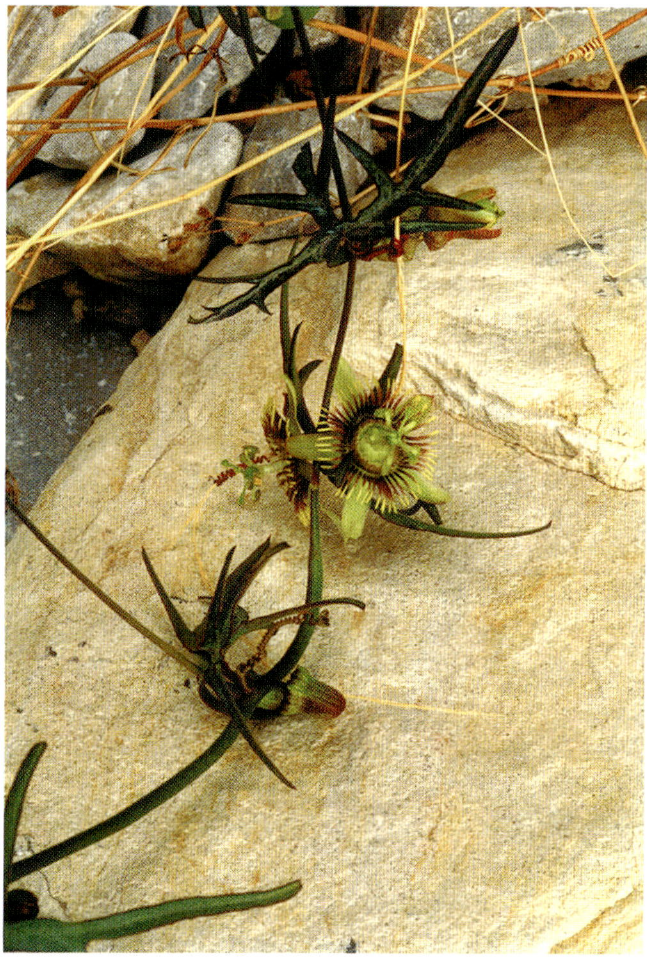

Also known as birdwing passionflower and batwing passionflower, this trailing and climbing vine produces glorious pale yellowish-green flowers three-fourths inch across. The solitary or paired flowers have an elaborate and unusual architecture: they have fleshy oblong sepals but no petals; a crown (corona) consisting of an inner and outer set of threadlike, yellow-tipped filaments that ring a membranous, plaited lid; and a protruding stalk supporting the reproductive parts of the flower.

In folklore the genus's intricate blossoms symbolize the Passion of Christ. The corona represents the crown of thorns, the five stamens the five wounds, the three styles the nails, and the ten petal-like parts the ten faithful apostles.

The dark green, clawlike leaves are strongly veined, deeply cut into two or three usually narrow but very irregular lobes that can be oblong, linear, or wedge-shaped. A quarter-inch wide, the spherical, berrylike, pulpy fruit turns purple to black as it matures. The delicious fruits of many passionflower species are cultivated for human consumption.

This exquisite herb is widely dispersed from northern Mexico (Coahuila to Tamaulipas) into the Rio Grande Plains, Edwards Plateau, and Trans-Pecos regions of Texas and west into New Mexico. Climbing over shrubs and boulders, or sprawling across rocky canyon washes, it blooms from May to October on limestone soils in the eastern Trans-Pecos (Brewster through Pecos and Terrell counties to Del Rio).

Although not well known in the park, spreading-lobe passionflower trails through a few off-the-beaten-path sites in the Dead Horse Mountains. The photo was taken in early October in Passionflower Canyon east of the Ernst Basin backcountry campsite.

Note: In the American South, some *Passiflora* species were known as maypops because children stomped on the ripe fruits to make them pop. The genus name *Passiflora* is from the Latin *passio* (passion) and *floris* (flower), for "passionflower." The species name *tenuiloba* refers to the leaves' "slender lobes." This species was described by George Engelmann in 1850 and first collected by Ferdinand Lindheimer on the Llano River in the Texas Hill Country.

Mexican Pinyon Pine

Pinus cembroides ssp. *cembroides*

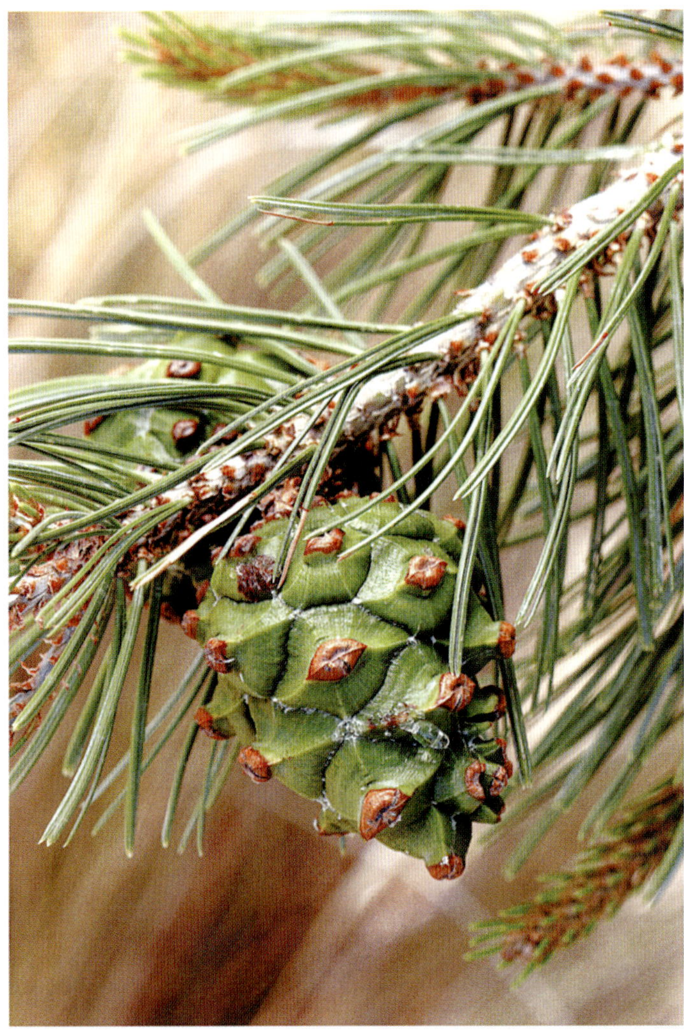

Also known as pinyon and pino piñonero, Mexican pinyon pine is a relatively small tree averaging about twenty feet high with a short trunk, many diffuse branches, and a rather cone-shaped crown. The bark may be noticeably reddish-brown, thin, and scaly. Spreading outward and curving upward in small bundles (of two or three) near the ends of branches, the narrow, evergreen pine needles are one to two inches long and blue- or grayish-green.

Separate male and female cones appear on the same tree. The male pollen cones are quite small, less than two-fifths inch long. The oval, female seed cones, usually maturing in two years, are at first fleshy but later hard and reddish-brown. No more than one and one-half inches long, they bear scales shaped like slightly flattened or compressed pyramids and contain very hard-shelled but edible nuts. The nuts may be eaten raw or roasted and are highly prized in Mexico and parts of the Southwest.

Mexican pinyon pine thrives at elevations above forty-seven hundred feet on arid rocky slopes and in canyons in pinyon-juniper woodlands, especially in igneous soils. In the United States this subspecies is concentrated in Trans-Pecos mountains (primarily in Jeff Davis and Brewster counties but also Presidio and Terrell counties). It has a much broader range in northern Mexico and throughout the Sierra Madre.

Mexican pinyon pine is abundant in the Chisos, from Green Gulch and the Basin to the highest elevations—at the top of Lost Mine Peak, on Mount Emory, and along the South Rim Trail. It also occurs in the Dead Horse Mountains and the northern Rosillos Mountains.

Note: The genus name *Pinus* is the ancient Latin name for pine. The species name *cembroides* means "like *cembra,*" implying a resemblance to the Swiss stone pine, *Pinus cembra.* This species was described by German botanist Joseph Zuccarini in 1832, from collections taken near Mexico City in 1827 by the Bavarian explorer Baron Karwinski. Collecting in Brazil and Mexico from 1826 to 1832 and 1840 to 1843, Karwinski sent live plants and specimens to the Saint Petersburg Botanical Gardens and also to Zuccarini in Munich.

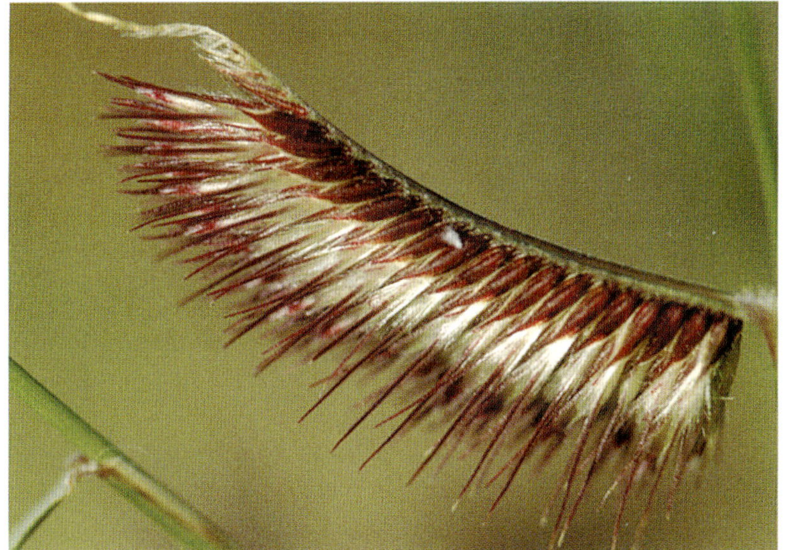

Chino Grama

Bouteloua ramosa

Chino grama is a clumped perennial grass that grows from a somewhat woody and knotty base with many erect or bending stems up to two feet high. The stems are thickly branched and very leafy near the base. One to two inches long, the mostly flat leaves curl inward at the tip.

You may not think of grass as beautiful until you take a closer look. This grass has comblike flower spikes about one inch long near the stem tip. Each spike usually has two branches, each with twenty to sixty tiny spikelets that are tightly clustered along one side of the stem like the teeth of a comb. Hairy, needlelike bristles extend from the spikelets; they are projecting veins of the bracts enclosing each flower. This grass blooms from May to October.

Chino grama may be the dominant vegetation in desert mountains and foothills, on dry gravelly and open rocky slopes, in limestone and igneous soils, often with lechuguilla. In the United States this grass covers large stretches near the Rio Grande in Trans-Pecos Texas (from Hudspeth County to Terrell County). It is also present in northern Mexico (from Chihuahua to Nuevo León and south to Zacatecas).

In the park chino grama dominates some areas surrounding the Chisos, especially to the west on and near Burro Mesa and Ward Spring and near Smoky Creek to the south. It also spreads outside the Chisos to the Dead Horse Mountains.

Note: The genus *Bouteloua* is named for the nineteenth-century Spanish brothers Claudio and Esteban Boutelou. Claudio was director of the Madrid Botanical Garden, and Esteban was the royal gardener to King Philip V of Spain. The species name *ramosa* means "branched," referring to the bushy stems. This species was collected by botanical explorer Greenleaf Nealley in 1888 in the Chinati Mountains of Presidio County and described by agrostologist George Vasey in 1890. From 1887 to 1889, Nealley collected plants for the U.S. Department of Agriculture in less explored areas of southwestern and western Texas.

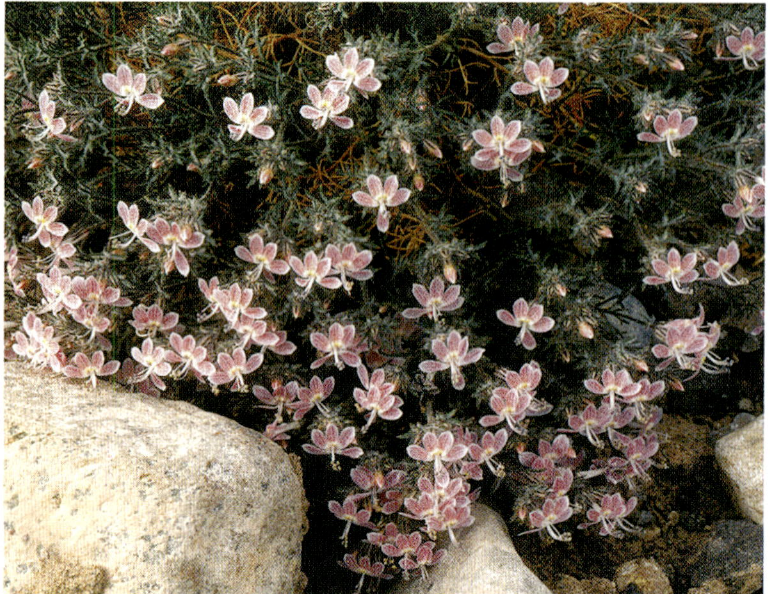

Havard Ipomopsis

Ipomopsis havardii

Havard ipomopsis, or Havard standing cypress, is a densely shaggy, miniature shrub less than eight inches high with highly dissected leaves. The leaves, perhaps one-half inch long, are divided into three to five thin, almost threadlike segments, which are slightly spiny at the tips. Covered with long, grayish-white hairs and sometimes sticky glands that collect blowing sand and dirt, this phlox usually has a distinctly gritty appearance.

From spring to fall, small flowers may completely cover the spreading stems, forming compact little flower mounds or bouquets. The irregular flowers have slender tubes that expand abruptly into spoon-shaped lobes only one-eighth inch long with unusual color markings: white near the base and edges and otherwise pink, or streaked with thin pink (or pinkish-red) stripes so close together that they appear pink. Another prominent feature is the long, protruding stamens.

Havard ipomopsis grows at lower elevations in sandy soils and open sun on dry hills and the shoulders of roads, on silty flats, or in sandy washes. This dwarf shrub is confined to Brewster and Presidio counties in the Big Bend and to northeastern Chihuahua and western Coahuila in Mexico. A delightful miniature that, strangely, has escaped the attention of gardeners, it is classified as vulnerable in Texas and worldwide.

This ipomopsis has a very spotty distribution in the park. It has been recorded on Dog Flats and Tornillo Flat, in the Rosillos foothills and near Roy's Peak in the north, and along the River Road at Smoky Creek and Mariscal Mine in the south.

Note: The genus name *Ipomopsis,* from the Greek *opsis* (likeness), probably means "resembling the genus *Ipomoea,*" an allusion to the similar flowers. This species was described by Asa Gray in 1883 (as *Loeselia havardii*) and named in honor of Valery Havard, who had collected it near Presidio in 1881.

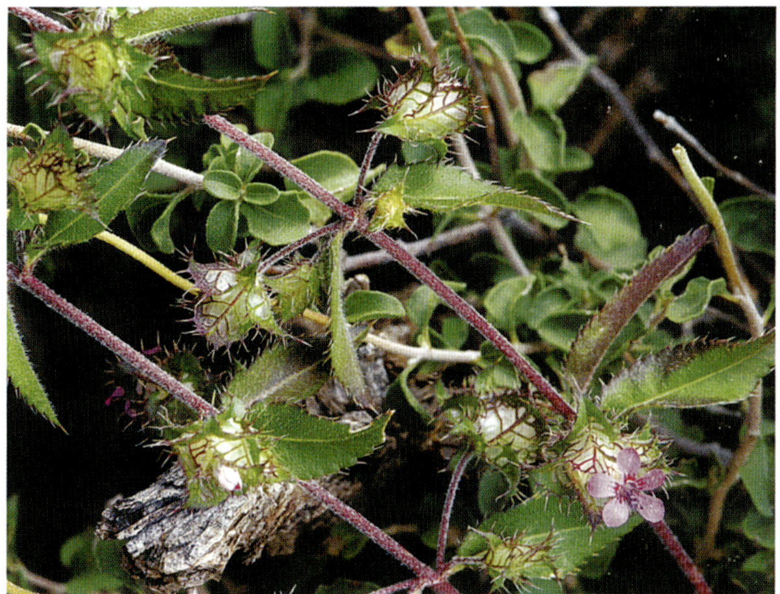

Papercup Loeselia

Loeselia greggii

ALSO KNOWN AS Chisos Mountain false calico and gordo lobo, papercup loeselia is a sometimes woody perennial no more than eighteen inches high with mottled pink flowers, bristle-tipped leaves, and odd papery bracts. The spreading low branches bear lancelike leaves about one inch long with sharp but flexible, bristly teeth along the margins.

The pink flowers, one-half inch wide, resemble a calico print: the five pink lobes have patches of white at the base and a ring of darker pink or red near the throat. Each flower rests above thin, bristle-tipped bracts that are divided by dark veins into windowlike, membranous compartments. The thin-walled bracts have the consistency of a paper cup.

This perennial is known exclusively from rocky slopes of the Chisos Mountains and northern Mexico (from Chihuahua to Nuevo León and south to Durango). It is secure globally but critically imperiled and especially vulnerable to extirpation in Texas (NatureServe 2005).

Blooming mostly in the fall, papercup loeselia has been documented in Upper Green Gulch, at Panther Pass, and on the Laguna Meadow Trail. Probably the best place to find this unusual phlox is on the Lost Mine Trail in late September.

Note: The genus *Loeselia* is named for Johannes Loeselius (1607–55), German professor of medicine at Königsberg. Species *greggii* was described in 1882 by Sereno Watson and named for botanical pioneer Josiah Gregg. This phlox was long regarded as *Hoitzia* (later *Loeselia*) *scariosa,* a species that had been collected by Belgian botanist Henri Galeotti in the Mexican states of Puebla and Michoacán and described in 1845 by him and Belgian taxonomist Martin Martens. But that species has been identified as *Loeselia coerulea,* a closely related and wider ranging Mexican species from which *Loeselia greggii* is considered distinct (Henrickson and Johnston, forthcoming).

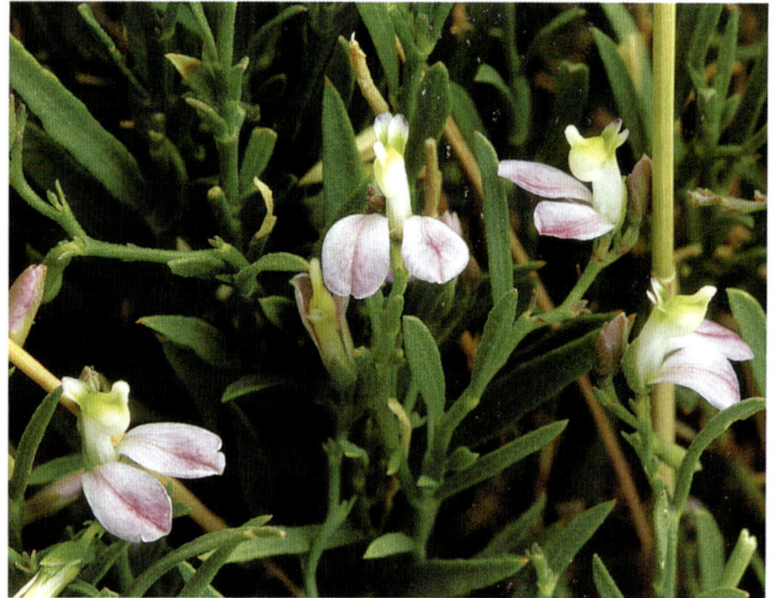

Rock Milkwort

Polygala lindheimeri var. *parvifolia*

ROCK MILKWORT is an erect or trailing perennial herb from four inches to more than twelve inches high and wide but sometimes flimsy and straggling. The stems may be densely leafy, bearing narrowly oblong or elliptic, often hairy leaves up to one and one-half inches long with short, abrupt tips. Basal leaves can be noticeably wider than the upper leaves.

Blooming from spring to fall, the tiny colorful flowers have an interesting architecture. Two large winglike "petals," white blushed with pink or rose, are not petals at all but rather colored sepals. Two of the three actual petals are minuscule, like little ears with dark rose tips. The more prominent lower petal (the keel) is creamy white tinged with greenish-yellow and shaped like a small pouch with a conspicuous, protruding beak.

Rock milkwort inhabits rocky slopes among boulders or in rock crevices and desert scrub, grasslands, and woodlands. Common in the Big Bend (in Brewster and Presidio counties), this herb is scattered throughout the Edwards Plateau, north-central Texas, and the Panhandle into southwestern Oklahoma. It also crosses into southwestern New Mexico, southeastern Arizona, and Mexico (from Chihuahua to Nuevo León and south to Zacatecas).

In the park rock milkwort is encountered in the Chisos as well as desert mountains. It has been identified at Sam Nail Ranch, Chisos Pens, Lone Mountain, and Fresno Spring in the Chisos foothills; near the outlying Chilicotal, Talley, and Mariscal mountains; and in the Rosillos and Dead Horse mountains.

Note: The genus name *Polygala,* from the Greek *polys* (much) and *gala* (milk), stems from the belief that consumption of some species increased animal milk production.

This species was named for Ferdinand Lindheimer in 1850, but variety *parvifolia* (small leaf) was collected by Cyrus Pringle in the Santa Rita Mountains of Arizona in 1884 and described by milkwort specialist William Wheelock in 1891.

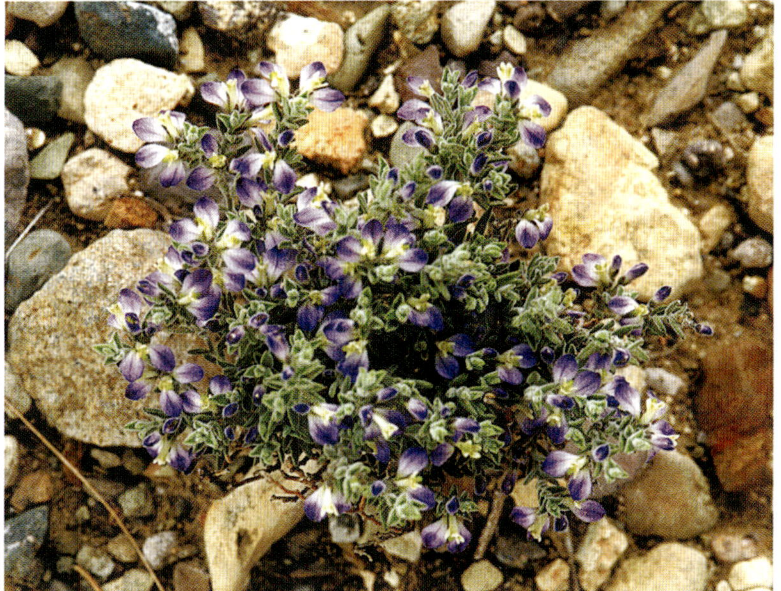

Glandleaf Milkwort

Polygala macradenia

GLANDLEAF MILKWORT is a low perennial herb or subshrub up to eight inches high with many densely leafy and hairy stems extending from a woody base. The slender, weak stems are mostly erect but crowded in compact clumps or mats. Hairy gray-green leaves only one-fourth inch long alternate up the stems. Thick and often folded, the blades are narrowly oblong or lance-shaped. Both leaves and stems are dotted with shiny, greenish-yellow to brown, pustular glands, which may produce a black, oily residue at the base of the plants.

Blooming mostly in spring and summer, the delicate flowers are no more than one-fourth inch long in shades of bluish-purple, greenish-yellow, and white. Each bloom has five sepals and three petals. The two lateral sepals (called wings) are relatively large, bluish-purple and white, and petal-like; the prominent middle petal (the keel) is greenish-yellow with a distinct beak at the tip.

Glandleaf milkwort forms dense clumps on dry, gravelly desert flats and rocky slopes in Chihuahuan Desert scrub habitats, mostly on limestone soil at elevations below forty-five hundred feet. In Texas this perennial is widespread throughout the Trans-Pecos and reaches into the extreme western Edwards Plateau and south along the Rio Grande to Laredo. It is also resident in southern New Mexico, central Arizona, and throughout northern Mexico (Baja California to Tamaulipas and south to Guanajuato and Hidalgo in central Mexico).

In the park glandleaf milkwort is most evident on desert flats: Tornillo Flat, Dog Flats, and Dagger Flat. It is also plentiful from the Rosillos Mountains east to the Dead Horse Mountains and on arid hills surrounding the Chisos (for example, Burro Mesa and Chilicotal Mountain).

Note: The species name *macradenia* means "with large glands." Asa Gray described this species in 1852 in *Plantae Wrightiana* from a specimen collected by Charles Wright in 1849.

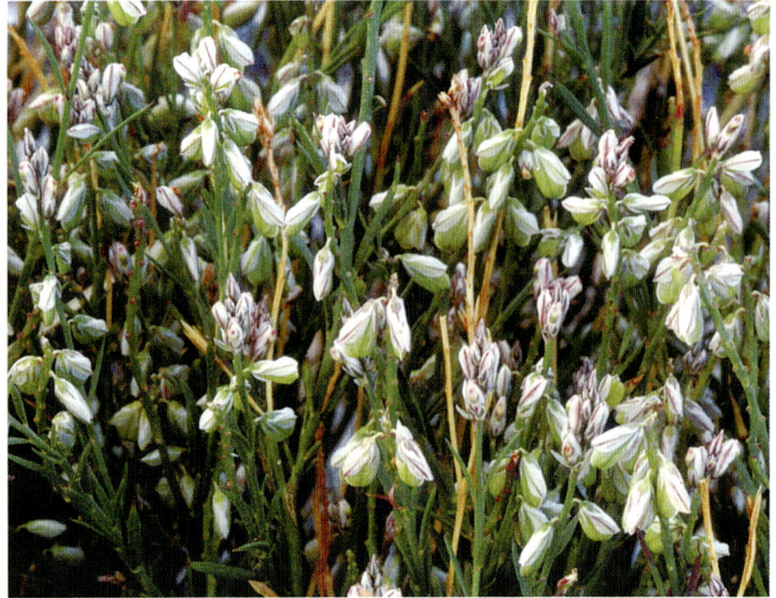

Broom Milkwort

Polygala scoparioides

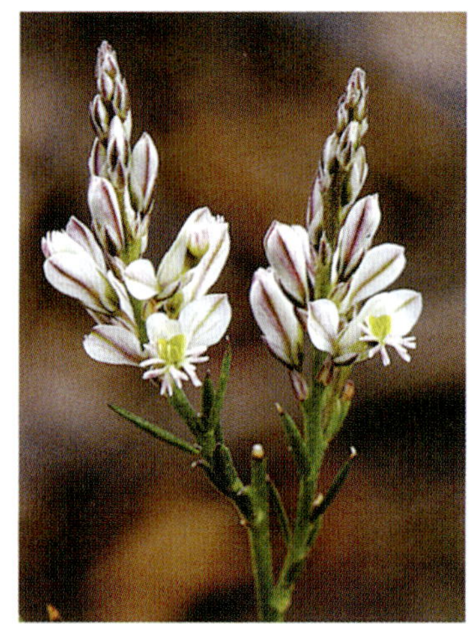

Broom milkwort is a low but upright perennial herb perhaps four to twelve inches high with many crowded stems that may collectively resemble little brooms. Often angled or ribbed and slightly hairy, the stems bear very narrow, alternate leaves, usually less than one-half inch long and linear in shape.

Minuscule and often unnoticed, the flowers form on short stalks in two-inch spikelike clusters. Appearing from spring to fall, the tiny blooms have an unusual structure: each blossom has five sepals and three petals but only two large, lateral petal-like sepals (wings) and a central petal (the keel) are prominent. The wings are white with a rose-red midstripe; the keel sports a conspicuous apical crest with a greenish-yellow eye.

Broom milkwort appears on limestone hills and gravelly flats in desert scrub habitats and sometimes on gypseous clay soils or along roadsides throughout Trans-Pecos Texas from Del Rio to El Paso. Its range extends west into southern New Mexico and central Arizona and south into northern Mexico (primarily in Chihuahua and Coahuila but also western Nuevo León and less commonly farther south to Guanajuato).

This low herb is far reaching throughout the park: in the Chisos foothills, the Dead Horse and Rosillos mountains, and the Grapevine Hills. It especially occupies clay flats—for example, along the Terlingua Ranch Road in association with wild onion, button gilia, and red dome blanketflower.

Note: The species name *scoparioides* means "broomlike." This species was collected by Cyrus Pringle in 1884 in the Santa Rita Mountains of Arizona. It was described in 1893 by Swiss botanist Robert Chodat, a professor of botany at the University of Geneva who is perhaps best known for his studies of the flora of Paraguay.

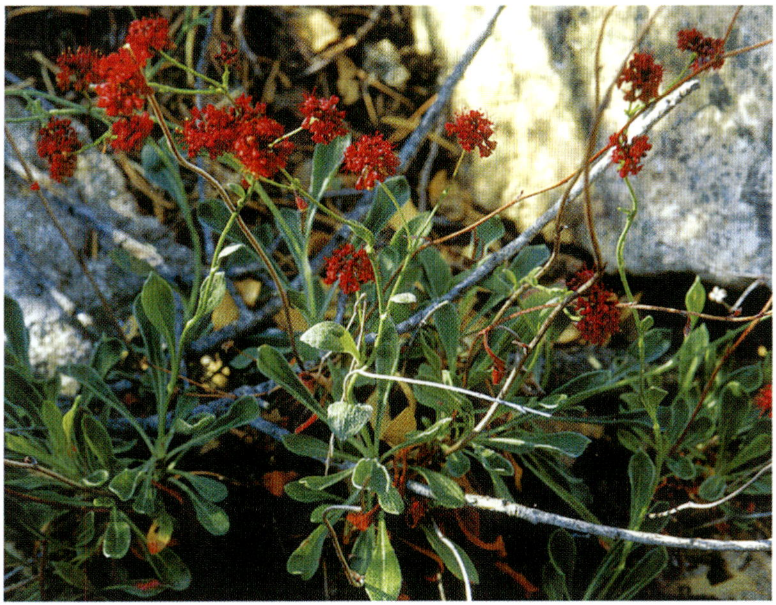

Chisos Mountain Buckwheat

Eriogonum hemipterum var. *hemipterum*

Chisos Mountain buckwheat is an erect perennial herb up to two feet high with mostly basal leaves around a usually single flowering stem. The basal leaves are elliptic or spatula-shaped, up to three inches long, with tapering bases and winged stalks. Leaves along the stem are fewer and smaller. Both leaves and stems are cloaked with stiff, straight hairs.

Flower clusters up to nearly one foot long form near the top of the stem, often stimulated by soaking summer rains. Dainty maroon or wine-red flowers appear on slender stalks that may approach three inches in length. Lacking petals, the flowers consist largely of six petal-like sepals, in two groups of three each, and conspicuous, protruding stamens that may be longer than the tiny flowers. As the floppy leaves age, they too turn deep red and match the flowers.

In the United States, Chisos Mountain buckwheat can be found only in the Chisos Mountains of southern Brewster County, where it is most prevalent in the oak-juniper-pinyon woodlands. This buckwheat also spills into the Sierra del Carmen in Coahuila, Mexico. Classified as vulnerable globally, it is considered imperiled in Texas.

Walk the Laguna Meadow Trail from summer to fall, especially July to September, and you will likely come across these attractive maroon flowers. This perennial is also accessible in the Basin and along Oak Creek.

Note: The genus name *Eriogonum* is from the Greek *erion* (wool) and *gonu* (knee or joint), referring to the hairy joints of the stem. This buckwheat was collected by Cornelius Müller in 1931 in Oak Canyon of the Chisos Mountains. Paul Standley, botanist with the Field Museum in Chicago and author of numerous works on the flora of Central America and Mexico, described it (as *Eriogonum hieracifolium* forma *atropupureum*) in 1936 as a form of another species, hawkweed buckwheat.

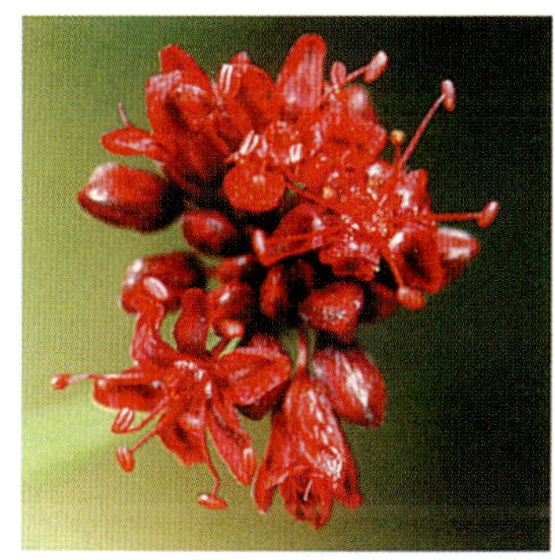

Orange Flameflower

Phemeranthus aurantiacus

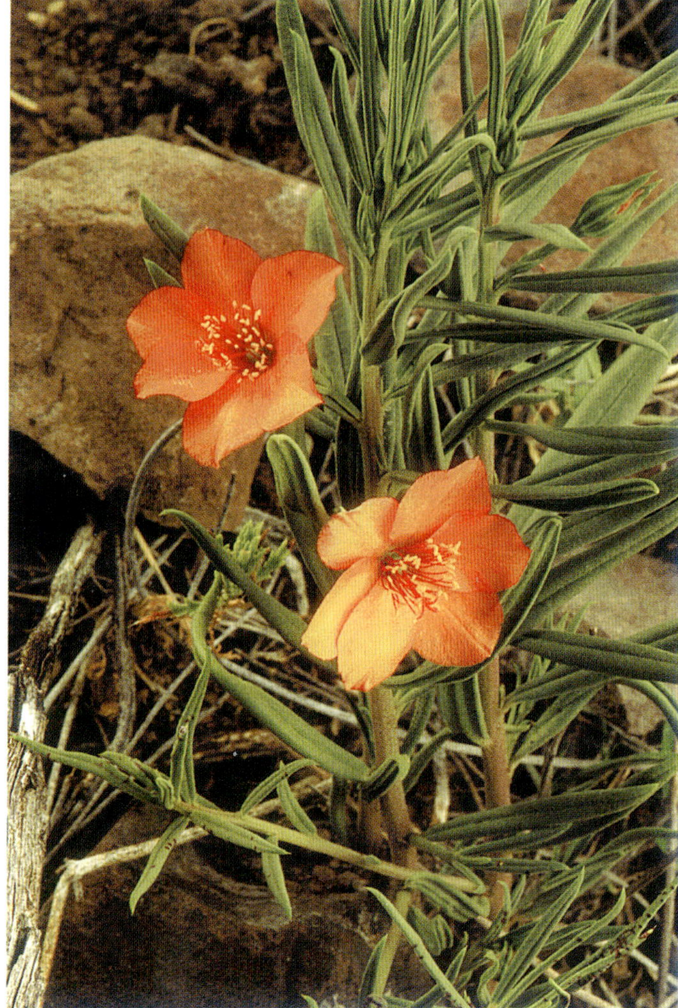

ORANGE FLAMEFLOWER is a succulent perennial herb usually six to fourteen inches high with notably fleshy stems and leaves. Flimsy and spreading or more often sturdy and erect, the stems grow from large, tuberlike roots. Seldom more than two inches long, the dark green, stalkless leaves are linear or narrowly lance-shaped, seemingly swollen or dilated, and curling along the edges and at the tip.

Blooming from late spring to fall, the lush flowers are about one inch across, yellow or deep orange to reddish-orange. Mostly solitary, the fleeting blossoms open for a few hours in the afternoon, only in bright sun, and soon wither. Each bloom has five or more thin petals and twenty to thirty red stamens. They reflect the late afternoon light like glowing lanterns.

Orange flameflower does best on rocky bluffs and open grassy slopes and in gravelly and sandy arroyos at elevations up to six thousand feet. In Texas this herb ranges throughout the Trans-Pecos (except in Reeves and Pecos counties) to Del Rio, northeast to Austin, and into parts of north-central Texas and the southern Panhandle. It is also widely distributed in southern New Mexico, southern Arizona, and northern Mexico (from Chihuahua to Nuevo León and south to San Luis Potosí).

In the park orange flameflower springs up quickly and fleetingly in the Chisos foothills. It has been confirmed west of the Chisos at Onion Spring, to the east at Crown Mountain, to the south at Dominguez Spring, in the Basin, and also in the Rosillos Mountains. Plants with yellow flowers and very slender stems and leaves were once considered to be a distinct species (*Talinum angustissimum*).

Note: The genus name *Phemeranthus* is presumably from the Greek *ephemeros* (living for one day) and *anthos* (flower), referring to the ephemeral blooms. The species name *aurantiacus* means "orange." This species was collected by Ferdinand Lindheimer in the Texas Hill Country in 1848 and described (as *Talinum aurantiacum*) by George Engelmann in 1850. In 2001 Robert Kiger, director of the Hunt Institute for Botanical Documentation and a Carnegie Mellon professor, placed this species in the genus *Phemeranthus*.

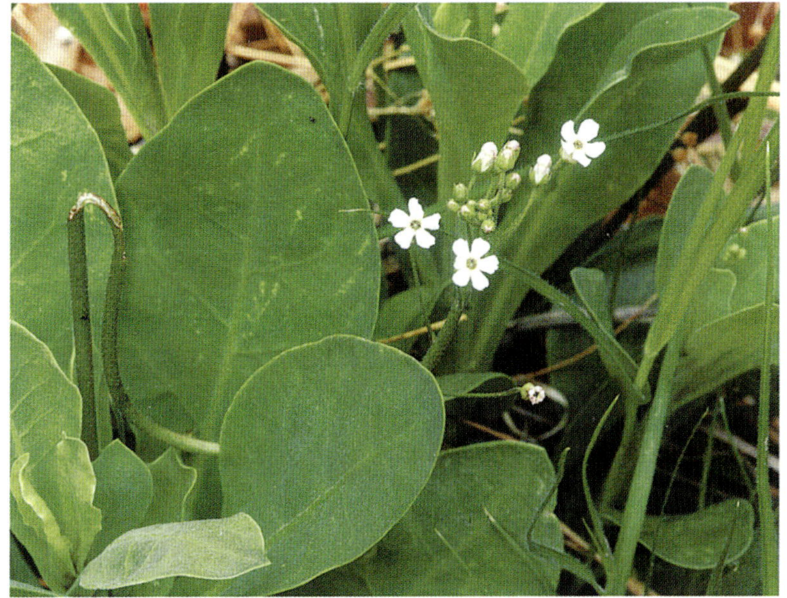

Limerock Brookweed

Samolus ebracteatus var. *cuneatus*

Limerock brookweed is a slightly fleshy perennial herb under two feet high with clumped, leafy stems and large broad leaves. Shaped like wide spoons or spatulas up to six inches long, the leaves form bulging basal rosettes and also grow along the stems.

From spring to fall, the flowers bloom in elongated clusters (racemes) on thin bare stalks that may shoot out obliquely high above the leaves. Each white, often pink-tinged flower has a short tube opening into five rounded but slightly notched lobes.

In the Big Bend, you are confronted with plants that have adapted in elaborate ways to the desert environment. It is surprising to find plants that rely on a steady water source, like limerock brookweed, proliferating at oases isolated from the surrounding desert expanse. This herb fills moist nooks by springs, creeks, waterfalls, and rivers.

Limerock brookweed is dispersed from Kansas throughout the Panhandle and northern Texas to Austin, southwest to Eagle Pass, and across the Trans-Pecos into New Mexico. It is also native to much of Mexico as far as Colima on the west coast and Baja California.

In the park search for this herb at springs (Oak, Mule Ears, McKinney, Fresno), pour-offs (below the Window), and sandy banks (Boquillas Canyon, Hot Springs).

Note: The genus name *Samolus* may be Celtic and refer to medicinal properties: plants were placed in cattle troughs to cure disease in livestock. *Ebracteatus* means "without bracts," and *cuneatus* refers to the "wedge-shaped" leaves. This variety was described in 1897 (as *Samolus cuneatus*) by John Small, curator of the New York Botanical Garden, and had been collected by botanist A. Arthur Heller in Kerr County in the Texas Hill Country in 1894. Heller and his wife, Gertrude, built a large herbarium of more than ten thousand specimens, many from the Sierra Nevada.

Creeping Cliff Brake

Pellaea intermedia

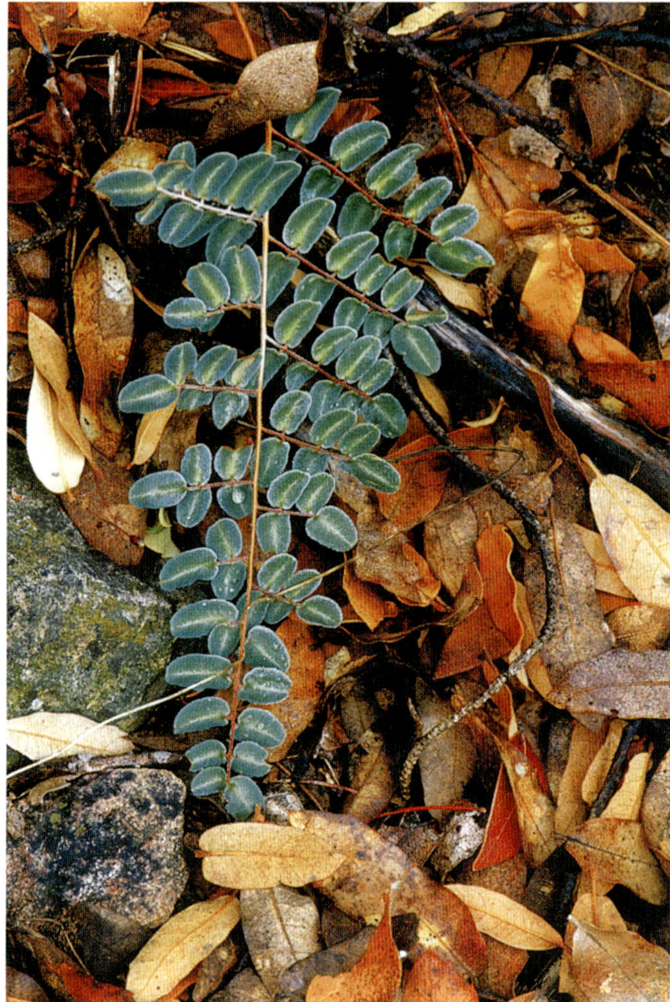

CREEPING CLIFF BRAKE, or intermediate cliff brake, grows from slender, creeping underground roots (rhizomes) that form aboveground leafy stems sometimes as much as twenty inches long. The leaves (fronds) are scattered along the stems and are usually bipinnate (twice divided into smaller leaflets or segments that are separately attached to the leaf axis). A single frond can be up to ten inches long and eight inches wide. The main leaf axis and the smaller leaf stalks are normally tan and relatively straight.

Each thick, leathery segment is oval to oblong, typically with a rounded or blunt tip and rounded base. The individual segments are dark green to gray-green, often with a distinctly bluish cast. They are neither lobed nor toothed. The fertile segments that produce spores in summer and fall are curled under along the edges and have distinctive white borders. Spore cases, usually containing thirty-two spores, are attached to the underside of the leaves.

This fern sprawls over relatively dry habitats, on rocky ledges and talus slopes, and amidst large boulders in igneous and limestone soil, often in shade. Populations of creeping cliff brake span Trans-Pecos Texas, southern New Mexico, and central and southern Arizona. They also stretch across large portions of northern Mexico (from Sonora to Nuevo León and south to northern parts of Zacatecas and San Luis Potosí).

Creeping cliff brake is frequent throughout the Chisos, from Green Gulch and the Basin to Casa Grande and Pulliam Canyon, on trails to the higher mountains (Laguna Meadow, Lost Mine, Pinnacles), and in wooded Boot Canyon and near Boot Spring.

Note: The genus name *Pellaea* is derived from the Greek *pellos* (dark), probably referring to the characteristically dark leaf stalks. The botanical name of this species was published in 1869 by German botanist Friedrich "Max" Kuhn, a specialist in ferns, fungi, and lichens, but was originally proposed by Georg Mettenius, who died three years earlier. Mettenius, a professor of botany at Leipzig, was an expert on brake or maidenhair fern and algae. The fern had been collected by Charles Wright on mountains near El Paso in 1849.

Trans-Pecos Cliff Brake

Pellaea ternifolia

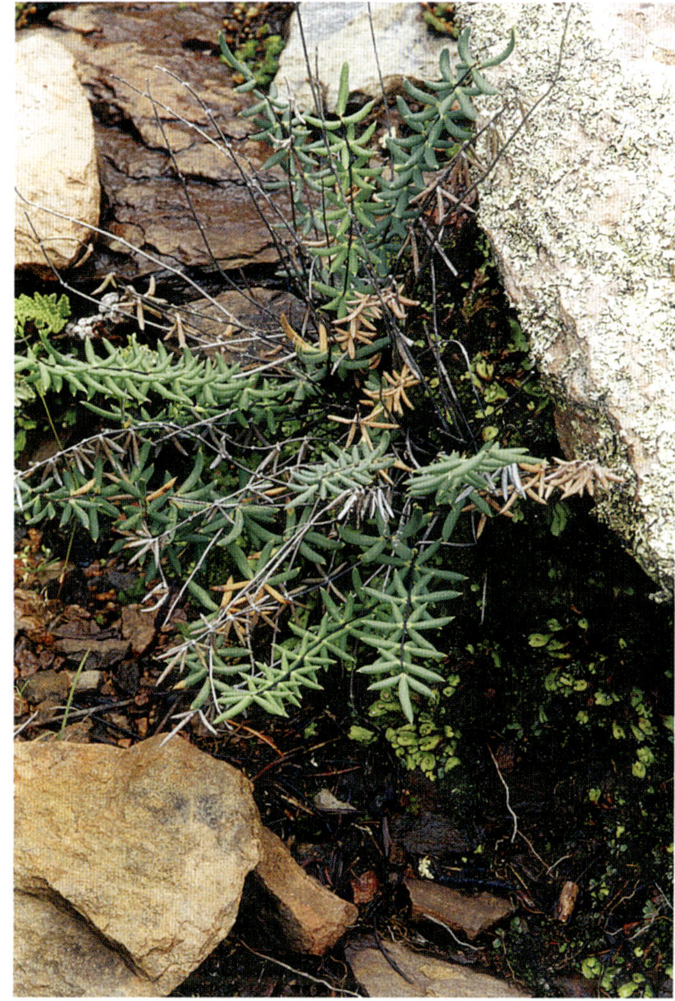

TRANS-PECOS CLIFF BRAKE, or ternate cliff brake, is a fern with relatively thick, dense stems, upright or sprawling, and many often tightly clustered fronds (compound leaves). On maroon or very dark brown and sometimes shiny stems, the fronds may occasionally reach sixteen inches long.

Narrowly lance-shaped in outline, each blade is divided into numerous leathery leaflets that are in turn three-parted into linear segments. The grayish-green segments, up to one and one-half inches long, lack leaf stalks and grow at sharp angles to the midrib of the blade. Spore cases containing sixty-four spores are attached on long stalks to the undersides of the fertile segments, which are curled inward and conspicuously whitish along the margins.

Trans-Pecos cliff brake lodges in shady and moist pockets on rocky, wooded slopes, in igneous soil. In the United States, this cliff brake is seen only in Trans-Pecos Texas, southeastern Arizona, and Hawaii. In the Trans-Pecos, it is apparently restricted to the Chisos Mountains and the mountains of Presidio County. Although imperiled and very vulnerable to extirpation in Texas (NatureServe 2005), the fern is pervasive in Mexico, Central America, and Argentina and Chile.

In the park Trans-Pecos cliff brake has been reported in upper Pulliam Canyon, on the Emory Peak Trail, and above Boot Spring. Good examples of this robust fern are also located near the beginning of the Northeast Rim Trail. Three distinct subspecies of this fern have been recognized: two are present in the Chisos Mountains (*P. ternifolia* ssp. *ternifolia* and *P. ternifolia* ssp. *arizonica*), and a third has been reported in the Davis Mountains (*P. ternifolia* ssp. *villosa*).

Note: The species name *ternifolia* means "three-leaved," referring to a leaf composed of three leaflets. Antonio Cavanilles described this fern in 1802 (as *Pteris ternifolia*) from a specimen collected by Luis Née in Peru. French-born Née was a botanist on the Spanish Malaspina expedition (1789–94), which traveled to South America and up the Pacific coast to Alaska (looking for the Northwest Passage to the Atlantic) and on to the Philippines and Australia.

Tuber Windflower

Anemone tuberosa var. *texana*

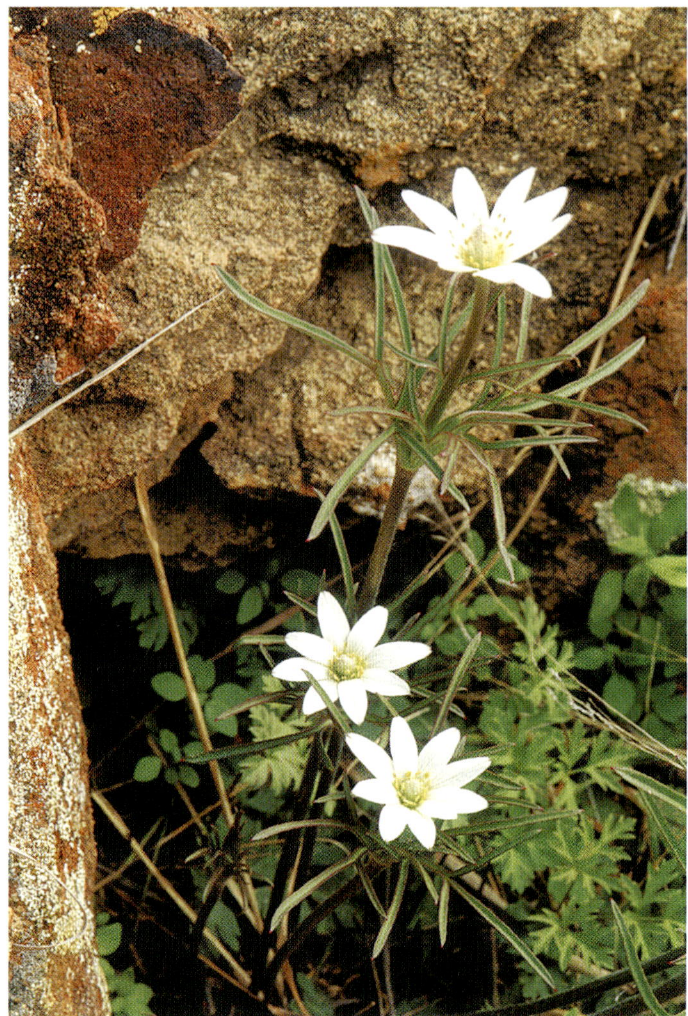

Tuber windflower, or desert anemone, is a perennial herb four to twelve inches high with a single erect, unbranched stem that grows from thick tuberous roots. On stalks about two and one-half inches long, a few leaves form a rosette at the stem base. The leaves are once or twice divided into three leaflets kidney-shaped in outline, and each one is deeply cleft or lobed and scalloped on the edges. Usually three leaflike bracts, split into thin linear segments, appear at the base of the flowering scape.

Perhaps the earliest spring bloomer, this herb produces a long flowering scape with one to three flowers. Each blossom consists of about eight to ten white to pink, petal-like sepals that encircle a cylindrical fruiting column. On a stalk as much as six inches long, the fruiting column gradually elongates as the colorful sepals wilt and fall.

Tuber windflower frequents grasslands and rocky slopes and canyons, rooting in rock crevices and at the base of cliffs, at shady sites in damp soil, and near creeks and springs. In the United States this species is wide ranging from southern parts of California, Nevada, and Utah to Texas. In Mexico it occurs in Baja California, Coahuila, and Nuevo León. The variety in the park (var. *texana*) is limited to the Edwards Plateau and eastern Trans-Pecos (also Culberson County) and rarely, adjacent Coahuila.

In the park this member of the crow's foot (Ranunculaceae) family was verified years ago on the east slopes of Pulliam Bluff, but Rough Spring just north of the Chisos is the best spot to view this intriguing perennial.

Note: The genus name *Anemone* may be from the Greek *anemos* (wind); Pliny and later herbalists thought the wind caused the flowers to bloom. The species name *tuberosa* refers to the tuberous roots. This species was collected by Cyrus Pringle in Arizona in 1884 and described by Per Rydberg in 1902. Rydberg wrote and illustrated books on the flora of the Rocky Mountains and Great Plains. Variety *texana* was described in 1995 by Marshall Enquist, author of *Wildflowers of the Texas Hill Country* (1987), and University of Texas botanist Bonnie Crozier; it was collected by them and B. L. Turner that same year near Del Rio.

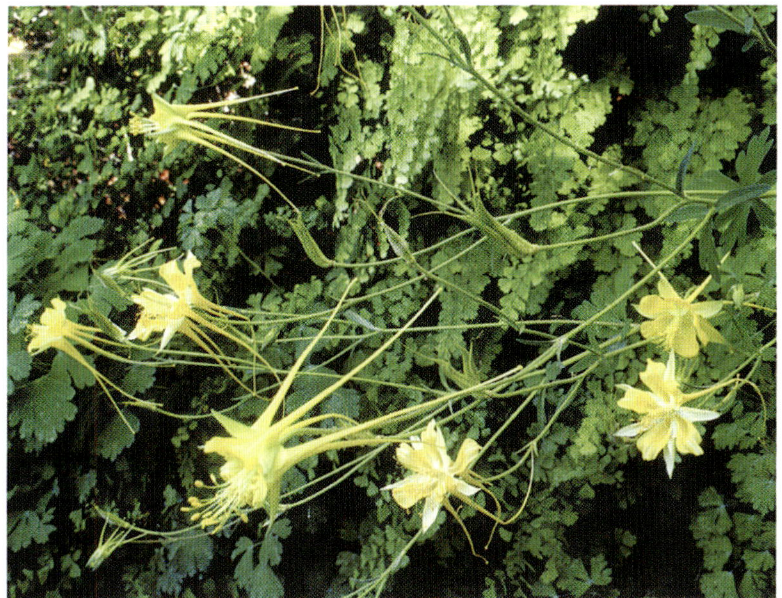

Longspur Columbine

Aquilegia longissima

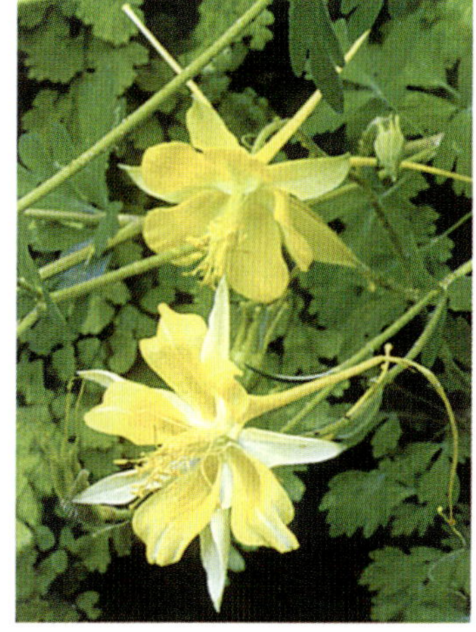

Longspur columbine is a perennial herb up to three feet high that often spreads to form large leafy clumps. Highly divided and fernlike, the long-stemmed basal leaves can exceed twelve inches in length with thin, deeply lobed leaflets up to two inches long.

This columbine is one of the most highly prized Big Bend plants. The pale yellow flowers have an elaborate architecture: the five spatula-shaped, rounded petals are no more than one inch long, but they extend backward to form slender, elongated spurs perhaps six times the petal length. Blooming mostly in spring and summer, the exquisite flowers are pleasantly lemon-scented but often slightly sticky.

Longspur columbine surrounds water sources, especially waterfalls and springs, and also dripping and seeping water in moist canyons. This delicate herb has a relatively circumscribed range: Brewster, Jeff Davis, and Presidio counties in the Trans-Pecos and Chihuahua, Coahuila, and Nuevo León in northern Mexico.

Globally longspur columbine is vulnerable throughout its range and in Texas imperiled and very vulnerable to extirpation. In the park it spreads across moist sites on west slopes of the Chisos and also in Pine Canyon at the waterfall, in secluded Maple Canyon, and on Pulliam Bluff.

Note: The genus name *Aquilegia* is perhaps from the Latin *aquila* (eagle) because of the likeness of the flower spurs to an eagle's talons, or it may be from *aquilegus* (water collector), alluding to the moisture collected in the hollow spurs. The species name *longissima* (longest) describes the long spurs. The name "columbine" is from the Latin word for "dove": the flower spurs of some species are said to resemble the heads of drinking doves. This species was collected by Edward Palmer in 1880 in Coahuila, Mexico, and described by Sereno Watson in 1882.

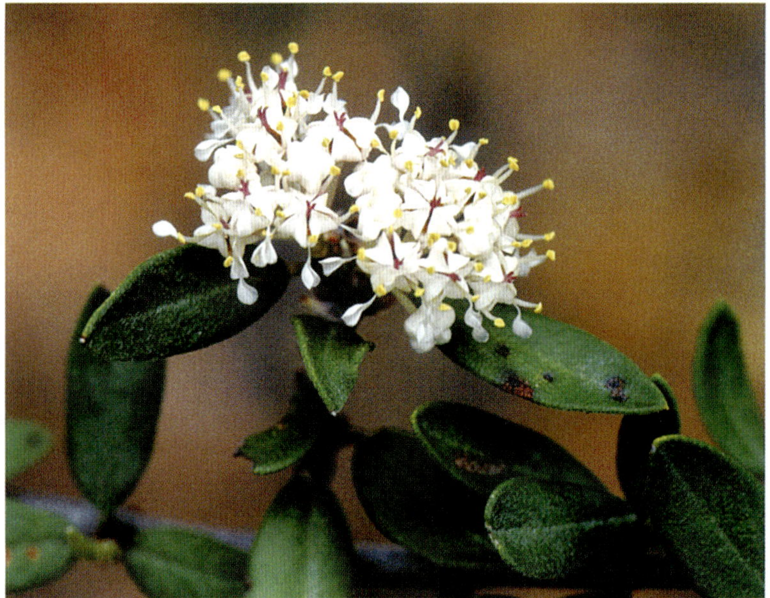

Desert Ceanothus

Ceanothus greggii

ALSO KNOWN AS buckbrush, desert ceanothus is a stubby shrub about three to four feet high with many rigid branches. Seldom more than one-half inch long, the dark green, oblong leaves are thick and leathery. The profuse white flower clusters contain small, convex flowers that are quite attractive, resembling tiny umbrella tops when they first bloom.

Desert ceanothus grows in Mexico from Oaxaca north to the Rio Grande, into parts of Trans-Pecos Texas, and throughout most of the Southwest, from New Mexico west to California and north to Utah and Nevada. The brilliant flower displays of this member of the buckthorn (Rhamnaceae) family brighten arid mountain slopes and canyons at elevations from thirty-five hundred to eight thousand feet.

In the park look for desert ceanothus in the Chisos Basin, along the Laguna Meadow Trail, and from Laguna West to Upper Cattail Canyon. Late March and early April are ideal times to find this pleasing little shrub in bloom.

According to Barton Warnock (1974), an excellent tea can be made from the leaves and flowers. A related species, New Jersey tea, was used as a tea substitute during the American Revolution. Flowers crushed in water make a fragrant skin lotion, and the fruits can be eaten by humans as well as birds. In Mexico the roots have been crushed to produce a red dye.

Note: The genus name *Ceanothus* is from the old Greek *keanothos,* a name used by Theophrastus for an unrelated plant. In 1853 Asa Gray named this species for Josiah Gregg, the frontier historian, trader, and plant collector. Gregg collected this buckthorn in 1847 near Saltillo, Mexico, at a famous battle site of the U.S.-Mexico War, Buena Vista. As many as eighty plants were named for Gregg, who sent specimens to George Engelmann in Saint Louis and Asa Gray at Harvard University.

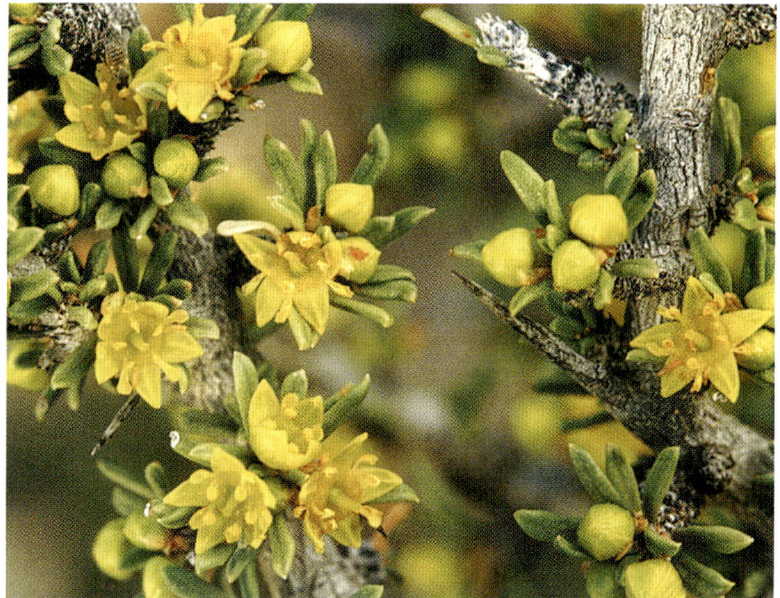

Javelina Bush

Condalia ericoides

Also known as little buckthorn and tecomblate, javelina bush is a squat shrub up to three feet high with convoluted spine-tipped branches. Narrowly oblong, gray-green leaves with curling margins grow in tight, overlapping clusters on short branchlets.

Dainty flowers with yellow petals are solitary or bunched in small groups on the spur branches. The fruit is a drupe (fleshy, one-seeded, with a hard stone enclosing a single seed), dark red to purplish-black at maturity. The other condalias in the park—Warnock condalia and green condalia—are much larger shrubs. Their flowers have sepals but no petals.

Usually in early spring, javelina bush blooms profusely throughout the Trans-Pecos, where it can be locally common. It also enters the western edge of the Edwards Plateau and, infrequently, the South Plains. A well-known plant of the desert Southwest, this low shrub is encountered from Texas west to Arizona and south into northern Mexico (from Chihuahua to Nuevo León and south to San Luis Potosí and Zacatecas).

Javelina bush prospers at lower elevations in desert scrub and grama grasslands and on gravelly slopes. Rather sparse in the park, it has been documented on Burro Mesa west of the Chisos and in the Rosillos Mountains. This photograph was taken in March on gravelly foothills of the Dead Horse Mountains near the north end of Old Ore Road.

Note: The genus *Condalia* was described in 1799 by Antonio Cavanilles and named for Spanish physician Antonio Condal. Condal was part of a team of naturalists led by Swedish botanist Pehr Löfling on José Iturriaga's 1754–61 Venezuelan expedition to explore the Orinoco River. The species name *ericoides* refers to the "heath-like" leaves. This species was collected by Charles Wright in 1849 in the Pecos River valley of western Texas and described by Asa Gray in 1852.

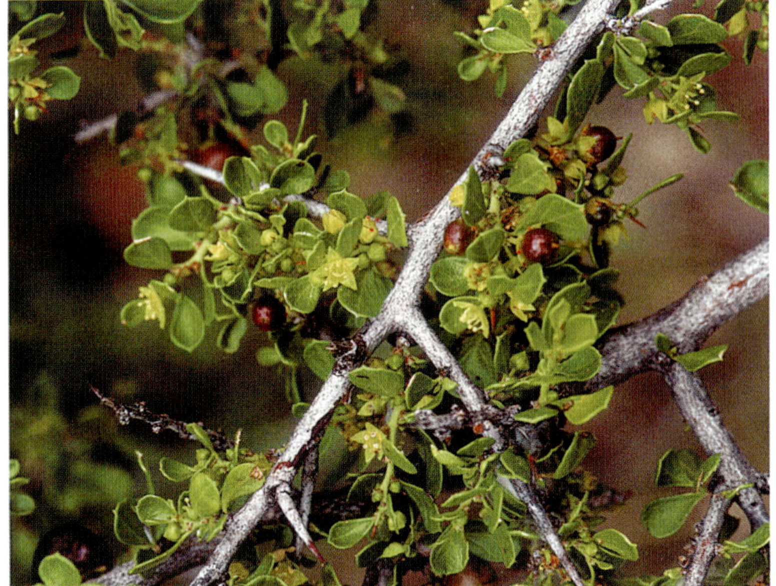

Green Condalia

Condalia viridis

GREEN CONDALIA, or green snakewood, is a shrub three to nine feet tall with rigid spreading branches and numerous short, spine-tipped branchlets. The bright green leaves are at most three-fourths inch long, usually smaller, with rounded or blunt tips. Relatively smooth and firm, the spoon-shaped leaves are often bunched in small clusters on the branchlets.

Tiny yellow flowers usually grow singly at the base of leaves on the spur branches. Lacking petals, each minute flower consists of five triangular petal-like sepals joined at the base into a disklike floral cup one-eighth inch wide. The flowers bloom in late spring and summer. Small globelike fruits form from the flowers. The fleshy, one-seeded fruits (drupes) are one-fourth inch long and dark maroon to black in color.

This evergreen snakewood inhabits gravelly washes and arroyos and dry limestone hills. Its distribution extends across the Edwards Plateau from San Antonio to Del Rio, into the Trans-Pecos to Presidio and Jeff Davis counties, and into northern Mexico (from Coahuila and Nuevo León south to San Luis Potosí).

In the park, green condalia can be observed on the limestone hills above Hot Springs by the Rio Grande and from the western slopes of the Chisos to Burro Mesa. It has been recorded near the mouth of Tornillo Creek, in the Dead Horse Mountains near the tunnel, and at Sam Nail Ranch. I have photographed this shrub in a gravelly wash below Trap Spring off the Mule Ears Spring Trail.

Note: The species name *viridis* refers to the brilliant green color of the leaves. This species was described by Ivan Johnston in 1939 from a specimen he collected in Coahuila in 1938. Johnston may be best known for his *Studies in the Boraginaceae,* a numbered series of thirty-one articles published from 1923 to 1961.

Havard Plum

Prunus havardii

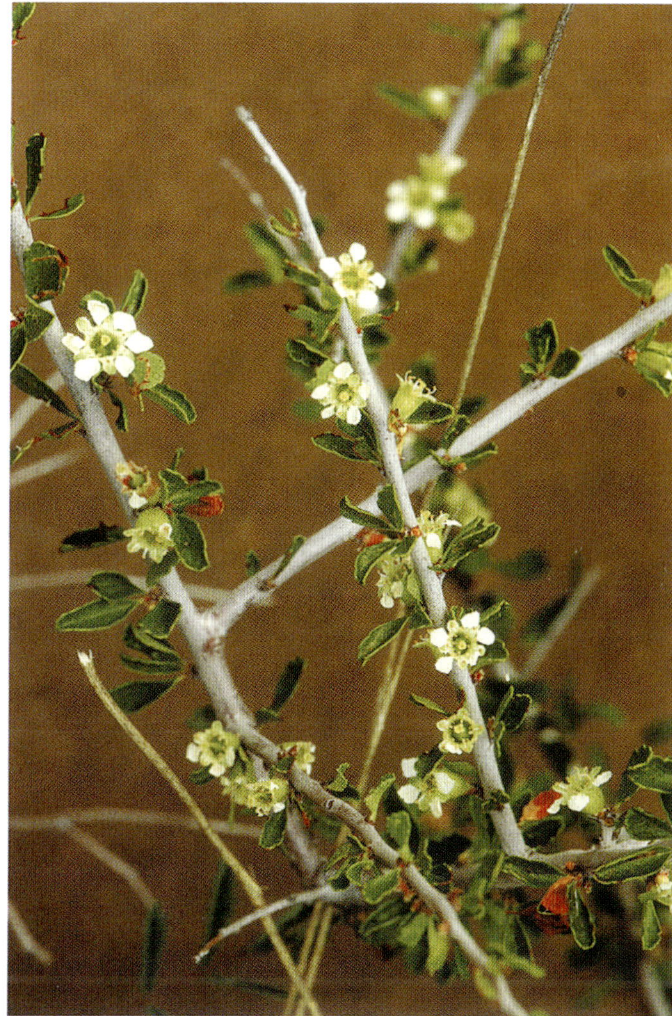

HAVARD PLUM is a stiffly branched shrub usually about three feet high, often rounded in shape and compact, and crowded with short, slender but rigid branchlets. Covered with pale gray bark, the branches may seem almost white. Small overlapping leaves, up to three-fourths inch long, form in tight groups along the spurs. On very short stalks, the spatula-shaped leaves are rounded, usually toothed at the tip, and curling at the base.

Blooming primarily in May and June, the white flowers are typically solitary and stemless at the base of leaves clustered on the branchlets. The flowers have spreading oval petals with a distinctly narrowed base. Very widely spaced, the petals seem to hang from the side of the bell-shaped floral cup. In July the flowers develop into globular, peach-colored fruits (fleshy, usually one-seeded drupes) less than one-half inch long.

Once considered rare, Havard plum is sparsely distributed on gravelly soils, along rocky slopes, and in canyons of the Trans-Pecos, where it is confined to southern Hudspeth, Presidio, and Brewster counties. Although it was once considered endemic to Trans-Pecos Texas, this shrub crosses into adjacent Chihuahua, Mexico, as well. It is classified as vulnerable in Texas and worldwide.

Havard plum is situated in the Chisos foothills and mountains at mid elevations and is probably most evident in Green Gulch and the Basin and along Oak Creek toward the Window. It sits atop rocky ledges near Ward Spring, at Rough Spring, and in Pine and Blue Creek canyons in the Chisos, and it also shows up in the Rosillos Mountains. I have also found this low shrub above McKinney Spring in the Dead Horse Mountains and in gravelly washes below the Grapevine Hills.

Note: The genus name *Prunus* is the ancient Latin name for a plum tree. This species was named for Valery Havard, who collected in the Chisos and Davis mountains in the 1880s. The species was described in 1913 (as *Amygdalus havardii*) by William Wight, horticultural collector with the U.S. Department of Agriculture and a specialist in South American plants.

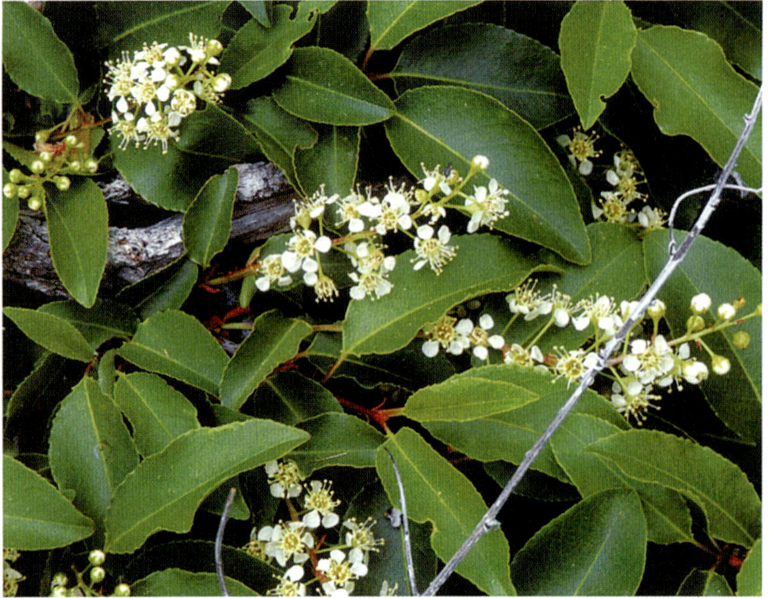

Southwestern Chokecherry

Prunus serotina ssp. *virens* var. *virens*

ALSO KNOWN AS black cherry and capulin, southwestern chokecherry can be a tree attaining a height of forty feet or more, but it is usually much shorter. It has rather smooth, gray bark with ruddy-brown branchlets. On occasion, this species will form dense brushy thickets. The glossy leaves, about two inches long, are lance-shaped or broadly elliptic and very finely toothed along the edges. They are dark green on top with a distinctly leathery texture.

The brilliant white flowers are crowded on slender spikes up to six inches long. Each flower has five spreading, spoon-shaped petals and many yellow stamens. Together, they are quite a show. The fruit is a fleshy, one-seeded spherical fruit (drupe) that turns red and then dark purple at maturity. According to Barton Warnock (1977), the fruits make very good wine and jelly.

Usually blooming in late spring and fruiting in summer, this small tree occupies mountainous areas of the Trans-Pecos from the Guadalupe Mountains to the Chisos Mountains. This variety is also widespread in New Mexico, Arizona, and much of Mexico (south to Guerrero and west to Baja California). It flourishes in closed canyons, arroyos, and along creeks.

In the park southwestern chokecherry is concentrated in the Chisos Mountains—in Upper Green Gulch and in Mouse, Pine, and Blue Creek canyons—and higher on the north slopes of Mount Emory.

Note: The species name *serotina* means "late blooming or fruiting." The subspecies name *virens* means "living," referring to the semi-evergreen leaves. This subspecies was collected by Paul Standley in the Organ Mountains of New Mexico in 1906 and described by him and botanist Elmer Wooton (as *Padus virens*) in 1913. Standley, while with the U.S. National Herbarium, published *Trees and Shrubs of Mexico* (1922).

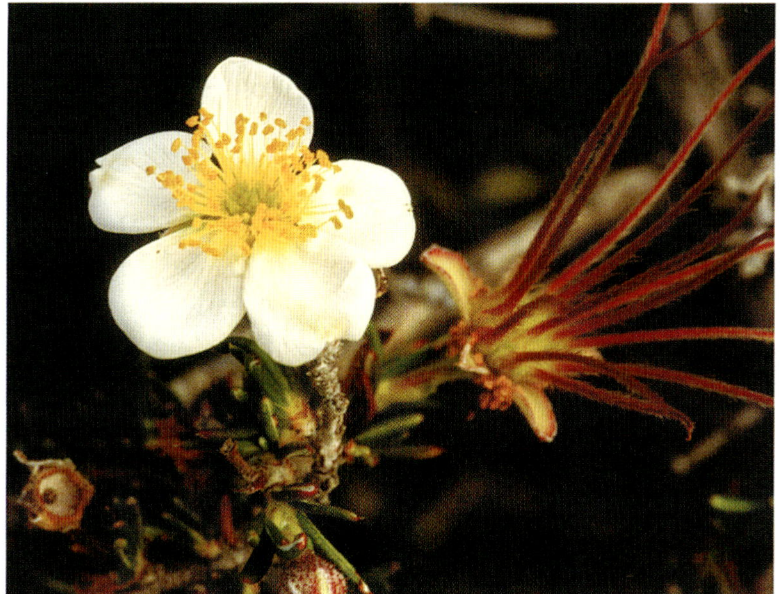

Heath Cliffrose

Purshia ericifolia

HEATH CLIFFROSE is a small shrub up to three feet high with dark wrinkled bark and sometimes contorted branches. The tiny, leathery leaves are linear and sharp-pointed, often crowded together at the ends of branches. According to Barton Warnock (1970), Native Americans used the leaves to cleanse wounds and induce vomiting.

The pure white flowers, an inch or more wide, are fragrant and solitary, with five broadly rounded petals. They bloom at the ends of short stalks. The distinctive fruits (achenes) are covered with shaggy hairs and equipped with a plumelike tail that helps them to disperse.

Flowering in summer and fall, heath cliffrose is known only from Brewster and Presidio counties in the Big Bend and adjacent Chihuahua and Coahuila in Mexico. It prefers rocky limestone habitats among boulders, in crevices, and on the edges of cliffs.

In the park heath cliffrose has been identified at sites in the Dead Horse Mountains: above Roy's Peak, near McKinney Spring, and northeast of Dagger Flat (Powell 1998). I have also come across this shrub at Passionflower Canyon in the Dead Horse Mountains and on Mesa de Anguila in the park's extreme southwest. The gnarled, windswept branches may be twisted into fantastic shapes.

Note: The genus *Purshia* is named for Frederick Pursh, horticulturalist at the Royal Botanical Gardens in Dresden who came to the United States in 1799 and participated in two extensive explorations. After publishing an important study of North American plants, Pursh went to Canada in 1816, collecting specimens for a Canadian flora, but his herbarium was destroyed in a fire, and he died in Montreal a destitute drunk. The species name *ericifolia* means "with Erica-like (heath-like) leaves." This species was collected by Charles Parry in 1852 in New Mexico "above the mouth of the Rio Pecos" and described (as *Cowania ericaefolia*) by John Torrey in 1853.

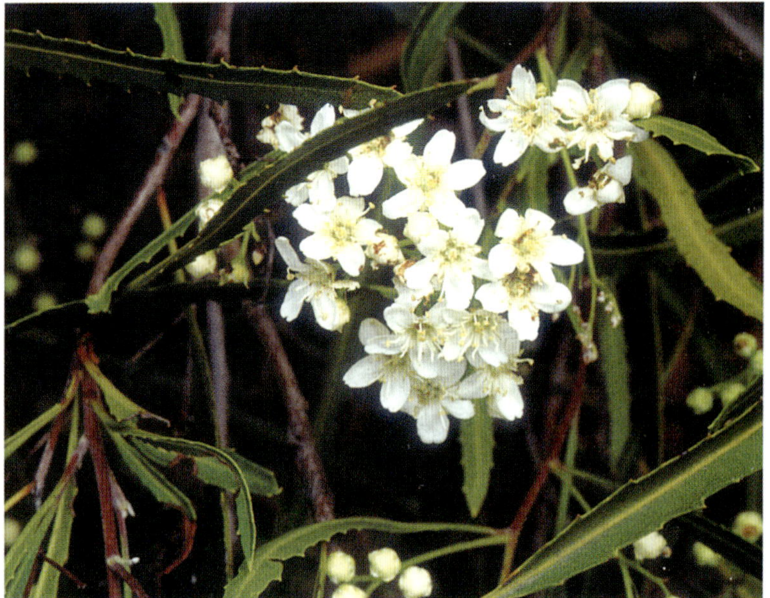

Palo Prieto

Vauquelinia corymbosa ssp. *angustifolia*

Also known as slimleaf vauquelinia and guauyul, palo prieto is an evergreen shrub that can reach fifteen feet in height. Young branches are smooth and light-colored, but older branches are darker with furrowed or cracked bark. The long, narrow leaves have sharp teeth on the edges. Noticeably tough and leathery, the blades are a highly reflective dark green.

Crowded into bouquetlike clusters, the white flowers steal the show. Each flower consists of five oblong petals set off by up to twenty-five yellow, eyelashlike stamens. The stamens and white petals collect light and shine like spotlights in the sun. The fruit is a five-celled oval capsule covered with densely matted woolly hairs.

In very late spring and summer, palo prieto blooms on steep rocky slopes and in mountain canyons. Geographically, it is relatively isolated in the Chisos and Dead Horse mountains of Brewster County, the mountains of Presidio County, and the adjacent Mexican states of Chihuahua and Coahuila. Subspecies *angustifolia* is critically imperiled in Texas (NatureServe 2005).

In the park, palo prieto thrives in the Basin, along the Window Trail, and below the Window in Oak Creek Canyon. It has also been noted in Pine Canyon, on the east slopes of Crown Mountain, at Sam Nail Ranch, and between Blue Creek and Elephant Tusk in the Chisos and in Telephone Canyon in the Dead Horse Mountains.

Note: In 1807 Baron von Humboldt and Aimé Bonpland named the genus *Vauquelinia* for Louis Nicolas Vauquelin, a French chemist who discovered chromium and beryllium and isolated the first amino acid in 1806. The species name *corymbosa* refers to the flat-topped flower clusters (corymbs). Subspecies *angustifolia* (narrow-leaved) was collected by Cyrus Pringle in Chihuahua, Mexico, in 1885 and described by Per Rydberg in 1908 as a distinct species (*Vauquelinia angustifolia*).

Scarlet Bouvardia

Bouvardia ternifolia

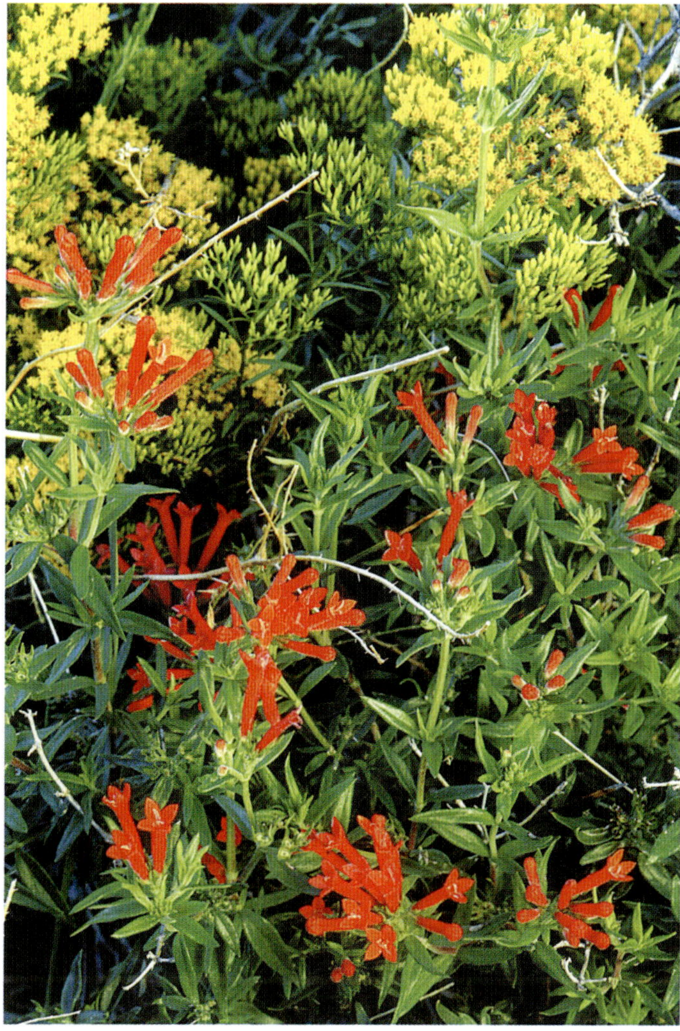

ALSO KNOWN AS trompetilla, clavillo, and mirto, scarlet bouvardia is an erect, extensively branched shrub usually less than three feet high with slender, leafy, pale-green or white-barked stems. The leaves form in small, ascending bunches of three or four along the stems. Very short-stalked, the blades are highly variable, from one-half inch to more than four inches long, linear or narrowly elliptic, often with pointed, tapering tips.

From late spring until the first frost, orange-red to brilliant scarlet flowers appear in small clusters at the branch tips. Well designed to attract hummingbirds, each blossom has a slender tube up to one and one-half inches long that opens into four short, egg-shaped lobes.

This low shrub fares well in brushy canyons and woodlands, on rocky slopes, and at the base of boulders at elevations above thirty-five hundred feet. In Texas scarlet bouvardia is present only in the Trans-Pecos mountains (in primarily Brewster and Presidio but also Culberson and southern Hudspeth counties). Its range reaches west to southwestern New Mexico and southeastern Arizona as well as deep into Mexico (from Sonora east to Tamaulipas and south to Oaxaca).

In the park scarlet bouvardia is prominent in Green Gulch and the Basin and common at other locations in the Chisos (from Ward Spring, Oak Creek, and Pine canyons and the south slope of Pulliam Bluff to upper Boot Canyon). It also spills into the Rosillos foothills.

Extracts of scarlet bouvardia roots have been used in Mexico as a natural remedy for dysentery and other ailments and as an antitoxin against scorpion poisoning.

Note: The genus *Bouvardia* in 1807 was named in honor of Charles Bouvard, director of the Jardin du Roi in Paris and personal physician to King Louis XIII of France. This species was described in 1797 (as *Ixora ternifolia*) by Antonio Cavanilles. The drawing in Cavanilles' famous work, *Icones Plantarum,* was of a plant grown in Madrid from Mexican seed. Diederich von Schlectendal, director of the botanical gardens at Halle, Germany, and editor of the journal *Linnaea,* reclassified the species as *Bouvardia ternifolia* in 1854.

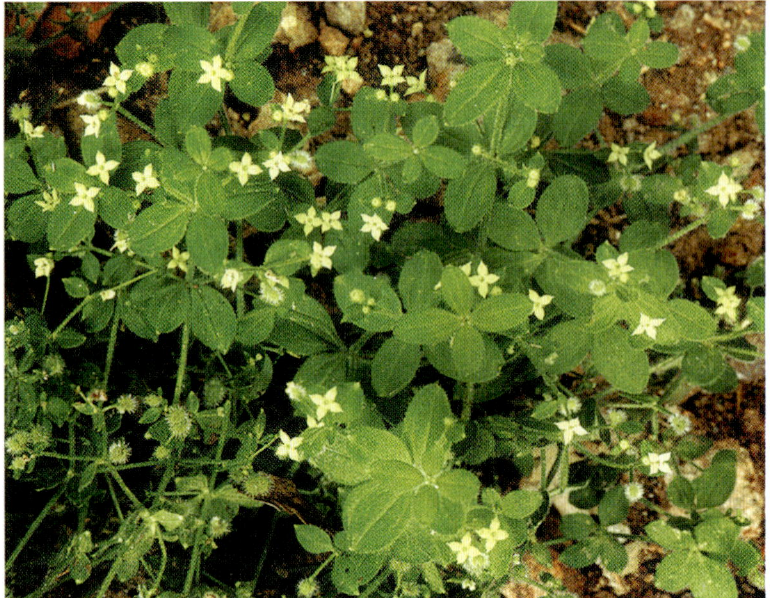

Bristly Bedstraw

Galium uncinulatum

Bristly bedstraw is aptly named: the entire plant is hairy, even the flowers and the fruits. This member of the madder (Rubiaceae) family is a weak perennial herb about one foot high with slender square stems. Long and slight, the stems may be nearly prone or slanting and widely spreading to form leafy canopies. Oval or broadly elliptic, the stemless leaves are clustered in groups of four along the stems. One-half inch or more long, they have a round but abrupt tip.

The dainty, pale yellow flowers usually have four united lobes. They appear on erect, stiff stalks from short, lateral branchlets and are sometimes conspicuously elevated above the leaves. The globose, paired fruits have distinctive hooked bristles at the tips.

Bristly bedstraw covers moist, shady sites and rocky and gravelly soils in canyons and on wooded slopes. In Texas this herb is limited to Presidio and Brewster counties in the Big Bend and to Val Verde and Kinney counties to the east along the Rio Grande. It occurs farther west in Arizona and Baja California and is plentiful throughout much of Mexico to Costa Rica and Panama.

In the park bristly bedstraw has been reported only in the Chisos Mountains, mostly at high elevations. It is easy to locate below the Pinnacles, along the trail to Emory Peak, and in Boot Canyon.

Note: The genus name *Galium* is from the Greek *gala* (milk): the flowers of *G. verum* and other species were used to curdle milk for cheese. Because some dried *Galium* plants have a pleasant fragrance, early settlers used bedstraw to bind straw for mattresses. The species name *uncinulatum* refers to the fruit's hooked bristles. This species was collected by Jean Louis Berlandier near Tampico, Mexico, and described by Augustin de Candolle in 1830.

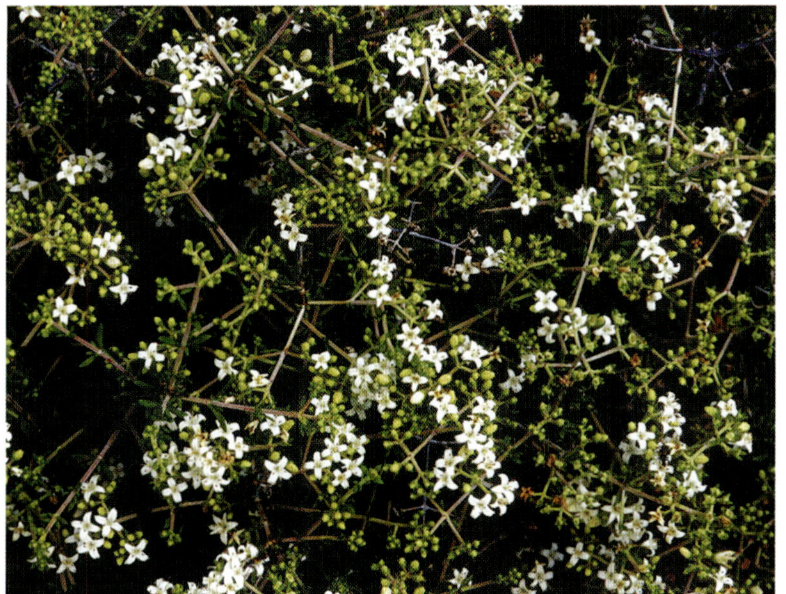

Fascicled Bluet

Hedyotis intricata

ALSO KNOWN AS tangle bluets, fascicled bluet is a woody shrub less than two feet high with intricate, rigid branches. The spreading branches bear paired, stemless leaves, less than one-half inch long, in compact, overlapping clusters. They are linear in shape and often curled under along the margins.

Blooming from summer to fall, the small white flowers are funnel-shaped with four recurved lobes and a noticeably hairy throat. They appear in few-flowered clusters, forming canopies near the tops of the stems.

Fascicled bluet spreads over foothills and mountains, rocky slopes of canyons, and exposed ridges at higher elevations. It is distributed from Brewster and Presidio counties in the southern Trans-Pecos to southern New Mexico and northern Mexico (Nuevo León to Chihuahua and south to northern Durango).

In the park this member of the madder (Rubiaceae) family is centered in the Chisos, from Green Gulch and the Basin to the South Rim. It has been seen atop Pulliam Bluff, at the base of Bailey Peak, near the Window, at Ward Spring, and above Lower Juniper Spring, but it may be most conspicuous in Blue Creek Canyon. Outside the Chisos, fascicled bluet frequents higher elevations of the Dead Horse Mountains.

Note: The genus name *Hedyotis* is from the Greek *hedys* (sweet) and *otos* (ear), but the significance is unknown. The species name *intricata* means "tangled." This species was described by Asa Gray in 1882 (as *Houstonia fasciculata*). Syntypes (specimens jointly representing the type) were collected by Edward Palmer (in Coahuila, Mexico, in 1880), John Bigelow (near Presidio, Texas, in 1852), and George Vasey (in the Organ Mountains of New Mexico in 1881). Vasey, later curator of the U.S. National Herbarium, had accompanied John Wesley Powell on his expedition by raft through the Grand Canyon in 1868.

Dutchman's Breeches

Thamnosma texana

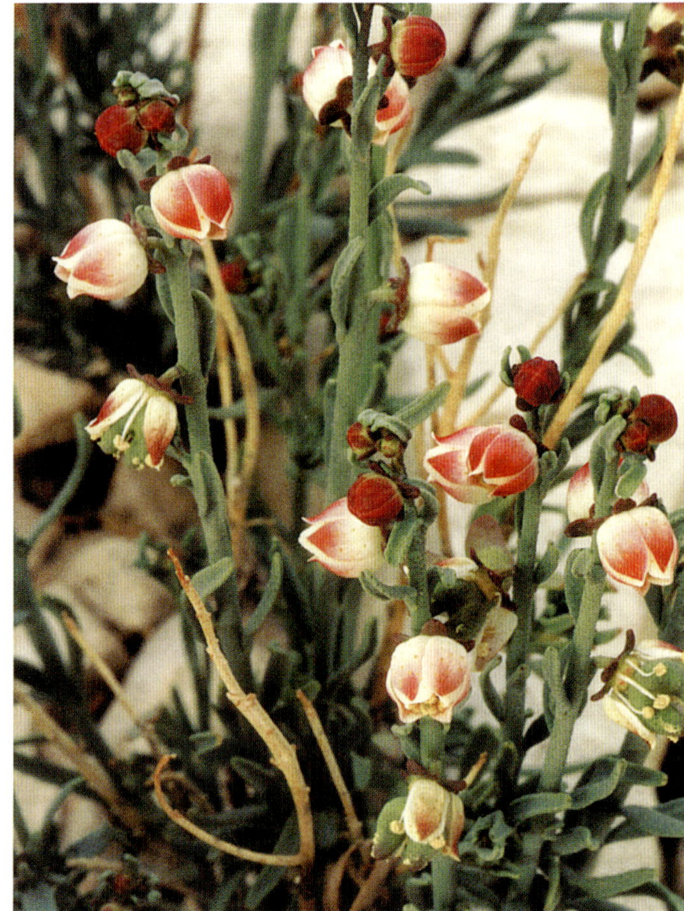

Also known as Texas desertrue, ruda del monte, and toronjil, this dwarf citrus is a perennial herb up to one foot high with sprawling, gangly stems. Dutchman's breeches is one of two citrus species in the park (the other is a small tree of the Chisos with no obvious resemblance, the hoptree, *Ptelea trifoliata*). This herb appears in desert scrub, and its relation to citrus trees seems a stretch, but the alternate, linear leaves are dotted with oil glands, and when the leaves are crushed a lemony (or to some, a turpentinelike) odor is readily apparent.

Half-closed, the urn-shaped flowers hang from upper stems on naked stalks. Blooming in spring and summer, the four-petaled flowers are multicolored with blotches of red, purple, and white and a yellow center. But this perennial is memorable for its amusing fruits that are shaped like a boy's breeches. Usually with two protruding lobes, the inflated, gland-dotted capsules look like pantaloons with the legs sticking upward.

On gravelly and rocky soils below forty-five hundred feet, Dutchman's breeches is wide-ranging throughout southern and central Texas to the South Plains and across the Trans-Pecos to Arizona. It grows extensively in northern Mexico (from Chihuahua to Tamaulipas and south to San Luis Potosí).

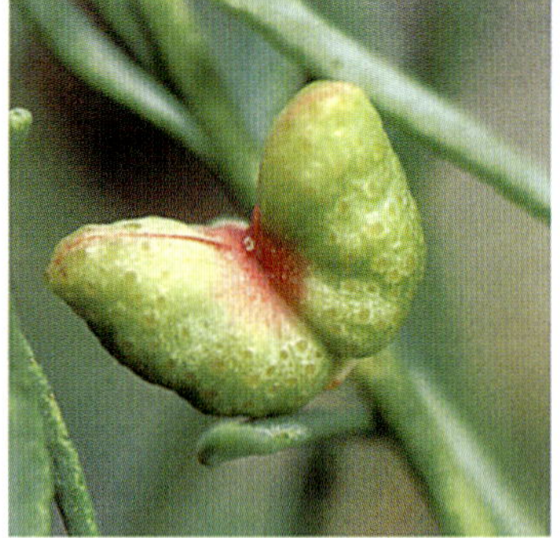

In the park this herb is scattered north of the Chisos—near Panther Junction, Hannold Draw, and Lone and Nugent mountains—and to the southeast near Talley and Chilicotal mountains and Glenn Spring. It also sprawls across many sites in the Dead Horse Mountains.

Note: The genus name *Thamnosma* is from the Greek *thamnos* (shrub) and *osme* (scent), for "strong-scented bush." John Torrey and John Frémont described the *Thamnosma* genus in 1845 from a specimen collected during Frémont's expedition to the Rocky Mountains in 1842. Frémont was the first botanizer of the Sierra Nevada. Asa Gray described this species (as *Rutosma texana*) in 1849.

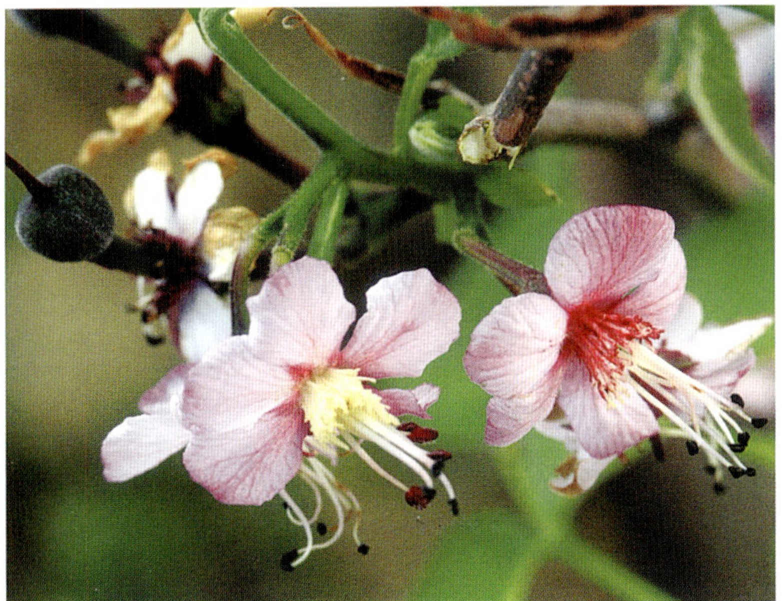

Mexican Buckeye

Ungnadia speciosa

Also known as monilla, this member of the soapberry (Sapindaceae) family is called "buckeye" because its seeds resemble *Aesculus* buckeye seeds. Mexican buckeye is either an extensively branched, spreading shrub six to ten feet high or a small tree up to thirty feet tall, both with thin, pale gray bark. The compound, alternate leaves, up to one foot long, are divided into one to three pairs of large leaflets and a single terminal leaflet. Each leaflet is narrowly egg- or lance-shaped, two to almost five inches long, with rounded teeth along the edges.

In spring fragrant pink flowers with seven to ten protruding stamens form in lateral clusters along the branches. Often appearing before the leaves, each blossom has four or five petals, broadly rounded with a narrow, hairy base (claw). The fruits are three-chambered, leathery pods up to two inches wide. The pale green (and then later brown) pods contain three shiny black seeds, each with a roundish white scar or eye (buckeye).

This tree lines rocky canyons, positioned along creeks, near springs, or at the base of cliffs on limestone and igneous soils. It is dispersed throughout the Trans-Pecos and Edwards Plateau into parts of north-central and northeastern Texas and the Coastal Prairies. Populations stretch across southern New Mexico and northern Mexico (from northeastern Chihuahua to Tamaulipas).

In the park Mexican buckeye is most prevalent in the Chisos: in the Basin, below the Window, in Pine and Pulliam canyons, and at Boot Spring. It also enters the Rosillos and Dead Horse Mountains.

Note: The genus *Ungnadia* may be named for Baron David von Ungnad, the Austrian ambassador to Constantinople who introduced the horsechestnut tree to Europe in 1581. *Speciosa* (showy) is the only species in the genus. It was described in 1833 by Austrian botanist and linguist Stephen Endlicher, who later discovered the western hemlock and named the genus *Sequoia* for the Cherokee who created an alphabet of his native language.

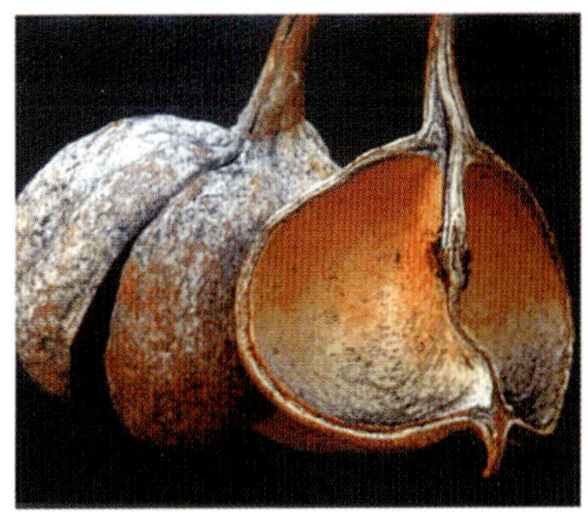

Twistleaf Paintbrush

Castilleja mexicana

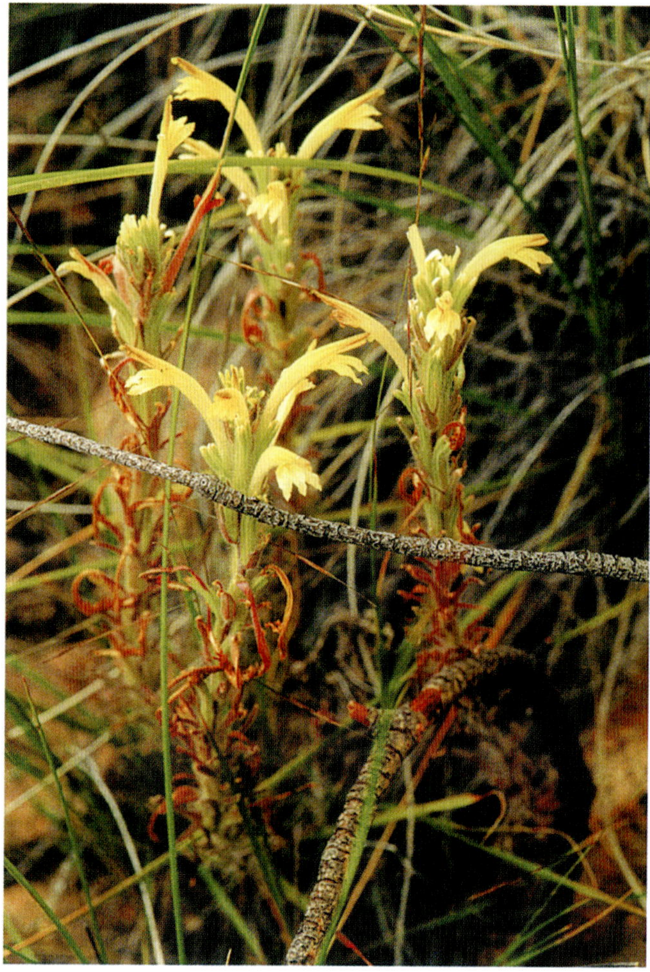

Also known as Mexican Indian paintbrush, twistleaf paintbrush is an annual or short-lived perennial with straight upright stems three to ten inches high. The mostly solitary stems may be densely covered with rough hairs. All leaves are narrowly linear or lance-shaped, and the upper are deeply divided, usually into three segments, which can be conspicuously curled or twisted, hence the common name "twistleaf."

Blooming from spring to early fall, lemon-yellow flowers form in spikes near the stem tips. Many paintbrushes have bracts (reduced leaves) below the flowers that are more striking and colorful than the flowers themselves, but in this species the bracts are mostly green and inconspicuous. Protruding well beyond the bracts, each two-lipped flower as much as two inches or more long has a hoodlike upper lip and a shorter lower lip with markedly flaring lobes. Arching and downward-curving, the flowers may bear slightly sticky, glandular hairs.

Twistleaf paintbrush usually comes up in grasslands and on open grassy and rocky slopes in the mountains. In the Trans-Pecos, this herb is restricted to the Guadalupe, Davis, and Chinati mountains and various sites in Brewster County, but it is less obscure in northern Mexico (from Chihuahua to Tamaulipas and south to San Luis Potosí and Aguascalientes).

The plant is rare in Big Bend National Park. This photograph was taken in Upper Cattail Canyon in September. Once considered a separate species (*Castilleja tortifolia*), this paintbrush is now subsumed under the name *Castilleja mexicana.*

Note: In 1781 this genus was named for Domingo Castillejo (1744–93), professor of botany at Cádiz, Spain. The species was described (as *Orthocarpus mexicanus*) in 1882 by British botanist William Hemsley from specimens collected by John Coulter in Zacatecas and Johann Schaffner in San Luis Potosí, Mexico. Hemsley published an early study of the fauna and flora of Mexico and Central America (1879–88). University of Chicago botanist Coulter collected extensively in Mexico and Texas and authored *Botany of Western Texas* (1891–94). German-born Schaffner became a physician, pharmacist, and plant collector in San Luis Potosí, where he developed an extensive herbarium.

Boquillas Silverleaf

Leucophyllum candidum

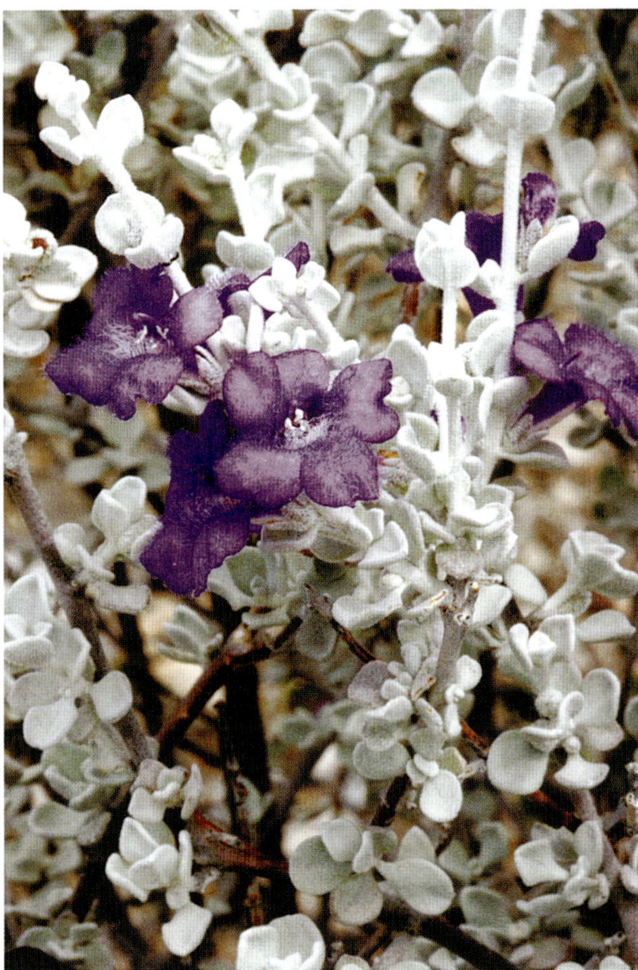

Also known as violet silverleaf and cenizo, Boquillas silverleaf is a small shrub less than four feet high, easily recognized by the silvery hairs that blanket the leaves and stems. Smaller, more compact, and paler in color than the more common purple sage, it has densely crowded, stiff branches and almost white leaves one-half inch long and wide. Alternate and sometimes opposite near the stem tips, the short-stalked, oval leaves appear in small bundles along the branchlets.

Boquillas silverleaf blooms in response to moisture, especially after good rains in late summer and early fall. Then it puts on a spectacular display of purple to dark violet flowers, which burst into bloom above the silvery leaves and color the landscape for miles. Like the leaves and stems, these flowers are also distinctly hairy, sometimes punctuated with orange or yellow blotches in the throat.

In the United States Boquillas silverleaf has a very circumscribed range: the gravelly and rocky limestone hills in Big Bend National Park and Black Gap Wildlife Refuge of southern Brewster County. It is more far reaching in northern Mexico (from Chihuahua to Coahuila and south to Durango and Zacatecas).

In the park this shrub inhabits portions of the Chisos foothills from Lone and Nugent mountains to Glenn Spring, east to Hot Springs and Boquillas Canyon, and north into the Dead Horse Mountains. It has also been observed at the north end of Mariscal Mountain.

Note: The genus name *Leucophyllum* refers to the grayish-white (*leuco*) leaf (*phylum*) of this shrub. The species name *candidum* is Latin for "shiny white." This species was described by Ivan Johnston in 1941 from a specimen he collected in Coahuila in 1940. A now synonymous species (*Leucophyllum violaceum*) was collected by Barton Warnock in 1937 between Lone and Nugent mountains, near Panther Junction. Warnock graduated from Sul Ross State University in Alpine that same year. After getting his doctorate at the University of Texas, he returned to Sul Ross in 1946 and taught botany there for the next thirty-three years. Warnock discovered many new Trans-Pecos species, and more than a dozen were named after him.

Purple Sage

Leucophyllum frutescens

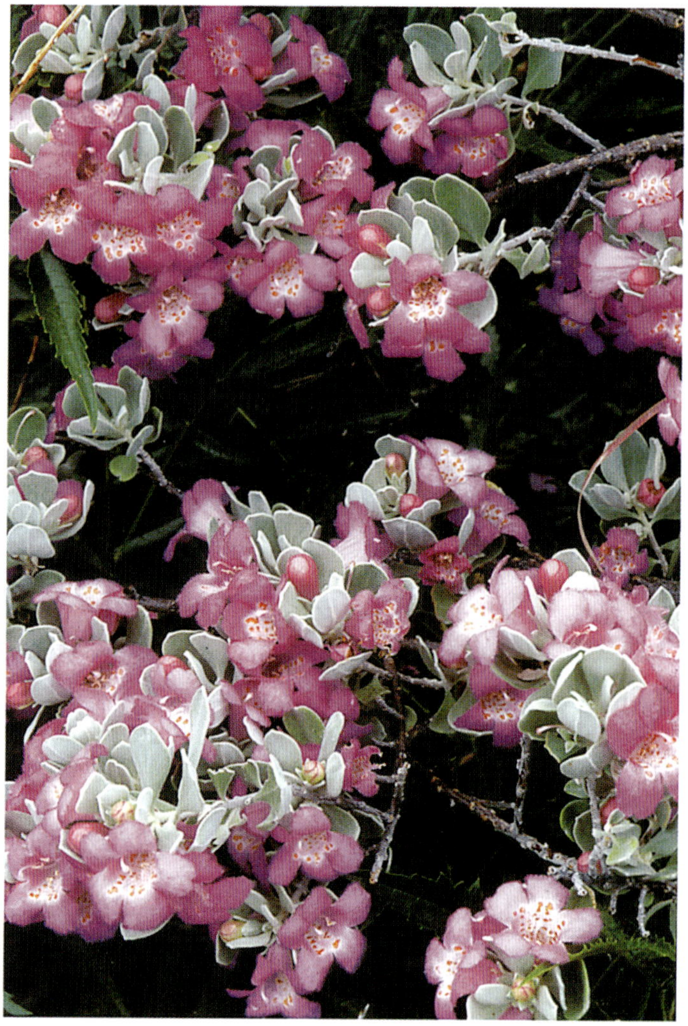

Also known as cenizo, Texas silverleaf, and barometer bush, purple sage is the largest and most common of three cenizo (*Leucophylllum*) species, with the largest, most spectacular flowers. This rounded evergreen shrub with widely spreading branches may reach six feet high or more. Stems (when young) and leaves are covered with dense, silvery-gray hairs, giving this cenizo (Spanish for "ashy") a distinctly gray-green or silvery-green appearance. Up to one inch long, the fuzzy, nearly stemless leaves are spoon-shaped or oblong, broadly rounded at the tip and wedge-shaped at the base.

Mostly rose-colored or lavender-pink flowers form in leafy clusters at the ends of branches. Up to one inch long, the flowers are bell-shaped, with a broad, inflated throat and five petal-like lobes. Blossoming in summer and fall, each flower has a bearded lower throat with a conspicuous, orange-dotted white blotch. Since they appear rather suddenly after summer rains, they were mistakenly thought to be barometers of coming storms.

Purple sage occupies low limestone hills, desert arroyo and scrub habitats, and chaparral and brush. In the United States this shrub is limited to Texas, from the southern and eastern Trans-Pecos to the Texas Hill Country and south to the Lower Rio Grande Valley. It is also encountered in northern Mexico, especially from Coahuila east to Tamaulipas.

In the park purple sage is most prominent in the Chisos foothills and in areas just west and north of the Chisos, from Paint Gap to Grapevine Hills, east to Lone Mountain and toward Boquillas Canyon. It is also present in the Rosillos and Dead Horse Mountains.

Note: The species name *frutescens* means "shrubby." Jean Louis Berlandier collected this species near Monterrey, Mexico, and described it (as *Terania frutescens*) in 1832 in a pamphlet chronicling new plants he discovered while working with the Mexican Boundary Commission.

Baccharisleaf Penstemon

Penstemon baccharifolius

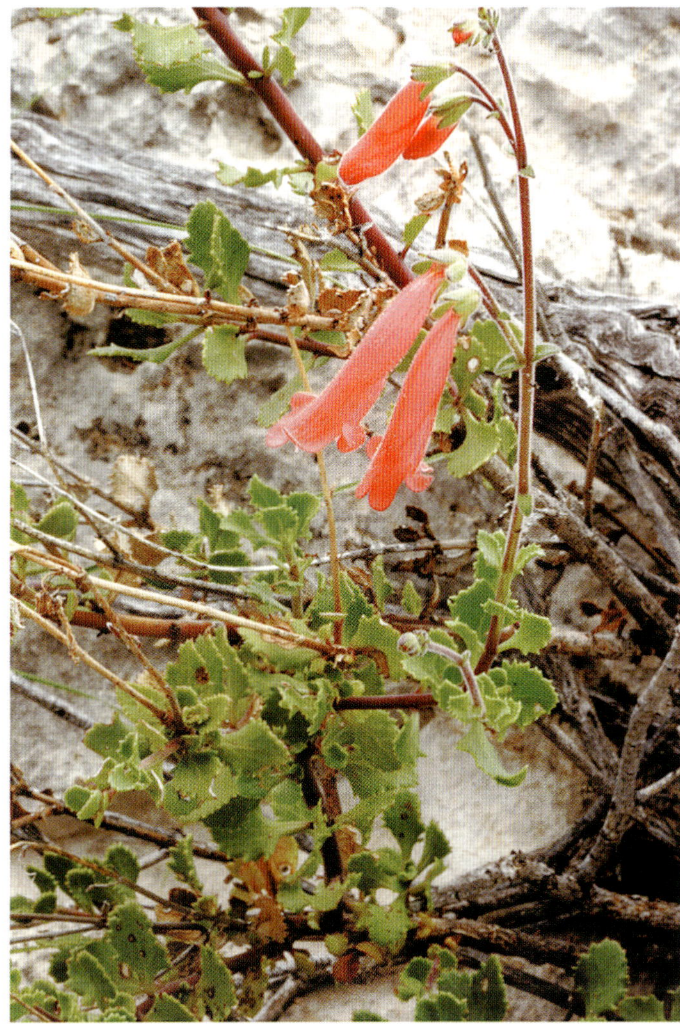

BACCHARISLEAF PENSTEMON, or cutleaf penstemon, is an attractive subshrub up to sixteen inches high with erect or sprawling stems that are distinctly woody at the base and densely hairy in the upper reaches. Usually less than one inch long, the thick leaves are spoon-shaped, often with tiny teeth along the rounded upper edge. Frequently, these yellowish-green leaves grow in tight bundles along the stems.

Like most beardtongues, *P. baccharifolius* has tubular flowers with a two-lobed upper lip and a three-cleft lower lip which may be distinctly hairy or "bearded." In this case, the scarlet flowers appear on a three-flowered stalk. They are often thickly covered with gland-tipped, sticky hairs, and a conspicuous white band may mark the throat at the base of the lower lip.

Frequently, this uncommon subshrub takes root on limestone cliffs, in rock crevices, and on the rocky beds of limestone canyons from Bandera in the Edwards Plateau west to Presidio County in the Trans-Pecos and south into the Sierra del Carmen and elsewhere in Coahuila, Mexico.

In Big Bend National Park, baccharisleaf penstemon is confined to remote canyons in the Dead Horse Mountains. I have witnessed these plants blooming in early October in Passionflower Canyon, not far from the Ernst Basin backcountry campsite. Depending on rainfall, they may bloom from late spring through early fall.

Note: The genus name *Penstemon,* from the Greek *pente* (five) and *stemon* (stamen), refers to the flower's five stamens (the fifth one sterile). The species name *baccharifolius* signifies "with leaves like those of genus *Baccharis.*" This species was described in 1852 by Sir William Jackson Hooker, author of a famous flora of North America (*Flora Boreali-Americana,* 1833–40). Hooker was a Glasgow University botanist and editor of the *Journal of Botany* and the *Botanical Magazine.* After serving for many years as director of Kew Gardens, he was succeeded by his son, the equally distinguished botanist Joseph Dalton Hooker. The type specimen was collected by Charles Wright on the Graham boundary survey in 1851–52.

Havard Penstemon

Penstemon havardii

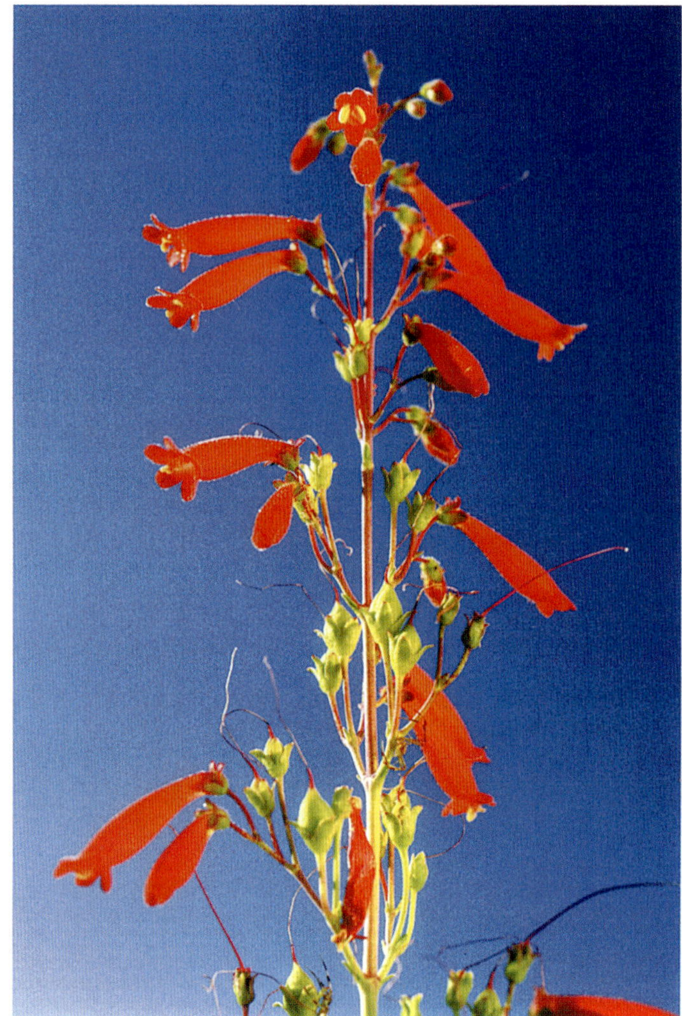

HAVARD PENSTEMON is an erect perennial herb usually three to four feet high with several sturdy, unbranched stems rising high above large, spatula-shaped basal leaves. These lower leaves are thick and fleshy, green and hairless but coated with a waxy white powder that gives them a distinctly grayish cast. The smooth, stalkless leaves along the stem are highly variable—egg- to lance-shaped to narrowly oblong—and as much as three inches long and one and one-half inches wide.

Bright red flowers appear in small, stalked clusters on a long spike. About one inch long, the tubular flowers are two-lipped like other flowers of the figwort (Scrophulariaceae) family. In this penstemon, the two-lobed upper lip extends beyond the three-cleft lower lip, forming a kind of hood. Havard penstemon blooms from spring to fall, but it is most prolific in late spring and early summer. The narrow tube and red color of the flower lures hummingbirds.

In the United States this imposing herb resides exclusively in the mountains of Jeff Davis, Presidio, and Brewster counties, but it is also found in the northern parts of Coahuila and Chihuahua in adjacent Mexico. Frequently, this perennial is situated in drainages leading out of mountain foothills, on rocky and gravelly hillsides, and also in sandy soils.

In the park Havard penstemon puts on showy displays to the west and north of the Chisos from Sam Nail Ranch to Government Spring and up Green Gulch toward the Basin. It is also unmistakable in Blue Creek and Oak Creek canyons, between Lone and Nugent mountains and near Pummel Peak, and in the Rosillos foothills.

Note: Species of this genus acquired the name "beardtongue" because in many species a single sterile stamen is cloaked with fuzzy hairs and rests flat, like a tongue, in the flower's tubular throat. The family is sometimes called foxglove presumably because the flowers, native to Europe, were reminiscent of the fingers of a glove and shared habitat with the burrows of foxes. Asa Gray named this penstemon for Valery Havard, who collected it in the Guadalupe Mountains of west Texas in 1882.

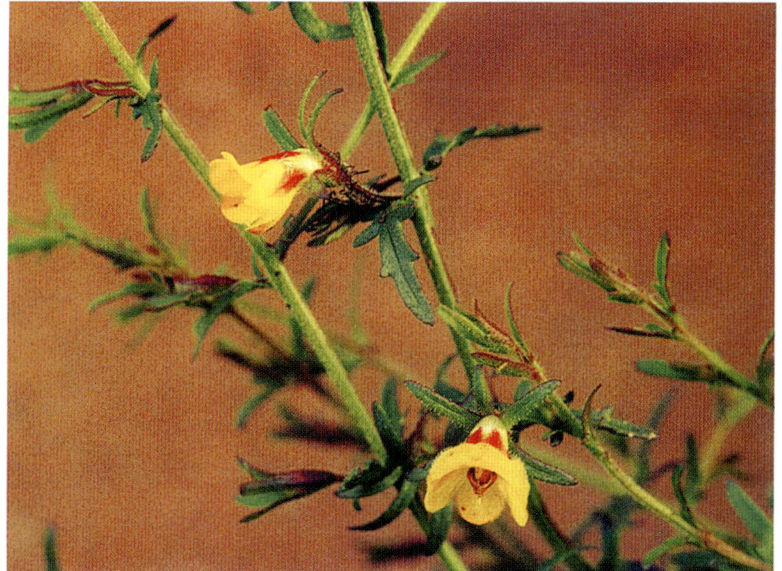

Limpia Seymeria

Seymeria scabra

Limpia seymeria is an annual less than two feet high with ascending but weak, branching, and arching stems. The surfaces of the stems and leaves are notably rough and covered with bristly, gland-tipped hairs, and the stems turn a conspicuous dark maroon with age. Most of the small leaves are divided into narrow linear segments.

The bell-shaped flowers, one-half inch wide, are almost always hanging or nodding. Curling back at the tips, the two-lipped, five-lobed yellow blossoms develop distinctive ruddy markings on their outer surface, similar in color to the maroon stems. Narrowly globose with a pointed tip, the reddish-black fruits release many seeds.

Limpia seymeria is native to the mountains of the Trans-Pecos and adjacent Mexico. Its distribution extends from the Guadalupe Mountains southeast to the Chinati and Chisos mountains and into northern Mexico (Chihuahua east to Nuevo León). Blooming in summer or fall, depending on rainfall, this figwort does best on dry rocky slopes and outcroppings high in the mountains.

In the park, check for limpia seymeria especially in August on Emory Peak and Toll Mountain, and in Upper Pine and Upper Cattail canyons. I have photographed this herb on the Emory Peak Trail and along Boot Canyon. On occasion you can encounter it at much lower elevations, even in Upper Green Gulch on the road into the Basin.

Note: This genus was described by Frederick Pursh in 1814 in his famous *Flora Americae Septentrionalis* and named for Henry Seymer (1745–1800), an English naturalist and owner of a large botanical garden of exotic plants. The species name *scabra* (rough), from Latin, refers to the stems and leaves. This herb was collected in 1849 by Charles Wright near Limpia Canyon in the Davis Mountains and described by Asa Gray in 1858.

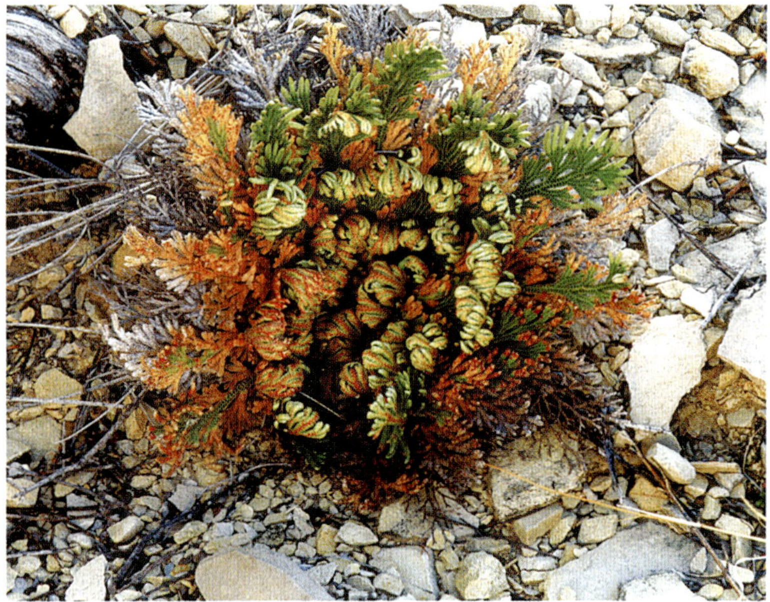

Flower of Stone

Selaginella lepidophylla

Flower of stone (flor de piedra) is also known as siempre viva, doradilla, and resurrection plant. This mosslike plant curls into a desiccated ball when dry but "resurrects" with moisture, opening into a bright green, flat rosette of slender, leafy stems up to ten inches wide. Broadly egg-shaped or triangular, the scalelike leaves overlap like miniature roof shingles. They are bristleless with a round or short, rigid tip, and their margins are transparent and fringed with hairs.

Spikemoss reproduces by spores instead of seeds. It has been known as little clubmoss because the fertile leaves form four-sided conelike structures (strobili) shaped like little clubs at the branch tips. The cones contain two types of thin-walled spore cases: one with up to four large female megaspores and the other with numerous male microspores.

Flower of stone may form large stands at north-facing sites, on dry rocky ledges, and in rock crevices and among boulders, especially on limestone and in shade. It spreads across the Trans-Pecos from Del Rio and Ozona to El Paso, west into New Mexico, and south throughout much of Mexico (from Baja California to Tamaulipas and south to Oaxaca).

In the park this fern is widespread in desert and foothills. It occurs on the Rio Grande (near Boquillas and Santa Elena canyons) and in the Chisos (Ward Spring, Blue Creek Canyon) as well as the Rosillos and Dead Horse mountains.

Note: The genus name *Selaginella* is the diminutive of *Selago,* an old name for *Lycopodium,* a genus of similar plants. The species name *lepidophylla,* from the Greek *lepis* (scale) and *phyla* (leaf), refers to the scaly leaves. This species was described in 1830 (as *Lycopodium lepidophyllum*) by Sir William Jackson Hooker and Robert Greville, an Edinburgh botanist and botanical illustrator, in their famous work *Icones Filicum,* which included 240 hand-colored plates of ferns.

Crucifixion Thorn

Holacantha stewartii

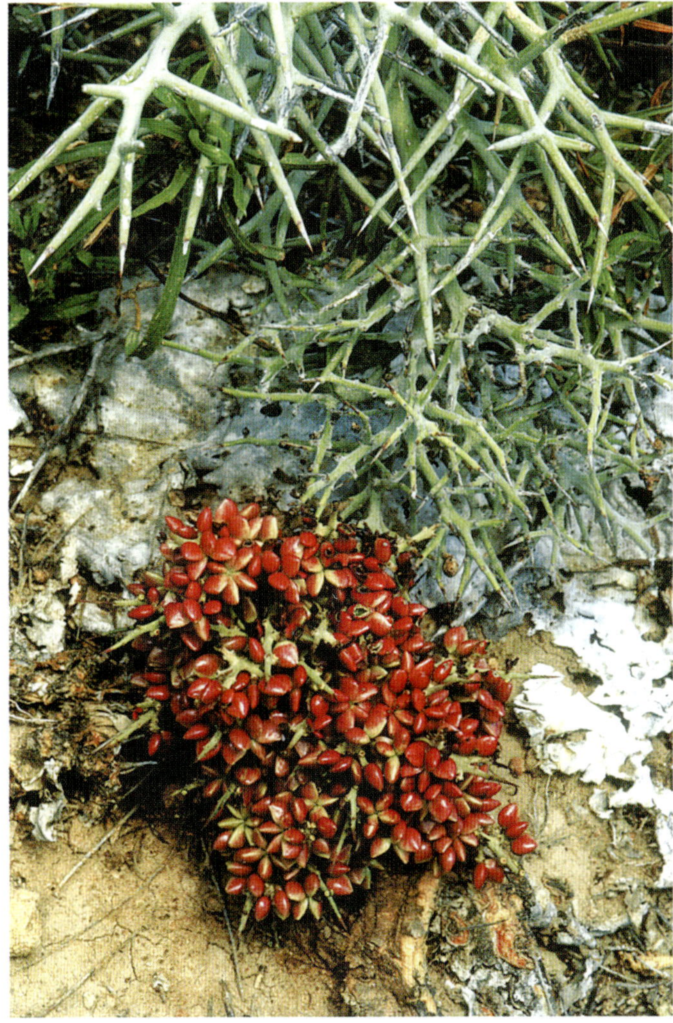

CRUCIFIXION THORN, or Stewart holacanth, looks like a crown of thorns up to three feet high, sprawled across the ground. Seldom seen, the tiny scalelike leaves soon fall, leaving only leafless, spine-tipped branches. The pale green stems of this sturdy shrub may form a dense, thorny maze many feet across.

Male and female flowers bloom on separate plants. Growing in small clusters, the unusual, fleshy flowers have five to eight petals that are rust colored with yellow margins. Crucifixion thorn is usually recognized by its distinctive clusters of shiny, scarlet fruits. Each fruit, consisting of up to nine individual, flattened nutlets (mericarps), may fade to brown and persist for several years.

Blooming in spring and summer, this member of the simarouba (Simaroubaceae) family grows only in Big Bend at disjunct locations in Brewster and southern Presidio counties and in northern Mexico from Coahuila to San Luis Potosí. It colonizes large expanses at the base of gravelly and clay hills and on desert flats.

In the park this shrub has been confirmed in the Paint Gap Hills, at Dagger Flat, on creosote flats between Glenn Spring and Mariscal Mine, west of Sublett Ranch on the Rio Grande, and south of Government Spring. It is often interspersed with allthorn (*Koeberlinia spinosa*), a similar, thorny plant (but with greenish-white flowers and shiny black berries).

Note: The genus name *Holacantha* is from Greek words meaning "all thorn." In 1942 California botanist Cornelius Müller named this species for Robert M. Stewart, his host at Santa Elena in the Mexican state of Coahuila, where the type specimen was collected in 1940. Müller discovered a number of rare plants while botanizing in the park, including curly coral root, Texas purplespike, and Chisos oak.

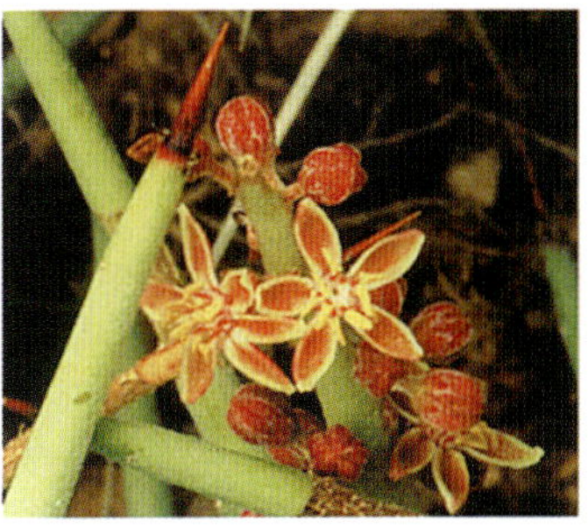

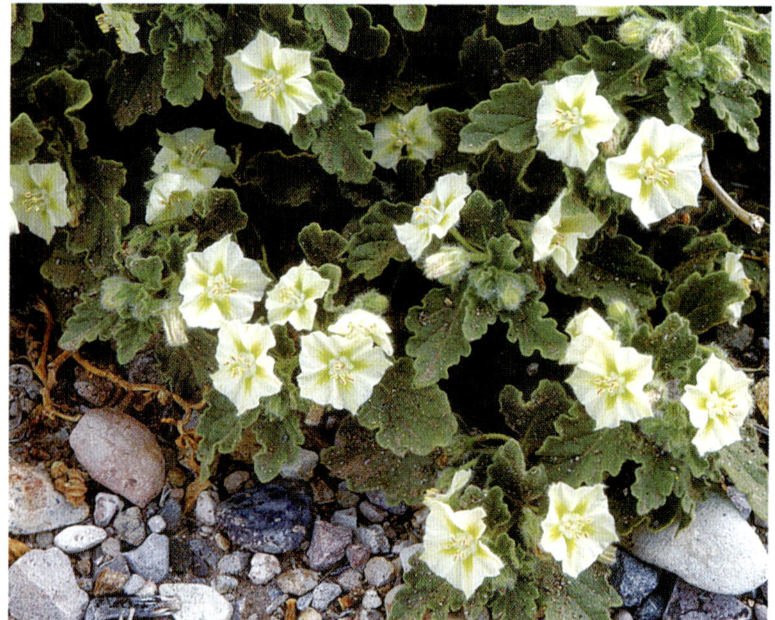

Shaggy False Nightshade

Chamaesaracha villosa

Shaggy false nightshade, or woolly false nightshade, is a sticky, odiferous perennial herb four to twelve inches high, extensively branched from woody roots. The thin, rough stems form rounded mounds or spread low to the ground. Occasionally two inches long or more, the dark green leaves are broadly egg-shaped, blunt-tipped, and conspicuously scalloped and wavy on the margins. Both the thick blades and the stems are cloaked with long, shaggy hairs and dotted with stalked glands.

In spring and fall, pale greenish-yellow flowers, three-fourths inch across, usually appear singly or in pairs at nodes along the stems. Each wheel-shaped bloom has five lobes united into a single plaited border. White cottony hairs project from the flower's throat.

Shaggy false nightshade proliferates along roadsides, in desert scrub habitats, on gravelly limestone, and in sandy arroyos. In the United States this perennial is mostly limited to Trans-Pecos Texas, especially near the Rio Grande, from southern Hudspeth County to Del Rio. It also reaches into northern Mexico (eastern Chihuahua, Coahuila, and northeastern Durango).

In the park this herb is most plentiful in the Dead Horse Mountains (at Hot Springs, Boquillas Canyon, and to the north near Roy's Peak and Dog Canyon). It is also scattered in the desert and foothills (at Santa Elena Canyon, Burro Mesa, and Glenn Spring).

Note: *Chamaesaracha,* from the Greek *chamai* (low, dwarf) and *Saracha* (a South American genus), implies "dwarf *Saracha.*" The species name *villosa* means "villous." This species was collected by Edward Palmer in Coahuila, Mexico, in 1880 and described in 1893 by Per Rydberg. Palmer got his start in botanical collecting on the Page expedition to Paraguay, worked as assistant surgeon at army outposts after the Civil War, led a survey of California flora in 1891, and collected for more than thirty years in the southwestern United States and Mexico.

Silvery Wolfberry

Lycium puberulum var. *berberioides*

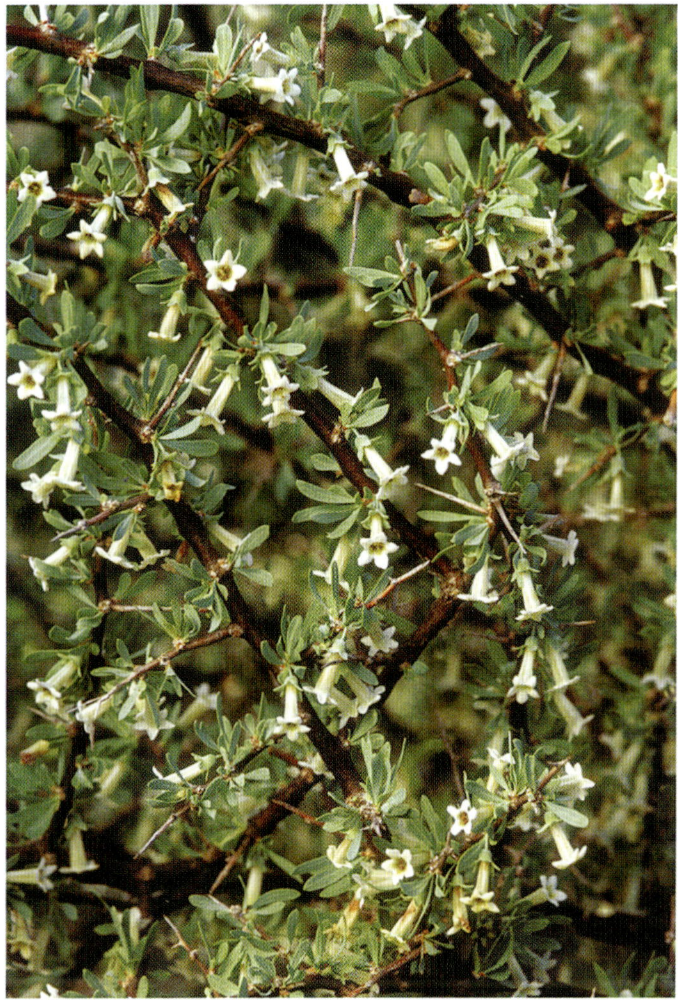

Silvery wolfberry is a spreading shrub usually three to four feet high with long, needlelike thorns and very distinctive chocolate to reddish-purple branches. Stemless or short-stemmed, the leaves are narrowly oblong or narrowly spatulate, from one-fourth inch to one and one-half inches long, and grouped at the nodes of branches. They are green and smooth but covered with a waxy bloom that gives the entire shrub a conspicuous gray or silvery cast.

The greenish-white flowers, no more than one-half inch long, are funnel-shaped with a short, five-lobed border that spreads abruptly and often curves backward. The inside of the flower tube is yellowish-green. Solitary or in pairs, the blooms appear in spring at the leaf clusters, and abundant oval berries cover the branches in summer. The dry, hard, slightly bitter berries are about one-fourth inch wide and resemble miniature tomatos. Pale orange and waxy, they are delicacies for birds and other wildlife.

In the United States silvery wolfberry is isolated in extreme southern Brewster and southeastern Presidio counties of the Big Bend. It covers gravelly hills, clay and stony flats, desert grasslands, and creosote scrub habitats.

In the park this shrub prospers on clay flats south of Persimmon Gap and near upper Tornillo Creek, at Crown Mountain to the east of the Chisos and Chisos Pens on Cottonwood Creek to the west, and on the southern boundary near Sublett Ranch on the Rio Grande. This photograph was taken east of Sam Nail Ranch in June.

Note: The genus name *Lycium,* from the Greek *lykion,* was used by Dioscorides for a thorny shrub, probably from Lycia in Asia Minor (in present-day Turkey). The species name *puberulum* (finely hairy) describes the leaves. Variety name *berberioides* ("like the genus *Berberis*") refers to the similarity of the grayish, glaucous leaves. This variety was collected in 1964 by Howard Gentry and Craig Hanson in the park three miles south of Persimmon Gap and described in 1965 by Donovan Correll as a distinct species (*Lycium berberioides*). Gentry was a widely known authority on agaves and the flora of the Río Mayo valley of Sonora, Mexico.

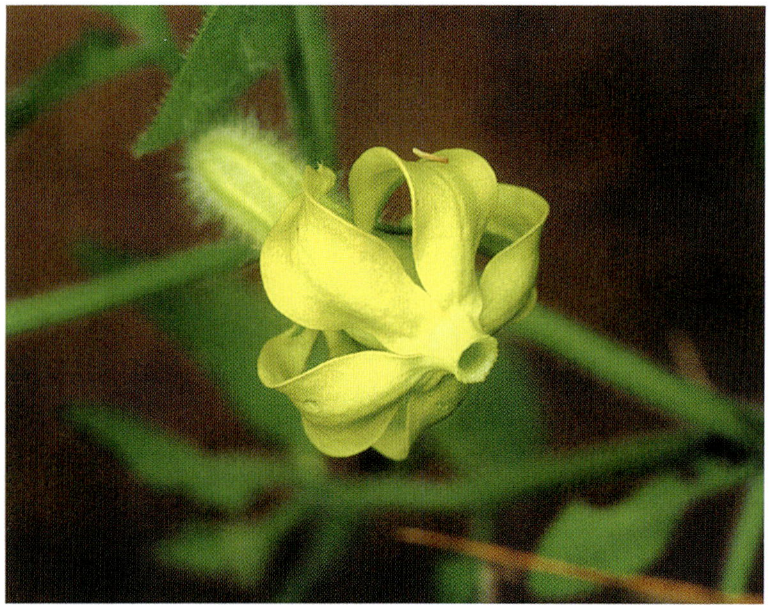

Puckering Nightshade

Nectouxia formosa

PUCKERING NIGHTSHADE is a slightly foul-smelling perennial herb less than two feet high. The erect or oblique stems are coated with hairs of various types, some tipped with glands that produce sticky secretions. Heart-shaped at the base and pointed at the tips, the egg-shaped leaves are also sticky-hairy.

In spite of the odor and stickiness, this nightshade is best known for its unusual trumpet-shaped flowers. The slender flower tubes abruptly expand at their apex into five narrow lobes that spread backward more like yellow ribbons than petals. Because of the small tubular crown in the center, the bizarre flowers really do seem to pucker.

The only U.S. location of puckering nightshade is high in the Chisos Mountains in Big Bend National Park, where it is closely monitored by the National Park Service. The herb is more common at moist, shady, rocky, wooded sites in northern Mexico (from Chihuahua to Nuevo León and south to Zacatecas) and farther south in Hidalgo and Oaxaca.

Puckering nightshade spills across the talus slopes of Mount Emory, where it blooms in summer and fall depending on rainfall. The photograph shows this rare perennial blooming in late July near the top of Mount Emory.

Note: The genus *Nectouxia* is named for French botanist Hyppolite Nectoux, who traveled with Napoleon to Egypt and published a chronicle of the expedition in 1808. The species name *formosa* in Latin is "beautifully formed." Both the genus and species were collected by Baron von Humboldt and Aimé Bonpland in about 1803 near Real del Monte, a silver mining town in the Mexican state of Hidalgo, and described by Carl Kunth. After describing Humboldt and Bonpland plants in Paris from 1815 to 1828, Kunth moved his large herbarium to Berlin and became assistant director of the Berlin Botanical Garden.

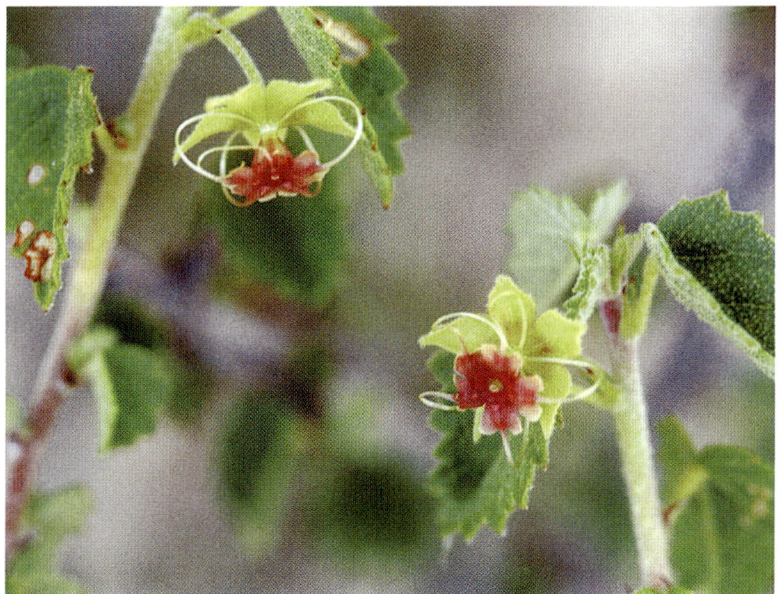

Dense Ayenia

Ayenia microphylla

Dense ayenia, or yerba del cáncer, is an often densely branched, low subshrub seldom two feet high. The rather thick, white-hairy stems branch mostly from the base and arch outward. Round or broadly egg-shaped, the leaves droop noticeably from long, downy stalks. Up to one inch long, the blades are coarsely toothed and coated with felty hairs, especially on the lower surface.

Blooming primarily in summer and fall, dark red or maroon flowers less than one-eighth inch across form on slender, hairy stalks. Pendent like the leaves, the curious flowers have five hoodlike petals, each with two rounded lobes at the tip and an arching, threadlike claw. With five greenish-yellow, pointed sepals above the petals, the blooms seem to dangle like tiny parachutes.

This weak shrub frequents desert scrub habitats and dry rocky and gravelly slopes at elevations below forty-four hundred feet. In Texas dense ayenia is heavily concentrated in southern Brewster and Presidio counties and is sparse elsewhere in the Trans-Pecos. Its range also spans portions of New Mexico, southern Arizona, and northern Mexico (from eastern Chihuahua to Coahuila and south to northern Durango and Zacatecas).

In the park dense ayenia is seen, rather infrequently, in the Dead Horse Mountains (at Boquillas Canyon, near Hot Springs and the tunnel, and at McKinney Spring) and the outskirts of the Chisos (near Pine Canyon, Burro Mesa, and Alamo Spring).

Note: In 1756 Carolus Linnaeus named the genus *Ayenia* for Louis de Noailles, the Duc d'Ayen. Noailles, pupil of botanist Bernard de Jussieu, was instrumental in Jussieu's appointment as designer of Louis XV's botanical garden near Versailles, Petit Trianon. The species name *microphylla* refers to the "small leaves." This species was collected by Charles Wright near El Paso in 1852 and described by Asa Gray that same year.

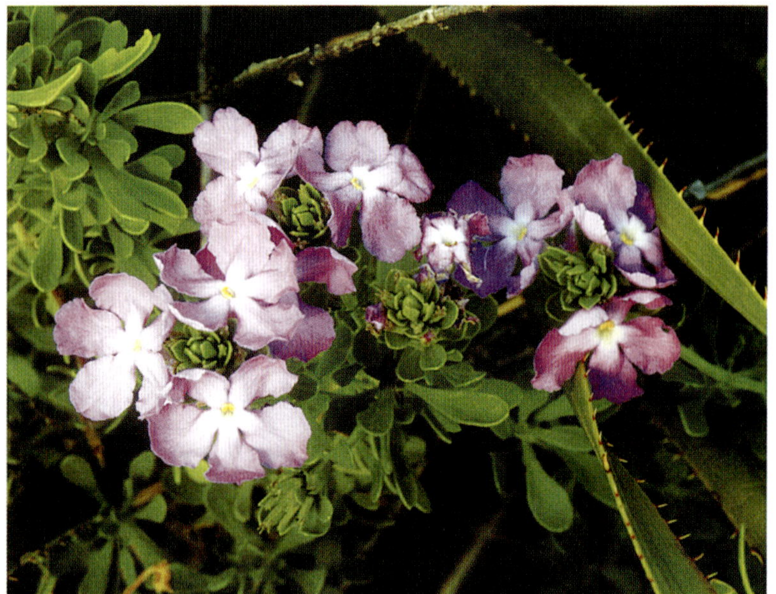

Spoonleaf Bouchea

Bouchea spathulata

Spoonleaf bouchea is an upright, extensively branched shrub up to two feet high and wide, with sturdy, very leafy, and densely hairy branches. Narrowly spatula-shaped and tapered at the base, the fleshy leaves can easily reach one inch in length. The thick blades are often clustered in groups of two or three and tightly crowded along the stems.

Lavender flowers as much as one inch wide appear in groups of three in compact spikes. Each flower is a slender tube which abruptly flares into five irregular, wavy lobes. The flower is supported at the base by prominent, reduced leaves or bractlets. Look for the blooms from late May through summer and occasionally into fall. The dry, oblong fruit splits into two one-seeded nutlets (mericarps) at maturity.

Spoonleaf bouchea sprawls over dry, rocky hillsides and gravelly slopes and also in canyons of desert mountains in limestone soil. Classified as critically imperiled in Texas (NatureServe 2005), it is known exclusively from the Dead Horse Mountains of southern Brewster County and parts of Coahuila and Chihuahua, Mexico.

In the park this shrub appears at widely separated locations from Tornillo Creek east to the mouth of Heath Canyon. It has been recorded in the hills above Boquillas Canyon, along the Old Ore Road, and in Telephone Canyon. I photographed this member of the vervain (Verbenaceae) family at the entrance to Passionflower Canyon, where a healthy population is flourishing.

Note: In 1832 Adelbert von Chamisso, at the Royal Botanical Garden in Berlin, named this genus for his coworkers, the German botanists and gardeners Carl and Peter Bouché. The species name *spathulata* (spatulate) refers to the leaves. This species was collected by Charles Parry (with Bigelow, Wright, and Schott) on the U.S.-Mexico boundary survey in "the Great Cañon of Mount Carmel, Rio Grande" and described by John Torrey in 1859.

Mejorana

Lantana macropoda

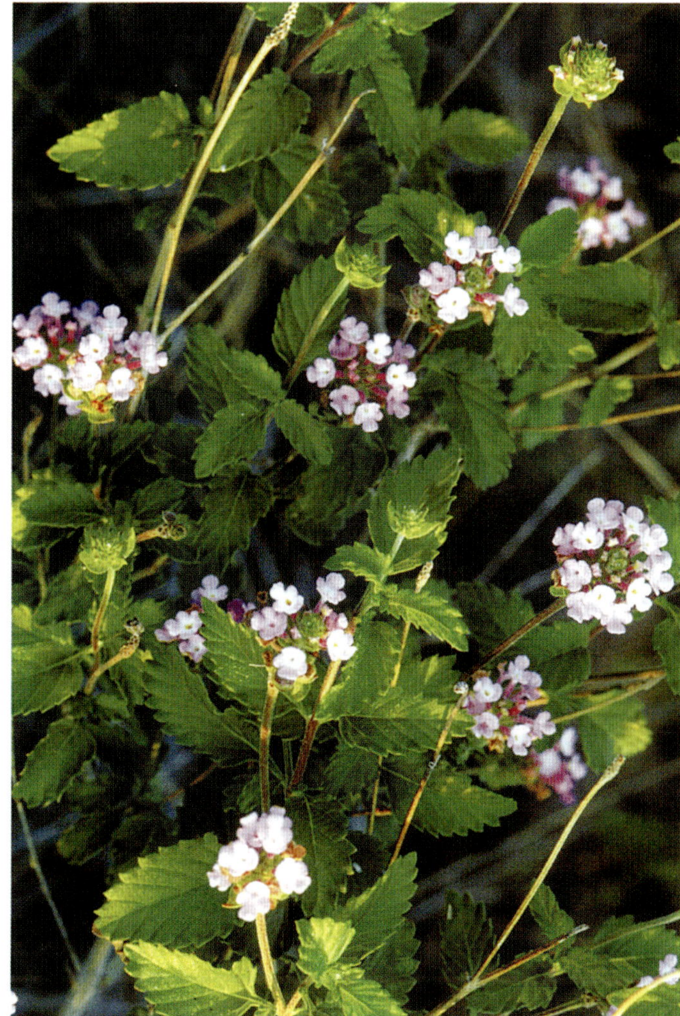

Also known as desert lantana, veinyleaf lantana, hierba negra, and yerba del Cristo, mejorana is an erect, aromatic shrub two to four feet high with many slender gray branches. Egg-shaped to oblong, the paired leaves are densely hairy and pointed at the tip, with sharp, broad, triangular teeth along the margins. These leaves may reach one and one-half inches long.

In spring and also in summer and fall after rains, compact flower clusters appear on elongated stalks up to four inches long. A single headlike cluster contains many irregularly shaped flowers in varying shades of pink, white, lavender, and purple, some with yellow at the throat. Each flower consists of a very slender tube that opens abruptly into a border (limb) with four broadly rounded, irregular lobes.

Mejorana fills dry rocky and gravelly hills and flats, desert washes, and roadsides, usually in full sun. In Texas this shrub is widely distributed in the southeastern Trans-Pecos, along the Rio Grande from Presidio to Del Rio. It also shows up in southern New Mexico, southern Arizona, and northern Mexico (from Sonora to Tamaulipas south to San Luis Potosí). A related, often considered synonymous species, *Lantana achyranthifolia,* ranges from South and Central America through Mexico into the Rio Grande Plains.

In the park mejorana is abundant in parts of the Chisos foothills near Lone and Nugent mountains, in Pine Canyon, and at Paint Gap and Oak Creek. It is also present near Persimmon Gap to the north, Hot Springs to the east, and Smoky Creek to the south.

Note: The genus name *Lantana* is Latin for *Viburnum,* a genus with similar flowers. The species name *macropoda* (large-stalked) refers to the long flower stalks. This species was collected by Charles Wright in 1849 and described by John Torrey in 1859. The *achyranthifolia* species was described in 1829 in a herbarium catalog of the Museum of Natural History in Paris. The description is attributed to René Desfontaines, museum director and author of *Flora Atlantica* (1798–99), an account of more than fifteen hundred northern Africa species previously unknown to science.

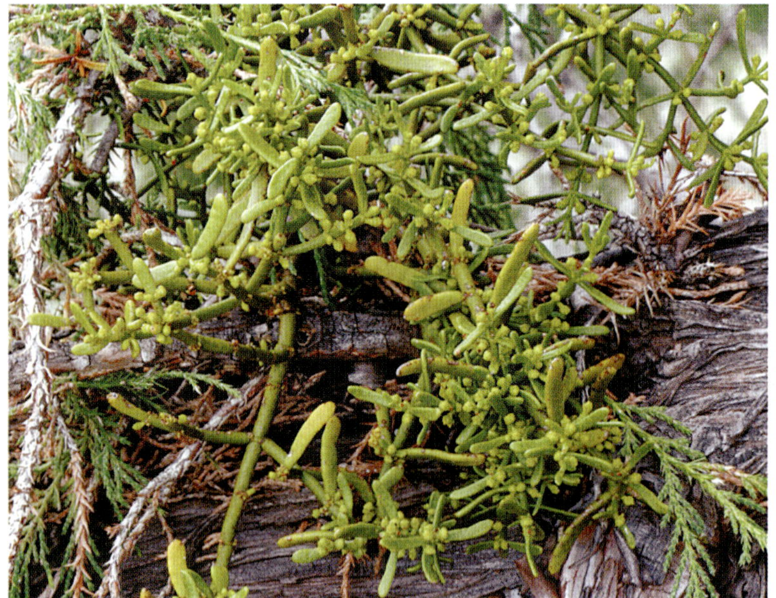

Rough Mistletoe

Phoradendron hawksworthii

Also known as hawksworth mistletoe, pink tree-thief, muerdago, and injerto, rough mistletoe is parasitic on juniper trees. This evergreen, with jointed, smooth cylindrical stems, embeds its roots in juniper branches to obtain nutrients. Fleshy and stemless, the one-inch-long, paired leaves are linear and flattened on the upper surface with a rounded tip.

These parasites may ultimately kill the junipers they infect, but the parasitic process does have some beneficial effects. Mistletoe attracts new bird species to a tree: the birds eat and disperse mistletoe and juniper seeds, encouraging propagation of the tree.

Embedded in thin, cylindrical spikes, the yellow unisexual flowers are only one-sixteenth inch wide. The male spike has six to ten minute flowers, the female spike two flowers. In this species, the spikes usually consist of a single segment or joint. The scalelike flowers bloom in summer. Up to one-fourth inch wide, the smooth fruits are semitranslucent, white to pink berries (fleshy, one-seeded drupes).

Rough mistletoe is dispersed across the Trans-Pecos from El Paso to Del Rio and into the western Edwards Plateau. It also enters juniper-pinyon woodlands in southern New Mexico and Coahuila, Mexico. In the park, look for this scarce mistletoe in Green Gulch and the Basin. Globally and in Texas it is classified as vulnerable throughout its range.

Note: The genus name *Phoradendron* may be from the Greek *phor* (thief) and *dendron* (tree), or "tree thief," because mistletoes rob nutrients from trees. In 1970 Utah botanist and mistletoe specialist Delbert Wiens named this species (as *Phoradendron bolleanum* ssp. *hawksworthii*) for his friend Frank Hawksworth, who collected it in the Chisos Mountains near the Basin campground in 1967. Hawksworth, an expert on dwarf mistletoes, was a plant pathologist for the Rocky Mountain Research Station in Fort Collins, Colorado.

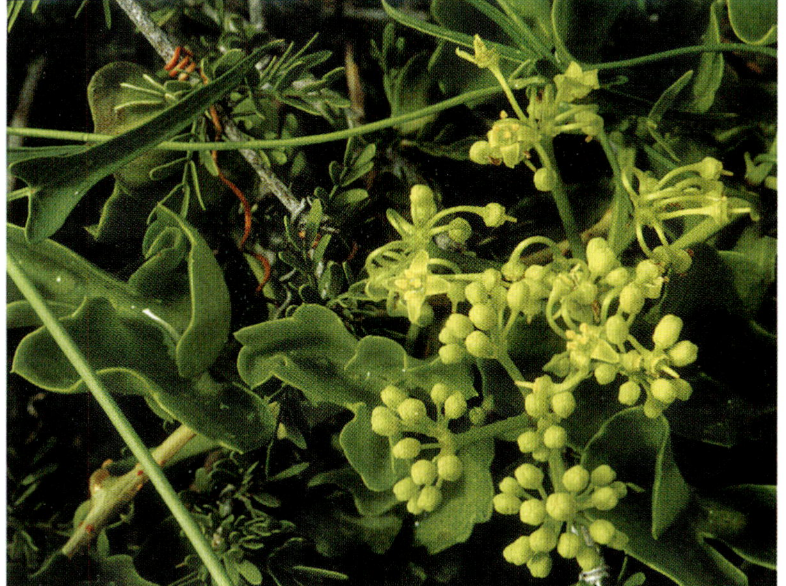

Ivy Treebine

Cissus trifoliata

Also known as marine ivy, cow-itch, and hierba del buey, ivy treebine is a sturdy perennial vine with climbing tendrils. The warty stems, growing from a tuberous base, can reach a length of thirty feet. Highly variable, the leaves are at first thick and succulent; later they dry and fall. About as long as they are wide, they may be scalloped and wavy along the margins, three-lobed, or divided into three distinct leaflets.

Blooming from May to September, greenish-yellow flowers are borne in rounded clusters. Each small flower consists of four petals, which sometimes spread sharply backward, and conspicuous yellow anthers. The fruits mature to black-colored berries.

Ivy treebine occupies diverse habitats: dense thickets, open woodlands, sandy banks, and saline soils. Its distribution stretches from Florida to Arizona, north to Kansas and Missouri, and south to Mexico and the Caribbean. The vine spreads across much of Texas from the Lower Rio Grande valley and Gulf Coast to Oklahoma and from the Hill Country west to Fort Davis and Presidio in the Trans-Pecos.

In the park ivy treebine thrives at springs (Fresno, Ward, Mule Ears) and along creeks (Alamo, Smoky). It has also been reported near the Rosillos Mountains, at Sam Nail Ranch, and on the east slopes of Crown Mountain. Superb examples of this robust vine climb over trees and shrubs at the mouths of Blue and Smoky creeks.

Note: The genus name *Cissus,* from the Greek *kissos* (ivy), refers to the twining habit. The species name *trifoliata* connotes "three-leaved." This species was discovered by scientist and collector Sir Hans Sloane on his 1687–89 Jamaican voyage and documented in his two-volume natural history (1707, 1725). Carolus Linnaeus, professor of medicine at the University of Uppsala, used Sloane's text and drawings to describe this species in his *Species Plantarum* in 1753.

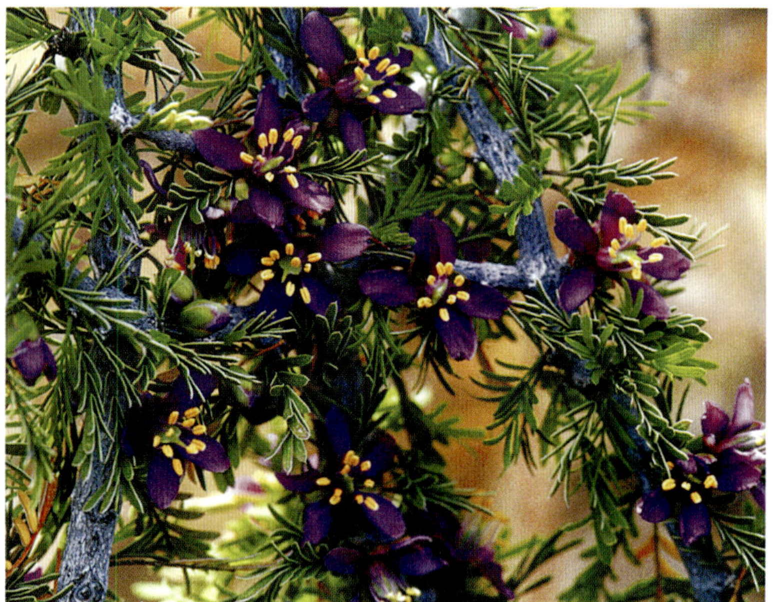

Guayacán

Guaiacum angustifolium

GUAYACÁN is a dense evergreen shrub or small tree with an obvious trunk and short, thick branches. Very scaly gray or black bark covers the stubby, sometimes gnarled branches. Usually two to eight feet high, this shrub may form dense, brushy thickets many feet across. The compound, dark green leaves, seldom more than one inch long, are often tightly bundled along the branches. Each leaf has four to eight pairs of narrowly oblong leaflets. To preserve moisture, the leathery leaflets close at night and partly fold in hot sun.

Blooming profusely in spring and sometimes into fall, vivid purple flowers form singly or in clusters on the branchlets. Each fragrant flower has five petals set off by ten protruding yellow anthers. The fruits are as attractive as the flowers. Each fruit is a tough, leathery capsule, usually two-lobed and heart-shaped, with a thin wing on each side and a beaked tip. The capsule splits open to reveal two shiny red, fleshy seed covers (arils).

Guayacán is situated in arroyos and brushland by streams and desert springs. The shrub occurs only in Texas in the United States, from the southern Trans-Pecos to Austin and south to Brownsville, especially in a band along the Rio Grande from Presidio to the coast. It inhabits northern Mexico from Chihuahua to Tamaulipas. Guayacán is pervasive in much of the park, from desert washes to foothills: at Persimmon Gap, Boquillas Canyon, Hot Springs, and Rooney's Place and near Burro Spring and Terlingua Abaja.

Note: Guayacán was known as soapbush in Mexico because the root bark was used to wash wool. The genus name *Guaiacum* is from a Native American word, *guaiac,* for a South American species (*G. officinale*). The species name *angustifolium* denotes "narrow-leaved." This species was described by George Engelmann in 1848. The type specimens were collected by Josiah Gregg in Coahuila, Mexico, and Ferdinand Lindheimer on the Pedernales River in the Texas Hill Country.

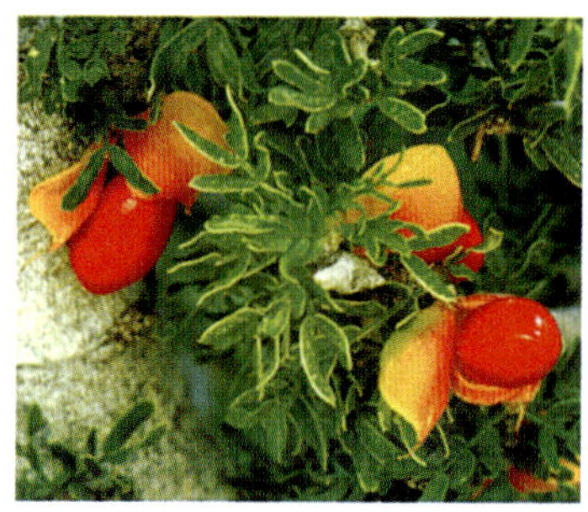

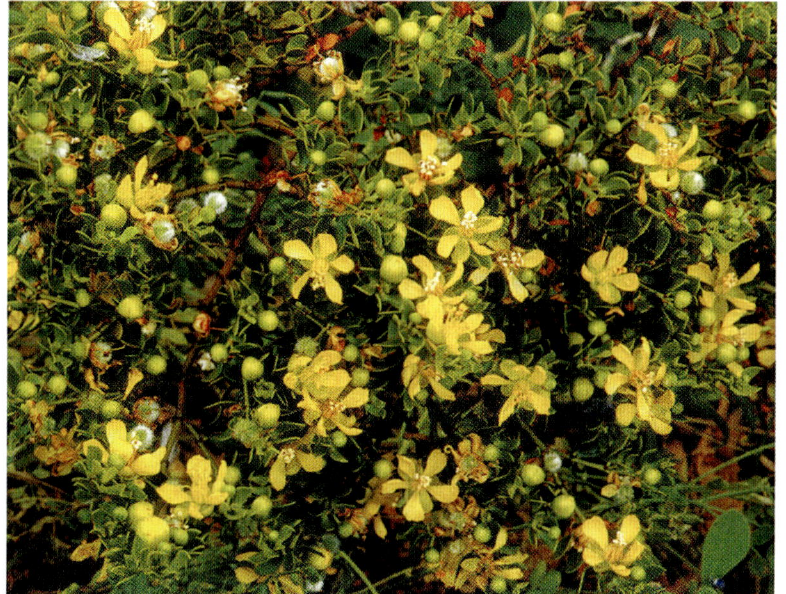

Creosote Bush

Larrea tridentata

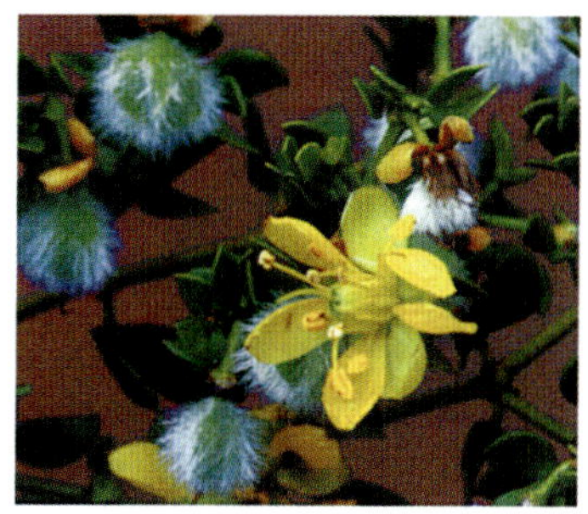

Also known as gobernadora, hediondilla (little stinker), and guame, creosote bush produces a heady creosote odor, especially after rains. Probably the most common shrub of southwestern deserts, it is typically three to five feet high with many slender, gray or black stems that radiate from the base and sometimes arch and droop. Resins coating the stems and leaves produce the creosote smell, reflect light, and prevent water loss. The dark green or yellowish-green leaves are leathery and waxy, each with usually two small leaflets united at the base. They contain chemicals that deter leaf eaters.

This evergreen is a quintessential study in desert survival. The crafty shrub has a shallow, extensive root system that collects rainfall and also produces a toxin that deters competing plants. It is one of the longest-living plants, with some shrubs dated to thousands of years in age.

Solitary, bright yellow, five-petaled flowers appear in spring and often much later during the summer monsoon season. They develop into round, five-lobed, hairy fruits with a fuzzy or fluffy appearance. At maturity, the fruits split into five one-seeded nutlets.

This ubiquitous desert shrub prefers desert flats in sandy, gravelly, and clay soils from California, Utah, and Nevada to Texas and northern Mexico (from Baja California to Tamaulipas and south to Zacatecas and San Luis Potosí). It is an occupant of the Sonoran and Mojave as well as Chihuahuan deserts. In Texas creosote grows throughout the Trans-Pecos and much of the Rio Grande Plains. At lower elevations in the park, it is nearly omnipresent, from Persimmon Gap and Dagger Flat to Hot Springs, Smoky Creek, and the Maverick Badlands.

Note: The genus *Larrea* is named for Juan Antonio de Larrea (1730–1803), a prominent patron of science in Valladolid, Spain. The species name *tridentata* (three-toothed) may refer to the three-angled seeds. This species was described in 1824 by Augustin de Candolle (as *Zygophyllum tridentatum*) from a drawing by Martín de Sessé y Lacasta and José Mociño during their botanical expedition to New Spain (1788–1803).

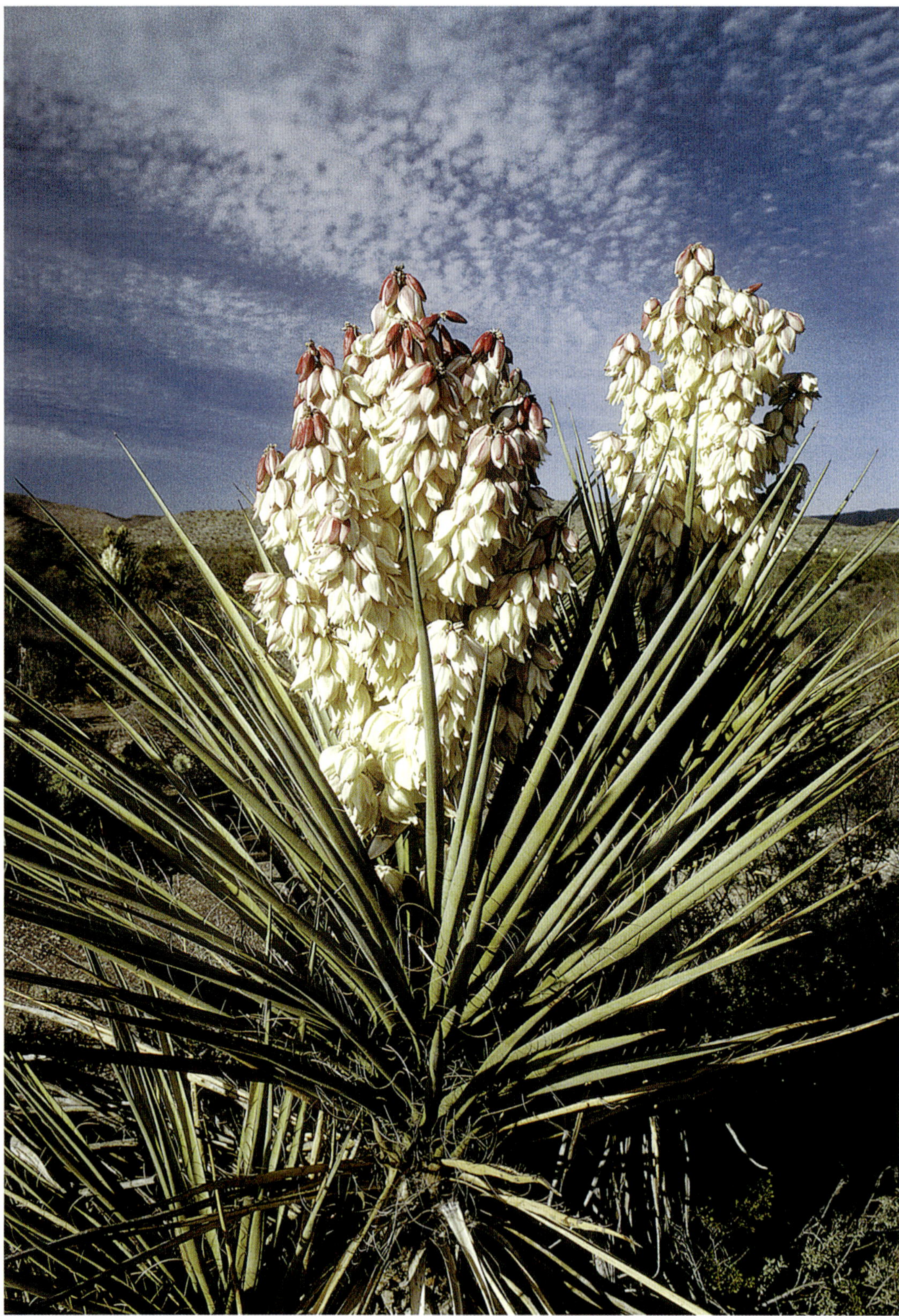

Spanish dagger

Appendixes

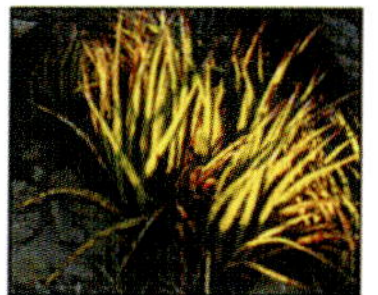

APPENDIX A

Critically Imperiled, Imperiled, and Vulnerable Plants

	Common Name	*Botanical Name*	*Global Status*	*Texas Status*	**Source*
1	Chaffey's pincushion	*Coryphantha chaffeyi*	imperiled	critically imperiled	NC
2	Duncan's pincushion	*Coryphantha duncanii*	critically imperiled/ imperiled	critically imperiled/ imperiled	NC
3	Whiskerbrush pincushion	*Coryphantha ramillosa*	imperiled/ vulnerable	imperiled/ vulnerable	NC
4	Silverlace cactus	*Coryphantha sneedii* var. *albicolumnaria*	imperiled/ vulnerable	imperiled/ vulnerable	NC
5	Chisos hedgehog	*Echinocereus chisoensis* var. *chisoensis*	critically imperiled	critically imperiled	NC
6	Texas claret-cup	*Echinocereus coccineus* var. *paucispinus*	vulnerable	vulnerable	NC
7	Mariposa cactus	*Echinomastus mariposensis*	imperiled	imperiled	NC
8	Warnock's cactus	*Echinomastus warnockii*	vulnerable	vulnerable	NC
9	Boke's button cactus	*Epithelantha bokei*	vulnerable	vulnerable	NC
10	Big Bend cholla	*Opuntia imbricata* var. *argentea*	critically imperiled	critically imperiled	NC
11	Golden-spined prickly pear	*Opuntia azurea* var. *aureispina*	critically imperiled	critically imperiled	NC
12	Warnock justicia	*Justicia warnockii*	vulnerable	vulnerable	NC
13	Fleshy tidestromia	*Tidestromia carnosa*	vulnerable	imperiled	NC
14	Arizona cockroach plant	*Haplophyton crooksii*	secure	critically imperiled	NS

*NC = Nature Conservancy NS = NatureServe

	Common Name	Botanical Name	Global Status	Texas Status	*Source
15	Soft twinevine	*Sarcostemma torreyi*	apparently secure	vulnerable	NS
16	Sandlot brickellbush	*Brickellia lemmonii* var. *conduplicata*	apparently secure	vulnerable	NS
17	Veronicaleaf brickellbush	*Brickellia veronicifolia*	not ranked	critically imperiled	NS
18	Turner thistle	*Cirsium turneri*	vulnerable	vulnerable	NC
19	Chisos Mountain brickellbush	*Flyriella parryi*	vulnerable	imperiled	NC
20	Rayless rockdaisy	*Perityle aglossa*	vulnerable	vulnerable	NC
21	Twobristle rockdaisy	*Perityle bisetosa* var. *scalaris*	critically imperiled	critically imperiled	NC
22	Slimlobe rockdaisy	*Perityle dissecta*	imperiled	imperiled	NC
23	Heartleaf rockdaisy	*Perityle parryi*	apparently secure	vulnerable	NS
24	Roundleaf stevia	*Stevia ovata* var. *texana*	secure	critically imperiled	NS
25	Threeflower goldenweed	*Xylothamia triantha*	apparently secure	critically imperiled	NS
26	Big Bend hophornbeam	*Ostrya virginiana* var. *chisosensis*	imperiled	critically imperiled	NC
27	Cory cryptantha	*Cryptantha palmeri*	apparently secure	vulnerable	NS
28	Green gromwell	*Lithospermum viride*	apparently secure	imperiled	NS
29	Texas largeseed bittercress	*Cardamine macrocarpa* var. *texana*	imperiled	imperiled	NC
30	Lyreleaf twistflower	*Streptanthus carinatus* ssp. *carinatus*	vulnerable	vulnerable	NC
31	Cutler twistflower	*Streptanthus cutleri*	imperiled	imperiled	NC
32	Texas thelypody	*Thelypodium texanum*	vulnerable	vulnerable	NC
33	Bigpod bonamia	*Bonamia ovalifolia*	critically imperiled	critically imperiled	NC
34	Creeping rockvine	*Bonamia repens*	vulnerable	imperiled	NC
35	Blue morning glory	*Ipomoea lindheimeri*	apparently secure	vulnerable	NS
36	Havard stonecrop	*Sedum havardii*	imperiled	imperiled	NC
37	Wright stonecrop	*Sedum wrightii*	apparently secure	vulnerable	NS
38	Stonecrop	*Villadia squamulosa*	apparently secure	vulnerable	NS
39	Smooth bur cucumber	*Sicyos glaber*	vulnerable	vulnerable	NC
40	Arizona cypress	*Cupressus arizonica*	apparently secure	critically imperiled	NS
41	Three-tongue spurge	*Chamaesyce chaetocalyx* var. *triligulata*	critically imperiled	critically imperiled	NC
42	Zigzag croton	*Croton pottsii* var. *thermophilus*	imperiled	critically imperiled	NC
43	Littleleaf brongniart	*Brongniartia minutifolia*	imperiled	critically imperiled	NC

	Common Name	Botanical Name	Global Status	Texas Status	*Source
44	Parry caesalpinia	*Pomaria melanosticta*	apparently secure	vulnerable	NS
45	Chisos oak	*Quercus graciliformis*	critically imperiled	critically imperiled	NC
46	Narrowleaf fendlerbush	*Fendlera linearis*	vulnerable	critically imperiled	NC
47	Mearns mockorange	*Philadelphus mearnsii*	apparently secure	imperiled	NS
48	Littleleaf mockorange	*Philadelphus microphyllus*	secure	critically imperiled	NS
49	Havard nama	*Nama havardii*	apparently secure	vulnerable	NS
50	Mat nama	*Nama torynophyllum*	apparently secure	critically imperiled	NS
51	Hairy hedeoma	*Hedeoma mollis*	vulnerable	vulnerable	NC
52	Mountain sage	*Salvia regla*	apparently secure	critically imperiled	NS
53	Stalkflower nesaea	*Nesaea longipes*	imperiled/vulnerable	imperiled	NC
54	Purple gaymallow	*Batesimalva violacea*	imperiled	critically imperiled	NC
55	Littleleaf moonpod	*Acleisanthes parvifolia*	vulnerable	vulnerable	NC
56	Hidalgo ladies tresses	*Deiregyne confusa*	vulnerable	not ranked	NS
57	Giant helleborine	*Epipactis gigantea*	vulnerable	vulnerable	NC
58	Giant coral root	*Hexalectris grandiflora*	apparently secure	imperiled	NS
59	Curly coral root	*Hexalectris revoluta*	critically imperiled/ imperiled	critically imperiled	NC
60	Texas purplespike	*Hexalectris warnockii*	imperiled/vulnerable	imperiled	NC
61	Wendt's malaxis	*Malaxis wendtii*	imperiled	critically imperiled	NC
62	Havard ipomopsis	*Ipomopsis havardii*	vulnerable	vulnerable	NC
63	Papercup loeselia	*Loeselia greggii*	secure	critically imperiled	NS
64	Chisos Mountain buckwheat	*Eriogonum hemipterum* var. *hemipterum*	vulnerable	imperiled	NC
65	Trans-Pecos cliff brake	*Pellaea ternifolia*	secure	imperiled	NS
66	Longspur columbine	*Aquilegia longissima*	vulnerable	imperiled	NC
67	Havard plum	*Prunus havardii*	vulnerable	vulnerable	NC
68	Palo prieto	*Vauquelinia corymbosa* ssp. *angustifolia*	apparently secure	critically imperiled	NS
69	Silvery wolfberry	*Lycium puberulum* var. *berberioides*	vulnerable	vulnerable	NS
70	Spoonleaf bouchea	*Bouchea spathulata*	apparently secure	critically imperiled	NS
71	Mejorana	*Lantana macropoda*	secure	critically imperiled	NS
72	Rough mistletoe	*Phoradendron hawksworthii*	vulnerable	vulnerable	NC

*NC = Nature Conservancy NS = NatureServe

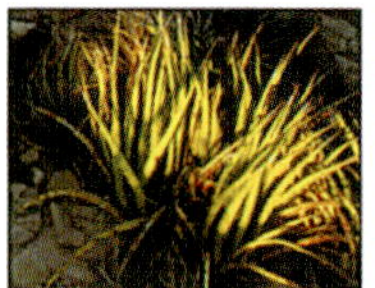

APPENDIX B

Selected Plants by Park Location

Location	*Selected Plants*
Alamo Creek	Ivy treebine, mat nama, thickleaf drymary
Alamo Spring to Burro Spring	Dense ayenia, trumpetflower
Avery Canyon	Bearded dalea, lyreleaf twistflower
Bailey Peak	Crestrib morning-glory, fascicled bluet, Mexican clammyweed, pearl netleaf milkweed, showy menodora, scarlet ladies tresses, Texas purplespike, Trans-Pecos carlowrightia
Black Dike	Candle cholla
Blue Creek Canyon	Arizona carlowrightia, Arizona cockroach plant, Chisos red oak, dwarf anisacanth, fascicled bluet, flower of stone, Havard penstemon, Havard plum, palo prieto, Parry ruellia, Roemer acacia, southwestern chokecherry, talayote, Texas purplespike
Blue Creek Canyon (Upper)	Chaffey's pincushion cactus, Chisos oak, spreading snakeherb
Blue Creek (mouth)	Earlobe mustard, ivy treebine, mat nama
Bois d'Arc Spring	Silver ponyfoot
Boot Canyon	Alligator juniper, Arizona cypress, blue morning-glory, bristly bedstraw, creeping cliff brake, curly coral root, drooping juniper, eggleaf silktassel, Gray bean, limpia seymeria, littleleaf mockorange, Mexican squawroot, mountain sage, roundleaf stevia, scarlet bouvardia, scarlet ladies tresses, Texas largeseed bittercress, Texas purplespike, veronicaleaf brickellbush, white giant hyssop, Wright stonecrop
Boot Creek (above Boot Spring)	Coahuila scrub oak, littleleaf mockorange, mountain sage, smooth bur cucumber, stonecrop, Texas madrone, Trans-Pecos cliff brake, Wendt's malaxis, white giant hyssop
Boot Spring (or nearby)	Chaffey's pincushion cactus, Chisos prickly pear, Coahuila scrub oak, creeping cliff brake, Gray bean, Havard agave, licorice marigold, littleleaf mockorange, Mexican buckeye, Mexican squawroot

Location	*Selected Plants*
Boquillas Canyon (or nearby)	Arizona carlowrightia, blind prickly pear, Boquillas silverleaf, Cory Dutchman's pipe, Cutler twistflower, dense ayenia, desert myrtlecroton, flower of stone, guayacán, limerock brookweed, Mexican navelseed, narrowleaf moonpod, Parry caesalpinia, shrubby tidestromia, silverlace cactus, soft twinevine, spiny-fruited prickly pear, spoonleaf bouchea, Texas thelypody, three-tongue spurge, Trans-Pecos senna, Wright dalea, yellow rocknettle, zigzag croton
Boquillas to Mariscal Canyon	Texas false agave, woollyflower spurge
Boquillas Flats	Leafy heliotrope, strawberry hedgehog cactus
Boquillas Flats (hills above)	Zigzag croton
Boulder Meadow (nearby)	Bigelow bristlehead
Burro Mesa	Chino grama, dense ayenia, dwarf fairy duster, green condalia, heartleaf rockdaisy, narrowleaf fendlerbush, New Mexico dalea, ocotillo, shaggy false nightshade, Texas rainbow, tree cholla, trumpetflower
Burro Spring (or nearby)	Cory Dutchman's pipe, guayacán, narrowleaf moonpod, Parry ruellia
Campground (or Pulliam) Canyon	Chisos Mountain brickellbush, creeping cliff brake, eggleaf silktassel, hairy hedeoma, Mearns mockorange, Mexican buckeye, Mexican starwort, mountain sage, pearl netleaf milkweed, rock betony, Texas purplespike, Trans-Pecos cliff brake
Casa Grande	Alligator juniper, bigelow bristlehead, Chisos Mountain brickellbush, creeping cliff brake, eggleaf silktassel, hairy hedeoma, Havard giant hyssop, mountain sage, rock betony, splitleaf brickellbush, stemless aletes, tailleaf pericome, Texas milkweed, Trans-Pecos spiderwort, veronicaleaf brickellbush
Castolon (or nearby)	Margined rockdaisy, rusty hedgehog cactus, Turner mimosa
Castolon to Santa Elena Canyon	Bearded dalea, bladder sage, fleshy tidestromia, slimlobe globeberry, three-flower goldenweed, woollyflower spurge
Castolon to Smoky Creek	Bearded dalea, bladder sage, Cory cryptantha
Castolon to Johnson Ranch	Berlandier flax, woollyflower spurge, yellow rocknettle
Cattail Canyon (Upper)	Bigelow beggarticks, Coahuila scrub oak, desert ceanothus, licorice marigold, limpia seymeria, Mexican squawroot, New Mexico ponyfoot, roundleaf stevia, roughstem hawkweed, scarlet ladies tresses, Texas flatsedge, Texas largeseed bittercress, twistleaf paintbrush, veronicaleaf brickellbush, viscid stevia
Chilicotal Spring	Lyreleaf twistflower
Chilicotal Mountain (or nearby)	Candelilla, Dutchman's breeches, glandleaf milkwort, narrowleaf fendlerbush, rock milkwort, Wright dalea
Chinese Wall Trail	James nailwort, littleleaf mockorange, roundleaf stevia, veronicaleaf brickellbush
Chisos Basin	Alligator juniper, Bigelow beggarticks, black dalea, blue morning-glory, Chisos Mountain buckwheat, Chisos prickly pear, Chisos pricklypoppy, Chisos red oak, creeping cliff brake, crestrib morning-glory, desert ceanothus, drooping juniper, eggleaf silktassel, fascicled bluet, hairy hedeoma, Havard agave, Havard plum, heartleaf rockdaisy, Mexican buckeye, Mexican pinyon pine, mountain sage, New Mexico ponyfoot, palmleaf thoroughwort, palo prieto, pearl netleaf milkweed, rough mistletoe, sandlot brickellbush, scarlet bouvardia, soft twinevine, splitleaf brickellbush, spreading snakeherb, Texas kidneywood, Texas madrone, Texas milkweed, Trans-Pecos spiderwort, Wright ageratina

Location	*Selected Plants*
Chisos Foothills	Allthorn, Arizona carlowrightia, beargrass, black dalea, broom milkwort, catclaw cactus, cob cactus, desert sumac, knotweed leafflower, leatherstem, lechuguilla, naked brittlestem, New Mexico dalea, orange flameflower, Parry ruellia, plume tiquilia, Pringle swallow-wort, propellerbush, purple sage, rusty hedgehog cactus, Trans-Pecos carlowrightia, tree cholla, trumpetflower
Chisos Mountains	Arizona cypress, black dalea, bottomwhite rocktrumpet, bristly bedstraw, Chisos oak, Chisos red oak, Havard stonecrop, Hidalgo ladies tresses, Michoacán ladies tresses, roughstem hawkweed, stemless aletes, stonecrop, Wright stonecrop
Chisos Mountains (west slopes)	Crested coral root, giant helleborine, longspur columbine, tarbush, Texas purplespike, Trans-Pecos carlowrightia, Warnock justicia, woven-spine pineapple cactus
Chisos Pens (or nearby)	Bearded dalea, littleleaf moonpod, margined rockdaisy, rock milkwort, rough mortonia, silvery wolfberry, stalkflower nesaea
Colima Trail	Curly coral root, drooping juniper
Cottonwood Campground (nearby)	Earlobe mustard, thickleaf drymary
Croton Spring (or nearby)	Narrowleaf moonpod, Trans-Pecos senna
Crown Mountain	Big Bend hophornbeam, narrowleaf fendlerbush, New Mexico ponyfoot, orange flameflower, palo prieto, pearl netleaf milkweed, silver ponyfoot, silvery wolfberry, soft twinevine, spreading snakeherb
Dagger Flat	Creosote bush, crucifixion thorn, desert olive, feather dalea, giant dagger, glandleaf milkwort, knotweed leafflower, manyflowered broomrape, slimlobe globeberry, tarbush, Thompson yucca, Warnock's cactus
Dagger Flat (above)	Guayule, heath cliffrose, Texas cone cactus
Dagger Flat (road to)	Allthorn, Cutler twistflower, littleleaf leadtree
Dead Horse Mountains	Big Bend devil cholla, Boke's button cactus, Boquillas silverleaf, candelilla, catclaw cactus, desert olive, Dutchman's breeches, glandleaf milkwort, golf ball cactus, leafy heliotrope, living rock cactus, Mariposa cactus, Mexican navelseed, Parry caesalpinia, plume tiquilia, round copperleaf, sea urchin cactus, shaggy false nightshade, shrubby tidestromia, Spanish dagger, stemless perezia, Thompson yucca, Warnock justicia, Warnock's cactus, woolly butterflybush, Wright dalea, zigzag croton
Dead Horse Mountains (northern)	Giant dagger
Dead Man's Cut	Warnock justicia
Desert-Mountain Overlook	Big Bend bluebonnet
Devil's Den	Stemless perezia, Texas kidneywood
Dodson Trail (west end)	Arizona cockroach plant, creeping rockvine, Mexican clammyweed, Parry caesalpinia
Dog Canyon	Cutler twistflower, desert olive, Mexican navelseed, narrowleaf moonpod, stemless perezia, Texas kidneywood
Dog Canyon (nearby)	Horse-crippler cactus, shaggy false nightshade, Texas selenia
Dog Flats	Glandleaf milkwort, greeneye heliotrope, Gregg keelpod, Havard ipomopsis, Heyder's pincushion cactus, horse-crippler cactus, leafy heliotrope, mesa greggia, naked brittlestem, New Mexico dalea, plume tiquilia, scurfy mallow, shaggy stenandrium, silvery wolfberry, soft heliotrope, tarbush, Texas selenia, thickleaf drymary

Location	Selected Plants
Dominguez Spring (or nearby)	Arizona carlowrightia, desert yaupon, Emory mimosa, orange flameflower, slimlobe globeberry, Texas false agave, woolly butterflybush
Dripping Spring	Emory mimosa
Dugout Wells (or nearby)	Allthorn, black dalea, desert sumac, Emory mimosa, mesa greggia, stinging cevallia
Elephant Tusk	Palo prieto
Emory Peak Trail	Big Bend hophornbeam, blue morning-glory, Chaffey's pincushion cactus, Chisos prickly pear, Chisos red oak, drooping juniper, licorice marigold, limpia seymeria, littleleaf mockorange, Mexican pinyon pine, Mexican starwort, mountain sage, puckering nightshade, rock betony, stemless aletes, tailleaf pericome, Texas flatsedge, Texas madrone, Texas milkweed, Trans-Pecos cliff brake, Trans-Pecos spiderwort, viscid stevia, Wright ageratina
Ernst Tinaja (or nearby)	Cory Dutchman's pipe, Cutler twistflower, Parry caesalpinia, Wright dalea, yellow rocknettle
Ernst Tinaja (above)	Creeping rockvine, woolly dogweed
Fisk Canyon	Berlandier lobelia
Fresno Creek	Berlandier lobelia, stemless perezia
Fresno Spring	Berlandier lobelia, ivy treebine, limerock brookweed, Mexican clammyweed, rock milkwort
Gano Spring (or nearby)	Allthorn, desert willow, Emory mimosa, margined rockdaisy, slimlobe globeberry, stinging cevallia
Glenn Draw (above)	Edwards Nicollet
Glenn Spring	Berlandier lobelia, Boquillas silverleaf, candelilla, desert willow, Edwards Nicollet, naked brittlestem, ocotillo, shaggy false nightshade, Texas false agave
Glenn Spring (nearby)	Dutchman's breeches, Parry caesalpinia, stalkflower nesaea, Texas kidneywood
Glenn Spring to Mariscal Mine	Crucifixion thorn
Government Spring (or nearby)	Cory cryptantha, Havard penstemon, ocotillo, Parry ruellia
Government Spring to Grapevine Hills	Cory cryptantha, Emory mimosa
Grapevine Hills (or nearby)	Broom milkwort, desert myrtlecroton, dwarf anisacanth, plume tiquilia, Havard plum, propellerbush, purple sage
Grapevine Spring	Desert willow, ocotillo
Gravel Pit	Big-needle pincushion cactus
Green Gulch	Alligator juniper, black dalea, Chisos red oak, desert sumac, desert yaupon, dwarf fairy duster, feather dalea, Havard agave, Havard penstemon, Havard plum, littleleaf leadtree, Mexican pinyon pine, Mexican starwort, nipple cactus, pretty dodder, rough mistletoe, scarlet bouvardia, soft twinevine, Texas claret-cup cactus, Texas kidneywood, Texas madrone, tree cholla
Green Gulch (Upper)	Bigelow bristlehead, Chisos prickly pear, drooping juniper, mountain sage, papercup loeselia, roughstem hawkweed, sandlot brickellbush, southwestern chokecherry, tailleaf pericome, Texas largeseed bittercress, Texas milkweed, veronicaleaf brickellbush
Hannold Draw (nearby)	Dutchman's breeches
Harte Ranch	Havard nama, Mexican clammyweed, pearl netleaf milkweed, Pringle swallowwort, scurfy mallow, slimlobe globeberry, soft heliotrope, stinging cevallia, thickleaf drymary, woolly dogweed

Location	Selected Plants
Hot Springs (or nearby)	Bearded dalea, Boquillas silverleaf, candelilla, dense ayenia, desert willow, Edwards Nicollet, golf ball cactus, green condalia, guayacán, hairy tubetongue, Havard nama, leatherstem, narrowleaf moonpod, shaggy false nightshade, shrubby tidestromia, spiny-fruited prickly pear, stemless perezia, stinging cevallia, Texas thelypody, Trans-Pecos senna, Wright dalea, yellow rocknettle
Hot Springs to San Vicente	Texas thelypody
Johnson Ranch (or nearby)	Chisos pricklypoppy, woollyflower spurge
Juniper Canyon	Crested coral root, black dalea, Chisos red oak, desert yaupon, drooping juniper, New Mexico dalea, silver ponyfoot, Texas kidneywood, Texas madrone
Juniper Spring (Lower)	Fascicled bluet, longpetal echeveria, Pringle swallow-wort, soft twinevine
Juniper Spring (Upper)	Chisos oak, Texas purplespike
K-Bar Ranch	Desert yaupon, slimlobe globeberry
La Clocha	Golf ball cactus
Laguna Meadow	Alligator juniper, blue morning-glory, Chisos pricklypoppy, Coahuila scrub oak, New Mexico ponyfoot, red cyphomeris, roughstem hawkweed, viscid stevia
Laguna Meadow (nearby)	Bottomwhite rocktrumpet
Laguna Meadow Trail	Black dalea, Chisos Mountain buckwheat, Chisos red oak, creeping cliff brake, desert ceanothus, drooping juniper, eggleaf silktassel, mountain sage, New Mexico ponyfoot, palmleaf thoroughwort, plateau rocktrumpet, roundleaf stevia, sandlot brickellbush, spreading snakeherb, Wright ageratina
Laguna West	Desert ceanothus
Lone Mountain (or nearby)	Boquillas silverleaf, Dutchman's breeches, Havard penstemon, mejorana, mesa greggia, purple sage, rock milkwort
Lost Mine Trail	Alligator juniper, Bigelow bristlehead, Chisos prickly pear, Coahuila scrub oak, creeping cliff brake, drooping juniper, eggleaf silktassel, hairy hedeoma, James nailwort, Mexican pinyon pine, mountain sage, papercup loeselia, roughstem hawkweed, sandlot brickellbush, Texas claret-cup cactus, Texas largeseed bittercress, Texas milkweed, veronicaleaf brickellbush
Maple Canyon	Chisos Mountain brickellbush, hairy hedeoma, Havard giant hyssop, longspur columbine, Texas largeseed bittercress, Turner thistle
Mariscal Mountain (or nearby)	Big Bend cholla, Big Bend devil cholla, blackbrush acacia, Boquillas silverleaf, candelilla, Chisos pricklypoppy, living rock cactus, mesa greggia, Potts' mammillaria cactus, rock milkwort, sea urchin cactus, Wright dalea
Marufo Vega Trail	Zigzag croton
Maverick Badlands	Fleshy tidestromia, littleleaf moonpod, margined rockdaisy, naked brittlestem, trumpetflower
McKinney Spring (or nearby)	Candelilla, dense ayenia, desert myrtlecroton, desert willow, Havard plum, heath cliffrose, limerock brookweed, narrowleaf moonpod, stemless perezia, Texas kidneywood, tubular slimpod, Warnock justicia, woolly butterflybush, Wright dalea, yellow rocknettle
Mesa de Anguila	Bladder sage, Cory cryptantha, Cory Dutchman's pipe, heath cliffrose, Mexican navelseed, Potts' mammillaria cactus, stinging cevallia, Texas false agave, tubular slimpod, Turner thistle

Location	*Selected Plants*
Mouse Canyon	Southwestern chokecherry
Mule Ears Spring Trail	Catclaw cactus, ivy treebine, limerock brookweed
Muskhog Spring (or nearby)	Candelilla, guayule, leafy heliotrope, round copperleaf, shrubby tidestromia
Northeast Rim Trail	Trans-Pecos cliff brake
Nugent Mountain (nearby)	Boquillas silverleaf, Dutchman's breeches, Havard penstemon, manyflowered broomrape, mejorana
Nugent Mountain to Glenn Spring	Desert yaupon
Oak Creek Canyon	Bigelow beggarticks, Chisos oak, Chisos red oak, desert olive, desert sumac, feather dalea, Havard penstemon, Havard plum, knotweed leafflower, mejorana, palo prieto, Parry ruellia, plateau rocktrumpet, Roemer acacia, scarlet bouvardia, splitleaf brickellbush, talayote, tarbush, Texas kidneywood, Texas madrone, Trans-Pecos carlowrightia, trumpetflower
Oak Creek Canyon (Lower)	Arizona snakecotton, crested coral root, crestrib morning-glory, dwarf anisacanth, New Mexico ponyfoot, woven-spine pineapple cactus
Oak Spring	Bearded dalea, giant fishhook cactus, limerock brookweed, showy menodora, soft twinevine
Old Maverick Road	Edwards Nicollet
Old Ore Road	Creeping rockvine, golf ball cactus, guayule, spoonleaf bouchea, Warnock justicia, zigzag croton
Old Ore Road (north end)	Javelina bush
Onion Spring	Arizona snakecotton, James nailwort, leafy heliotrope, orange flameflower, stemless perezia
Paint Gap Hills (or nearby)	Black dalea, crucifixion thorn, Emory mimosa, hairy tubetongue, mejorana, purple sage, trumpetflower
Panther Canyon	Berlandier flax, talayote
Panther Junction (nearby)	Dutchman's breeches, nipple cactus
Panther Junction to Basin turnoff	Pretty dodder
Panther Junction to Boquillas Canyon	Blind prickly pear
Panther Junction to Persimmon Gap	Big Bend bluebonnet, escobilla butterflybush, hairy tubetongue, Torrey ephedra
Panther Pass	Chisos prickly pear, papercup loeselia
Panther Spring (nearby)	James nailwort
Passionflower Canyon	Baccharisleaf penstemon, Cutler twistflower, heath cliffrose, rayless rockdaisy, red cyphomeris, spoonleaf bouchea, spreading-lobe passionflower, Turner thistle, twobristle rockdaisy
Persimmon Gap (or nearby)	Berlandier flax, big-needle pincushion cactus, cob cactus, desert olive, feather dalea, guayacán, leatherstem, lyreleaf twistflower, manyflowered broomrape, mejorana, mesa greggia, plume tiquilia, round copperleaf, rusty hedgehog cactus, shaggy stenandrium, strawberry cactus
Pine Canyon (or nearby)	Big Bend hophornbeam, Chisos red oak, Gray bean, hairy hedeoma, Havard plum, longspur columbine, Mexican buckeye, mountain sage, palo prieto, plateau rocktrumpet, rock betony, Roemer acacia, roundleaf stevia, scarlet bouvardia, southwestern chokecherry, Texas kidneywood, Texas largeseed bittercress, Texas madrone, Texas milkweed, Texas purplespike, Trans-Pecos spiderwort

Location	Selected Plants
Pine Canyon (Upper)	Giant coral root, Gray bean, limpia seymeria, Trans-Pecos chickweed, Turner thistle
Pinnacles Trail	Big Bend hophornbeam, Bigelow bristlehead, black dalea, bristly bedstraw, Chisos prickly pear, Chisos red oak, Gray bean, hairy hedeoma, mountain sage, palmleaf thoroughwort, plateau rocktrumpet, rock betony, sandlot brickellbush, spreading snakeherb, stemless aletes, stonecrop, Texas claret-cup cactus, Trans-Pecos chickweed, Wright ageratina
Pinnacles Trail backcountry campsite (nearby)	Bigelow bristlehead
Pulliam Bluff (or nearby)	Drooping juniper, eggleaf silktassel, fascicled bluet, green gromwell, hairy tubetongue, longspur columbine, Mexican starwort, rock betony, scarlet bouvardia, slimlobe rockdaisy, soft twinevine, Trans-Pecos chickweed, tuber windflower
Pulliam Peak	Mexican squawroot
Pummel Peak (or nearby)	Chisos prickly pear, Havard penstemon, narrowleaf fendlerbush
Rattlesnake Mountains	Bladder sage, littleleaf moonpod, naked brittlestem, woollyflower spurge
Rio Grande	Arizona carlowrightia, bearded dalea, bigpod bonamia, Big Bend cholla, Big Bend devil cholla, desert willow, Duncan's pincushion cactus, earlobe mustard, flower of stone, glory of Texas cactus, golden-spined prickly pear, hairy tubetongue, Havard nama, living rock cactus, mesa greggia, propellerbush, shrubby tidestromia, silverlace cactus, strawberry hedgehog cactus, Texas false agave, Texas thelypody, Trans-Pecos senna, whiskerbrush pincushion cactus, woollyflower spurge
River Road	Big-needle pincushion cactus, Chisos pricklypoppy, Havard ipomopsis, Havard nama, Mexican navelseed
River Road (east end)	Blackbrush acacia, manyflowered broomrape
River Road (west end)	Manyflowered broomrape, ocotillo
Rooney's Place	Guayacán
Rosillos Mountains (or nearby)	Broom milkwort, flower of stone, giant fishhook cactus, glandleaf milkwort, guayule, Havard agave, Havard ipomopsis, Havard plum, javelina bush, lyreleaf twistflower, Mexican buckeye, Mexican pinyon pine, orange flameflower, Parry caesalpinia, Pringle swallow-wort, propellerbush, red cyphomeris, Roemer acacia, scarlet bouvardia, showy menodora, slimlobe rockdaisy, Thompson yucca, Trans-Pecos carlowrightia, Trans-Pecos spiderwort, Turner thistle
Rough Run Creek	Glory of Texas cactus
Rough Spring (or nearby)	Berlandier lobelia, Cory cryptantha, Havard plum, Parry ruellia, tuber windflower
Roy's Peak (or nearby)	Havard ipomopsis, heath cliffrose, shaggy false nightshade
Sam Nail Ranch (or nearby)	Black dalea, desert myrtlecroton, Emory mimosa, green condalia, Havard penstemon, silvery wolfberry, tarbush
Sam Nail Ranch to Wilson Ranch Overlook	Dwarf fairy duster
San Vicente (or nearby)	Blackbrush acacia, naked brittlestem, Trans-Pecos senna
San Vicente to Glenn Spring	Leafy heliotrope, littleleaf brongniart, Texas thelypody
San Vicente to Mariscal Canyon	Littleleaf brongniart

Location	Selected Plants
Santa Elena Canyon (or nearby)	Arizona carlowrightia, bladder sage, Cory Dutchman's pipe, earlobe mustard, flower of stone, Havard nama, heartleaf rockdaisy, lyreleaf twistflower, mesa greggia, naked brittlestem, narrowleaf moonpod, Parry caesalpinia, shaggy false nightshade, slimlobe rockdaisy, Trans-Pecos senna, woolly butterflybush, woollyflower spurge
Santiago Mountains	Thompson yucca
Sierra Quemada	Arizona carlowrightia, Texas false agave
Smoky Creek	Candle cholla, Chino grama, Emory mimosa, Havard ipomopsis, ivy treebine, littleleaf moonpod, manyflowered broomrape, mejorana, Trans-Pecos senna, Turner mimosa
Smoky Creek (mouth)	Ivy treebine
Smoky Creek to Johnson Ranch	Desert olive
Solis (or nearby)	Big Bend devil cholla, big-needle pincushion cactus, blackbrush acacia, New Mexico dalea
Sotol Vista (or nearby)	Pringle swallow-wort, smooth sotol, trumpetflower
South Rim Trail	Beargrass, Bigelow beggarticks, Chaffey's pincushion cactus, Chisos red oak, Coahuila scrub oak, eggleaf silktassel, fascicled bluet, giant fishhook cactus, licorice marigold, Mexican pinyon pine, mountain sage, stonecrop, Texas flatsedge, Texas madrone, Trans-Pecos spiderwort
Sublett Ranch (or nearby)	Crucifixion thorn, silvery wolfberry
Sue Peaks (or nearby)	Leafy heliotrope, littleleaf leadtree
Talley Mountain (or nearby)	Dutchman's breeches, lyreleaf twistflower, narrowleaf moonpod, pearl netleaf milkweed, rock milkwort, woollyflower spurge
Telephone Canyon	Creeping rockvine, palo prieto, Parry caesalpinia, rough mortonia, round copperleaf, showy menodora, spoonleaf bouchea, trumpetflower, Warnock justicia, woolly butterflybush, Wright dalea
Terlingua Abaja (or nearby)	Fleshy tidestromia, guayacán, threeflower goldenweed
Terlingua Creek	Bearded dalea, propellerbush
Terlingua Creek (mouth)	Havard nama
Terlingua Ranch Road	Berlandier lobelia, escobilla butterflybush, greeneye heliotrope, Gregg keel-pod, soft heliotrope
Toll Mountain	Big Bend hophornbeam, limpia seymeria
Tornillo Creek	Havard nama, littleleaf moonpod, margined rockdaisy, yellow rocknettle, zigzag croton
Tornillo Creek (near mouth)	Earlobe mustard, green condalia, mat nama
Tornillo Creek Bridge (south)	Big Bend bluebonnet
Tornillo Creek to Heath Canyon	Spoonleaf bouchea
Tornillo Flat	Berlandier flax, feather dalea, glandleaf milkwort, Heyder's pincushion cactus, horse-crippler cactus, leafy heliotrope, Texas selenia, threeflower goldenweed, Torrey ephedra
Tornillo Flat to Persimmon Gap	Berlandier flax
Trap Spring	Green condalia
Tuff Canyon (or nearby)	Big Bend bluebonnet, Turner mimosa, woollyflower spurge

Appendix B (continued)

Location	*Selected Plants*
Tunnel (or nearby)	Cutler twistflower, dense ayenia, green condalia, margined rockdaisy, Mexican navelseed, rayless rockdaisy, spiny-fruited prickly pear, zigzag croton
Ward Spring	Beargrass, blue morning-glory, Chino grama, desert myrtlecroton, fascicled bluet, flower of stone, Havard plum, ivy treebine, longpetal echeveria, Pringle swallow-wort, red cyphomeris, rough mortonia, scarlet bouvardia, soft twine-vine, talayote, Trans-Pecos carlowrightia, Warnock justicia
Wasp Spring	Narrowleaf moonpod
West entrance to park	Ocotillo
Wilson Ranch	Arizona carlowrightia
Window Trail	Chisos prickly pear, Cory Dutchman's pipe, dwarf anisacanth, heartleaf rockdaisy, palo prieto, pearl netleaf milkweed, sandlot brickellbush, splitleaf brickellbush
Window (nearby or at base)	Fascicled bluet, hairy hedeoma, Havard plum, limerock brookweed, Mexican buckeye, purple gaymallow, scarlet ladies tresses
Woodson's Camp	Turner mimosa

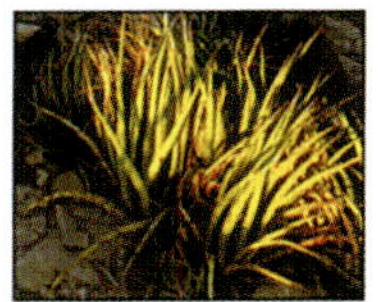

APPENDIX C

Photo Equipment and Techniques for Plant Photography

Cameras, Film, and Lenses

For many years, my primary camera has been an old Nikon F3 35mm film camera, with Fuji Velvia film, and macro lenses designed for high-quality close-up photography: a Nikon 105mm macro lens and a Nikon 55mm macro lens, and extension tubes for more extreme close-up images. When not hiking long distances, I often carry a Mamiya RZ Pro II medium format film camera, with Fuji Velvia film, and with normal and macro lenses and extension tubes. My first digital camera, a Canon EOS20D with a 100mm macro lens, was purchased in early 2005, and although it has proven a valuable addition, it has not replaced my film cameras.

Tripod

A sturdy tripod is mandatory. I use one with a specially purchased short center column to get the camera as close as possible to the ground.

Lens Filters

Polarizing filters are helpful, especially when photographing skies, to increase color saturation. Split or graduated neutral density filters are essential for landscape photography, especially early and late in the day, to reduce illumination from the sky relative to the foreground. A warming filter is often required in cool, drab light.

Reflectors and Umbrellas

Although cumbersome, thirty-inch Westcott reflectors are ideal for blocking wind, reflecting light, and filtering (softening) direct light. I also employ Photoflex white nylon umbrellas to keep rain off the camera and me and as an added means of blocking wind and reflecting light.

Flash Equipment

I don't use flash equipment. I prefer to rely on natural light and work with reflectors for subtle lighting effects.

Camera Bag

For several years, I have worn a series of Lowepro photo backpacks, which carry all my equipment (often two camera bodies plus lenses) except the tripod. My backpack also includes three liters of water.

Other Equipment

Wimberley PLAMPs (plastic adjustable arms) are my preferred tools for holding plants in place. I also carry a hard clipboard, which I place behind a plant when the natural background is too busy or confusing; a small memo pad, to make notes about species characteristics, plant habitats, and locations; and a collapsible ruler to measure the size of leaves and flowers, the height of plants, etc., for subsequent plant identification.

Techniques

For this type of close-up photography, mirror lock-up to eliminate camera shake is critical. In addition, maximum depth of field for sharp focus is usually preferred. I try to capture many views of the same subject to get the best composition and, above all, exercise PATIENCE.

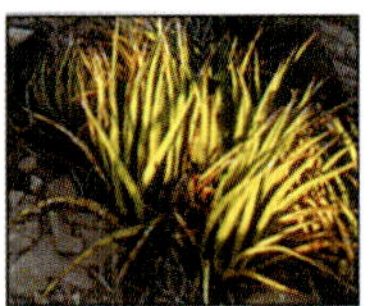

APPENDIX D

A Note on Botanical Nomenclature

In specifying scientific and common names for plants featured in this book, an effort was made to be consistent with well-known and respected works on the area's flora. In the choice of scientific (botanical) names for trees and shrubs, cacti, grasses, and ferns, A. M. Powell's four works were followed, with only a few exceptions (Powell 1994, 1998; Powell and Weedin 2004; Yarborough and Powell 2002).

In three cases, scientific names from recent genera revisions were used: for Parry caesalpinia, *Pomaria melanosticta* instead of *Caesalpinia parryi,* and for narrowleaf moonpod and littleleaf moonpod, *Acleisanthes angustifolia* and *Acleisanthes parvifolia* instead of *Selinocarpus angustifolius* and *Selinocarpus parvifolius.* In three other cases, recent species treatments in the *Flora of the Chihuahuan Desert Region* (Henrickson and Johnston, forthcoming) were followed: for narrowleaf fendlerbush, *Fendlera linearis* instead of *Fendlera rigida;* for mejorana, *Lantana macropoda* rather than *Lantana achyranthifolia;* and for ivy treebine, *Cissus trifoliata* rather than *Cissus incisa.*

In the choice of scientific names for herbaceous plants, the *Flora of the Chihuahuan Desert Region* was usually followed, with the following several exceptions. In five cases, scientific names cited in *Atlas of the Vascular Plants of Texas* (Turner et al. 2003) were used: palmleaf thoroughwort (*Conoclinium dissectum*), Chisos Mountain brickellbush (*Flyriella parryi*), creeping rockvine (*Bonamia repens*), Berlandier flax (*Linum berlandieri* var. *filifolium*), and Wendt's malaxis (*Malaxis wendtii*). For woolly dogweed (*Thymophylla micropoides*), the genus name *Thymophylla* was used, following Powell; for three-tongue spurge (*Chamaesyce chaetocalyx* var. *triligulata*), the genus name *Chamaesyce* was used, following Turner. Manyflowered broomrape was referred to as a subspecies, *Orobanche ludoviciana* ssp. *multiflora,* rather than a variety, following a 2001 article by H. L. White and W. C. Holmes in the journal *Sida.*

More latitude was exercised in the choice of common names for plants, but wherever possible widely used common names and especially names known locally were employed. For trees and shrubs, cacti, grasses, and ferns, common names cited by Powell were generally used, with a few exceptions. For *Yucca faxoniana,* the common name giant dagger was chosen because it is a name so frequently used in the park. For *Echinocereus enneacanthus* var. *enneacanthus,* strawberry hedgehog was employed (instead of strawberry), to avoid using the same common name applied to *Echinocereus stramineus* var. *stramineus.* For *Glandulicactus uncinatus* var. *wrightii,* catclaw cactus was used (instead of eagle-claw cactus) because eagle claw is a name often applied to another cactus species, *Echinocactus horizonthalonius,* which occurs in the park. For *Thamnosma texana,* Dutchman's breeches was chosen because it is too apt a name not to use.

Other exceptions were relatively minor—a matter of syntax or personal preference. The term "rockdaisy" was used in the common name for all plants of the genus *Perityle* (e.g., heartleaf rockdaisy); the term "brickellbush" was applied to all plants of the genus *Brickellia* (e.g., splitleaf brickellbush). Common names were usually not joined together into a single word (e.g., desert yaupon rather than desertyaupon); and names were usually not joined by hyphens (e.g., sea urchin instead of sea-urchin).

For the remaining 109 mostly herbaceous plants, local sources of common names were again followed. Seventy-two of these common names appeared in one of Barton Warnock's three wildflower studies of the region (1970, 1974, 1977), and another 17 were cited in the National Park Service's checklist of plants known to occur in the park (Big Bend National Park, Division of Science and Resource Management 1996). Generally, only when no common names were cited in the above publications were other sources utilized. The remaining 20 common names were drawn from the U.S. Department of Agriculture *Plants Database, Wild Orchids of Texas* (Liggio and Liggio 1999), the Nature Conservancy of Texas, NatureServe, and a Web site maintained by the U.S. Geological Survey, *Rare and Threatened Plants of Big Bend National Park.*

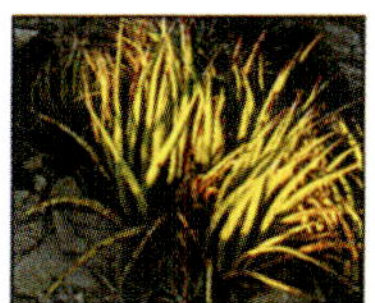

GLOSSARY

achene—Small, dry, one-seeded fruit, not opening at maturity, with the seed free from the fruit wall.
acorn—One-seeded fruit of an oak; nut seated in or partly surrounded by woody cuplike base.
alluvial—Deposited by flowing water, as in soil, silt, or sediment.
alternate—On one side and then the other, not opposite or whorled; having a single leaf or branch at a node.
annual—A plant that matures, flowers, fruits, and dies in a single year; lasting one year or one season.
anther—Upper, pollen-bearing part of the stamen.
anthocarp—Structure in which the fruit is united with the base of the perianth, as in Nyctaginaceae.
apex—Tip, such as the end of a leaf; the highest point.
appendage—Any attached supplementary or subordinate part.
appressed—Flat against, pressed close.
areole—Small raised or depressed, cushionlike area from which spines, flowers, and other organs grow.
aril—Growth partially or wholly enveloping the seed; a usually fleshy, brightly colored seed cover.
armed—Bearing spines, thorns, prickles, or similar structures.
aromatic—Fragrant or pungent but usually implying a pleasant odor.
arroyo—Channel or gully formed by water; a small water course, often dry.
ascending—Rising, sloping, or trending upward; usually somewhat less than straight or fully erect.
attenuate—Tapering gradually to a slender point.
awn—A slender, terminal bristle or stiff, hairlike appendage.
axil—Upper angle between a leaf and the stem to which it is attached.
banner—Upper petal or standard of a pea or papilionaceous flower.
barbed—With barbs or sharp, usually rigid, reflexed or hooked points.
basal—At or near the base.
beak—A projecting, usually narrowed, firm tip, as in a fruit.
bearded—With long, stiff hairs.
berry—Small, fleshy, pulpy fruit with several to many seeds, not splitting open at maturity.
biennial—A plant maturing and dying in two years, usually flowering and fruiting in second year.
bipinnate—Twice divided into smaller leaflets or segments that are separately attached to the leaf axis.
bisexual—Having both stamens and pistils; with both male and female reproductive parts on the same plant.
blade—An expanded part, often used in reference to a leaf.
bract—Reduced or modified leaf subtending a flower or flower cluster.
bristle—Stiff hair or hairlike projection.
bur—Fruit with spines or prickles; a prickly seed covering.
calcareous—Containing calcium carbonate; derived from limestone, as in calcareous soil.
caliche—White or yellowish calcium carbonate soil; a crust of calcium carbonate or decomposed limestone.

calloused—Abnormally thickened or toughened.

calyx—External, usually green outer whorl or floral envelope of the flower; collectively the sepals.

capsule—Dry fruit composed of more than one carpel, splitting open at maturity.

carpel—A simple pistil, or one member of a compound pistil.

catkin—Usually slender, drooping, deciduous cluster of unisexual flowers that lack petals.

chaparral—Dense, low vegetation often composed of thorny shrubs with relatively thick, leathery, evergreen leaves.

cholla—Cactus with cylindrical stem segments.

clasping—Refers to a leaf whose base surrounds or partly surrounds the stem.

claw—Narrowed base of some petals or sepals.

cleft—Cut or divided partway, as in a leaf.

cleistogamous—Mostly small, self-fertilizing flowers, usually not opening.

column—Columnar structure formed by united stamens, as in Malvaceae, or united stamen and style in Orchidaceae.

composite head—Compact flower cluster with many small flowers surrounded by an involucre, as in Asteraceae.

compound—Composed of two or more similar parts.

compound leaf—Leaf separated into two or more leaflets.

compressed—Flattened laterally.

cone—Inflorescence of flowers or fruits with overlapping scales, as in Pinaceae.

corolla—Collective term for the petals of a flower; the inner, usually colored, floral envelope.

corona—Crownlike outgrowth of the corolla in some flowers, as in Asclepiadaceae.

corymb—Flat-topped or convex flower cluster with individual flower stalks of unequal length, the outer the longest.

crest—Elevation or ridge on a plant part.

cyathium—Flower cluster of Euphorbiaceae, with a flowerlike, cup-shaped involucre enclosing the real flowers.

deciduous—Falling; said of plants that shed leaves seasonally, or of leaves that fall at the end of a growing season.

depressed—Sunken or flattened.

dioecious—With staminate and pistillate flowers borne on separate plants.

disjunct—Occurring in two or more widely separated geographic regions.

disk—Fleshy outgrowth of the receptacle within the perianth; central part of flower head in Asteraceae.

disk floret—Small, tubular flower in the central part or disk of a flower head in the Asteraceae.

dissected—Deeply divided or cut into segments.

dominant species—The most prominent species in a plant community.

drupe—Fleshy, indehiscent fruit, usually with a hard stone enclosing a single seed.

elliptic—Shaped like an ellipse; a flattened circle; longer than broad, widest in the middle, narrowing to the ends.

elliptical—Elliptic.

endemic—Restricted to or native to a specific geographic region.

evergreen—With leaves throughout the year; not deciduous.

family—Grouping of related organisms, consisting of one or more related genera.

fertile—With functional reproductive organs.

fibrous—Possessing fibers; with woody fibers.

filament—The slender stalk of the stamen, supporting the anther.

floral cup—Cuplike structure containing perianth or other floral parts.

floret—Small or diminutive flower, usually in a dense flower cluster, as in Asteraceae.

follicle—Dry, single-chambered fruit that splits open lengthwise along one seam.

frond—Leaf of a fern, including the blade and the petiole.

fruit—Matured ovary of a flower, containing the seeds.

genera—Plural of genus.

genus—Category or grouping of plants consisting of one or more related species.

gland—Organ that secretes a fluid.

glandular—Having glands, often at the end of a hair.

glaucous—Covered with a white, blue, or gray, waxy or powdery, usually easily removable, coating or bloom.

globose—Spherical in shape.

globular—Globose; nearly orbicular.

glochid—Minute barbed, easily removed, modified spine or bristle, as in Cactaceae.

glutinous—Covered with a sticky or gluey exudation.

habit—General appearance or manner of growth of a plant.

habitat—Place where the plant lives; its environment.

head—Dense cluster of stemless or short-stemmed flowers arising from a common point, as in Asteraceae.

herbaceous—Like an herb; not woody; often with stems that die back at the end of the growing season.

herbarium—Collection of dried and

pressed plants; place maintaining such a collection for research and scientific study.

horn—Appendage shaped like a horn; a hard protrusion.

hybrid—Cross between two species.

igneous—Resulting from intense heat; solidified from a molten condition; volcanic.

incised—Cut, usually deeply and sharply, as in the edges of leaves.

indehiscent—Not splitting open at maturity.

inflated—Bladdery, swollen, expanded.

inflorescence—Flower cluster; or the typical arrangement of flowers on a plant.

involucre—Bracts, or circle of bracts, below or around a flower or flower cluster.

joint—Articulation, as in a node on a stem; an internode or space between two nodes.

keel—Longitudinal ridge, shaped like the keel of a boat; the two lower, united petals of a papilionaceous flower.

lance-shaped—Several times longer than wide, broadest toward the base, tapering to the tip.

latex—Milky, sticky, whitish juice or sap of some plants.

lax—Loosely arranged.

leaflet—Division or segment of a compound leaf.

legume—Dry fruit or seedpod, usually splitting along two seams into two valves; any plant of the Fabaceae family.

linear—Long, narrow, with parallel sides.

lip—Upper or lower division of a two-lipped flower or calyx; petal of an orchid.

lobe—Division of an organ, such as part of a leaf or flower.

margin—Edge, as in the margin of a leaf.

mat—Dense, low cover of foliage.

megaspore—Larger spore that produces the female gametophyte.

membranous—Thin, usually soft, layer; like a membrane.

mesic—With moderate moisture.

microspore—Smaller spore that produces the male gametophyte.

midrib—Central vein or rib of a leaf.

montane—Growing in the mountains.

needle—Stiff linear leaf, as in Pinaceae.

net-veined—Reticulated; with a network of crossed veins.

node—Point on the stem where a leaf or branch is attached.

nut—Dry, one-seeded fruit with a hard shell, not splitting open.

nutlet—A small, dry nut or nutlike fruit.

oblong—Several times as long as wide with roughly parallel sides.

obovate—Egg-shaped but wider at the upper end.

odd-pinnate—Pinnate leaf with a single, terminal leaflet.

opposite—On opposing sides, as in opposite sides of the stem; two at a node.

oval—Broadly elliptic, rounded at the ends.

ovary—Swollen basal ovule-bearing portion of the pistil of the flower that becomes the fruit.

ovate—Egg-shaped, with the widest end at the base and the more pointed end at the top.

papilionaceous—Butterfly-shaped flower of many Fabaceae plants; with a banner, wing (two), and keel petals.

parasite—Organism growing on and obtaining nourishment from a host; usually lacking chlorophyll.

pedicel—Stalk of a single flower.

pendulous—Hanging or dangling; pendent.

perennial—A plant living for several years or more.

perianth—Outer floral envelope, consisting of calyx and corolla, excluding stamens and pistils.

petal—A usually colorful part of the corolla; flower part within the sepals, peripheral to stamens and pistils.

petiole—Leaf stalk.

pinna—Major division of a pinnately compound leaf.

pinnae—Plural of pinna.

pinnate—With divisions arranged on either side of a common axis, as in leaflets on a common petiole.

pistil—Female, ovule-bearing organ of the flower (ovary, style, stigma) that matures into fruit.

pod—Any dry fruit or seedpod that splits open when mature.

pollen—Microscopic male spores borne by the anther of the stamen; usually fine, powdery grains.

prickle—Sharp spiny or thorny outgrowth.

prickly pear—Cactus with flat stem joints.

prostrate—Lying flat on the ground.

raceme—Simple flower cluster, with each stalked flower growing singly off an elongated stem.

radial spine—Spine that radiates from the edge of a spine cluster, mostly parallel to the stem surface.

ray floret—Outer, petal-like, often strap-shaped flower, typically on the perimeter of a central disk in the Asteraceae.

receptacle—The enlarged stem tip bearing the flower parts; or in Asteraceae, bearing the florets in the head.

recurved—Curving backward or downward.

reflexed—Turned backward or downward.

revolute—Rolled or curled backward along the edges or from the tip.

rhizome—Horizontal underground stem that generates new stems above its nodes and new roots below.

rib—Longitudinal ridge on a surface, as in a rib on the stem of a cactus; prominent vein of leaf.

rosette—Circular, typically crowded cluster of radiating leaves, usually at or near ground level.

samara—Dry, winged, indehiscent fruit.

scale—Small, thin, flattish or appressed structure, usually a vestigial leaf or bract.

scape—Leafless flower stalk rising from ground level.

schizocarp—Dry fruit splitting at maturity into two or more one-seeded compartments (mericarps).

scrub—Vegetation stunted due to lack of moisture.

scurfy—With small, loose, branlike scales or crust on the epidermis.

segment—Subdivision; part of a divided leaf or frond.

sepal—One of the usually green, leafy, outer parts of a flower, beneath the petals; leaflike segment of a calyx.

serrated—With small, sharp teeth along the margin; saw-toothed.

sheath—Tubular structure surrounding a plant part; such as thin tissue enclosing part of a stem or spine.

shrub—Woody plant usually smaller than a tree, usually with several stems from base.

silique—Slender, dry seed capsule of Brassicaceae, with two sides opening to expose a central partition.

solitary—Single.

spatulate—Shaped like a spatula; wider and rounded at the tip, gradually narrowed to the base.

species—Division of a genus, with related, similar individuals reproducing among themselves.

spike—Unbranched, usually elongated cluster of flowers, each without individual stalks or pedicels.

spikelet—Small spike; part of a grass flower cluster, with one or two scalelike, basal bracts and one or more florets.

spine—Sharp-pointed, stiff, typically woody projection; usually a modified leaf.

spinescent—Ending in a sharp point; bearing a spine or spines.

spore—Small, usually one-celled, reproductive structure, especially of some nonflowering plants, such as ferns.

spur—Short branch or branchlet or a slender extension at the base of a petal or sepal.

stamen—Male organ of a flower that bears the pollen in the anther; the filament and anther collectively.

stem—Aboveground plant axis, often branching.

sterile—Without male or female organs; barren; unable to produce fruit, seed, spores, or other reproductive organs.

stigma—The tip of the pistil of a flower that receives the pollen.

strobilus—Cone-shaped structure of overlapping or closely spaced, spore-bearing appendages, as in Selaginellaceae.

style—Tubular, usually slender, upper portion of the pistil above the ovary and below the stigma.

subspecies—Taxonomic classification between species and variety; often, a geographically distinct group.

subtend—To be closely below or adjacent to; to underlie.

succulent—Fleshy, juicy, containing or storing water.

talus—Typically smaller rocks, usually on steep mountainous slopes; rock fragments, often at the base of a cliff.

taproot—Stout, central, vertical root.

taxonomy—Study of the hierarchical classification of life forms.

tendril—Slender, coiling structure, arising from stem or leaf, that enables climbing or attachment for support.

terminal—At the tip; at the end of a stem.

ternate—In groups of three in a cluster; having three parts, as in leaflets.

thorn—Sharp, woody structure; a modified branch; a spiny projection.

throat—Opening of a flower with fused or united petals.

tooth—Small marginal lobe, often with reference to the edge of a leaf or petal.

tree—Woody plant with one main stem or trunk, usually larger than a shrub.

tuber—Short, thickened, fleshy, usually underground stem.

tubercle—Small, wartlike structure or knobby protrusion; in Cactaceae, a nipplelike or ridgelike projection.

tubular—Shaped like a tube or hollow cylinder.

tuft—Clump.

two-lipped—With upper and lower lobes.

type—Specimen or drawing on which a plant name and description is based.

umbel—Flower cluster with flowers on stalks arising from a common point, like spokes of an umbrella.

unisexual—Male or female; flowers with stamens or pistils but not both.

valve—One of the parts into which a

dehiscent capsule splits at maturity.

variety—Subdivision of a species or subspecies, with differences from the rest of the species or subspecies.

vein—Threadlike ridge in a leaf; a strand or bundle of conducting vascular tissue.

villous—With long, soft, shaggy, unmatted hairs.

viscid—Adhesive; sticky.

whorl—Circular arrangement of three or more similar organs radiating from a common point, as in a whorl of leaves.

wing—Thin, flat membrane bordering a leaf stalk, stem, fruit, or seed; also, lateral petal of many Fabaceae flowers.

winged—With thin, papery wings or membranous extensions.

woolly—With long, soft, closely interwoven, or matted hairs.

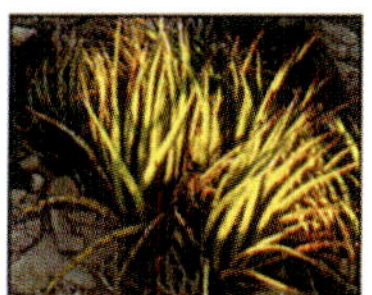

SOURCES

I want to specifically indicate and gratefully acknowledge the sources that proved most current, reliable, and informative in the preparation of this guide.

Introductory Information on Big Bend National Park

For broad general information on the Big Bend country, no source was more useful than *Big Bend: Official National Park Handbook* (National Park Service 1983). This handbook contains a wealth of interesting facts about everything relating to the Big Bend experience. Barney Nelson's compilation of nature writings on the Big Bend, *God's Country or Devil's Playground* (Nelson 2002), was a valuable compendium of human perceptions of Big Bend over time. Ro Wauer's *Naturalist's Big Bend* (Wauer 1980) provided a concise summary of Big Bend history and an excellent analysis of the park's ecological zones and related plant communities.

The Big Bend: A History of the Last Texas Frontier (Tyler 1975) and *A Most Singular Country* (Gomez 1995) were primary sources on human settlement of the region. The key resource on the park's geologic history and natural features was Ross Maxwell's *The Big Bend of the Rio Grande* (Maxwell 1968). Detailed accounts of the origins of the national park were found in *The Story of Big Bend National Park* (Jameson 1996) and *Big Bend Country: A History of Big Bend National Park* (Maxwell 1985).

Botanical Information

For botanical information on plant physiography and blooming periods, *The Manual of the Vascular Plants of Texas* (Correll and Johnston 1979) was regularly consulted. *The Flora of the Chihuahuan Desert Region* (Henrickson and Johnston, forthcoming) proved an invaluable update to the Correll and Johnston work. The Web site of *Flora of North America North of Mexico* (Flora of North America Editorial Committee 1993+) was another significant and timely source.

For descriptions of trees and shrubs, *Trees, Shrubs and Woody Vines of the Southwest* (Vines 1960) was helpful after forty-five years in publication. *Trees and Shrubs of the Trans-Pecos and Adjacent Areas* (Powell 1998) was a more current and critically important work covering the trees and shrubs of Big Bend National Park.

Other sources were important for specific types of plants, such as cacti, orchids and ferns. *Cacti of the Trans-Pecos and Adjacent Areas* (Powell and Weedin 2004) was the most reliable source of detailed botanical information on cacti. *Wild Orchids of Texas* (Liggio and Liggio 1999) was often consulted on the botany of Big Bend orchids. Only a few ferns and grasses are included in this book, but *Ferns and Fern Allies of the Trans-Pecos and Adjacent Areas* (Yarborough and Powell 2002) was the primary source of botanical data on ferns, and *Grasses of the Trans-Pecos and Adjacent Areas* (Powell 1994) was the principal authority on grasses.

For many years, Barton Warnock's

three books—*Wildflowers of the Big Bend Country, Texas* (1970); *Wildflowers of the Davis Mountains and Marathon Basin, Texas* (1977); and *Wildflowers of the Guadalupe Mountains and the Sand Dune Country, Texas* (1974)—provided the only photographic documentation of Big Bend plants. These references were my constant companions.

Geographic Distribution of Plants

The U.S. Department of Agriculture Natural Resources Conservation Service online *Plants Database* was a starting point for data on plant distribution by state in the United States, and the *Flora of North America* Web site was another useful resource. The invaluable *Atlas of the Vascular Plants of Texas* (Turner et al. 2003) became my principal authority for the distribution of plants by county in Texas. The online *Herbarium Specimen Browser* of Texas A&M University and the online *Flora of Texas Database* of the University of Texas at Austin were also consulted.

The Flora of the Chihuahuan Desert Region was the primary source for most of the detailed information on the geographic range of species in Mexico. Web sites of the Missouri Botanical Garden, the New York Botanical Garden, and the Harvard University Herbaria were also consulted for detailed data on geographic locations where specimens were collected, especially in the United States and Mexico.

The Nature Conservancy's *Annotated List of the G3/T3 and Rarer Plant Taxa of Texas* (Carr 2005) proved a reliable source of geographic distributions both globally and in Texas and was the primary authority for a plant's conservation status. The *NatureServe Explorer* Web site was also helpful on geographic distributions by state, and for the conservation status of plants globally and by state in the United States.

Other authorities were examined where appropriate: for the distribution of trees and shrubs, *Trees and Shrubs of the Trans-Pecos and Adjacent Areas;* for cacti, *Cacti of the Trans-Pecos and Adjacent Areas;* for orchids, *Wild Orchids of Texas;* for ferns, *Ferns and Fern Allies of the Trans-Pecos and Adjacent Areas;* and for grasses, *Grasses of the Trans-Pecos and Adjacent Areas.*

Plant Locations in the Park

Big Bend National Park botanist Joe Sirotnak provided me with a plants database, which listed specimens of plants collected in the park by date and location, and this database was a crucial resource for locating the more uncommon plants. Over the years, every location where I photographed plants was documented, and that personal knowledge is also reflected in this book. In addition, *Trees and Shrubs of the Trans-Pecos and Adjacent Areas* frequently lists specific locations of plants within the park, as do Barton Warnock's three books.

The occurrence of specific species within Big Bend National Park is documented in *The Vascular Flora of Big Bend National Park, Texas* (Big Bend National Park, Division of Science and Resource Management 1996) and in *Inventory of Vascular Flora, Big Bend National Park* (Clelland 2001). Both are useful checklists of plants known to occur in the park.

Notes

The note in the final paragraph of each individual plant description in this book required extensive research on the origin and meaning of botanical names, the first descriptions and collections of plants, and the lives of botanists and plant collectors.

For the meaning of plant names, the old but still valuable *Gray's Manual of Botany* (Fernald 1970) and *Trees, Shrubs and Woody Vines of the Southwest* were regularly consulted. The most helpful recent works on scientific names and etymologies were the *CRC World Dictionary of Plant Names* (Quattrocchi 2000) and *The Names of Plants* (Gledhill 2002). Online sources reviewed (often with healthy skepticism) include the *Flora of North America* Web site; *California Plant Names: Latin and Greek Meanings and Derivations* (Charters 2005), and *Botanary—The Botanical Dictionary* (Dave's Garden 2005).

Data on the first descriptions and collections of plants, and on designated type specimens, were obtained principally through the online resources of herbaria. Those Web sites consulted most frequently were the Missouri Botanical Garden's *VAST (VAScular Tropicos) Nomenclatural Database,* the New York Botanical Garden's *International Plant Science Center Virtual Herbarium,* the Harvard University Herbaria's *Harvard University Herbaria Database, Index of Botanical Specimens,* and the *International Plant Names Index,* a joint project of the Royal Botanic Gardens, Kew, the Harvard University Herbaria, and the Australia National Herbarium.

Historical data on botanists and plant collectors were derived from many online resources and a handful of books. *Naturalists of the Frontier* (Geiser 1948) was particularly helpful, especially on the botanical collector Charles Wright. *Pioneer Naturalists* (Evans 1993) was a more current and very entertaining work on numerous botanists, including George Engelmann, Josiah Gregg, and Edwin James.

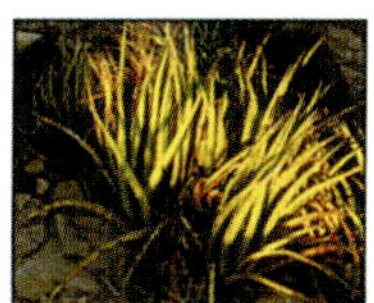

REFERENCES

Ajilvsgi, G. 1984. *Wildflowers of Texas.* Fredericksburg, TX: Shearer Publishing.

Anderson, E. F. 2001. *The Cactus Family.* Portland, OR: Timber Press.

Anthony, M. S. 1956. The *Opuntiae* of the Big Bend Region of Texas. *American Midland Naturalist* 55(1): 225–56.

Arnberger, L. P. 1982. *Flowers of the Southwest Mountains.* Tucson, AZ: Southwest Parks and Monuments Association.

Beatty, J. J. 1964. *Plants in His Pack: A Life of Edward Palmer.* New York: Random House.

Big Bend National Park, Division of Science and Resource Management. 1996. *The Vascular Flora of Big Bend National Park, Texas.* Panther Junction, TX: National Park Service.

Bowers, J. E. 1987. *100 Roadside Wildflowers of the Southwest Woodlands.* Tucson, AZ: Southwest Parks and Monuments Association.

———. 1989. *100 Desert Wildflowers of the Southwest.* Tucson, AZ: Southwest Parks and Monuments Association.

———. 1993. *Shrubs and Trees of the Southwest Deserts.* Tucson, AZ: Southwest Parks and Monuments Association.

Brown, P. M. 2003. *Wild Orchids of North America, North of Mexico.* Gainesville: University Press of Florida.

Carr, W. R. 2005. *An Annotated List of the G3/T3 and Rarer Plant Taxa of Texas* (working draft). Austin: Texas Conservation Data Center, The Nature Conservancy of Texas.

Carter, J. L. 1997. *Trees and Shrubs of New Mexico.* Silver City, NM: Mimbres Publishing.

Charters, M. L. 2005. *California Plant Names: Latin and Greek Meanings and Derivations.* Sierra Madre, CA: Michael L. Charters. http://www.calflora.net/botanical-names/index.html.

Clary, K. H. 1997. Phylogeny, Character Evolution, and Biogeography of Yucca L. (Agavaceae) as Inferred from Plant Morphology and Sequences of the Internal Transcribed Spacer (ITS) Region of the Nuclear Ribosomal DNA. Ph.D. dissertation, University of Texas.

Clelland, R. 2001. *Inventory of Vascular Flora, Big Bend National Park.* Panther Junction, TX: Big Bend Natural History Association.

Coffey, T. 1993. *The History and Folklore of North American Wildflowers.* Boston, MA: Houghton Mifflin Company.

Coombes, A. J. 1985. *Dictionary of Plant Names.* Portland, OR: Timber Press.

Correll, D. S., and M. C. Johnston. 1979. *Manual of the Vascular Plants of Texas.* Richardson: University of Texas at Dallas.

Dave's Garden. 2005. *Botanary—The Botanical Dictionary.* San Antonio, TX: Dave's Garden. http://davesgarden.com/botany/.

Dodge, N. N. 1985. *Flowers of the Southwest Deserts.* Tucson, AZ: Southwest Parks and Monuments Association.

Enquist, M. 1987. *Wildflowers of the Texas Hill Country.* Austin, TX: Lone Star Botanical.

Evans, D. B. 1998. *Cactuses of Big Bend National Park.* Austin: University of Texas Press.

Evans, H. E. 1993. *Pioneer Naturalists: The Discovery and Naming of North American Plants and Animals.* New York: Henry Holt.

Everitt, J. H., D. L. Drawe, and R. I. Lonard. 1999. *Field Guide to the Broad-Leaved Herbaceous Plants of South Texas.* Lubbock: Texas Tech University Press.

———. 2002. *Trees, Shrubs and Cacti of South Texas.* Lubbock: Texas Tech University Press.

Fernald, M. L. 1970. *Gray's Manual of Botany.* New York: D. Van Nostrand.

Fischer, P. C. 1989. *70 Common Cacti of the Southwest.* Tucson, AZ: Southwest Parks and Monuments Association.

Flora of North America Editorial Committee. 1993+. *Flora of North America North of Mexico.* New York: Flora of North America Association. http://hua.huh.harvard.edu/FNA/.

Flora of Texas Consortium. 2000. *Flora of Texas Database.* Austin: University of Texas at Austin.

Foster, S., and C. Hobbs. 2002. *A Field Guide to Western Medicinal Plants and Herbs.* Boston: Houghton Mifflin.

Gehlbach, Frederick R. 1981. *Mountain Islands and Desert Seas: A Natural History of the U.S.-Mexican Borderlands.* College Station: Texas A&M University Press.

Geiser, S. W. 1948. *Naturalists of the Frontier.* Dallas, TX: Southern Methodist University.

Gledhill, D. 2002. *The Names of Plants.* Cambridge, UK: Cambridge University Press.

Gomez, A. R. 1995. *A Most Singular Country: A History of Occupation in the Big Bend.* Provo, UT: Charles Redd Center for Western Studies, Brigham Young University.

Goyne, M. A. 1991. *Life among the Texas Flora: Ferdinand Lindheimer's Letters to George Engelmann.* College Station: Texas A&M University Press.

Harvard University Herbaria. 2001. *Harvard University Herbaria Database, Index of Botanical Specimens.* Cambridge, MA: President and Fellows of Harvard College. http://cms.huh.harvard.edu/databases/specimen_index.html.

Henrickson, J., and M. C. Johnston. Forthcoming. *The Flora of the Chihuahuan Desert Region.*

Jaeger, E. C. 1969. *Desert Wild Flowers.* Stanford, CA: Stanford University Press.

Jameson, J. 1996. *The Story of Big Bend National Park.* Austin: University of Texas Press.

Keenan, P. E. 1998. *Wild Orchids across North America: A Botanical Travelogue.* Portland, OR: Timber Press.

Kirkpatrick, Z. M. 1992. *Wildflowers of the Western Plains.* Austin: University of Texas Press.

Kurz, D. 1999. *Ozark Wildflowers: A Field Guide to Common Ozark Wildflowers.* Helena, MT: Falcon Publishing.

Ladd, D. 1995. *Tallgrass Prairie Wildflowers.* Helena, MT: Falcon Publishing.

Lamb, E., and B. Lamb. 1974. *Colorful Cacti of the American Deserts.* New York: Macmillan Publishing Company.

Leake, D. V., J. B. Leake, and M. L. Roeder. 1993. *Desert and Mountain Plants of the Southwest.* Norman: University of Oklahoma Press.

Liggio, J., and A. O. Liggio. 1999. *Wild Orchids of Texas.* Austin: University of Texas Press.

Loughmiller, C., and L. Loughmiller. 1994. *Texas Wildflowers.* Austin: University of Texas Press.

MacLeod, W. 2002. *Big Bend Vistas: A Geological Exploration of the Big Bend.* Alpine: Texas Geological Press.

Madison, V. D., and H. C. Stillwell. 1958. *How Come It's Called That? Place Names in the Big Bend Country.* Albuquerque: University of New Mexico Press.

Mattiza, D. B. 1993. *100 Texas Wildflowers.* Tucson, AZ: Southwest Parks and Monuments Association.

Maxwell, R. A. 1968. *The Big Bend of the Rio Grande.* Austin: Bureau of Economic Geology, University of Texas at Austin.

———. 1985. *Big Bend Country: A History of Big Bend National Park.* Big Bend National Park, TX: Big Bend Natural History Association.

McDougall, W. B., and O. E. Sperry. 1951. *Plants of Big Bend National Park.* Washington, DC: U.S. Government Printing Office.

McVaugh, R. 1956. *Edward Palmer: Plant Explorer of the American West.* Norman: University of Oklahoma Press.

Missouri Botanical Garden. 2005. *VAST (VAScular Tropicos) Nomenclatural Database.* Saint Louis: Missouri Botanical Garden. http://mobot.mobot.org/W3T/Search/vast.html.

National Park Service. 1983. *Big Bend: Official National Park Handbook.* Washington, DC: U.S. Department of the Interior.

———. 2006. *Big Bend National Park.* U.S. Department of the Interior, Washington, DC. http://www.nps.gov/bibe/.

NatureServe. 2005. *NatureServe Explorer, An Online Encyclopedia of Life.* Arlington, VA. http://www.natureserve.org/explorer/.

Nelson, B., ed. 2002. *God's Country or Devil's Playground.* Austin: University of Texas Press.

New York Botanical Garden. 2005. *International Plant Science Center Virtual Herbarium.* Bronx, NY. http://sciweb.nybg.org/science2/vii2.asp.

Niehaus, T. F. 1984. *A Field Guide to Southwestern and Texas Wildflowers.* Boston, MA: Houghton Mifflin.

Novovitch, Barbara. 2000. "Clear Days Less Frequent in Texas Park." *Washington Post,* October 27, 2000. Posted at Rio Grande/Rio Bravo Basin Coalition, http://www.rioweb.org.

Powell, A. M. 1994. *Grasses of the Trans-Pecos and Adjacent Areas.* Austin: University of Texas Press.

———. 1998. *Trees and Shrubs of the Trans-Pecos and Adjacent Areas.* Austin: University of Texas Press.

Powell, A. M., and J. F. Weedin. 2004. *Cacti of the Trans-Pecos and Adjacent Areas.* Lubbock: Texas Tech University Press.

Quattrocchi, U. 2000. *CRC World Dictionary of Plant Names: Common Names, Scientific Names, Eponyms, Synonyms, and Etymology.* Boca Raton, FL: CRC Press.

Quinn, Meg. 2000. *Wildflowers of the Desert Southwest.* Tucson, AZ: Rio Nuevo.

Richardson, A. 1990. *Plants of the Rio Grande Delta.* Austin: University of Texas Press.

Rose, F. L., and R. W. Strandtmann. 1986. *Wildflowers of the Llano Estacado.* Dallas, TX: Taylor.

Royal Botanic Gardens, Kew, the Harvard University Herbaria, and the Australia National Herbarium. 2004. *International Plant Names Index.* http://www.ipni.org.

Sanders, J. 2003. *The Secrets of Wildflowers.* Guilford, CT: Lyons Press.

Silverthorne, E. 1996. *Legends and Lore of Texas Wildflowers.* College Station: Texas A&M University Press.

Spellenberg, R. 1998. *National Audubon Society Field Guide to North American Wildflowers, Western Region.* New York: Alfred A. Knopf.

Stearn, W. T. 1992. *Stearn's Dictionary of Plant Names for Gardeners.* Portland, OR: Timber Press.

———. 2004. *Botanical Latin.* Portland, OR: Timber Press.

Taylor, R. B., J. Rutledge, and J. G. Herrera. 1999. *A Field Guide to Common South Texas Shrubs.* Austin: Texas Parks and Wildlife.

Taylor, R. J. 1998. *Desert Wildflowers of North America.* Missoula, MT: Mountain Press.

Texas A&M University Bioinformatics Working Group. 2005. *Digital Flora of Texas, Herbarium Specimen Browser.* College Station. http://www.csdl.tamu.edu/FLORA/tracy2/main1.html.

Tull, D., and G. O. Miller. 1991. *A Field Guide to Wildflowers, Trees and Shrubs of Texas.* Houston, TX: Gulf.

Turner, B. L. 1959. *The Legumes of Texas.* Austin: University of Texas Press.

Turner, B. L., H. Nichols, G. Denny, and O. Doron. 2003. *Atlas of the Vascular Plants of Texas.* Fort Worth: Botanical Research Institute of Texas.

Tveten, J., and G. Tveten. 1993. *Wildflowers of Houston and Southeast Texas.* Austin: University of Texas Press.

Tyler, R. C. 1975. *The Big Bend: A History of the Last Texas Frontier.* Handbook 128. Washington, DC: National Park Service, U.S. Department of the Interior.

USDA, NCRS. 2004. *Plants Database.* U.S. Department of Agriculture, Natural Resources Conservation Service. Baton Rouge, LA: National Plant Data Center. http://plants.usda.gov.

Vines, R. A. 1960. *Trees, Shrubs and Woody Vines of the Southwest.* Austin: University of Texas Press.

———. 1984. *Trees of Central Texas.* Austin: University of Texas Press.

Warnock, B. H. 1970. *Wildflowers of the Big Bend Country, Texas.* Alpine, TX: Sul Ross State University.

———. 1974. *Wildflowers of the Guadalupe Mountains and the Sand Dune Country, Texas.* Alpine, TX: Sul Ross State University.

———. 1977. *Wildflowers of the Davis Mountains and Marathon Basin, Texas.* Alpine, TX: Sul Ross State University.

Wauer, R. H. 1980. *Naturalist's Big Bend.* College Station: Texas A&M University Press.

Welch, M. 2002. *The Administrative History of Big Bend National Park.* Washington, DC: National Park Service, U.S. Department of the Interior.

Wells, D. 1997. *100 Flowers and How They Got Their Names.* Chapel Hill, NC: Algonquin Books.

Weniger, D. 1978. *Cacti of the Southwest.* Austin: University of Texas Press.

West, S. 2000. *Northern Chihuahuan Desert Wildflowers.* Helena, MT: Falcon Publishing.

White, H. L., and W. C. Holmes. 2001. Validation of the name *Orobanche ludoviciana* ssp. *multiflora* (Orobanchaceae). *Sida* 19(3): 623–24.

Whitson, P. D. 1974. *The Impact of Human Use upon the Chisos Basin and Adjacent Lands.* Scientific Monograph 4. Washington, DC: National Park Service, U.S. Department of the Interior.

Wislizenus, A. 1848. *Memoir of a Tour to Northern Mexico, Connected with Col. Doniphan's Expedition, in 1846 and 1847.* Washington, DC: Tippin & Streeter.

Wrede, Jan. 1997. *Texans Love Their Land: A Guide to 76 Native Texas Hill Country Plants.* San Antonio, TX: The Watercress Press.

Yarborough, S. C., and A. M. Powell. 2002. *Ferns and Fern Allies of the Trans-Pecos and Adjacent Areas.* Lubbock: Texas Tech University Press.

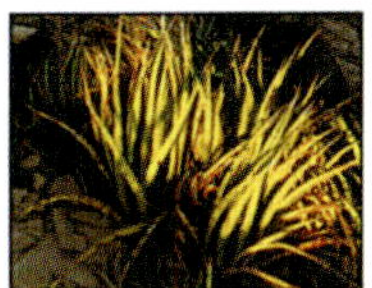

INDEX

Page numbers in bold refer to images.

abori, 171
Acacia
  angustissima, 12
  farnesiana, 8
  greggii, 9
  neovernicosa, 9
  rigidula, 8, **185**
  roemeriana, 12, **186**
Acalypha
  hederacea, 177
  monostachya, **177**
Acanthaceae, 79–86
acanthus (family), 79–86
Acer grandidentatum, 16
Acleisanthes
  angustifolia, 9, **226**, 304
  parvifolia, 9, **227**, 304
A'Court, Mary Gibbs, 102
Acourtia runcinata, **102**
Agarito, 12
Agastache
  micrantha var. *durangensis*, 213
  micrantha var. *micrantha*, **213**
  pallidiflora ssp. *neomexicana* var. *havardii*, **214**
Agavaceae, 36–42
agave (family), 36–42
Agave
  glomerulifolia, 12
  havardiana, 12, **36**
  lechuguilla, 10, **37**
Ageratina wrightii, **103**
agrillo, 90
Alamo Creek, 151, 211, 285
Alamo Spring, 130, 281
albarda, 205
Aletes acaulis, **91**
alicoche, 55
alligator juniper, 13, 15, **171**
Aloysia gratissima, 12
allthorn, 10, 168, **212**, 277
althorn (family), 212
amaranth (family), 87–89
Amaranthaceae, 87–89
Ames, Blanche, 237
Ames, Oakes, 231, 237
Amson, John, 92
Amsonia longiflora var. *longiflora*, **92**
Anacardiaceae, 90
Anemone tuberosa var. *texana*, **256**
angel hair, 173
Anisacanthus
  insignis, 79
  linearis, **79**
  puberulus, 79
Anthony, Margery, 71, 73, 76, 78
Apache Canyon Trail, 54
Apache plume, 12
Apaches, 16–17
Apiaceae, 91
Apocynaceae, 92–95
Aquilegia longissima, **257**
Arbutus xalapensis, 14, **176**
Arch Camp, 38
Argemone chisosensis, **242**
Ariocarpus fissuratus var. *fissuratus*, 10, **43**
Aristolochia coryi, **96**
Aristolochiaceae, 96
Arizona
  carlowrightia, **80**
  cockroach plant, **93**
  cottonwood, 8
  crested coral root, 236
  cypress, 15, **170**
  pine, 15
  snakecotton, **87**
Arundo donax, 8
Asclepiadaceae, 97–101
Asclepias texana, **97**
Asteraceae, 102–127
atempatli, 93
athel, **8**
Atriplex
  acanthocarpa, 11
  canescens, 11
  obovata, 11
Aublet, Jean Baptiste, 100
Avery Canyon, 142, 192
Ayenia microphylla, **281**
azafrán, 147

Baccharis salicifolia, 8
baccharisleaf penstemon, **273**
Bailey Peak, 81, 100, 156, 162, 230, 232, 237, 267
balsam apple, 168
Barneby, Rupert C., 198
barometer bush, 272
Basin. *See* Chisos Basin
Basin and Range physiographic province, 3
Basin campground. *See* Chisos Basin campground
Basin turnoff. *See* Chisos Basin turnoff
Bates, David M., 224
Batesimalva violacea, 16, **224**
batwing passionflower, 243
Bauhin senna, 201
beaked yucca, 41
bearded dalea, **192**
bearded swallow-wort, 98
beargrass. 11, 12, 38, **39**
beebrush, 12
beech (family), 202–204
beetle spurge, 182
Benson, Lyman, 60–61, 93
Bentham, George, 94, 117, 147, 182, 185, 188, 195, 197
Berlandier, Jean Louis, 51, 111, 126, 147, 149, 173, 219, 266, 272
Berlandier flax, **219**, 304
Berlandier lobelia, **149**
Berlandier wolfberry, 11
  Bernardia obovata, **178**

Betulaceae, 128
bicolor mustard, 139
Bidens bigelovii var. *bigelovii*, **104**
Big Bend
  bluebonnet, 11, 165, **196**
  cholla, 9, 48, **71**
  Cory cactus, 48
  devil cholla, 9, **73**
  hophornbeam, 13, 14, **128**
  Last Frontier, the, 20
  life on the edge, 21
  mystique, 18, 20
  Official National Park Handbook, 313
  purplish prickly pear, 10, **75**, 76
  silverleaf, 12
  state of mind, 20
  State Park, 19
Big Bend National Park, xi, 3, 4, 7–8, 18–20
big nipple Cory cactus, 47
Bigelow, John, 41, 50, 66, 77, 91, 101, 104, 108, 115, 119, 134, 197, 215, 218, 227, 267, 282
Bigelow beggarticks, **104**
Bigelow bristlehead, **108**
bigtooth maple, 15, 16
big-needle pincushion cactus, **47**, 48
Bignoniaceae, 129–130
bigpod bonamia, **158**
binomial system, 22
biological island, Chisos Mountains, 18
birch (family), 128
birchleaf buckthorn, 15–16
birdwing passionflower, 243
birthwort (family), 96
biscuit cactus, 44
bittersweet (family), 154–155
Biznaga de chilitos, 66
Biznaga de limilla, 62
black cherry, 262
black dalea, 12, **190**
Black Dike, 9, 72
blackbrush, 111
blackbrush acacia, 8, **185**
bladder sage, 8, **216**
Blake, Sidney F., 127
blanco viznagita, 58
blind prickly pear, 10, 77
Blue Creek, 229, 264
  Canyon, 79–80, 85, 99, 186, 203, 237, 261–262, 267, 274, 276
  Canyon, Lower, 14
  Canyon, Upper, 44, 82, 202, 237
  mouth of, 141, 211, 285
  Trail, 160
  trailhead, 93
blue morning-glory, **163**
bluebell (family), 149
Bois d'Arc Spring, 160
Boke, Norman, 61
Boke's button cactus, 10, **61**, 65
Bonamia
  ovalifolia, **158**
  repens, **159**, 304
Bonamy, François, 158
Bonpland, Aimé, 107, 123–124, 148, 153, 160, 174, 176, 230, 264, 280
Boot Canyon, 15–16, 44, 107, 163, 166, 170–172, 190, 199, 209, 217, 232, 235, 237, 240, 254, 266, 275
Boot Canyon, upper, 206, 265
Boot Creek, 87, 150, 169, 213
Boot Creek, Upper, 16
Boot Spring, 36, 76, 87, 97, 113, 125, 150, 162, 167, 169, 176, 204, 209, 213, 228, 232, 238, 240, 254–255, 269
Boot Spring Trail, 91, 123, 138, 213
Boquillas, Mexico, 6, 17, 77–78, 89, 193
Boquillas button cactus, 61
Boquillas Canyon, 5–9, 12, 49, 55, 77–78, 80, 96, 101, 109, 130, 136, 139, 143, 145–146, 178–180, 182, 191, 200–201, 221, 223, 226, 253, 271–272, 276, 278, 281–282, 286
Boquillas Flats, 55, 132, 180
Boquillas silverleaf, 12, **271**
borage (family), 131–137
Boraginaceae, 131–137
Botanical Nomenclature, International Code of, 23
botanical nomenclature, laws of, 23
Botanary—The Botanical Dictionary (Web site), 314
bottomwhite rocktrumpet, **94**
Bouché, Carl, 282
Bouché, Peter, 282
Bouchea spathulata, **282**
Boulder Meadow, 108
boundary ephedra, 11
Boutelou, Claudio, 245
Boutelou, Esteban, 245
Bouteloua ramosa, 11, **245**
Bouvard, Charles, 265
Bouvardia ternifolia, 14, **265**
bowl flax, 219
Brack, Steven, 74
Brassicaceae, 138–145
Brewer, William H., 240
Brickell, John, 105
Brickellia
  conduplicata, 106
  laciniata, **105**
  lemmonii var. *conduplicata*, **106**
  shineri, 112
  veronicifolia, **107**
  veronicifolia var. *petrophila*, 107
bristly bedstraw, **266**
Britton, Nathaniel, 44, 68
broadpod twistflower, 143
Bromeliaceae, 146
Brongniart, Adolphe Theodore, 187
Brongniartia minutifolia, 8, **187**
broom milkwort, **250**
broomrape (family), 240–241
brown-flowered cactus, 57
buckbrush, 258
buckthorn (family), 258–260
Buddle, Adam, 147
buddleja (family), 147–148
Buddleja
  marrubiifolia, **147**
  scordioides, **148**
Buddlejaceae, 147–148
bunched Cory cactus, 48
bur marigold, 104
Burro Mesa, 54, 70, 120, 130, 188, 191, 205, 207, 227, 245, 249, 259–260, 278, 281
Burro Spring, 85, 96, 130, 226–227, 286
burrobrush, 8

Cabeza de Vaca, Alvar Nuñez, 16
cacao (family), 281
Cactaceae, 43–78
cactus (family), xi, 43–78
cactus, illegal trade in, 18
Caesalpinia parryi, 200, 304
California Plant Names (Web site), 314
Calliandra
  herbacea, 188
  humilis var. *humilis*, **188**
caltrop (family), 286–287
Campanulaceae, 149
Campground Canyon, 208. *See also* Pulliam Canyon
candelilla, 10, 18, **181**
candle cholla, 9, **72**
candlewood, 205
cane cholla, 70
canyon spiderwort, 157
capul, 155
capulin, 262
Cardamine macrocarpa var. *texana*, **138**
Carlowrightia
  arizonica, **80**
  serpyllifolia, **81**
Carneros yucca, 40
Carphochaete bigelovii, **108**
Carr, William R., x, 314
Caryophyllaceae, 150–153
Casa Grande, 39, 91, 97, 105, 107–108, 112, 116, 156–157, 171, 206, 214–215, 217–218, 254
Castilleja
  mexicana, **270**
  tortifolia, 270
Castillejo, Domingo, 270
Castolon, 3, 8, 18, 57, 62, 88, 121, 127, 131, 140, 168, 182, 192, 198, 212, 216, 219, 221
catalpa (family), 129–130
catclaw acacia, 9, 186
catclaw cactus, 13, 60, **63**, 304
catclaw mimosa, 12
cattail, 8
Cattail Canyon, 14
Cattail Canyon, Upper, 104, 107, 113, 123–125, 138, 161, 174, 204, 232, 240, 258, 270, 275
Cavanilles, Antonio José, 123, 125, 129, 156, 177, 200, 217, 220, 255, 259, 265
Ceanothus greggii, 14, **258**
cedro blanco, 170
Celastraceae, 154–155
Celtis
  laevigata var. *reticulata*, 14
  pallida, 12
cenizo, 12, 271–272
century plant, 36
Cerastium axillare, **150**
Cercocarpus montanus, 14

Cevallia sinuata, **220**
Chaffey, Elswood, 44
Chaffey's pincushion cactus, 15, **44**
Chamaesaracha villosa, 278
Chamaesyce
chaetocalyx var. *chaetocalyx*, 179
chaetocalyx var. *triligulata*, **179**, 304
Chamisso, Adelbert von, 172, 282
chaparro prieto, 185
chatterbox orchid, 233
chautle, 43
checkerbark juniper, 171
Chihuahua Trail, 17
Chihuahuan Desert, xi, 6–7, 10–12, 18–19, 37–38, 40, 57, 75, 88, 103, 111, 122, 128, 139, 181, 199–200, 205, 216, 249, 287
Chilicotal Mountain, 85, 181, 193, 207, 248–249, 268
Chilicotal Spring, 142
Chilopsis linearis ssp. *linearis*, 8, **129**
Chimneys Trail, 7
Chinese Wall Trail, 107, 123, 152, 209
chino grama, 11, 18, 21, 38, 56, **245**
Chisos
agave, 12
Basin, 3, 13–14, 36, 44, 76, 81–82, 90, 97–106, 110–111, 119–120, 129, 132, 138, 155, 157, 161–163, 171–172, 176, 190, 194, 203, 206, 215, 217, 219, 228, 235, 242, 244, 251–252, 254, 258, 261, 264–265, 267, 269, 274–275, 284
Basin campground, 284
Basin turnoff, 66, 173
bluebonnet, 196
coral root, 235
foothills, 6, 9, 11–13, 18, 37–38, 46, 50, 63, 70, 79–81, 85, 87, 90, 98, 122, 130, 137, 139, 159, 163–164, 173, 184, 190–193, 200–201, 210, 212, 248, 250, 252, 260–261, 271–272, 283
hedgehog cactus, 10, **52**
Mountain brickellbush, **112**, 304
Mountain buckwheat, **251**
Mountain false calico, 247
Mountain hedgehog, 52
Mountains, 3–4, 6, 11, 14–16, 18–19, 36, 44, 52, 56, 62, 66, 75–76, 82, 87, 90–91, 94, 98–99, 101, 103, 107, 109, 111–113, 119–121, 124–125, 128, 135, 138, 153, 160, 163, 165–166, 170–172, 174, 178, 185, 193, 199, 202–204, 206–208, 213–214, 217–219, 223–224, 228–232, 234–235, 237–240, 244–245, 247, 251, 254–255, 262, 266, 275, 280, 284
Mountains (west slopes), 14, 16, 58, 81, 84, 111, 233, 236–237, 257
oak, 14, **202**, 277
Pens, 121, 154, 192, 222, 227, 248, 279
prickly pear, 15, **76**
pricklypoppy, **242**
red oak, 13, **203**
tribe, 16
woodlands, 13–16, 44, 53
Chloracantha spinosa, 8
Chodat, Robert H., 250
Choisy, Jacques D., 173
cinnabar ladies tresses, 232
Cirsium turneri, **109**
Cissus
incisa, 285, 304
trifoliata, **285**, 304
citrus (family), 268
Clary, Karen, 40–41
clavillo, 265
Clayton, John, 92, 213
Clelland, R., xi, 314
Cleomaceae, 156
cleome (family), 156
cliff thistle, 109
club cholla, 9, **73**
clumped dog cholla, **73**
coachwhip, 205
Coahuila scrub oak, 14, **204**
cob cactus, 13, 44–45, 49, **50**, 67
cola de alacrán, 133
Coldenia greggii, 137
Colima Trail, 160, 172, 235
Comanche War Trail, Great, 17
Comanches, 16–17
Commelinaceae, 157
common button cactus, 61
common hoptree, 268
common reed, 8
Condal, Antonio, 259
Condalia
ericoides, **259**
viridis, **260**
warnockii, 10
conejitos, 196
confused ladies tresses, 231
Conoclinium dissectum, **110**, 304
Conopholis alpina var. *mexicana*, **240**
Convolvulaceae, 158–163
corona de Cristo, 212
Correll, Donovan, 123, 128, 150, 235, 237, 279, 313
correosa, 90
Cory cryptantha, **131**
Cory dutchman's pipe, **96**
Cory, Victor, 96, 131, 138, 143, 145
Coryphantha
chaffeyi, 15, **44**
duncanii, **45**
echinus, **46**
macromeris var. *macromeris*, **47**
ramillosa, 9, **48**
sneedii var. *albicolumnaria*, **49**
sneedii var. *sneedii*, 49
tuberculosa var. *tuberculosa*, 13, 44–45, 49, **50**
tuberculosa var. *varicolor*, 50
Cottonwood Campground, 141, 151
Cottonwood Creek, 154, 222, 227, 279
Coulter, John M., 91, 121, 270
Coulter, Thomas, 72, 188
cow-itch, 285
coyonostyle, 70
coyote candles, 70
coyote willow, 8
Crassulaceae, 164–167
creeping cliff brake, **254**
creeping rockvine, **159**, 304
creosote (bush), 8, 10, 37, 43, 47, 52, 57, 59–60, 71, 111, 187, **287**
creosote flats, 9, 71, 277
creosote scrub, 279
crested coral root, **236**
crestrib morning-glory, **162**
Cretaceous period, 5
cristate, 61
critically imperiled (species), xi, 14, 44–45, 52, 71, 74, 93, 107, 118, 123, 127–128, 135, 158, 170, 179–180, 187, 202, 207, 209, 211, 217, 224, 235, 238, 247, 264, 282, 291–293
Crooks, D. M., 93
Croton
pottsii var. *pottsii*, 180
pottsii var. *thermophilus*, **180**
Croton Spring, 201, 226
crowded heliotrope, 132
Crown Mountain, 15, 82, 100–101, 128, 160–161, 207, 252, 264, 279, 285
crow's foot (family), 256–257
Crozier, Bonnie, 256
crucifixion thorn, 277
Cryptantha palmeri, **131**
Cucurbitaceae, 168–169
Cupressaceae, 170–172
Cupressus arizonica, 15, **170**
curly coral root, **235**, 277
Cuscuta indecora, **173**
Cuscutaceae, 173
Cutak, Ladislaus, 48
cutleaf brickellbush, 105
cutleaf penstemon, 273
Cutler, Hugh C., 143
Cutler twistflower, **143**
Cylindropuntia (section), 70–72
Cynanchum
pringlei, **98**
racemosum var. *unifarium*, **99**
Cyperaceae, 174
Cyperus seslerioides, **174**
Cyphomeris gypsophiloides, **228**
cypress (family), 170–172

Dagger Flat, 11, 14, 40–41, 60, 68, 111, 115, 143, 168, 184, 189, 195, 212, 229, 241, 249, 263, 277, 287

Dale, Samuel, 191
Dalea
formosa, 12, **189**
frutescens, 12, **190**
neomexicana, **191**
pogonathera var. *pogonathera*, **192**
pogonathera var. *walkerae*, 192
wrightii var. *wrightii*, **193**
Dasylirion leiophyllum, 11, **38**
Davis, A. R., 48
de Candolle, Alphonse, 23, 149, 204
de Candolle, Augustin, 23, 51, 68, 72, 111, 126, 149, 164, 204, 266, 287
Dead Horse Mountains, 3, 6, 9–11, 40–43, 46, 57, 59–63, 65, 68, 73, 84, 87, 89, 92, 96, 98, 102–103, 109, 115, 117–118, 121, 130–132, 136–137, 143, 146–147, 155, 159, 164, 168, 177–178, 180–181, 184, 193–195, 200, 223, 228–230, 243–245, 248–250, 259–261, 263–264, 267–269, 271–273, 276, 278, 281–282
Dead Man's Cut, 84
Deiregyne confusa, **231**
dense ayenia, **281**
Deppe, Ferdinand, 171
desert
anemone, 256
candle, 38
catalpa, 129
ceanothus, 14, **258**
Christmas cholla, 9, 72
hackberry, 12
lantana, 21, 283
marigold, 196
myrtlecroton, **178**
olive, 12, **229**
pavement, 7
savior, 164
sumac, 12, 14, **90**
turtleback, 122
velvet, 122
willow, 8, **129**
yaupon, 12, **155**, 305
Desert-Mountain Overlook, 196
Desfontaines, René L., 283
Devil's Den, 102, 194
devil's head, 51
devil's pincushion, 51
devil's pitchforks, 104
devil's rope, 70
d'Iberville, Pierre Le Moyne, 168
Dichondra
argentea, **160**
brachypoda, **161**
Dichromanthus cinnabarinus, **232**
dike, 6
Dioscorides, Pedanius, 85, 138, 279
Diospyros texana, 12
Dobie, J. Frank, 18, 19
dodder (family), 173
Dodson Trail, 156
Dodson Trail, west end, 93, 159, 200
Dog Canyon, 51, 64, 102, 136, 140, 143, 194, 226, 229, 278
Dog Canyon Trail, 86
dog cholla, 52, 71, 73
Dog Flats, 11, 64, 86, 111, 114, 122, 132–134, 137, 139–140, 143–145, 151, 191, 210, 225, 246, 249
dog turd cholla, 73
dogbane (family), 92–95
Dominguez Spring, 80, 146–147, 155, 168, 197, 252
Don, David, 102
Doniphan, Col. Alexander, 47
doradilla, 276
Douglas, David, 133, 206, 233
Douglas fir, 15, 233
dragon's blood, 10, 183
Dripping Spring, 197
drooping juniper, 13, 15, **172**
Drummond, Thomas, 185
dry whiskey, 43
Drymaria pachyphylla, **151**
Dugout Wells, 90, 139, 190, 197, 212, 220
Duncan, Carl, 242
Duncan, Frank, 45
Duncan's pincushion cactus, **45**, 54
Duncan's snowball, 45
Durango senna, 201
dutchman's breeches, **268**, 304
dwarf anisacanth, **79**
dwarf fairy duster, **188**
dwarf oak, 204
Dyschoriste
decumbens, 82
schiedeana var. *decumbens*, **82**

eagle-claw cactus, 63, 304
earlobe mustard, **141**
early bloomer, 58
Echeveria strictiflora, **164**
Echeverría y Godoy, Atanasio, 164
Echinocactus
horizonthalonius, 304
texensis, 10, **51**
Echinocereus
chisoensis var. *chisoensis*, 10, **52**
coccineus var. *paucispinus*, 15, **53**
dasyacanthus, 10, **54**
enneacanthus var. *brevispinus*, 55
enneacanthus var. *enneacanthus*, **55**, 304
stramineus var. *stramineus*, 55, **56**, 304
viridiflorus var. *russanthus*, 13, 46, **57**
Echinomastus
intertextus, **58**
mariposensis, 58, **59**
warnockii, 58, **60**
Echols, Lt. William, 20
Edwards, L. A., 114, 192
Edwards Nicollet, **114**
Edwards' hole-in-the-sand plant, 114
eggleaf silktassel, 14, **206**
Ehrenberg, Carl A., 172
Ehrenberg adder's mouth, 238
El Despoblado, 16, 20
El Pico, 6
Elephant Tusk, 264
Elliott, Stephen, 105
Emory mimosa, 12, **197**, 198
Emory oak, 13–14, 202–203
Emory Peak, 3, 13, 16, 44, 91, 97, 103, 116, 125, 128, 153, 157, 162–163, 172, 176, 203, 209, 217–218, 235, 240, 244, 262, 266, 275, 280
Emory Peak Trail, 76, 113, 124, 174, 255, 275
Emory, Maj. William, 197
Emory boundary survey. *See* U.S.–Mexico boundary survey, Major William Emory's
Endlicher, Stephen, 269
enebro, 171
Engelmann, George, 43, 46–47, 50, 53–54, 56, 58, 63, 65–66, 69–70, 77, 90, 205–206, 243, 252, 258, 286, 314
Enquist, Marshall, 256
ephedra (family), 10–11, 175
Ephedra
aspera, 11
torreyana var. *powelliorum*, **175**
trifurca, 11
Ephedraceae, 175
Epipactis gigantea, **233**
Epithelantha
bokei, 10, **61**
micromeris var. *micromeris*, 61
Ericaceae, 176
Ericameria triantha, 127
Eriogonum hemipterum var. *hemipterum*, **251**
Ernst Basin backcountry campsite, 118, 243, 273
Ernst Tinaja, 96, 126, 143, 159, 193, 200, 221
Escobaria
albicolumnaria, 49
chaffeyi, 44
escobilla butterflybush, **148**
esperanza, 130
Esteve, Pedro Jaime, 123
Eucnide bartonioides, **221**
Eupatorium
greggii, 110
wrightii, 103
Euphorbia
antisyphilitica, 10, **181**
eriantha, **182**
Euphorbiaceae, 177–184
Evans, Walter H., 208
evergreen sumac, 14
Eysenhardt, Carl Wilhelm, 194
Eysenhardtia texana, 14, **194**

Fabaceae, 185–201
Fagaceae, 202–204
Fallugia paradoxa, 12
false mesquite, 188
fascicled bluet, 14, **267**
Faxon, Charles, 40
Faxon yucca, 40
feather dalea, 12, **189**, 190
feather plume, 189
featherleaf desert peony, 102
Fendler, Augustus, 70, 207, 209

Fendlera
linearis, **207**, 304
rigida, 304
Ferguson, David J., 76
Ferguson, William, 20
fern acacia, 12
Ferocactus
hamatacanthus var. *hamatacanthus*, **62**
hamatacanthus var. *sinua tus*, 62
Ferris, Roxana S., 242
fifth season, 7
figwort (family), 12, 270–275
firecracker bush, 14
Fisk Canyon, 149
flattened mammillaria, 64
flax (family), 219
fleshy tidestromia, 11, **88**
flor de mimbre, 129
flor de piedra, 276
flor de San Juan, 95
Flora of North America (Web site), 313–314
Flora of Texas Database, University of Texas, 314
Flora of the Chihuahuan Desert Region, 304, 313–314
Flourens, Marie Jean Pierre, 111
Flourensia cernua, 10, **111**
flower of stone, **276**
Flyr, David, 112
Flyriella parryi, **112**, 304
foothill basketgrass, 39
Forestier, Charles Le, 229
Forestiera angustifolia, 12, **229**
Fouquier, Pierre E., 205
Fouquieria splendens, 10, **205**
Fouquieriaceae, 205
four-o'clock (family), 226–228
fourwing saltbush, 11
foxtail cactus, 67
fragrant ash, 14
Fraxinus
cuspidata, **14**
greggii, 14
Frémont, John C., 77, 216, 268
Fresno Creek, 9, 102, 149
Fresno Spring, 149, 156, 248, 253, 285
Fries, Elias M., 113
Froelich, Joseph von, 87
Froelichia arizonica, **87**
Fryxell, Paul A., 224
fuzzy mammillaria, 65
Galeotti, Henri G., 228, 234, 247
Galium uncinulatum, **266**
Gano Spring, 121, 129, 168, 191, 197, 212, 220
Garay, Leslie, 231
Garry, Nicholas, 206
Garrya ovata ssp. *lindheimeri*, 14, **206**
Garryaceae, 206
gavia, 185
Gentry, Howard S., 279
giant
cane, 8
coral root, **234**
dagger, 11, **40**, 41, 60, 196, 304
fishhook cactus, **62**
helleborine, **233**
glandleaf milkwort, **249**
Glandulicactus uncinatus var. *wrightii*, 13, 60, **63**, 304
Glenn Draw, 114
Glenn Spring, 8, 12, 85, 114, 122, 129, 132, 145–146, 149, 155, 181, 187, 194, 200, 205, 222, 230, 268, 271, 277–278
Glenn Spring Road, 241
glory of Texas cactus, 9, **69**
gobernadora, 287
goldenball leadtree, 195
golden-spined prickly pear, 9, 48, **74**, 76, 78
golf ball cactus, 10, 45, 54, 59, **65**
golondrilla, 148
Gomez, A. R., 313
Goodding willow, 8
gordo lobo, 247
gourd (family), 168–169
Government Spring, 85, 131, 197, 205, 274, 277
Graham, Colonel James, 73
Graham boundary survey. *See* U.S.–Mexico boundary survey, Colonel Graham's
Graham dog cholla, **73**
grape (family), 285
Grapevine Hills, 12, 79, 81, 131, 137, 178, 197, 223, 250, 261, 272
Grapevine Hills Road, 140
Grapevine Spring, 129, 205
Grashoff, Jerold L., 123
grass (family), 245
Gravel Pit, 47
Graves, H. S., 203
Graves' oak, 13, 203
Gray, Asa, 37, 56, 70, 80–82, 86, 98, 100–108, 110, 112, 114–120, 122, 139, 141–142, 144, 152, 154–155, 163–164, 166, 168, 188, 190–193, 199, 207, 209–210, 213–214, 218, 222–223, 225, 230, 240, 246, 249, 258–259, 267–268, 274–275, 281
Gray bean, **199**
gray oak, 13–14, 235
green condalia, 259, **260**
green gromwell, **135**
Green Gulch, 13–14, 36, 53, 70, 87, 90, 101, 105, 107, 120, 153, 155, 162–163, 171, 173, 176, 188–190, 194–195, 203, 244, 254, 261, 265, 267, 274, 284
Green Gulch, Lower, 105
Green Gulch, Upper, 14, 76, 97, 105–106, 108, 112–113, 116, 138, 156, 172, 217, 247, 262, 275
green puccoon, 135
green snakewood, 260
Greene, Edward L., 135, 170
greeneye heliotrope, **133**
Greenman, Jesse M., 211, 234
Greenman's hexalectris, 234
Gregg, Josiah, 79, 83, 110, 137, 139, 144, 204, 247, 258, 286, 314
Gregg ash, 14
Gregg keelpod, **144**
Gregg tiquilia, 137
Greville, Robert K., 276
Griffiths, David, 87
Gronovius, Johannes F., 213
Grusonia (section), 73
Guaiacum angustifolium, 8, **286**
guame, 287
guauyul, 264
guayacán, 8, **286**
guayule, 11, 18, 68, **115**
guirambo, 94

Hagen, Stanley, 79
hairy false pennyroyal, 215
hairy hedeoma, **215**
hairy tubetongue, **83**
Hannold Draw, 268
Hanson, Craig A., 279
Haplophyton crooksii, **93**
Harte Ranch, 19, 98, 100, 126, 134, 151, 156, 168, 210, 220, 225
Hartweg, Carl Theodore, 94
Harvard University Herbaria (Web site), 314
harvest lice, 104
Havard, Valery, 36, 38, 146, 165, 187, 196, 201, 210, 214, 239, 246, 261, 274
Havard
agave, 12, **36**, 165
giant hyssop, **214**
ipomopsis, **246**
maguey, 36
nama, **210**, 211
penstemon, **274**
plum, 14, **261**
standing cypress, 246
stonecrop, **165**, 166
Hawksworth, Frank, 284
Hawksworth mistletoe, 284
heartleaf rockdaisy, **120**, 305
heath (family), 176
Heath Canyon, 282
Heath cliffrose, **263**
Heath Creek, 177
Hecht, Julius G., 146
Hechtia texensis, 10, **146**
Hedeoma mollis, **215**
hedgehog Cory cactus, 46
hediondilla, 287
Hedyotis intricata, 14, **267**
Heil, Kenneth D., 74
Heliotropium
confertifolium, **132**
glabriusculum, **133**
molle, **134**
Heller, A. Arthur, 97, 253
Hemsley, William B., 270
Henrickson, James, x, 82, 128, 304, 313
Herbarium Specimen Browser, Texas A&M University, 314
Hester, J. Pinckney, 45, 49, 59
Hexalectris
grandiflora, **234**
revoluta, 234, **235**
spicata var. *arizonica*, 236
spicata var. *spicata*, **236**
warnockii, **237**
Heyder, Edward, 64
Heyder's pincushion cactus, 10, **64**, 66
Hidalgo ladies tresses, **231**
Hieracium schultzii, **113**
hierba

de la cucaracha, 93
de las escobas, 148
del buey, 285
del corazon, 192
del coyote, 156
negra, 283
hikuli mulato, 61
Hill, Robert T., 18
Hilsenbeck, Richard, 73, 82–83
Hoffmannseggia melanosticta var. *parryi*, 200
hojase, 111
hojasen, 111
Holacantha stewartii, **277**
Holmes, W. C., 304
honey mesquite, 8–9
Hooker, Joseph Dalton, 273
Hooker, Sir William Jackson, 219, 233, 273, 276
Hopffer, Carl, 51
horse-crippler cactus, 10, **51**
Hot Springs, 9, 65, 78, 89, 102, 114, 117, 121, 127, 129, 141, 145, 181, 183, 186, 189, 191–193, 201, 210, 220–221, 226, 228, 253, 260, 271, 278, 281, 283, 286–287
Hot Springs Trail, 8, 83
huisache acacia, 8
human influence, 16–18
Humboldt, Baron Alexander von, 107, 123–124, 148, 153, 160, 174, 176, 230, 264, 280
hydrangea (family), 207–209
Hydrangeaceae, 207–209
Hydrophyllaceae, 210–211
Hymenoclea monogyra, 8

Ibervillea tenuisecta, **168**
imperiled (species), xi, 11, 15–16, 44–45, 49, 88, 112, 119, 128, 135, 138, 143, 159, 165, 180,187, 208, 222, 224, 234–235, 237–238, 251, 255, 257, 291–293
indicator species, 10, 37, 111
injerto, 284
inkweed, 151
intermediate cliff brake, 254
International Plant Names Index (Web site), 314
invasive species, 18
Ipomoea
costellata, **162**
lindheimeri, **163**
Ipomopsis havardii, **246**
Iturriaga, José, 259
ivy treebine, **285**, 304

Jacob's staff, 205
James, Edwin, 133, 152, 189, 314
James nailwort, **152**
Jameson, J., 313
Janusia gracilis, 11, **223**
Jatropha
dioica var. *dioica*, 183
dioica var. *graminea*, 10, **183**
javelina bush, **259**
Jefea brevifolia, 12
Johnson Ranch, 140, 182, 219, 221, 229, 242
Johnston, Ivan, 88, 96, 131, 134, 159, 178, 260, 271
Johnston, Marshall C., 123, 128, 180, 304, 313
Juba II, King of Mauretania, 181
Jumanos, 16
junco, 212
Juniper Canyon, 176, 190, 194, 236, 242
Juniper Canyon, Lower, 160, 191
Juniper Canyon, Upper, 155, 172
Juniper Spring, Lower, 98, 101, 164, 267
Juniper Spring, Upper, 202, 237
Juniperus
deppeana var. *deppeana*, 13, **171**
flaccida, 13, **172**
pinchotii, 13
Jussieu, Bernard de, 178, 281
Justice, James, 84
Justicia
pilosella, **83**
warnockii, **84**

Karwinski von Karwin, Baron Wilhelm Friedrich, 104, 181, 212, 221, 244
K-Bar Ranch, 155, 168
Kiger, Robert W., 252
King, Robert, 112
Klein cholla, 72
Klotzsch, Johann F., 146, 180
knotweed (family), 251
knotweed leafflower, **184**
Koeberlin, Christoph Ludwig, 212
Koeberlinia spinosa, 10, 168, **212**, 277
Koeberliniaceae, 212
Kuhn, Friedrich, 254
Kunth, Carl S., 107, 123–124, 130, 148, 174, 194, 222, 280
Kuntze, Otto, 82

La Clocha, 65
La Llave, Pablo de, 232, 239
lacespine pincushion, 65
lady's leg, 176
Lagasca y Segura, Mariano, 220
Laguna Meadow, 14, 94, 103, 113, 124, 161, 163, 171, 204, 228, 235, 242
Laguna Meadow Trail, 14, 82, 95, 106–107, 110, 123, 161, 172, 188, 190, 206, 217, 247, 251, 254, 258
Laguna West, 258
Lajitas, 3, 5, 16, 46, 49, 63, 67, 69, 127, 198, 212
Lamiaceae, 213–218
Lantana
achyranthifolia, 283, 304
macropoda, 12, **283**, 304
Larrea, Juan Antonio Pérez Hernandez de, 287
Larrea tridentata, 10, **287**
Last Frontier, the, 20
lateleaf oak, 16
leafy heliotrope, **132**, 133
leatherstem, 10, **183**
Leaton, Ben, 17
lechuguilla, 10, 12, 15, **37**, 38, 43, 52, 56, 59–60, 65, 71, 86, 146, 187, 196, 245
Leclerc, Jules, 21
legume (family), 185–201
Lemmon, John, 106
Leucaena retusa, 14, **195**
Leucophyllum
candidum, 12, **271**
frutescens, 12, **272**
minus, 12
violaceum, 271
Levin, Rachel A., 226
Lexarza, Juan José Martínez de, 232, 239
licorice marigold, **125**
Liggio, J. and A. O., 305, 313–314
limerock brookweed, **253**
Limpia seymeria, **275**
Linaceae, 219
Lindheimer, Ferdinand, 51, 90, 99–100, 163, 186, 190, 194, 206, 229–230, 243, 248, 252, 286
Lindheimer morning-glory, 163
Lindheimer silktassel, 206
Linnaeus, Carolus (Carl von Linne), 22, 85, 125, 130, 141, 147, 160, 176, 191, 281, 285
Linum berlandieri var. *filifolium*, **219**, 304
Lippia graveolens, 8
Lithospermum viride, **135**
little buckthorn, 259
little chilis, 66
littleleaf
brongniart, 8, **187**
leadtree 14, **195**
mockorange, 15–16, **209**
moonpod, 9, **227**, 304
sumac, 90
live forever, 164
living rock cactus, 10, 22, **43**, 61
Lloyd, Francis, 68
Lloyd's mariposa, 59
Loasaceae, 220–221
l'Obel, Matthias de (Lobelius), 149
Lobelia berlandieri var. *brachypoda*, **149**
Loeselia greggii, 247
Loeselius, Johannes, 247
Löfling, Pehr, 259
Lone Mountain, 139, 191, 248, 268, 271–272, 274, 283
Long Expedition, 152, 189
long mamma, 47
longleaf ephedra, 11
longpetal echeveria, **164**
longspur columbine, 8, **257**
loosestrife (family), 222
Lost Mine Peak, 97, 112, 123, 138, 150, 153, 204, 206, 217–218, 240, 244
Lost Mine Trail, 53, 76, 106–108, 113, 152, 171–172, 202, 215, 237, 247, 254
love vine, 173
Lupinus havardii, 11, 165, **196**
Lycium
berlandieri, 11

puberulum var. *berberi-oides*, 11, **279**
lyreleaf twistflower, **142**
Lythraceae, 222

Macbride, James F., 201
madder (family), 265–267
madron, 176
madroño, 176
Mahonia trifoliolata, 12
maidenhair fern (family), 254–255
Malaspina Expedition, 125, 255
Malaxis
ehrenbergii, 238
wendtii, **238**, 304
mallow (family), 224–225
malpighia (family), 223
Malpighiaceae, 223
Malvaceae, 224–225
Malvella lepidota, **225**
Mammillaria
heyderi var. *heyderi*, 10, **64**
lasiacantha, 10, 45, **65**
meiacantha, 12, 63–64, **66**
pottsii, 9, **67**
manca caballo, 51
manyflowered broomrape, **241**, 304
Maple Canyon, 16, 109, 112, 138, 214–215, 257
margined rockdaisy, 120, **121**
Marie Antoinette, Queen of France, 136
marine ivy, 285
mariola, 11
Mariposa cactus, 58, **59**
Mariscal Canyon, 5, 16, 86, 139, 146, 187, 195
Mariscal cholla, 71
Mariscal Mine, 17, 187, 191, 246, 277
Mariscal Mountain, 3–5, 8–10, 17, 43, 46, 67, 71, 73, 181, 185, 193, 242, 248, 271
Marshall, William, 52
Martens, Martin, 228, 247
Marufo Vega Trail, 180
mat nama, **211**
Matelea reticulata, **100**
Maverick Badlands, 7, 11, 57, 88, 121–122, 130, 227, 287
Maverick Mountain, 69
Maxwell, Ross, 313
McKinney Spring, 84, 92, 98, 102, 129, 146–147, 178, 181, 193–194, 221, 226, 253, 261, 263, 281
McVaugh, Rogers, 183, 219
Mearns, Edgar A., 199, 208
Mearns mockorange, **208**
mejorana, 12, 21, **283**, 304
Menodora longiflora, **230**
Mesa de Anguila, 3, 5–6, 10, 46, 63, 67, 89, 92, 96, 109, 131, 136, 146, 216, 220, 263
mesa greggia, **139**, 144, 196
mesa sacahuista, 39
Mescalero Apaches, 16
mesquite, 8–10, 18, 47–48, 55, 71
Mettenius, Georg Heinrich, 254
Mexican
buckeye, 13–14, **269**
cancer-root, 240
clammyweed, **156**
devilweed, 8
Indian paintbrush, 270
navelseed, **136**
navelwort, 136
pinyon pine, 13, **244**
Plateau, 3, 7
silktassel, 206
squawroot, **240**
starwort, **153**
tarragon, 125
Mexican Boundary Commission, 111, 126, 272
Michoacán ladies tresses, **239**
milkweed (family), 97–101
milkwort (family), 248–250
Miller, Philip, 71, 178, 201
mimbre, 129
Mimosa
aculeaticarpa var. *biuncifera*, 12
emoryana var. *emoryana*, 12, **197**
turneri, 8, **198**
mint (family), 213–218
mirto, 265
Missouri Botanical Garden (Web site), 314
mistflower, 110
mistletoe (family), 284
Mociño, José Mariano, 156, 164, 287
Moench, Conrad, 87
moist Chisos woodland. *See* vegetation zones, moist Chisos woodland
monilla, 269
monsoon rains, summer, 3–4, 7, 88, 216, 287
Moore, John A., 79, 88, 178–179
Mormon tea, 10–11, 175
morning glory (family), 158–163
Morton, Samuel, 154
Mortonia
scabrella, 154
sempervirens ssp. *scabrella*, **154**
sempervirens ssp. *sempervirens*, 154
Moss Well, 150, 153
Mount Emory. *See* Emory Peak
mountain mahogany, 14
mountain sage, 14, **217**
mountain starwort, 153
mountain tailleaf, 116
Mouse Canyon, 262
Muehlenpfordt, F., 62, 64
muerdago, 284
muicle, 79
Mule Ears Spring, 253, 285
Mule Ears Spring Trail, 63, 260
Müller, Cornelius H., 202, 235, 237, 251, 277
Müller, M. T., 235
Muskhog Spring, 89, 115, 132, 177, 181, 227
Muskhog Spring Canyon, 230
mustard (family), 138–145

naked brittlestem, **122**
naked Indian, 176
naked turtleback, 122
Nama
havardii, **210**
torynophyllum, **211**
narrowleaf cottonwood, 8
narrowleaf fendlerbush, **207**, 304
narrowleaf forestiera, 229
narrowleaf moonpod, 9, **226**, 304
Nature Conservancy of Texas, 305, 314
NatureServe, 305, 314
Nealley, Greenleaf C., 121, 245
Nectoux, Hyppolite, 280
Nectouxia formosa, **280**
Née, Luis, 125, 255
Nees, Christian, 83
Nelson, Barney, 313
Neolloydia conoidea var. *conoidea*, **68**
Nerisyrenia camporum, **139**, 144
Nesaea longipes, **222**
Nesom, Guy L., 127
netleaf hackberry, 14
netted milkvine, 100
New Mexico dalea, **191**
New Mexico giant hyssop, 214
New Mexico ponyfoot, **161**
New York Botanical Garden (Web site), 314
Nicollet, Joseph N., 114
Nicolletia edwardsii, **114**
Nicotiana glauca, 8
nightshade (family), 11, 278–280
nipple beehive, 47
nipple cactus, 12, 63–64, **66**, 67
Noailles, Louis de, Duc d'Ayen, 281
Nolin, Abbé C.P., 39
Nolina erumpens, 11, **39**
Northeast Rim, 15
Northeast Rim Trail, 255
notched leadtree, 195
Nugent Mountain, 6, 12, 155, 241, 268, 271, 274, 283
Nuttall, Thomas, 140, 184, 241
Nyctaginaceae, 226–228

Oak Creek, 99, 104–105, 111, 130, 162, 184, 186, 189, 229, 251, 261, 283
Oak Creek Canyon, 39, 81, 85, 90, 95, 156, 176, 190, 194, 202–203, 251, 264–265, 274
Oak Creek Canyon, Lower, 58, 87, 161, 236
Oak Spring, 62, 79, 101, 192–193, 230, 253
Oak Spring Trail, 79, 162
obovateleaf saltbush, 11
ocotillo, 10, 15, **205**
ocotillo (family), 205
Old Maverick Road, 88, 114, 182
Old Ore Road, 65, 73, 84, 115, 159, 180, 227, 259, 282
Old Shag, 42
Oleaceae, 229–230
olive (family), 229–230
Omphalodes aliena, 136
one-gland clammyweed, 156

Onion Spring, 87, 102, 132, 152, 252
Opuntia (section), 74–78
Opuntia
aggeria, **73**
azurea var. *aureispina*, 9, **74**, 78
azurea var. *parva*, 10, **75**
chisosensis, 15, **76**
densispina, 9, **73**
dulcis, 78
imbricata var. *arborescens*, 12, **70**
imbricata var. *argentea*, 9, **71**
imbricata var. *imbricata*, 70
kleiniae, 9, **72**
leptocaulis, 9, 72
macrocentra var. *aureispina*, 74
rufida, 10, **77**
schottii var. *grahamii*, **73**
schottii var. *schottii*, **73**
spinosibacca, 9, **78**
orange flameflower, **252**
orchid (family), 231–239
Orchidaceae, 231–239
oreja de ratón, 160
Orobanchaceae, 240–241
Orobanche ludoviciana ssp. *multiflora*, **241**, 304
orpine (family), 164–167
Ostrya virginiana var. *chisosensis*, 14, **128**
Ownbey, Gerald B., 242
Ozark-Ouachita-Marathon mountain range, 5

Pacific Railroad Survey, 66, 77, 86, 91, 114, 140
Paint Gap Hills, 12, 83, 130, 190, 197, 272, 277, 283
palma samandoca, 40
Palmer, Edward, 80, 136, 219, 257, 267, 278
Palmer's catseye, 131
palmita, 41
palmleaf thoroughwort, **110**, 304
palo prieto, 14, **264**
pañalero, 155, 229
pancake pincushion, 64
Panther Canyon, 99, 219
Panther Junction, 3, 20, 66, 77, 148, 173, 175, 196, 268, 271
Panther Pass, 76, 247
Panther Spring, 152
Papaveraceae, 242
paperbagbush, 216
papercup loeselia, **247**
park history, 18–19
Parks, Harris B., 131, 143
Paronychia jamesii, **152**
Parry, Charles C., 85, 101, 108, 112, 117–118, 120, 133–134, 158, 226, 263, 282
Parry caesalpinia, **200**, 304
Parry ruellia, **85**
Parry's rockdaisy, 120
parsley (family), 91
Parthenium
argentatum, 11, **115**
incanum, 11
Passiflora tenuiloba, **243**
Passifloraceae, 243
passionflower (family), 243
Passionflower Canyon, 109, 117–118, 143, 228, 243, 263, 273, 282
Pavón Jimenez, José, 137
pearl netleaf milkweed, **100**
Pellaea
intermedia, **254**
ternifolia, **255**
Penstemon
baccharifolius, **273**
havardii, **274**
peonia, 102
Pérez, Lorenzo, 102
Pericome caudata, **116**
Perityle
aglossa, **117**
bisetosa var. scalaris, **118**
dissecta, **119**
parryi, **120**
vaseyi, 120, **121**
Persimmon Gap, 5, 38, 47, 50, 56–57, 86, 101, 122, 131, 134, 137, 139, 142, 144–145, 148, 177, 183, 189, 195, 219, 225, 229, 241–242, 279, 283, 286–287
Persimmon Gap Road, 7, 83, 175, 196
Persoon, Christian H., 137
Petit-Thouars, Louis du, 159
Petrogenia repens, 159
peyote, 61
peyote cimarron, 43
Phaseolus grayanus, **199**
Phemeranthus aurantiacus, **252**
Philadelphus
mearnsii, **208**
microphyllus, 16, **209**
phlox (family), 246–247
Phoradendron hawksworthii, **284**
Phragmites australis, 8
Phyllanthus polygonoides, **184**
pinabete, 170
Pinaceae, 244
pine (family), 244
Pine Canyon, 6, 15–16, 95, 97, 99, 106, 112, 123, 128, 138, 156–157, 163, 176, 184, 186, 194, 199, 203, 215, 217–218, 237, 257, 261–262, 264–265, 269, 281, 283
Pine Canyon, Upper, 16, 109, 150, 199, 234, 275
pineapple (family), 10, 146
pineapple cactus, 54, 58–60
pink (family), 150–153
pink tree-thief, 284
Pinnacles, the, 91, 232, 266
Pinnacles Mountains, 81, 96
Pinnacles Trail, 53, 76, 82, 91, 95, 103, 106, 108, 110, 128, 150, 167, 190, 199, 215, 217–218, 232, 254
Pinnacles Trail backcountry campsite, 108
pino piñonero, 244
Pinus
arizonica var. *stormiae*, 15
cembroides ssp. *cembroides*, 13, **244**
pinyon, 13, 171, 240, 244
pitaya, 55
Pitcock Rosillos Mountain Ranch, 19
plants, the naming of, 22–23
Plants Database, U.S. Department of Agriculture, 305, 314
plateau rocktrumpet, 94, **95**
Pliny the Elder (Caius Plinius Secundus), 139, 175, 256
plume tiquilia, 12, **137**
Plumier, Charles, 130
Poaceae, 245
Poiret, Jean L. M., 229
Polanisia uniglandulosa, **156**
Polemoniaceae, 246–247
Polygala
lindheimeri var. *parvifolia*, **248**
macradenia, **249**
scoparioides, **250**
Polygalaceae, 248–250
Polygonaceae, 251
Pomar, Jaime Honorato, 200
Pomaria melanosticta, **200**, 304
Pompadour, Madame de, 39
Pope, Capt. John, 140
poppy (family), 242
Populus
angustifolia, 8
fremontii ssp. *mesetae*, 8
tremuloides, 16
Portulacaceae, 252
Potts, Frederick, 67
Potts, John, 67
Potts' mammillaria cactus, 9, **67**
Powell, A. Michael, 74–75, 78, 118, 175, 304, 313–314
Powell, Shirley, 75, 220
pretty dodder, **173**
primrose (family), 253
Primulaceae, 253
Pringle, Cyrus, 40, 81, 98, 106–107, 167, 207, 224, 248, 250, 256, 264
Pringle swallow-wort, **98**
propellerbush, 11, **223**
Prosopis
glandulosa, 8
pubescens, 8
Prunus
havardii, 14, **261**
serotina var. *virens*, 14, **262**
Psathyrotes scaposa, **122**
Pseudotsuga menziesii, 15
Ptelea trifoliata, 268
Pteridaceae, 254–255
Ptolemy II Philadelphus, 209
puckering nightshade, **280**
Pulliam Bluff, 83, 101, 119, 135, 150, 153, 172, 206, 218, 256–257, 265, 267
Pulliam Canyon, 16, 100, 112, 153, 206, 217–218, 237, 254, 269
Pulliam Canyon, Upper, 215, 255
Pulliam Peak, 240
Pulliam Ridge, 6
Pummel Peak, 76, 207, 274
purple gaymallow, 16, **224**
purple sage, 12, 75, 271, **272**
Purpus, Carl A., 211
Pursh, Frederick, 263, 275
Purshia ericifolia, **263**

purslane (family), 252

quaking aspen, 15–16
Quercus
emoryi, 14
graciliformis, 14, **202**
gravesii, 13, **203**
grisea, 14
intricata, 14, **204**
tardifolia, 16
quicksilver mining, 17

Radley, Frank, 52
Ragsdale, Kenneth, 20
rain shadow desert, 7
Ralston, Barbara, 73
Ranunculaceae, 256–257
rat's-tail succulent, 167
Rattlesnake Mountains, 7, 122, 182, 216, 227
rayless peritYle, 117
rayless rockdaisy, **117**, 120
red cyphomeris, **228**
redberry juniper, 13
Rehder, Alfred, 207
resurrection plant, 276
Rhamnaceae, 258–260
Rhamnus betulifolia, 16
rhinoceros cactus, 46
Rhus
microphylla, 12, **90**
trilobata, 14
virens, 14
Richard, Achille, 234
Ringgold expedition, 56, 166
Río Bravo del Norte, 5, 43
Río Conchos, 5
Rio Grande, 3–9, 11, 13, 16–19, 21, 38, 43, 45, 47–50, 55, 62–65, 69, 71–74, 76–78, 80, 83–84, 86, 89, 95, 101, 108–109, 114, 117, 120, 122, 126, 129, 132, 134, 139, 141, 145–147, 151, 154, 158, 177–179, 182–183, 185, 187, 195, 198, 201, 203, 210–212, 216, 219–221, 223, 226–227, 229, 241–242, 245, 249, 258, 260, 266, 276–279, 282–283, 286
Rio Grande, diminished flow of, 18
Rio Grande prickly pear, 74
Rio Grande Village, 5, 6, 8, 78, 117
Rio Grande Wild and Scenic River, 19
river floodplain-arroyo. *See* vegetation zones, river floodplain-arroyo
River Road, 8, 10, 47, 136, 140, 198, 210, 219, 242, 246
River Road, east end, 185, 241
River Road, west end, 205, 241
Robinson, Benjamin L., 106
Robinson, Harold, 112
rock betony, **218**
rock brush, 194
rock hedgenettle, 218
rock milkwort, **248**
rockslide daisy, 116
Roemer, Carl Ferdinand, 186
Roemer acacia, 12, **186**
Rollins, Reed, 138
Rooney's Place, 286
Rosaceae, 261–264
rose (family), 261–264
Rose, Joseph N., 44, 68, 91, 165, 224
Rosillos Mountains, 3, 7, 19, 36, 39, 41, 62, 81, 98, 101, 109, 111, 115, 119, 126, 133–134, 142, 148–149, 151, 157, 164, 173, 186, 191, 200, 223, 228, 230, 244, 246, 248–250, 252, 259, 261, 265, 269, 272, 274, 276, 285
Rosillos Peak, 3
Ross Maxwell Road, 11, 188
rough mistletoe, **284**
rough mortonia, **154**
Rough Run Creek, 69
Rough Spring, 85, 131, 149, 256, 261
roughstem hawkweed, **113**
round copperleaf, **177**
roundleaf stevia, **123**
royal sage, 217
Roy's Peak, 246, 263, 278
rubber plant, 11, 18, 115, 183
Rubiaceae, 265–267
ruda del monte, 268
Ruelle, Jean, 85
Ruellia parryi, **85**
Ruiz Lopez, Hipólito, 137
rusty hedgehog cactus, 13, 46, **57**
Rutaceae, 268
Rydberg, Per Axel, 256, 264, 278
sabino, 172
saffron, 147
Salazar, Gerardo A., 238
Salazar Ylarregui, Jose, 216
Salazaria mexicana, 8, **216**
Salix
exigua, 8
gooddingii, 8
Salm-Dyck, Prince Joseph, 67
Salvia regla, 14, **217**
Sam Nail Ranch, 12, 111, 178, 188, 190, 197, 248, 260, 264, 274, 279, 285
Samolus ebracteatus var. *cuneatus*, **253**
San Vicente, 8, 16, 122, 132, 185, 187, 201
sandlot brickellbush, **106**
sandpaperbush, 154
sangre de drago, 183
Santa Elena Canyon, 3, 5, 8–9, 11, 19, 80, 88, 96, 119–120, 122, 127, 139, 141–142, 145, 147, 173, 182, 186, 192, 196, 200–201, 210, 216, 219, 226, 276, 278
Santiago Mountains, 4, 7, 41, 136
Sapindaceae, 269
Sarcostemma torreyi, **101**
Sargent, Charles, 40
scarlet bouvardia, 14, **265**
scarlet ladies tresses, **232**
scented lippia, 8
Schaeffer, Jacob C., 155
Schaefferia cuneifolia, 12, **155**
Schaffner, Johann W., 113, 270
Schauer, Sebastian, 200
Scheele, Georg Heinrich Adolf, 99, 186, 194
Scheer, Friedrich, 67
Schiede, Christian Julius Wilhelm, 82, 171
Schlectendal, Diederich Franz Leonhard von, 172, 265
Schott, Arthur, 43, 65, 69, 73, 101, 108, 134, 282
Schott dog cholla, **73**
Schott Tower, 6
Schultz-Bipontinus, Carl Heinrich, 113
screwbean mesquite, 8
Scrophulariaceae, 12, 270–275
scurfy mallow, **225**
sea urchin cactus, 22, **46**, 52, 305
sedge (family), 174
Sedum
havardii, **165**
wrightii, **166**
seepwillow, 8
Selaginella lepidophylla, **276**
Selaginellaceae, 276
Selenia dissecta, **140**
Selinocarpus
angustifolius, 226, 304
parvifolius, 227, 304
Senna
bauhinioides, 201
pilosior, **201**
sensitive plants, xi
Sessé and Mociño expedition, 102, 156, 164, 219, 287
Sessé y Lacasta, Martín de, 156, 164, 287
Seymer, Henry, 275
Seymeria scabra, **275**
Shafer, John, 42
shaggy false nightshade, **278**
shaggy stenandrium, **86**
Shiner's brickellbush, 112
Shinners, Lloyd, 219
Shirley's nettle, 220
shorthorn jefea, 12
showy menodora, **230**
shrub desert. *See* vegetation zones, shrub desert
shrubby tidestromia, 11, **89**
Sicyos glaber, **169**
Siebold, Philipp Franz von, 212
siempre viva, 276
Sierra
del Carmen, 5–6, 16–17, 39, 208, 234, 251, 273
Maderas del Carmen, 39, 76, 128, 224
Madre, 3, 5, 67, 244
Madre Occidental, 3, 5, 7
Madre Oriental, 3, 6–7, 224, 227
Quemada, 6–7, 80, 146, 227
silktassel (family), 206
silver ponyfoot, **160**
silverlace cactus, **49**, 67, 71
silverspine cholla, 71
silvery wolfberry, 11, **279**
simarouba (family), 277
Simaroubaceae, 277
Simpson, Beryl, 200
Sirotnak, Joe, 314
Sisymbrium auriculatum, **141**
skeleton goldeneye, 11
skunkbush sumac, 14
slender janusia, 223

slender oak, 202
slimleaf vauquelinia, 264
slimlobe globeberry, **168**
slimlobe rockdaisy, **119**
Sloane, Sir Hans, 285
Small, John K., 253
Smithers, W.D., 18
Smoky Creek, 8, 72, 85, 131, 159, 192, 197–198, 201, 216, 227, 229, 241, 245–246, 283, 285, 287
smooth bur cucumber, **169**
smooth sotol, 11–12, **38**
Sneed, J.R., 49
snowcone nipple cactus, 49
soapberry (family), 269
soapbush, 286
soaptree yucca, 11
soft heliotrope, **134**
soft twinevine, **101**
Solanaceae, 11, 278–280
Solis, 47, 73, 185, 191
Sonoran bean, 199
Sophora secundiflora, 14
Sotol Vista, 11, 38, 98, 130
sotol-grassland. *See* vegetation zones, sotol-grassland
South Rim, 6, 14, 62, 169, 176, 204, 206, 217, 267
South Rim Trail, 39, 44, 104, 125, 157, 167, 174, 203, 244
southwestern brickellbush, 106
southwestern chokecherry, 13–14, **262**
soyate, 41
Spach, Edouard, 103
Spanish dagger, 11, **42**
Spanish exploration, 16–17
spectaclepod, 144
spiderwort (family), 157
spikemoss (family), 276
spine mound cactus, 56
spiny-fruited prickly pear, 9, **78**
splitleaf brickellbush, **105**, 305
spoonleaf bouchea, **282**
spreading snakeherb, **82**
spreading-lobe passionflower, **243**
Sprengel, Kurt P.J., 184
spurge (family), 177–184
Stachys bigelovii, **218**
Stairstep Mountain, 118
stairstep rockdaisy, 118
stalkflower heimia, 222
stalkflower nesaea, **222**
Standley, Paul C., 87, 151, 161, 169, 199, 251, 262
star cactus, 43
Stellaria cuspidata, **153**
stemless aletes, **91**
stemless Indian parsley, 91
stemless perezia, **102**
Stenandrium barbatum, **86**
Stenorrhynchos michuacanum, **239**
Sterculiaceae, 281
Steudel, Ernst Gottlieb von, 171
Stevia ovata var. *texana*, **123**
Stevia viscida, **124**
Stewart, Robert M., 277
Stewart holacanth, 277
Steyermark, Julian, 79, 88, 178–179
stickleaf (family), 220–221
sticktights, 104
stiff fendlerbush, 207
stinging cevallia, **220**
stinging serpent, 220
stonecrop, **167**
strangle vine, 173
strawberry cactus, 55, **56**, 68, 304
strawberry hedgehog cactus, **55**, 304
strawpile hedgehog, 56
stream orchid, 233
Streptanthus
carinatus ssp. *carinatus*, **142**
cutleri, **143**
platycarpus, 143
Study Butte, 75, 88, 127
Sublett Ranch, 277, 279
Sudworth, George B., 203
Sue Peaks, 3, 132, 195
sumac (family), 90
sunflower (family), 102–127
sweet prickly pear, 78
Synthlipsis greggii, **144**
Szlachetko, Dariusz, 238

Tagetes
lucida, 125
micrantha, **125**
tailleaf pericome, **116**
talayote, **99**
Talinum
angustissimum, 252
aurantiacum, 252
Talley Mountain, 100, 142, 182, 226, 248, 268
tamarisk, 8, 18
Tamarix aphylla, 8
Tamayo, Roberto, 238
Tamayorkis (genus), 238
tangle bluets, 267
taperleaf, 116
Tarahumara, 43, 61, 171
tarbush, 10, **111**
tasajillo, 9, 72
tascate, 172
Tecoma
stans, **130**
stans var. *angustata*, 130
tecomblate, 259
Telephone Canyon, 84, 130, 147, 154, 159, 177, 193, 200, 230, 264, 282
Telosiphonia
hypoleuca, **94**
macrosiphon, **95**
Terlingua, 17, 20, 45, 59, 67, 127
Abaja, 9, 11, 18, 88, 127, 286
Creek, 5, 18, 145, 168, 192, 210, 223
Ranch Road, 133–134, 144, 148–149, 250
stickbush, 127
ternate cliff brake, 255
Terra Incognita, 20
Tertiary period, 6
Texan candyleaf, 123
Texas
Canyons State Park, 19
claret-cup cactus, 15, **53**
cone cactus, **68**
desertrue, 268
false agave, 10, **146**
flatsedge, **174**
kidneywood, 14, **194**
largeseed bittercress, **138**
madrone, 13–14, **176**
milkweed, **97**
mountain laurel, 14
persimmon, 12
pride, 69
purplespike, **237**, 277
rainbow cactus, 10, **54**, 60
selenia, **140**
silverleaf, 272
thelypody, **145**
Thamnosma texana, **268**, 304
Thelocactus bicolor var. *bicolor*, 9, **69**
Thelypodium
texanum, **145**
wrightii, 145
Theophrastus of Eresus, 71, 90, 233, 258
thickleaf drymary, **151**
Thompson, Charles, 41
Thompson yucca, 11, **41**
Thornber, John, 87
Threatened (species), 9–10, 48, 52, 59, 236, 305
threeflower goldenweed, 11, **127**
three-tongue spurge, **179**, 304
Thymophylla micropoides, **126**, 304
tickbush, 154
Tidestrom, Ivar, 88
Tidestromia
carnosa, 11, **88**
suffruticosa, 11, **89**
Tiquilia greggii, 12, **137**
Toll Mountain, 128, 275
torch cactus, 57
Tornillo Creek, 9, 121, 180, 210, 221, 282
Tornillo Creek, Lower, 227
Tornillo Creek, mouth of, 141, 211, 260
Tornillo Creek, Upper, 11, 175, 227, 279
Tornillo Creek Bridge, south, 196
Tornillo Flat, 11, 18, 51, 64, 73, 127, 132, 140, 142, 175, 189, 219, 246, 249
toronjil, 268
Torrey, John, 37, 39, 42, 85–86, 89, 91–92, 95, 101, 119, 132–134, 137, 140, 152, 157–158, 162, 175, 177, 189, 206, 215–216, 226–227, 229, 263, 268, 282–283
Torrey ephedra, **175**
Torrey jointfir, 175
Torrey twinevine, 101
Torrey yucca, 42
Townsend, Everett, 18
Tradescant, John, the younger and the elder, 157
Tradescantia
brevifolia, **157**
leiandra, 157
Trans-Pecos
carlowrightia, **81**
chickweed, **150**
cliff brake, **255**
senna, **201**
spiderwort, **157**
Trap Spring, 260

tree cholla, 12, **70**, 71–72
tree tobacco, 8
Trelease, William, 36, 38, 41
trompetilla, 265
tronadora, 130
trumpetflower, **130**
tuber windflower, **256**
tubercled saltbush, 11
tubular bluestar, 92
tubular slimpod, **92**
Tuff Canyon, 182, 196, 198
tunnel, the, 78, 84, 117, 121, 136, 143, 180, 260, 281
Turk's head, 62
Turner, B. L., 84, 102, 109, 198, 256, 304, 314
Turner mimosa, 8, **198**
Turner thistle, **109**
twinpod, 230
twistleaf paintbrush, **270**
twobristle rockdaisy, **118**, 120
Tyler, R. C., 313
type (or type specimen), 23, 51, 70, 75, 78, 86, 91, 162–163,175, 237, 267, 273, 277, 286, 314
Typha domingensis, 8

uña de gato, 186
Ungnad, Baron David von, 269
Ungnadia speciosa, 14, **269**
U.S. Geological Survey, 305
U.S.–Mexico
  boundary survey, 39, 41, 43, 50, 56, 63, 65–66, 69, 73, 85, 89, 104, 108, 112, 115–118, 120, 133–134, 155, 158, 197, 215–216, 218, 225–226, 273, 282
  boundary survey, Colonel Graham's, 39, 56, 63, 73, 89, 116, 155, 225, 273
  boundary survey, Major William Emory's, 41, 43, 65, 85, 115, 197, 218
  boundary survey, report, 69, 104, 134, 162
  boundary survey, second, 199, 208
  War, 17, 47, 114, 192, 258

vara dulce, 194
varicolor cob cactus, 50
varnish bush, 111
Vasey, George S., 121, 245, 267
Vasey's rockdaisy, 121
Vauquelin, Louis Nicolas, 264
Vauquelinia corymbosa ssp. *angustifolia*, 14, **264**
vegetation zones, 8
  moist Chisos woodland, 8, 15–16
  river floodplain-arroyo, 8–10
  shrub desert, 3, 7–8, 10–13, 18
  sotol-grassland, 8, 11–13
  woodland, 8, 13–15
veinyleaf lantana, 283
velas de coyote, 70
Verbenaceae, 282–283
veronicaleaf brickellbush, **107**
vervain (family), 282–283
Victorio, Mescalero Apache Chief, 17
Viguiera stenoloba, 11
Villada, Manuel M., 167
Villadia squamulosa, **167**
Vines, R. A., 313–314
violet silverleaf, 271
Viscaceae, 284
viscid acacia, 9
viscid candyleaf, **124**
viscid stevia, 124
Vitaceae, 285
volcanic activity, 6
vulnerable (species), xi, 49, 53, 60–61, 84, 88, 109, 112, 117, 120, 142, 145, 159, 166, 169, 207, 210, 215, 222, 227, 233, 237, 246, 251, 257, 261, 284, 291–293

Wagstaff, Robert M., 18
walking-stick cholla, 70
Walter, Thomas, 92, 236
Ward Spring, 39,81, 84, 98–99, 101, 132, 154, 163–164, 178, 193, 228, 245, 261, 265, 267, 276, 285
Warnock, Barton, 60, 84, 109, 118, 127, 135, 143, 159, 178, 225, 235, 237, 258, 262–263, 271, 305, 313–314
Warnock condalia, 10, 259
Warnock justicia, **84**
Warnock's cactus, 58–59, **60**, 68
Wasp Spring, 226, 227
waterleaf (family), 210–211
Watson, Sereno, 146, 167, 175, 187, 196, 240, 247, 257
Wauer, R. H., 313
wax plant, 10, 18, 181
Webb, Walter Prescott, 19–20
Weedin, James F., 74–75, 78, 304, 313–314
weeping juniper, 13, 172
Wendt, Tom, 175, 238
Wendt's adder's-mouth, 238
Wendt's malaxis, **238**, 304
Weniger, Del, 57
Wheeler, Louis C., 179
Wheelock, William, 248
Whipple, Lt. Amiel, 66, 91
whiskerbrush pincushion cactus, 9, 22, **48**
whiskered barrel, 62
White, H. L., 304
white column cactus, 49
white giant hyssop, **213**
Wiens, Delbert, 284
Wight, William, 261
wild petunia, 85
Willdenow, Carl, 153, 160
Williams, Louis O., 234
Wilson Ranch, 80
Wilson Ranch Overlook, 188
Window, the, 62, 96, 162, 186, 215, 232, 253, 261, 264, 267, 269
Window Trail, 76, 79, 100, 105–106, 120, 264
Wislizenus, Adolph (or Frederick A.), 47, 54–55, 83, 205
witches' shoelaces, 173
wolf flower, 196
woodland. *See* vegetation zones, woodland
Woodson's Camp, 198
woolly butterflybush, **147**
woolly dogweed, **126**, 304
woolly false nightshade, 278
woollyflower spurge, **182**
Wooton, Elmer O., 151, 161, 169, 199, 262
woven-spine pineapple cactus, **58**
Wright
  ageratina, **103**
  dalea, **193**
  snakeroot, 103
  stonecrop, **166**
  thelypody, 145
  thoroughwort, 103
Wright, Charles, 37, 39, 46, 56, 63, 73, 80, 86, 89, 92, 95, 101, 103, 105, 108, 110, 116, 120, 122, 132–134, 139, 141–142, 149, 154–155, 162–164, 166, 168, 175, 191, 193, 195, 213, 222–223, 225, 240, 249, 254, 259, 273, 275, 281–283, 314

Xylothamia triantha, 11, **127**

Yarborough, S. C., 304, 313–314
yellow bells, 130
yellow rocknettle, **221**
yerba
  de Alonso Garcia, 189
  de chivato, 116
  del cancer, 281
  del cenizo, 137
  del Cristo, 283
yucca, 10–11, 37–38, 40–42
Yucca
  baccata var. *macrocarpa*, 42
  carnerosana, 40
  faxoniana, 11, **40**, 304
  rostrata, 41
  thompsoniana, 11, **41**
  torreyi, 11, **42**
  treculeana, 42

zigzag croton, **180**
Zuccarini, Joseph Gerhard, 181, 212, 221, 244
Zygophyllaceae, 286–287